T0328332

AN INTRODUCTION TO BIOLOGICAL MEMBRANES

AN INTRODUCTION TO BIOLOGICAL MEMBRANES

COMPOSITION, STRUCTURE AND FUNCTION

SECOND EDITION

WILLIAM STILLWELL

Professor Emeritus, Department of Biology,
Indiana University-Purdue University Indianapolis,
Indianapolis, Indiana, USA

AMSTERDAM • BOSTON • HEIDELBERG • LONDON
NEW YORK • OXFORD • PARIS • SAN DIEGO
SAN FRANCISCO • SINGAPORE • SYDNEY • TOKYO
Academic Press is an imprint of Elsevier

Academic Press is an imprint of Elsevier
125 London Wall, London EC2Y 5AS, United Kingdom
525 B Street, Suite 1800, San Diego, CA 92101-4495, United States
50 Hampshire Street, 5th Floor, Cambridge, MA 02139, United States
The Boulevard, Langford Lane, Kidlington, Oxford OX5 1GB, United Kingdom

Notices

Knowledge and best practice in this field are constantly changing. As new research and experience broaden our understanding, changes in research methods, professional practices, or medical treatment may become necessary.

Practitioners and researchers must always rely on their own experience and knowledge in evaluating and using any information, methods, compounds, or experiments described herein. In using such information or methods they should be mindful of their own safety and the safety of others, including parties for whom they have a professional responsibility.

To the fullest extent of the law, neither the Publisher nor the authors, contributors, or editors, assume any liability for any injury and/or damage to persons or property as a matter of products liability, negligence or otherwise, or from any use or operation of any methods, products, instructions, or ideas contained in the material herein.

British Library Cataloguing-in-Publication Data
A catalogue record for this book is available from the British Library

Library of Congress Cataloging-in-Publication Data
A catalog record for this book is available from the Library of Congress

ISBN: 978-0-444-63772-7

For information on all Academic Press publications visit our website at https://www.elsevier.com/

Publisher: Sara Tenney
Acquisition Editor: Jill Leonard
Editorial Project Manager: Pat Gonzalez
Production Project Manager: Julia Haynes
Designer: Matt Limbert

Typeset by TNQ Books and Journals
www.tnq.co.in

Contents

8. From Lipid Bilayers to Lipid Rafts

9. Basic Membrane Properties of the Fluid Mosaic Model

10. Lipid Membrane Properties

11. Long-Range Membrane Properties

12. Membrane Isolation Methods

13. Membrane Reconstitution

II
MEMBRANE BIOLOGICAL FUNCTIONS

14. Membrane Biogenesis: Fatty Acids

15. Membrane Biogenesis: Phospholipids, Sphingolipids, Plasmalogens, and Cholesterol

16. Membrane Biogenesis: Proteins

17. Moving Components Through the Cell: Membrane Trafficking

18. Membrane-Associated Processes

19. Membrane Transport

20. Bioactive Lipids

21. Lipid-Soluble Vitamins

22. Membrane-Associated Diseases

23. Membranes and Human Health

24. Cell Death, Apoptosis

Preface

Membranes are highly complex, dynamic structures that are fundamental to life. It is likely that life could not have begun if the first collections of primitive proteins, nucleic acids, and so on were not encapsulated inside a primal, semipermeable membrane. Every living organism on planet Earth is now and always has been surrounded by a plasma membrane. Although biological membranes are more similar than they are different in their compositions and structures, it is their differences that account for diverse membrane functions. Great progress in understanding how membranes work continues to be made, but it is likely that membranes will remain the next great frontier in the life sciences.

Membrane studies can be roughly divided into three arenas—composition, structure, and function. The first edition of this book, *An Introduction to Biological Membranes: From Bilayers to Rafts*, focused on composition and structure, while this second edition, *An Introduction to Biological Membranes: Composition, Structure, and Function*, is a greatly expanded version of the first edition, integrating many aspects of complex biological membrane functions with the composition and structural components featured in the first edition.

A single membrane is composed of hundreds of proteins and thousands of lipids, all in constant flux. Every aspect of membrane structural studies involves parameters that are very small and very fast. Parameters of interest to this book range in size from microns (μ, 10^{-6} m) to Angstroms (Å, 10^{-10} m) and in time from milliseconds (ms, 10^{-3} s) to picoseconds (ps, 10^{-12} s). Both size and time ranges are so vast that multiple instrumentations must be used, often simultaneously. As a result, a variety of highly specialized and esoteric biophysical methodologies are often utilized.

Like the first edition, the second edition is presented in a single voice (ie, I am the sole author). The purpose of this book is to address salient features of membranes as a broad introduction aimed at college seniors and beginning graduate students. It is written in the form of a textbook and hopefully does not get bogged down in overly detailed, technical minutia. As in the first edition, the emphasis is placed on what can be gleamed from a particular biophysical technique and not the sophisticated theory behind the instrumentation. In other words, the presented experiments can best be described as "biophysics light" and should be understandable by anyone interested in membranes. Choosing what should be included and what excluded from the book reflects my personal research experience that has been based on the biophysics and biochemistry of model membrane systems. There is also an emphasis placed on who actually were the pioneers of membrane studies. I have found that most contemporary membranologists have little idea how the problems they study originated. And often these early pioneers were quite interesting and even eccentric characters! The last chapter of this book, Chapter 25, is a timeline of the most important discoveries in membrane science from about 540 BC to the present.

Writing these two books has been a task that is beyond any one individual. It was certainly beyond me. No one can possibly be an expert in all aspects of membrane science. The field of membranes has just become too enormous for one individual to tackle. However, the other option of presenting the material as a large, multiauthored compendium results in an unevenly written book that is too technical and can only be understood by researchers who are already experts in their narrow field of study. While multiauthored compendiums have their place in membrane studies, they are essentially noncohesive collections of loosely related review articles that are written by experts, for experts. The authors approach aspects of membranes from very different backgrounds and writing styles. As a result, multiauthored compendiums are comprehensive but hard to read and are essentially "preaching to the choir." In contrast, my book is an introductory textbook aimed at a more general audience. It took me the past 11 years, writing 7 days a week, to finish. I am almost completely blind, type with one finger, and have no clerical help. And every day I think of additional things that should have been included.

William Stillwell
Sand Key, Florida
October 6, 2015

PART I

MEMBRANE COMPOSITION AND STRUCTURE

CHAPTER

1

Introduction to Biological Membranes

OUTLINE

1. WHAT IS A BIOLOGICAL MEMBRANE?

The *American Heritage Dictionary* defines a *membrane* as "A thin pliable layer of plant or animal tissue covering or separating structures or organs." The impression this description leaves is *static*—like plastic wrap covering a package of hamburger. By this definition, membranes are tough, impenetrable, and visible. Nothing can be further from the truth. The entire concept of *dynamic* behavior is missing from this definition, yet dynamics is what makes

An Introduction to Biological Membranes
http://dx.doi.org/10.1016/B978-0-444-63772-7.00001-4

membranes both essential for life and so difficult to study. Dynamic is characterized by "constant change, activity, or progress."

If we could somehow instantaneously freeze a membrane and learn the composition and location of each of the countless numbers of molecules composing the membrane and then instantly return the membrane back to its original unfrozen state for a microsecond before refreezing, we would find that the membrane had substantially changed while unfrozen. Although the total molecular composition would remain the same, the molecular locations and interrelationships would have been altered. Therefore, membranes must have both *static* and *dynamic* components. While "static" describes what is there, "dynamics" describes how the components interact to generate biological function. The first half of this book focuses on static membrane composition and structure, while the second half concerns dynamic membrane functions.

Every cell in the human body is a tightly packed package of countless membranes. The human body is composed of approximately 63 trillion cells (6.3×10^{13} cells), each of which is very small. For example, a typical liver cell would have to be five times larger to be seen as a speck by someone with excellent vision (it is microscopic). Each liver cell has countless numbers of internal membranes. If you could somehow open a single liver cell, remove all of the internal membranes, and sew them together into a planar quilt, the quilt would cover about 840 acres, the size of New York City's Central Park! And all of that is from one cell. Therefore, there is sufficient membrane area in a human body to cover the earth millions of times over.

All life on Earth is far more similar than it is different. Living organisms share a number of essential biochemical properties, collectively termed the "thread of life." Included in these essential properties is ownership of a surrounding plasma membrane (PM) that separates the cell's interior from its external environment. It is likely that all living things inhabiting planet Earth today arose from a single common ancestor more than 3.5 billion years ago. The first cell probably contained minimally a primitive catalyst (a pre-protein), a primitive information storage system (a pre—nucleic acid), a source of carbon (perhaps a primitive carbohydrate), and a primitive energy-transducing system, and this mixture had to be surrounded by a primitive PM that was likely made of polar lipids. Membranes were therefore an essential component of every cell that is alive today or has ever been alive. If life is found elsewhere in the universe, it would likely be surrounded by some type of PM.

With 3.5 billion years of biological evolution, the complexity of membranes in cells has greatly expanded from that of a simple surrounding PM to where they now occupy a large portion of a eukaryote's interior space. An electron microscopic picture of a "typical" eukaryotic (liver) cell is shown in Fig. 1.1 [1]. It is evident from the complexity of this micrograph that identifying, isolating, and studying membranes will be a difficult task.

2. GENERAL MEMBRANE FUNCTIONS

It is now generally agreed that biological membranes are probably involved somehow in all cellular activities. The most obvious function of any membrane is separating two aqueous compartments. For the PM, this involves separation of the cell contents from the very different extracellular environment. Membranes are therefore responsible for containment,

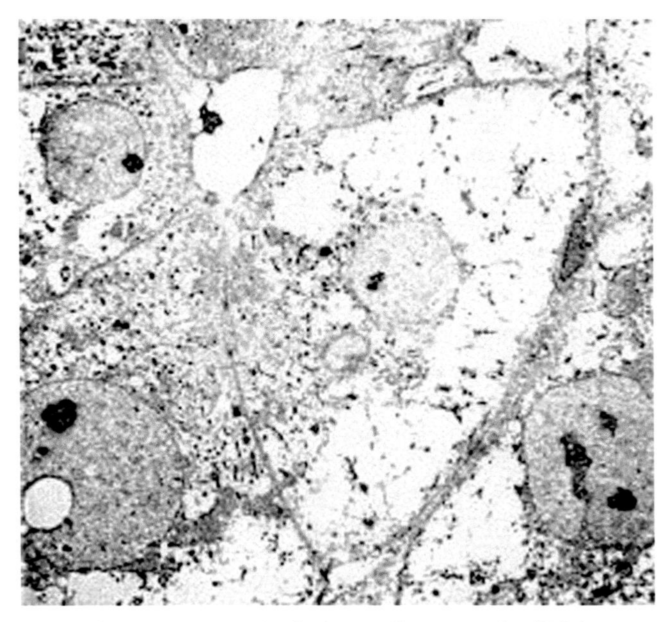

FIGURE 1.1 Transmission electron micrograph of a liver cell, a "typical cell" [1].

ultimately delineating the cell. Separation, however, cannot be absolute as the cell must be able to take up essential nutrients, gases, and solutes from the exterior while simultaneously removing toxic waste products from the interior. A biological membrane therefore must be selectively permeable, possessing the ability to distinguish many chemically different solutes and knowing in which direction to redistribute them. Biological membranes must house a variety of specific, vectorial transport systems (discussed in Chapter 19).

A characteristic of all living cells is the establishment and maintenance of transmembrane gradients of all solutes. Of particular interest are large ion gradients typically associated with the PM. Table 1.1 is a comparison of the mean concentration of selected ions inside and outside a typical mammalian cell and the magnitude of each gradient. To maintain gradients of this size, efficient energy-dependent transport systems must be used (discussed in Chapter 19). Vectorial transmembrane structure is required to generate these ion gradients.

TABLE 1.1 Transmembrane Ion Gradients of a "Typical" Mammalian Cell

Ion	Inside	Outside	Gradient
Na^+	140 mmol/L	10 mmol/L	14-fold
K^+	4.0 mmol/L	140 mmol/L	35-fold
Ca^{2+}	1.0 μmol/L	1.0 mmol/L	1000-fold
Cl^-	100 mmol/L	4.0 mmol/L	25-fold

In addition to transmembrane structure, it is now believed that biological membranes are composed of countless numbers of very small, transient, lateral lipid microdomains. Each of these domains is proposed to have a different lipid and resident protein composition. Thus, the activity of any membrane must reflect the sum of the activities of its many specific domains. One type of lipid microdomain, termed a "lipid raft," has received considerable recent attention as it is reputed to be involved in a variety of important cell signaling events. If the lipid raft story (discussed in Chapters 8 and 13) holds up, this new paradigm for membrane structure/function may serve as a model for other types of as-yet-undiscovered nonraft domains. Each of these domains might then support a different collection of related biochemical activities. Therefore, membranes have both transmembrane and lateral structures.

All membranes possess an extreme water gradient across their very thin (~5 nm) structure. In the membrane aqueous bathing solution, water concentration is about 55.5 mol/L water in water (1000 g of water per liter divided by 18, the molecular weight of water), while the membrane interior is very dry (<1 mmol/L water). The aqueous interface provides a charged or polar physical surface to help arrange related enzyme sequences, known as pathways, in one plane for increased efficiency (see Chapter 18 on Electron Transport and Oxidative Phosphorylation). In contrast, the dry interior provides an environment for dehydration reactions.

In addition to transport, biological membranes are a site of many other biochemical or physiological processes, including intercellular communication, cell–cell recognition and adhesion, cell identity and antigenicity, regulation (resident home of many receptors), intracellular signaling, and some energy transduction events.

3. EUKARYOTE CELL STRUCTURE

While the plasma (cell) membrane defines cell boundaries, internal membranes define a variety of cell organelles. In eukaryotes, the internal membranes also separate very different internal aqueous compartments resulting in compartmentation into membrane-bordered packets called organelles. Each organelle supports different sets of biological functions. Fig. 1.2 is a cartoon depiction of a "typical" animal (eukaryote) cell [2]. All of the important membrane-bound organelles are depicted. Throughout this book, specific examples demonstrating aspects of membrane structure or function will be selected from these membrane types. Following is a very brief description of the major cellular membranes. More detailed descriptions can be readily found in many cell biology [3–6] and biochemistry [7,8] textbooks.

3.1 Endomembrane System

With the exception of mitochondria, peroxisomes, and, in plants, chloroplasts, the intracellular membranes are suspended together in the cytoplasm, where they form an interconnected complex. Although each membrane type is unique and can be separated from one another (Chapter 12), all are related structurally, chemically, functionally, and developmentally. This strong interrelationship is referred to as the "endomembrane system" (Fig. 1.3 and Chapter 17). The endomembrane system divides the cell into many compartments,

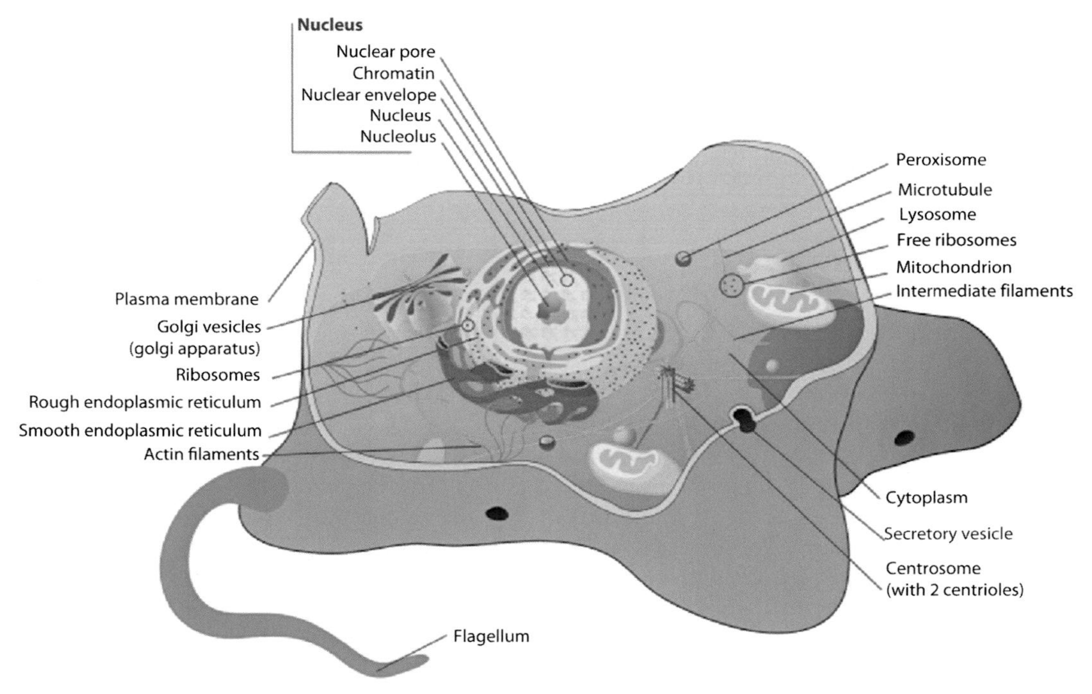

FIGURE 1.2 Cartoon depiction of the major components of an animal cell [1,2].

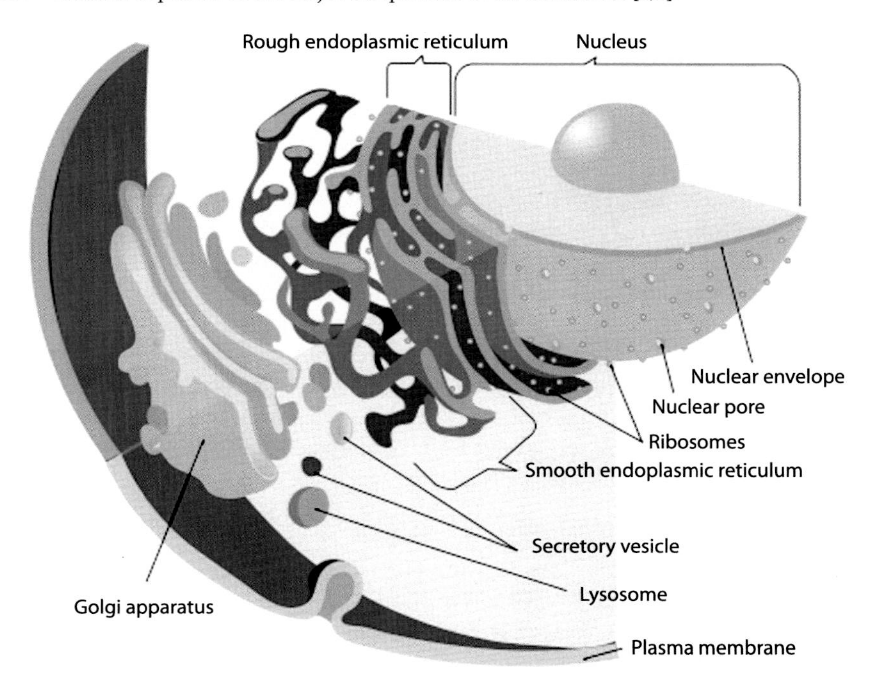

FIGURE 1.3 The endomembrane system of a "typical" animal cell. The various membranes and compartments are connected to one another structurally, chemically, functionally, and developmentally [1–9].

primarily organelles, that are distinct yet interconnected. The connected membranes include the nuclear envelope, the rough and smooth endoplasmic reticula (ERs), the Golgi apparatus, lysosomes, vacuoles, many types of small vesicles, and the PM [9]. The organelles are suspended in a crowded cytoplasm by a complex, interwoven protein mesh known as the cytoskeleton. Transport back and forth between these various membranes and compartments is continuous and under strict regulation [10]. It forms the foundation of cell trafficking (see Chapter 17).

3.2 Plasma Membrane

All cells are surrounded by a PM that separates the cell contents from the rest of the world [11]. The PM is the most dynamic and busiest of all cellular membranes, and more is known about the PM than about any other membrane. The majority of examples described in this book were obtained from PM studies. The PM defines the cell's boundary and its interaction with the environment. It is responsible for transporting nutrients into the cell while allowing waste products to leave. It prevents unwanted materials from entering the cell while keeping needed materials from escaping. It maintains the pH of the cytosol and preserves the proper cytosolic osmotic pressure. Proteins on the PM surface assist the cell in recognizing and interacting with neighboring cells. Other proteins on the PM allow attachment to the cytoskeleton and extracellular matrix, functions that maintain cell shape and fix the location of membrane proteins. The PM contains several characteristic functional structures that can be used to identify PM fractions. Tight junctions seal contacts between cells, while desmosomes are adhesion sites between adjacent cells. Gap junctions (Chapter 19) contain hexagonal arrays of pores that allow communication between adjacent cells. Caveolae and coated pits (see Chapter 19) are similarly shaped PM invaginations that are involved in cell signaling and solute uptake, respectively. PMs are indeed complex entities.

3.3 Cytoskeleton

In addition to a surrounding PM, all living cells have a cytoskeleton occupying their cytoplasm. The cytoskeleton is a type of cellular scaffolding that forms unusual structures, including flagella, cilia, and lamellipodia. They are also involved in the intracellular transport of cell vesicles and cell division. In 1903, Russian biologist Nikolai Koltsov (Fig. 1.4) proposed that the shape of cells was maintained by a network of cytoplasmic tubules that he termed the "cytoskeleton." We now know that eukaryotic cells contain three major types of cytoskeletal components: microfilaments, intermediate filaments, and microtubles. These components form complex interactions with other cell components and with one another and, importantly from the perspective of this book, are physically connected to the PM (Fig. 1.5) [12].

3.4 Nuclear Envelope (Membrane)

The nuclear envelope is a double membrane encompassing a perinuclear space [13]. This space is probably contiguous with the lumen of the ER. The envelope has large nuclear pores (about 600 Å) that allow passage of large RNA–protein complexes out of the nucleus into the cytoplasm and movement of regulatory proteins from the cytoplasm into the nucleus.

FIGURE 1.4 Nikolai Koltsov, 1872–1940. http://www.ecolife.ru/jornal/ecob/3-2002/koltzov.JPG.

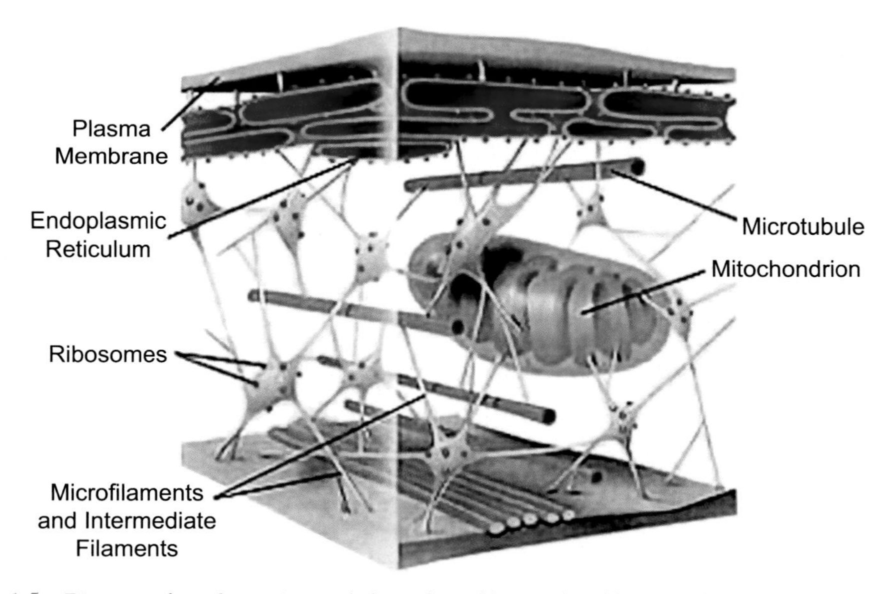

FIGURE 1.5 Diagram of a eukaryotic cytoskeleton. http://www.daviddarling.info/images/cytoskeleton.jpg.

3.5 Endoplasmic Reticulum

The ER is a complex network of cisternae or tube-like structures occupying a considerable percentage of the cell's internal volume [14]. The portion of the ER with attached ribosomes is known as the rough ER. It is the site for the biosynthesis of noncytoplasmic proteins that are either secreted or internalized into the lysosome or become PM proteins (see Chapters 16 and 17). The portion of ER devoid of ribosomes is known as the smooth ER. Functions of the smooth ER include sterol biosynthesis, drug detoxification, calcium regulation, and fatty acid desaturation. A specialized ER, the sarcoplasmic reticulum, has only one function: regulation of intracellular calcium levels. The ERs were first seen by Porter, Claude, and Fullam in 1945 [15].

3.6 Golgi Apparatus

The Golgi apparatus [16,17] is a series of stacked, disk-shaped tubules. The Golgi is named for its discovery in 1898 by Carmillo Golgi (see Fig. 17.4). For this, in 1906 Golgi was awarded one of the first Nobel Prizes. It is in the Golgi that posttranslational modification of glycoproteins synthesized originally in the ER and destined for secretion takes place (see Chapter 16). Other possible final destinations of proteins passing through the Golgi include incorporation into the PM or to the lysosome. Characteristic Golgi-resident enzymes involve sugar modification of proteins and include glucosidases and glycosyl transferases.

3.7 Lysosome

The lysosome contains some 40 hydrolytic enzymes whose function is to degrade macromolecules for reuse in the cell [18]. As a result, lysosomes are often referred to as the "cell's garbage disposal." The organelle was discovered in the early 1950s and in 1955 was named "lysosome" by Belgian cytologost Christian de Duve for its ability to lyse membranes (Fig. 1.6).

FIGURE 1.6 Christian de Duve (1917–2013). *Courtesy of Christian de Duve Institute of Cellular Pathology [25].*

3.8 Peroxisome

Peroxisomes contain a battery of oxidative enxymes that are involved in breakdown of small molecules [19]. Important peroxisomal enzymes include D–amino acid oxidase and catalase, the enzyme responsible for the degradation of dangerous peroxides. Peroxisomes are also involved in drug detoxification and the biosynthesis of essential ether phospholipids known as plasmalogens (discussed in Chapter 5). Peroxisomes were also discovered by Christian de Duve (Fig. 1.6) in 1967 [20]. For discovering lysosomes and peroxisomes, de Duve was awarded the 1974 Nobel Prize for Medicine.

3.9 Mitochondria

Mitochondria form the "powerhouse" of the cell, producing most eukaryotic cellular adenosine triphosphate. The processes of electron transport and oxidative phosphorylation (discussed in Chapter 18) are housed in the highly folded mitochondrial inner membrane (the cristae) [21]. The mitochondrial aqueous interior chamber, called the matrix, houses most of the enzymes involved in the Krebs cycle (terminal steps in sugar oxidation) and β-oxidation (fatty acid oxidation) [22]. The inner mitochondrial membrane is surrounded by a second membrane, the outer mitochondrial membrane, that is very different and far less dynamic than the inner membrane. Mitochondria have been defined as "semi-autonomous, self-replicating organelles," meaning that they grow and replicate independent of the cell in which they are housed. This is a vestige left over from their origin as freely living prokaryotes that took up refuge inside larger prokaryotes about 1.5 billion years ago. This concept, known as endosymbiont theory [23,24], was originally ridiculed but is now generally accepted. Mitochondria contain their own, albeit small, genome and code for a handful of mitochondrial membrane–protein components.

4. SIZE OF DOMAINS

Because membrane studies span a wide range of size and time domains (see Chapter 9), a variety of often esoteric instrumentations must be used. The studies addressed in this book will range in size from Angstroms (Å) to microns (μm): Å (10^{-10} m), nm (10^{-9} m), and μm (10^{-6} m).

We will first address the question of size by asking whether someone with excellent vision can actually see a membrane. A person with excellent vision can resolve two spots about 0.1 mm apart.

$$0.1\ \text{mm} = 10^{-4}\ \text{m} = 10^{2}\ \mu\text{m} = 10^{5}\ \text{nm} = 10^{6}\ \text{Å}$$

4.1 Can We See a Membrane?

Because a membrane bilayer is less than 50 Å thick, it would have to be about 20,000 times bigger than it is just to be seen as a speck by someone with excellent vision. Is it then possible that someone is able to see a much larger object, a cell?

4.2 Can We See a Cell?

Resolving power of the eye	100 μm	1 million Å
Liver cell	20 μm	200,000 Å
Erythrocyte	7 μm	70,000 Å
Bacteria	2 μm	20,000 Å
Mitochondrion	2 μm	20,000 Å

Even the relatively large liver cell would have to be about five times larer to even be seen as a speck by the human eye.

4.3 What Can Be Seen With a Light Microscope?

The resolving power of a good light microscope is wavelength dependent. Shorter wavelength (blue) light has a better resolving power than does longer wavelength (red) light. Although the entire visible light spectrum is narrow, from 380 to 750 nm, it is possible to push resolution of the light microscope down to about 300 nm, or 3000 Å. Because a biological membrane bilayer is about 50 Å thick, the membrane would have to be 60 times larger than it is to be detectable even with a light microscope.

Recent advances in "light technology" are predicted to soon have an enormous impact on the imaging of cells. The development of blue LEDs (light-emitting diodes) won the 2014 Nobel Prize in Physics for its Japanese inventors, Isamu Akasaki, Hiroshi Amano, and Shuji Nakamura. LED is a new type of light source that is much brighter and more efficient than its predecessors, incandescence and fluorescence. Light intensity of a regular tungsten light produces about 16 lumen of light/watt of input energy and a fluorescent bulb produces about 70 lumen/watt but an LED exceeds 300 lumen/watt! The energy output of a tungsten light bulb is about 96% heat, but an LED produces more than 50% light. LEDs are narrow-band light sources based on semiconductor principles. By varying the composition of the diode central gallium nitride (GaN) crystal, LEDs can produce light of wavelengths that cover the entire electromagnetic spectrum from ultraviolet to infrared. Millions of colors are possible. Blue LEDs in the presence of a phosphor and three-color LEDs can both generate white light. Light from a blue LED is particularly useful in cell imaging due to its relatively good resolving power and its sharp, highly focused beam. Other color LEDs are more dispersed. It is anticipated that blue LED light will greatly enhance the sharpness of cell images.

4.4 What Can Be Seen With an Electron Microscope?

Resolution of a membrane was not possible until discovery of the electron microscope in 1931 by Ernst Ruska (Fig. 1.7), who later won the 1986 Nobel Prize in Physics for this achievement. The electron microscope, however, did not become commonly used until the 1950s. Most electron microscopes that are used with biological samples have a resolution of ~10 Å and so can readily distinguish a biological membrane. Even higher resolution is possible in certain cases.

FIGURE 1.7 Ernst Ruska (1901–1988). *Courtesy of Siemens AG, http://www.siemens.com/history/en/personalities/scientists_and_engineers.htm.*

TABLE 1.2 Diffusion Distance for Water in Water

Time (s)	Distance
0.1	10 μm
1	100 μm
100	1 mm

It is evident that living cells, and therefore the membrane that surrounds them, are very small, making their study all the more difficult. It is now believed that if living cells are found elsewhere in the universe, they, too, will be small. The size of a living cell is limited by a universal constant: the rate of solute diffusion in water.

$$\text{Diffusion rate} = K(T/m)^{1/2}$$

where K is a constant that is related to the size of the solute, T is the temperature in K, and m is the molecular weight of the solute. Even diffusion of water (a tiny solute) in water is very low (Table 1.2).

Diffusion works well over very short differences but is far too slow to be effective over long distances. This limits the size of cells. Diffusion is further discussed in Chapter 9. Coupled with the small sizes associated with membranes are the rapid times involved in membrane dynamics. Size domains spanning the μm-to-Å range and time domains ranging from μs to ns necessitate the use of combinations of biophysical instrumentation introduced in Chapter 9.

5. BASIC COMPOSITION OF MEMBRANES

A striking feature of biological membranes is that while at first glance all biological membranes may appear to be similar in size, structure, and basic composition, they support very different functions. The conundrum of how membranes can be similar yet different forms the

TABLE 1.3 General Composition, Expressed as Percent by Weight, of Three Very Different Mammalian Membranes

Membrane	Lipid	Protein
Myelin sheath	80%	20%
PM	50%	50%
Mitochondrial inner membrane	25%	75%

central theme of this book. The structural similarity of all membranes is to a large extent due to the polar lipids that compose the lipid bilayer (see Chapters 4 and 5), while biochemical diversity is due to the proteins that reside in the bilayer (see Chapter 6). As a general rule, the more biochemical functions a particular membrane supports, the higher will be its protein content. Table 1.3 demonstrates this concept. The myelin sheath has perhaps the least biochemical functions of any mammalian membrane (it provides insulation for nerves), while the mitochondrial inner membrane has many functions including a major connected pathway (electron transport and associated oxidative phosphorylation). The PM falls between the two extremes. The many types of membrane lipids and proteins are discussed in Chapters 5 and 6, respectively.

6. SUMMARY

Understanding membrane structure and function is one of the major unsolved problems in life science. Membranes are intimately involved in almost all biological processes, including establishing and maintaining transmembrane gradients, compartmentalizing biochemical reactions into distinct functional domains, controlling transport into and out of cells, intercellular and intracellular communication, cell–cell recognition, and energy transduction events. What makes biological membranes so difficult to study is their small size (<10 nm in width) and compositional complexity, being composed of hundreds of proteins, thousands of lipids, and numerous surface carbohydrates, all in constant flux. A microscopic liver cell is so packed with internal membranes that the total membrane surface area of a single cell is about 840 acres, the size of New York City's Central Park. Remarkably, all of the cell membranes are related to one another structurally, chemically, functionally, and developmentally. A basic conundrum is that while at first glance all membranes appear to be similar in size, structure, and basic composition, they support very different functions.

Chapter 2 discusses the chronology of membrane studies from their murky beginnings centuries ago to the classic Gorter and Grendel experiment (1925) that proposed the membrane lipid bilayer.

References

[1] UCSF School of Medicine. Office of educational technology. Prologue histology resource. Cell Structure Lab.
[2] Davidson MW. Molecular expressions. Cell biology and microscopy. Structure and function of cells and viruses. Tallahassee (Florida): National High Magnetic Field Laboratory (NHMFL) Florida State University; 1995–2015.

[3] Lodish H, Berk A, Matsudaira P, Kaiser CA, Kreiger M, Ploegh H, Scott MP. Molecular cell biology. 6th ed. 2007.
[4] Pollard TD, Earnshaw WC, Lippincott-Schwartz J. Cell biology. 2nd ed. Elsevier; 2007. 928 pp.
[5] Becker WM, Kleinsmith LJ, Hardin J, Bertoni GP. The world of the cell. 7th ed. San Francisco: Benjamin Cummings; 2009. 881 pp.
[6] Alberts B, Johnson A, Lewis J, Raff M, Roberts K, Walter P. Molecular biology of the cell. 4th ed. Garland Science; 2007. 1358 pp.
[7] Nelson DL, Cox MM. Lehninger principles of biochemistry. 5th ed. New York (NY): W.H. Freeman & Co.; 2008. 1262 pp.
[8] Berg JM, Tymoczko JL, Stryer L. Biochemistry. 6th ed. New York (NY): W.H. Freeman & Co.; 2007. 1120 pages.
[9] Webster's Online Dictionary.
[10] Lippincott-Schwartz J, Phair RD. Lipids and cholesterol as regulators of traffic in the endomembrane system. Ann Rev Biophys 2010;39:559–78.
[11] Cooper GM. The cell: a molecular approach. 2nd ed. Sinauer Associates; 2000.
[12] Doherty GJ, McMahon HT. Mediation and consequences of membrane-cytoskeleton interactions. Ann Rev Biophys 2008;37:65–95.
[13] Chi YH, Chen ZJ, Jeang KT. The nuclear envelopathies and human diseases. J Biomed Sci 2009;16:96.
[14] Becker WM, Kleinsmith LJ, Hardin J, Bertoni GP. The world of the cell. San Francisco (CA): Benjamin Cummings; 2009. p. 333–9.
[15] Porter KR, Claude A, Fullam EF. A study of tissue culture cells by electron microscopy. J Exp Med 1945;81:233–46.
[16] Pavelk M, Mironov AA. The Golgi apparatus: state of the art 110 years after Camillo Golgi's discovery. Berlin: Springer; 2008.
[17] Glick BS. Organization of the Golgi apparatus. Curr Opin Cell Biol 2000;12:450–6.
[18] Luzio JP, Pryor PR, Bright NA. Lysosomes: fusion and function. Nat Rev Mol Cell Biol 2007;8:622–32.
[19] Wanders RJ, Waterham HR. Biochemistry of mammalian peroxisomes revisited. Annu Rev Biochem 2006;75:295–332.
[20] de Duve C. The peroxisome: a new cytoplasmic organelle. Proc R Soc Lond B Biol Sci 1969;173:71–83.
[21] Mannella CA. Structure and dynamics of the mitochondrial inner membrane cristae. Biochim Biophys Acta 2006;1763.
[22] McBride HM, Neuspiel M, Wasiak S. Mitochondria: more than just a powerhouse. Curr Biol 2006;16:R551.
[23] Sagan L. On the origin of mitosing cells. J Theor Bio 1967;14:255–74.
[24] Margulis L. Symbiosis in cell evolution. New York: W.H. Freeman; 1981. p. 452.
[25] Christian de Duve Institute of Cellular Pathology, http://www.icp.ucl.ac.be/about/deduve.htm.

CHAPTER

2

Membrane History

OUTLINE

The current concept of membrane structure is based on the fluid mosaic model outlined by Singer and Nicolson in 1972 (see Chapter 8) [1]. Not surprisingly, this model did not just spring to life in a fully developed form but instead was conceived slowly over centuries [2,3]. This chapter discusses the two seemingly unrelated but parallel historical paths taken: the study of oil on water and the study of the cell outer barrier (plasma membrane). These two approaches did not converge until the classic experiment of Gorter and Grendel in 1925 [4].

1. OIL ON WATER: INTERFACE STUDIES

Although not appreciated until the twentieth century, the study of membrane physical properties had its origin in prehistoric observations by ancient mariners of oils floating on the surface of water. Some of the most revered figures in the history of science have contributed to this field.

An Introduction to Biological Membranes
http://dx.doi.org/10.1016/B978-0-444-63772-7.00002-6

1.1 Pliny the Elder (AD 55)

The concept of the oily nature of a membrane had a murky origin in the prehistoric past. The first written description concerned a practice commonly used in ancient times of stilling water surfaces by the use of cooking oils. The practice was used by ancient seafarers and was described in the first century AD by the naturalist Pliny the Elder (Fig. 2.1) in his 37-volume encyclopedic series *Naturalis Historia.* In Book 2, Chapter 106, Pliny described the ancient common knowledge about the effect of oils on rough seas, stating "all sea water is made smooth by oil and so divers sprinkle oil on their face because it calms the rough element and carries light down with them." Pliny died in the Mt. Vesuvius eruption of AD 79 when he decided to sail his boat into Pompeii to get a better look. Later, it was reported that oyster fisherman poured oil onto the surface of rough seas to calm them so they could better see shellfish sitting on the bottom. In a similar fashion, ancient fishermen applied oil to better see schools of herring at long distances. When whaling ships returned to port with their catch, the entire harbor was found to be covered with oil, and the water was not choppy. Although the calming effect of oil on the sea was common knowledge in ancient times, it remained little more than an observation until the pioneering experiments of Benjamin Franklin.

Pliny the Elder, whose actual name was Gaius Plinius Secundus, was a prolific writer. His *Naturalis Historia* is a collection of just about everything he found interesting in his surroundings, including important descriptions of manufacturing processes for papyrus and purple

FIGURE 2.1 Pliny the Elder, AD 23–79. *Courtesy Prints and Photographs Division, Library of Congress, LC-USZ62.*

dyes and for mining of gold. Because Pliny was probably the most famous person to die in a volcanic eruption, current volcanologists use the term "Plinian" when referring to a sudden volcanic eruption and "ultra-Plinian" when describing an exceptionally large and violent eruption. The 1883 eruption of Krakatoa was ultra-Plinian. Although no real image or description of Pliny the Elder survives today, the highly imaginative nineteenth century portrait shown in Fig. 2.1 is often passed off as being authentic.

1.2 Benjamin Franklin (1772)

Benjamin Franklin (Fig. 2.2) was an early American statesman, an internationally renowned scientist, and a world-class tinkerer. Scientifically, he is best known as being one of the discoverers of electricity based on his courageous kite experiment in 1750. However, Franklin had an exceptional curiosity and dabbled in many other ventures, including creating the first public library and fire department, inventing a new musical instrument (the glass harmonica for which Mozart wrote a piece), accurately mapping the Gulf Stream, suggesting Daylight Savings Time, and becoming the "father of metrology" (the science of measurement). One of Franklin's less-known contributions involved "the stilling of waves by oil." Franklin's oil slick work is described in an interesting 1989 book, *Ben Franklin Stilled The Waves* by Charles Tanford [5]. Tanford is better known for his development of the

FIGURE 2.2 Benjamin Franklin, 1706–1790. *This 1777 portrait of Benjamin Franklin (1706–1790) is by Jean-Baptiste Greuze.*

"hydrophobic effect theory" that describes the physics behind membrane stabilization (discussed in Chapter 3).

Franklin's wave-stilling experiment was first done on a mission to London representing the American colony Pennsylvania. In a 1772 visit to the Lake District in northern England, he poured a single teaspoon of olive oil (primarily triolein) on a pond in Clapham Common. The oil rapidly spread, smoothing the water surface over an incredible half-acre! He recognized that the oil layer must be very thin because it displayed "prismatic colors" as described by Issac Newton in his book *Opticks*. Franklin also noted that if a similar drop of oil was placed on a polished marble table, it did not spread, indicating the importance of both oil and water. Franklin's observations were not published in a formal scientific paper written by him but instead appeared in extractions of personal letters written to friends in 1774. These letters appeared the same year in *Philosophical Transactions of the Royal Society of London*. Unfortunately, Franklin missed a golden opportunity when he failed to estimate how thin the oil slick must have been. He knew the volume of the olive oil as well as the lake area it covered. A simple calculation would have estimated the oil layer at about 10 Å, a good estimate of the molecular length of triolein. This would have been the first correct measurement of molecular size, a parameter that would have to wait another 120 years until Lord Rayleigh repeated Franklin's experiment.

Franklin's work on oils on water were extracted from a letter to a friend, Dr. William Brownrigg, dated November 7, 1773. The extract was published in *Philosophical Transactions of the Royal Society of London*, in 1774 [6]. A quote from this letter follows:

> At length being at Clapham where there is, on the common, a large pond, which I observed to be one day very rough with the wind, I fetched out a cruet of oil, and dropt a little of it on the water. I saw it spread itself with surprising swiftness upon the surface; ... the oil, though not more than a teaspoonful, produced an instant calm over a space several yards square, which spread amazingly, and extending itself gradually till it reached the lee side, making all that quarter of the pond, perhaps half an acre, as smooth as a looking glass.

1.3 Lord Rayleigh (Real Name John William Strutt) (1890)

Lord Rayleigh (Fig. 2.3) was a well-educated, gentleman scientist of very high profile at the turn of the nineteenth century. In 1904, he became one of the first Nobel Prize winners, winning a Nobel Prize in Physics for his discovery of the noble gas argon and was the first to explain the age-old question of why the sky is blue [6]. During his long and prolific career, he published more than 500 papers, primarily on optics and sound, and all were personally hand written! Although, as his title implies, he was wealthy, he did not use a secretary. In 1890 Rayleigh repeated the Franklin experiment, but he took it a step further. He precisely measured both the volume of olive oil added to his experimental water bath and the surface area over which it spread. From these measurements, he calculated the molecular size of triolein, the major component of olive oil, to be 16 Å.

1.4 Agnes Pockels (1891)

Perhaps the most compelling of the major players in the colorful history of membranes was Agnes Pockels (Fig. 2.4) [5]. Miss Pockels, as she was always referred to, overcame almost

FIGURE 2.3 Lord Rayleigh, 1842–1919. *John William Strutt, Lord Rayleigh (1842–1919), from Popular Science Monthly; vol. 25, 1884.*

FIGURE 2.4 Agnes Pockels, 1862–1935. *Courtesy of the Archive of Braunschweig Technical Universit.*

insurmountable odds to become one of the earliest women to make a lasting impact on science. She lived in Braunschweig in north central Germany and, as was typical of girls in the late nineteenth century, had no formal education. Her job was to cook, sew, and take care of the house. However, she had a keen interest in science and dabbled in her kitchen with pots and pans, sewing threads, and buttons to study the phenomenon of oil on water. In the late nineteenth century, it was not unusual for scientists to work in kitchen-based homemade laboratories. The corecipients of the 1906 Nobel Prize in Physiology or Medicine, Ramon y Cajal and Camilo Golgi, worked in their kitchens (see Chapter 18).

Pockels' now famous kitchen experiments began in 1880 when she was only 18. In these primitive conditions, she worked out the basic methods of making and measuring lipid monolayers. Her basic methods are still in practice today. The Langmuir Troughs used today closely resemble those she developed in the late nineteenth century, and the sewing buttons she used to measure surface tension later became the du Nuoy ring (see Chapter 3). In 1891, she gathered her results and sent an unsolicited letter to Lord Rayleigh. Despite her lack of credentials, Lord Reyleigh realized the novelty of her work and helped to get her paper published in *Nature* within 2 months (1891) [7]. It is hard to imagine that someone with no position or formal education could dabble in her kitchen and have a first scientific paper immediately published in the prestigious journal *Nature*.

Miss Pockels' methodology was so superior to Rayleigh's that he changed to her procedure. During her "career" of 35 years, she published 14 articles. She also repeated the Franklin experiment in 1892 and obtained a molecular size for triolein of 13 Å. Her science career faded with World War I and the post-war problems in Germany. Her life finally ended much as it began, as a homemaker.

The first description of Pockels' experiments were in the form of a letter from her to Lord Rayleigh dated January 10, 1891. This letter was later included in the journal *Nature*, published on March 12, 1891 [8]. The following quote describes what would decades later be known as a "Langmuir Trough," although it was first invented by Agnes Pockels.

> A rectangular tin trough, 70 cm long, 5 cm wide, 2 cm high, is filled with water to the brim, and a strip of tin about $1^1/_2$ cm wide laid across it perpendicular to its length, so that the under side of the strip is in contact with the surface of the water, and divides it into two halves. By shifting this partition to the right or left, the surface on either side can be lengthened or shortened in any proportion, and the amount of the displacement may be read off on a scale held along the front of the trough.

Soon after her death in 1935, Nobel Laureate Irving Langmuir, the "dean of surface chemists," stated that her methodology "laid the foundation for nearly all modern work with films on water."

1.5 Irwin Langmuir (1917)

At a time when nearly all world-renowned scientists were European, Irving Langmuir was born in America and did his work while employed by General Electric in Schenectady, New York (Fig. 2.5). While at GE, he became famous for developing the nitrogen-filled incandescent light bulb. Because of the large income the new bulb made for GE, he was allowed free reign to experiment on just about whatever he pleased, including oil—water interface physics.

FIGURE 2.5 Irving Langmuir, 1881–1957.

In 1917, Langmuir also repeated the Franklin experiment and reported the molecular size of triolein to be 13 Å [9]. Langmuir is credited with refining air–water interface studies into a precise science. For this contribution, he was awarded the 1932 Nobel Prize in Chemistry [10]. Langmuir was one of the first to understand that molecules were not just simple spheres. He envisioned the asymmetric molecular shape of surfactants (molecules that accumulate at the air/water interface) as having a dual personality – one region being hydrophobic and another being hydrophilic. In his 1917 *Journal of the American Chemical Society* article, he wrote "Oleic acid on water forms a film one molecule deep, in which the hydrocarbon chains stand vertically on the water surface with the COOH groups in contact with the water." He also was the first to predict that hydrocarbon chains are very flexible and that a fatty acid with a double bond (oleic acid) occupies a larger cross-sectional area and is more compressible than is a similar straight-chain saturated fatty acid (stearic acid). For his many contributions, Langmuir was also awarded the Franklin Medal in 1934, although, ironically, he did not mention Franklin in his acceptance speech for the medal, a later publication of the speech, or his seminal 1917 paper. The surface science journal *Langmuir* is named after him. Remarkably, Langmuir's iconic 1917 report was the only one he ever published on lipid monolayers!

Langmuir had a very long and, for the most part, distinguished career at GE. However, in 1947 he dabbled in salting clouds to effect rain. In collaboration with the U.S. military, they dropped 80 kg of dry ice into a hurricane that was safely offshore in the Atlantic Ocean. Unfortunately, the hurricane abruptly changed direction and came ashore, doing extensive damage in Georgia. To frustrate potential lawsuits, the U.S. military classified the data, burying evidence of the failed project for decades.

2. THE LIPID BILAYER MEMBRANE

Our current understanding of membrane structure, the fluid mosaic model, is based on a lipid bilayer (see Chapter 8). A lipid bilayer is essentially two oil–water monolayers, (described above), placed back to back. The first biological membrane to be appreciated, studied, and characterized is the plasma membrane. We now know that every cell currently living on planet Earth and probably every cell that has ever existed on our planet are surrounded by a plasma membrane. The plasma membrane is the barrier that separates the cell interior from the rest of the external environment. The study of biological membranes began in 1665 with Robert Hooke's discovery of the cork cell and subsequently followed a long and tortuous path. Outlined next are only a few early but important steps in this process.

2.1 William Hewson (1773)

The earliest experiments on the plasma membrane were osmotic studies on red blood cells. The first important investigator was the underappreciated Englishman William Hewson (Fig. 2.6), a good friend and colleague of Benjamin Franklin. In fact, in 1770 Franklin sponsored Hewson for Fellow of the Royal Society. Hewson's work involved microscopic studies of erythrocyte shape. He noted that on the addition of water, erythrocyte shape changed from flat (discoid) to spherical (globular). With too much water, erythrocytes simply dissolved, a process now known as hemolysis.

Hewson was the first to show osmotic swelling and shrinking of erythrocytes and, from this, deduced the existence of a cell (plasma) membrane as a structure surrounding a liquid protoplasm [11]. Unfortunately, this important membrane conclusion was largely ignored. Hewson, however, was highly respected and made major contributions to several fields. He has sometimes been referred to as the "father of hematology." He is given credit for isolating fibrin and for defining the lymphatic system, for which he was given the Copley Medal in 1769.

In 1998, there was renewed interest in Hewson but for an entirely different reason. While in London (1757–1762 and 1764–1775), Franklin stayed at a house at 36 Craven Street that is now the home of the Benjamin Franklin House Museum. In 1772, Hewson ran an anatomy school at this same location. In 1998, workmen restoring the Franklin museum dug up the remains of six children and four adults hidden below the home. The bones dated to the time Franklin had stayed at the house. Most of the bones showed signs of having been "dissected, sawn or cut." One skull had been drilled with several holes. The question arose, could this have been a very old crime scene? And what did Franklin know about the happenings?

FIGURE 2.6 William Hewson, 1739–1774. *Courtesy of the National Library of Medicine.*

2.2 C.H. Schultz (1836)

Although membranes are too thin to be seen directly by the human eye or even by light microscopy (Chapter 1), their presence can be detected as an invisible barrier by plasmolysis or by staining the outer membrane surface with various dyes. In 1836, C.H. Schultz used iodine to visualize the erythrocyte plasma membrane. He was then able to estimate the erythrocyte membrane thickness to be about 220 Å, only a little larger than the currently accepted measurement of about 100 Å. The importance of this very early date in membrane studies must be put in historical context. Schultz's work came only 8 years after Wohler's classic experiment synthesizing the first organic molecule (urea) by heating an inorganic molecule (ammonium cyanate). This was the beginning of the end for the theory of "vitalism," a concept that believed organic molecules were somehow different from inorganic molecules by possessing a "vital spirit" imparted by God. Three years later (1839), Theodor Schwann established the cell theory. Incidentally, Schwann's promising career, which included the first use of the term "membrane," ended by age 30 as he was ridiculed for having the audacity to suggest that alcohol fermentation was the result of living organisms. Schultz's use of iodine foreshadowed later experiments from the mid to late nineteenth century using "vital dyes" in the study of membranes.

2.3 Karl von Nageli (1855)

The botanist Karl von Nageli (Fig. 2.7) is usually given credit for laying the foundation for cell membrane osmotic studies on plant cells. Indeed, it is much easier to accurately measure

FIGURE 2.7 Karl von Nageli, 1817–1891.

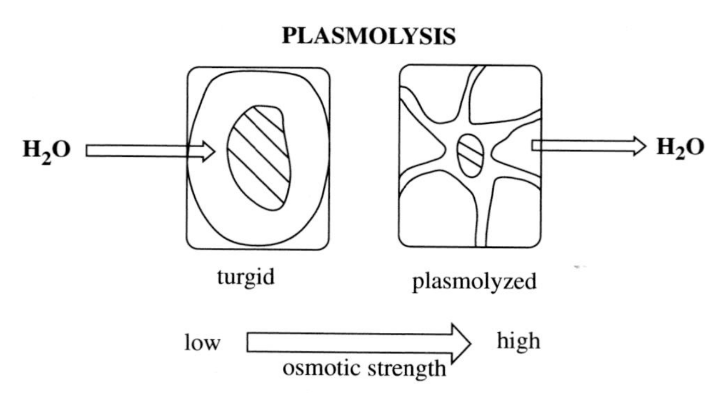

FIGURE 2.8 Plasmolysis in a plant cell. At a critical osmotic strength of the media, water leaves the cell and the plasma membrane pulls away from the cell wall. At this point the cell is said to be "plasmolyzed."

plasmolysis (Fig. 2.8) as a function of external osmotic strength with plant cells due to their thick and easily observable cell walls. Simple microscopic observations of plasmolysis on thin onion peels (only a few cells thick) are often used in high school biology classes. In a solution of sufficient salt or sugar, the cytoplasm is observed to pull away from the cell wall (Fig. 2.8). In animal cells, the plasma membrane barrier can be distinguished due to differences in light refraction across the membrane. Light travels slower in the dense cytoplasm than it does in the much less dense bathing solution. Using thin leaves of the common aquarium plant

Elodea, the location of the plasma membrane can be more readily observed than by light refraction. The outer plasma membrane barrier defines a track around which tiny green chloroplasts are carried via cytoplasmic streaming. This can be very easily detected with simple light microscopy.

In his day, von Nageli was a highly respected botanist. Among his discoveries were chromosomes and the function of many plant parts, including antheridia and spermatozoids of ferns. Unfortunately, he was stubborn and refused to accept new ideas proposed by others – and he believed in spontaneous generation! From 1866 to 1873, von Nageli exchanged letters with the Austrian monk Gregor Mendel, whom we now know as the "father of modern genetics." von Nageli did not appreciate or even understand Mendel's work and discouraged him from continuing. It was von Nageli who rejected the original paper by Mendel that, when rediscovered 40 years later, revolutionized genetics.

2.4 Wilhelm Pfeffer (1877)

Membrane theory, the first general theory of cell physiology, is often attributed to Wilhelm Pfeffer (Fig. 2.9). From his studies on the osmotic behavior of then invisible plasma membranes, he concluded the cell barrier must be thin and semipermeable. He proposed that the plasma membrane was similar to an artificial copper ferrocyanide membrane. Some 20 years later, Overton demonstrated that the permeability barrier (plasma membrane) was in fact lipid in nature.

FIGURE 2.9 Wilhelm Pfeffer, 1845–1920. *From Popular Science Monthly; vol. 50, 1896.*

2.5 Charles Ernest Overton (1899)

Charles Ernest Overton (Fig. 2.10) is credited with being the first true "membranologist." Although Overton was born in England, he worked in Germany and Sweden and published exclusively in German. Overton's major contribution to the understanding of membranes involved permeability studies [12,13]. In the late nineteenth century, it was believed that membranes were semipermeable, meaning at that time that they were only permeable to water. In sharp contrast, Overton found other molecules also crossed membranes, with neutral molecules crossing faster than charged molecules. Overton also showed that very different types of cells exhibited similar permeabilities, implying a common set of properties. He also found a parallel between a solute's membrane permeability and its solubility in olive oil. He concluded that membranes must include a lipid-like barrier. He also guessed, correctly, that membranes might contain cholesterol and lecithin (phosphatidylcholine), two molecules that were known at the time.

The permeability Overton measured for most tested solutes was "passive," meaning diffusion down the solute's concentration gradient by dissolving into and crossing the membrane (see Chapter 19). He also recognized the need for "uphill" (now called "active") transport against a gradient for some solutes that did not follow his lipid-solubility rule (see Chapter 19). Overton's work, published in 1899, was so far ahead of its time that it was not truly appreciated for decades. As an extension of his permeability studies, Overton later found an important relationship between an anesthetic's efficacy, its membrane permeability, and its solubility in olive oil (see Chapter 18). This correlation is now referred to as the Meyer—Overton theory (see Chapter 18) [14].

FIGURE 2.10 Charles Ernest Overton, 1855—1933. *Courtesy of Oregon State University Libraries Special Collections.*

2.6 Everet Gorter (1925)

During his entire and very long scientific career (born in 1881, he worked until he died in 1954), Evert Gorter (Fig. 2.11) was a highly respected Dutch pediatrician [15]. Gorter must have been a very busy man as he balanced two jobs: a paying one (pediatrician) and a nonpaying one (scientist). In addition, he had arthritis and was wheelchair bound.

In a 1925 article in the *Journal of Experimental Medicine*, Gorter and his research assistant, F. Grendel, published a very short report consisting of only five pages with no figures and one brief data table (shown in Fig. 2.12). This article [4] is now considered to be the most significant work ever done on membranes. However, as is often the case, this monumental insight was unappreciated when it was published and was only rediscovered many years later. Gorter and Grendel were the first to offer experimental proof that membranes are "lipid bilayers." They extracted erythrocyte membrane lipids with benzene and floated them on a Langmuir trough. On rapid evaporation of the benzene, a lipid monolayer was formed. They then used the earlier 1917 report by Langmuir [9] as a guide to compress the membrane lipid monolayer until it resembled a normal compression. Unfortunately, this was a guess. They knew how many extracted erythrocytes produced a measured monolayer area, and they calculated the surface area of a typical biconcave erythrocyte. Knowing each of these parameters, they discovered the area of a monolayer from extracted erythrocyte lipids was

FIGURE 2.11 Evert Gorter, 1881–1954. *Courtesy of Digitaal Wetenschapshistorisch Centrum (DWC).*

Animal	Amount of blood used for the analysis	No. of chromocytes per c.mm	Surface of one chromocyte	Total surface of the chromocytes (a)	Surface occupied by all the linoids of the chromocytes (b)	Factor a:b
	gm		*sq.μ*	*sq.m*	*sq.m*	
Dog A	40	8,000,000	98	31.3	62	2
	10	6,890,000	90	6.2	12.2	2
Sheep 1	10	9,900,000	29.8	2.95	6.2	2.1
	9	9,900,000	29.8	2.65	5.8	2.2
Rabbit A	10	5,900,000	92.5	5.46	9.9	1.8
	10	5,900,000	92.5	5.46	8.8	1.6
	0.5	5,900,000	92.5	0.27	0.54	2
Rabbit B	1	6,600,000	74.4	0.49	0.96	2
	10	6,600,000	74.4	4.9	9.8	2
	10	6,600,000	74.4	4.9	9.8	2
Guinea Pig A	1	5,850,000	89.8	0.52	1.02	2
	1	5,850,000	89.8	0.52	0.97	1.9
Goat 1	1	16,500,000	20.1	0.33	0.66	2
	1	16,500,000	20.1	0.33	0.69	2.1
	10	19,300,000	17.8	3.34	6.1	1.8
	10	19,300,000	17.8	3.34	6.8	2
	1	19,300,000	17.8	0.33	0.63	1.9
Man	1	4,740,000	99.4	0.47	0.92	2
	1	4,740,000	99.4	0.47	0.89	1.9

FIGURE 2.12 Data table from the Gorter and Grendel 1925 article indicating that there was enough lipid in the plasma membrane of erythrocytes to surround the cell exactly twice. This is the first experimental support for the existence of a lipid bilayer. *The table is taken from Gorter E, Grendel F. On bimolecular layers of lipids on the chromocytes of the blood. J Exp Med 1925;41:439–43.*

twice the surface area of the erythrocytes; hence, the membrane was a lipid bilayer. To prove the universal importance of their conclusion, they obtained erythrocytes from different animals, including humans, rabbits, dogs, guinea pigs, sheep, and goats. All specimens produced the same result. The report ends with: "It is clear that all our results fit in well with the supposition that the chromocytes (erythrocytes) are covered by a layer of fatty substances that is two molecules thick."

Although their conclusion that the lipid bilayer is the fundamental building block of membranes is now universally recognized to be at the heart of membrane structure, it is evident how lucky Gorter and Grendel were. The erythrocyte is perhaps the only eukaryotic cell that would give a ratio of monolayer area to cell surface area of 2. All other eukaryotic cells have extensive internal membranes that would significantly affect these measurements. The prediction that there is exactly enough lipid to surround a cell twice means the lipid bilayer had to constitute 100% of the membrane surface. There is no room for integral membrane proteins! Benzene is a poor solvent for extracting polar lipids, and Gorter probably extracted only about 70% of the membrane lipids. These authors did not compress their monolayer to biological pressures, and they miscalculated the surface area of an erythrocyte biconcave disc! All of these errors "conveniently" canceled out each other, resulting in a nice, interpretable ratio of 2. The only data table reported in their seminal report is shown in Fig. 2.12. Gorter went on to publish more than 50 additional articles, all on proteins. Grendel ceased publishing in 1929.

The year 1925 turned out to be a crucial year for the beginning of lipid bilayer studies. While Gorter and Grendel used the general, undefined term "lipoid" to describe the nature of the bilayer component, J.B. Leathes and H.S. Raper, in their 1925 book *The Fats* [16], suggested phospholipids might be essential structural elements of cell membranes. Also in 1925, the American Hugo Fricke used electrical impedance measurements to determine the thickness of the erythrocyte membrane [17]. His value of 33 Å was essentially correct, but he did not recognize it as a bilayer. This is surprising because his value is approximately twice the lipid monolayer values of 13–16 Å reported by Pockels, Rayleigh, and Langmuir.

In 1926, James B. Sumner (Fig. 2.13) reported isolation of the first enzyme, urease, an astonishing feat in its day [18]. For this work, Sumner was a corecipient of the 1946 Nobel Prize in Chemistry. Sumner accomplished his work despite having lost his left arm (he was left-handed) in a hunting accident when he was 17 years old. It is ironic that urea was the central compound in two of the major breakthroughs in biochemistry: Wohler's first synthesis of an organic molecule (urea) from an inorganic molecule (ammonium cyanate) in 1828 and Sumner's 1926 purification of urease. By the late 1920s, all pieces were in place to establish a realistic, working model for a biological membrane. This model was to first appear in a 1935 article by Davson and Danielli (discussed in Chapter 8).

3. SUMMARY

Membrane history has a very colorful past that followed two distinct paths. One path investigated lipid monolayers (oil on water), while the other monitored the plasma membrane of living cells. The first oil-on-water experiment is attributed to Benjamin Franklin in 1772. The Franklin experiment was greatly refined and quantified in the 1880s by Pockels, a most remarkable woman. Despite being an uneducated homemaker, at the age of 18 Pockels used old pots and pans, sewing threads, and buttons to measure lipid monolayer properties. Pockel's methodologies were further refined in 1917 by Langmuir. Paralleling the lipid monolayer studies were experiments done primarily on erythrocyte plasma membranes (hemolysis). At the end of the nineteenth century, Overton, the first true

FIGURE 2.13 James B. Sumner, 1887–1955. *Reproduced from http://chemistry.about.com/od/famouschemists/ig/Chemistry-Nobel-Prize-Photos/1946—James-B–Sumner-.htm.*

"membranologist," reported a relationship between solute solubility in olive oil and its membrane permeability. These early experiments led to the 1925 proposal of a membrane lipid bilayer by Gorter and Grendel.

Chapter 3 will discuss the most important molecule in life, water, and how it stabilizes membrane structure through the hydrophobic effect.

References

[1] Singer SJ, Nicolson GL. The Fluid Mosaic model of the structure of cell membranes. Science 1972;175:720–31.
[2] Eichman P. From the lipid bilayer to the Fluid Mosaic: A brief history of model membranes. SHiPS Resource Center for Sociology, History and Philosophy in Science Teaching; 1996.
[3] Ling GN. Life at the cell and below-cell level. The hidden history of a fundamental revolution in biology. Pacific Press; 2001.
[4] Gorter E, Grendel F. On bimolecular layers of lipids on the chromocytes of the blood. J Exp Med 1925;41:439–43.
[5] Tanford C. Ben Franklin stilled the waves. Duke University Press; 1989.
[6] Philos Trans R Soc London 1774;64:445–7.
[7] From Nobel Lectures, Physics 1901–1921. Amsterdam: Elsevier Publishing Company; 1967.
[8] Agnes Pockels A. Surface tension. Nature 1891;43:437–9.
[9] Langmuir I. The constitution and fundamental properties of solids and liquids. II. J Am Chem Soc 1917; 39:1848–906.
[10] From Nobel Lectures, Chemistry 1922–1941. Amsterdam: Elsevier Publishing Company; 1966.
[11] Kleinzeller A. William Hewson's studies of red blood corpuscles and the evolving concept of a cell membrane. Am J Physiol 1996;271:C1–8.

[12] Kleinzeller A. Ernest Overton's contribution to the cell membrane concept: a centennial appreciation. Am J Physiol Cell Physiol 1998;274:C13–23.
[13] Kleinzeller A. Membrane permeability – 100 years since Ernest Overton. Curr Top Membr 1999;48:1–22.
[14] Janoff AS, Pringle MJ, Miller KW. Correlation of general anesthetic potency with solubility in membranes. Biochim Biophys Acta 1981;649:125–8.
[15] Dooren LJ, Wiedemann HR. The pioneers of pediatric medicine: Evert Gorter (1881–1954). Eur J Pediatr 1986;145:329.
[16] Leathes JB, Raper HS. The Fats. In: Monographs on Biochemistry. 2nd ed. London: Longmans, Green and Co; 1925. 242 pp.
[17] Frlcke H. The electrical capacity of suspensions with special reference to blood. J Gen Physiol 1925;9:137–52.
[18] Sumner JB. The isolation and crystallization of the enzyme urease. J Biol Chem 1926;69:435–41.

CHAPTER

3

Water and the Hydrophobic Effect

OUTLINE

1. WATER: STRENGTH IN NUMBERS

What is so special about water? It is colorless, tasteless, odorless, and extremely abundant. At first glance it would appear that water is as bland and ordinary as any compound could possibly be. Yet water is the most studied material on Earth and has been for a very long time. It is evident from early historical records that water was known to be the most important biochemical required for life. Hidden by its deceptively unexciting cloak, water houses a plethora of subtle, but extraordinary properties. The basic properties of water can be found in any basic Biology or Biochemistry book, and are summarized nicely in Wikipedia [1].

Water is the most abundant molecule on the Earth's surface 70–75% of the Earth's surface is comprised of water, in both liquid and solid forms. The total volume of the Earth's water is approximately 1,360,000,000 km^3, with 97.2% in oceans, 1.8% as ice, 0.9% in ground water, 0.02% in fresh water and 0.001% as water vapor. The abundance of liquid water leaves one with the presumption that the fluid state is common on Earth. In fact, fluids are quite rare. The second most abundant fluid on this planet is petroleum, but even petroleum has its origins in water-based life.

An Introduction to Biological Membranes
http://dx.doi.org/10.1016/B978-0-444-63772-7.00003-8

From early Greek times through to modern space exploration it has been believed that water is essential for life as we know it. The Greek natural philosopher Thales of Miletus (c.640–548 BC) was reputedly the first to emphasize the fundamental importance of water, assigning it as the first principle of the cosmos, being "the origin of all things". Hippo of Samos (450 BC) perceived life as water. NASA now bases its search for life in the Universe on the assumption that without liquid water, life is not possible. Therefore the search for extraterrestrial life is predicated on the search for water.

Although not yet proven, it is likely water is abundant in all corners of the Universe [2,3]. Since H and O are among the most abundant elements in the Universe, it is reasonable to assume that water will be found everywhere as well. In our solar system, evidence of water has been found on the moon, Mercury, Mars, Neptune, Pluto, Jupiter's moons (Triton and Europa), and Saturn's moon (Enceladus) [4]. Of particular interest to the question of extraterrestrial life are Mars, Europa, and Enceladus.

In the eyes of the general public, it is and always has been Mars that is the most likely place to house alien life. Speculation about the existence of water and even life on Mars has existed since the first reports of canals on the red planet, which were viewed through telescopes by Angelo Secchi in 1858 and Giovanni Schiaparelli in 1877. Percival Lowell took the questionable descriptions of long straight channels and the unquestioned existence of Martian seasons to another level in the 1890s when he adopted the view that the canals were built by intelligent beings. Intelligent life on Mars soon became ingrained in the public conscience after the publication of *The War of the Worlds* by H.G. Wells in 1897.

In recent years, the United States has been joined by the Soviet Union, the European Space Agency, and India in launching successful ventures to the red planet. NASA's many Martian explorations have followed the trail of water in the search for extraterrestrial life (Table 3.1). At first, Mars appeared to be a dry desert with no pooled surface water (Viking 1 and Viking 2 in 1975). However, many subsequent landers, rovers, and orbiters demonstrated that there is actually a lot of water on Mars that is primarily located as frozen ice beneath the planet's surface. Much of the ice is associated with the South Polar Cap [5]. The total amount of frozen ice at or near the Martian surface is estimated to be an astonishing 5 million cubic kilometers, that if melted would cover the

TABLE 3.1 Successful Mars Landings

Name	Launch year	Rover
Viking 1	1975	Laboratory
Viking 2	1975	Laboratory
Pathfinder	1996	Sojourner
Spirit	2003	Spirit
Opportunity	2003	Opportunity
Phoenix	2007	Laboratory
Curiosity	2011	Curiosity

entire planet to a depth of 35 m! And even more ice is likely to be locked away in the deep subsurface [6].

Today, Mars cannot support any large body of liquid water due to its low atmospheric pressure (approximately 0.6% of Earth's) and low temperature (−63°C). However, strong geological evidence of fast flowing water can be found everywhere on the Martian surface. In addition, there are rocks and minerals that could only have formed in liquid water. The best potential locations for discovering life on Mars will be in subsurface environments [7].

Perhaps of even more importance than Mars, in the pursuit of extraterrestrial life, is Jupiter's moon Europa and Saturn's moon Enceladus. Europa is unlike other planetary moons. Its surface is smooth and uncratered, indicating the existence of a giant subterranean ocean made of water that is covered by a frozen ice crust. The frozen ice surface (−260°F) cracks as a result of Jupiter's enormous gravitational pull. The Jovian moon's surface is therefore covered with stress fracture lines that are colored, indicating the presence of slushy water that is extruded from the moon's interior. The extruded water refreezes on the surface, and likely contains organic solutes. The underlying water-ocean has been estimated to be perhaps as deep as 100 km. Warmed by Jupiter's gravitational field, Europa's interior is full of liquid water, which may in locations be of sufficient temperature to support life. Even if life did not have enough time to evolve on Europa, the moon may be a chemical evolution "laboratory", frozen in time.

Enceladus is a small moon revolving just outside Saturn's rings. Like Europa, Enceladus is home to a vast underground ocean. The moon's surface appears to be "fizzy like a soft drink and could be friendly to microbial life" [8]. Fig. 3.1 shows a Cassini spacecraft image of vaporous, icy jets emerging from fissures on Enceladus. If water is the most important biochemical for the development of life, Mars, Europa, and Enceladus are the places to explore.

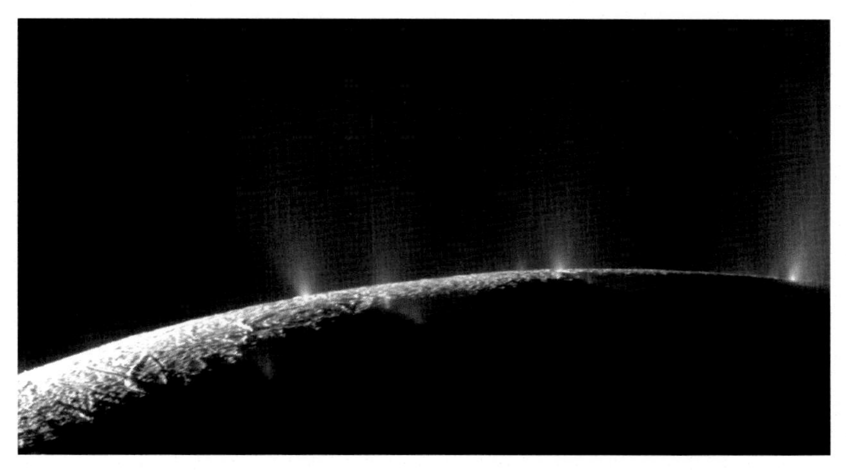

FIGURE 3.1 A Cassini spacecraft image of vaporous, icy jets emerging from fissures on Enceladus.

2. STRUCTURE OF WATER

The chemical composition of water, (two hydrogens and one oxygen), was discovered by Henry Cavendish in about 1781. Water is a polar molecule, having both positive and negative poles, and exists in a tetrahedral arrangement, extending out in all dimensions (Fig. 3.2). The partially positively charged hydrogens (each H is +0.41) are at an angle of 104.5° from each other with respect to the partially negatively charged oxygen (−0.82). The charge distribution produces a dipole moment of 1.85 D (Fig. 3.2). This polar structure is responsible for water's exceptionally strong self-attraction (cohesion) and numerous unusual properties that are responsible for its central role in life. Water is both small in size (MW 18) and highly mobile. The concentration of water in water (1000 $gl^{-1}/18\ gmol^{-1}$) is 55.5 M, making water both life's solvent, as well as being an essential reagent in life processes.

An important feature of water is its ability to readily form hydrogen bonds (H-bonds) with itself and other polar entities. H-bonding in water is the major driving force for membrane stability, as first suggested by Latimer and Rodebush [9]. As discussed in Chapter 2, the first quarter of the twentieth century was a critical time period that produced many seminal membrane studies. Included in this was a first understanding of the importance of water structure.

H-bonding occurs when an atom of hydrogen is shared between two electronegative atoms instead of residing on just one (Fig. 3.3). In the case of water, the hydrogen is shared unequally between two oxygens of adjacent waters. Water is unique in that it can form four H-bonds while other H-bonding molecules like HF, ammonia, and methanol can only form one H-bond. In fact, liquid water contains by far the densest H-bonding of any solvent. H-bonding accounts for the anomalous physical properties of water that are particularly important over a narrow temperature range (0–100°C), where water exists as a liquid and is therefore suitable for biological processes. For example, the similar compound H_2S that has much weaker H-bonding than water is a gas at biological temperatures.

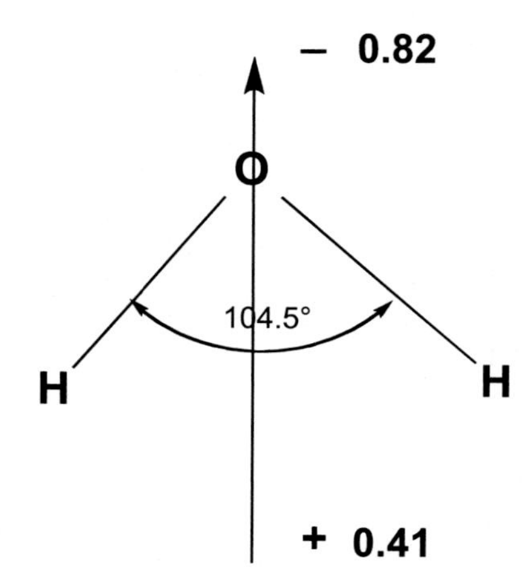

FIGURE 3.2 The structure of water.

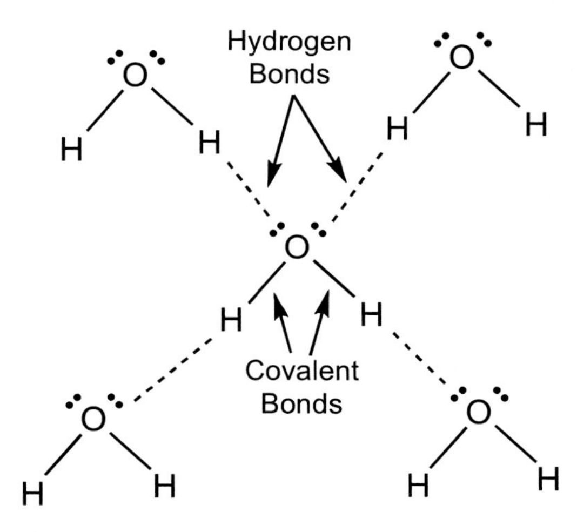

FIGURE 3.3 Hydrogen bonding in water.

At any one instant in time, a hydrogen is more closely associated with one water's oxygen (covalent bond length 0.101 nm) than the other water's oxygen to which it is H-bonded (bond length 0.175 nm). As a result the covalent bond (~492 kJ/mol) is stronger than is the H-bond (~23.3 kJ/mol). However, even the weaker H-bond is almost five times stronger than the average thermal collision fluctuation energy at 25°C, and both are far stronger than weak van der Waals interactions. The hydrogen residence is rapidly alternated between the two adjacent water oxygens, with an equilibrium time less than a femtosecond (10^{-15} s).

In the solid ice state, every water is H-bonded to four other water molecules producing an enormous, highly organized extended cluster. Upon melting, an initial assumption would be that the organized water clusters would be completely destroyed, hence resulting in a fluid. However, as shown in Table 3.2, this is not the case. The solid is transformed into a liquid as the result of breaking a surprisingly small number of H-bonds. Even near physiological temperature (40°C), water exists primarily in extended clusters, where most waters are still H-bonded to four other water molecules. It is not until water is in the gaseous state (100°C), that the extended clusters disappear.

TABLE 3.2 Size of Water Clusters as a Function of Temperature

Temperature (°C)	Number of water molecules in a cluster
0 (ice)	All (100)
0 (liquid)	91
40 (liquid)	38
100 (gas)	0

3. PROPERTIES OF WATER: NO ORDINARY JOE

The structure of water and its facility to form H-bonds dictates its extraordinary properties. For example, water is small, stable, abundant, and is the only pure substance on Earth that exists in all three states of matter, solid, liquid, and gas. The following list considers a few of water's special properties that are related to its central role in life [10–12]:

1. Most solids expand when they melt, but water expands when it freezes. In fact, water is the only substance on Earth where the maximum density of its mass does not occur when it becomes solidified. The density of liquid water (1.000 g/cm^3) is greater than solid ice (0.917 g/cm^3), causing ice to float in water and resulting in an approximate 9% increase in volume upon freezing. If ice were denser than liquid water, bodies of water would freeze from the bottom up, vastly reducing the amount of liquid water on Earth. Also, it is the expansion of water during freeze–thaw cycles that turns inhospitable rock into soil.
2. Water has a boiling point of almost 200°C higher than that expected, on the basis of boiling points of similar compounds. While the boiling point of H_2O is 100°C, that of H_2S is only −60°C. Therefore H_2S is a gas at biologically important temperatures and could not serve in place of water as "life's solvent".
3. Water's high latent heat of evaporation (2.270 kJ/kg) gives resistance to dehydration and provides considerable evaporation cooling.
4. Water has a melting point (0°C) at least 100°C higher than expected, on the basis of melting points of similar compounds, and also has a high heat of fusion (melting, 334 kJ/kg).
5. Water is an excellent solvent and is often referred to as a "miracle liquid" or the "universal solvent". Water is such a good solvent, that it almost always has some solute dissolved in it (approximately <5 ppb impurities). Almost every substance known has been found dissolved in water to some extent. Water efficiently dissolves salts like NaCl that have very limited solubility in most other liquids. Seawater for example is approximately 3.5%(w/v) salt. This property allows for facile transfer of nutrients necessary for life. Liquid water is an excellent solvent due to its polarity, high dielectric constant, and small size. In contrast, solid ice is a very poor solvent.
6. Although water has an unusually high viscosity [a measure of resistance to flow (0.001 Pa s at 20°C)], it is still 10^4–10^5 less viscous than the interior of a membrane.
7. Water conducts heat more easily than any other liquid, except mercury.
8. Water has an unusually high heat capacity. It takes more heat to raise the temperature of 1 g of water 1°C than any other liquid. Its specific heat is 1 cal/g°C or 4.186 J/g°C. As a result, water can store enormous amounts of thermal energy, minimizing the volume of body coolant required to remove metabolic heat. Due to its high heat capacity, high thermal conductivity, and abundance in cells, water is the major thermal regulator in life. Liquid water is truly special as it has over twice the specific heat capacity of ice or steam. Thermal properties of water are due to the ease of intramolecular H-bonding.
9. Water's pKa is 15.74 making the pH of unmodified water approximately 7.0, neither acidic nor basic. However, absolutely pure water is very hard to produce. CO_2 is

rapidly absorbed producing carbonic acid and dropping the pH of freshly distilled water to approximately 5.7.

10. Water is highly noncompressible. Even in a deep ocean at 4000 m, under a pressure of 4×10^7 PA there is only a 1.8% change in water's volume.
11. Pure water is an excellent electrical insulator and in fact does not conduct electricity at all! But water is such a good solvent it almost always has some solute dissolved in it, meaning that in a living cell where the salt concentration is high, water can readily conduct electricity.

4. SURFACE TENSION

Another important property of water, stemming from its structure, is surface tension [13]. Surface tension is of tremendous importance in the historical study of membranes (Chapter 2) and is essential in understanding membrane structure and stability. Water has the largest surface tension (72.8 dyn/cm at 25°C) of any common liquid, except mercury. For example, ethanol, a much weaker H-bonding molecule than water, has a surface tension of only 22 dyn/cm. Surface tension is responsible for holding raindrops together, for making ocean waves possible, and for the capillary action that brings water up a tree through the xylem. Water's surface tension is even high enough to support the weight of some insects like the Water strider. An extreme example is the Basilisk lizard that can actually "walk on water" for 10–20 min and so is often referred to as the "Jesus lizard" (Fig. 3.4).

Surface tension is a measure of the strength of the water surface film that reflects the strong intramolecular H-bond-driven cohesive forces of water. The force of cohesion (sticking

FIGURE 3.4 The Basilisk or "Jesus lizard" can "walk on water" due to water's high surface tension.

together) between water molecules is the same in all directions (Fig. 3.3). In bulk solution, every water molecule can be H-bonded to four other water molecules, but waters that are unfortunate to find themselves at the air–water or oil–water interface can only form two H-bonds. This is depicted in Fig. 3.5 for the case of adding hexane to water. The increase in number of H-bonds in the bulk water solution compared to the interface, make the bulk water structure more thermodynamically favorable. As a general rule, the more H-bonds that can be formed, the more stable the structure will be. Therefore water molecules at the air–water interface feel a net force of attraction that pulls them away from the interface and back into the bulk water. As a result, the entire liquid tries to take on a shape that has the smallest possible surface contact area with air or oil, therefore creating a sphere. Thus, ignoring gravitational affects, the shape of a raindrop would be a sphere. The stronger the H-bonds between molecules, the higher the surface tension will be. The intermolecular force of adhesion between water and a hydrocarbon that comprise the interior of a membrane is much lower than the intramolecular force of cohesion between the water molecules. Therefore water tries to minimize its surface contact area by separating from the hydrocarbon. This is known as the Hydrophobic Effect [10,11] and is the major stabilizing force for membrane structure (Section 5). Hydrobpobic means fear of (phobia) water (hydro).

Surface tension was one of the first quantitative measurements that was possible on a "model membrane" and so was heavily employed by the early monolayer experimental physicists, including Agnes Pockels and Irwin Langmuir (Chapter 2). Earliest measurements were done employing a Du Nouy Ring, named after the French physicist who developed the technique in the late 1890s. Actually, Agnes Pockels predated Du Nouy by using common shirt buttons! By this method, a platinum ring (now replaced by a platinum–iridium alloy) is attached to a sensitive tensiometer incorporating a precision micro balance. The ring is

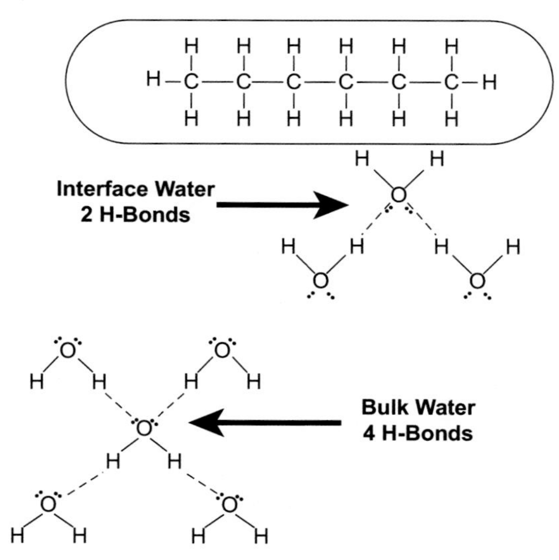

FIGURE 3.5 Dispersion of hexane into water is highly unfavorable due to the difference in H-bonds available to bulk-waters [4] compared to hexane-interface waters [2].

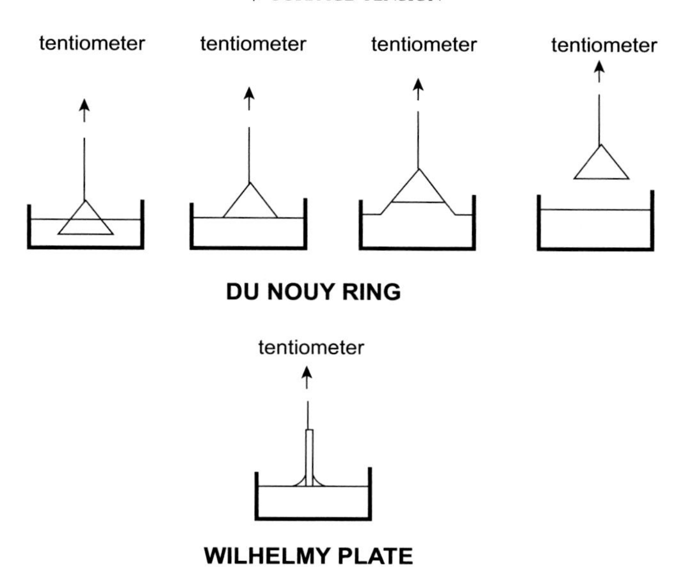

FIGURE 3.6 Du Nouy Ring and Wilhelmy Plate methods to measure surface tension.

immersed in the aqueous solution just beneath the air–water interface and the force required to pull the ring slowly through the interface is measured (Fig. 3.6). The force slowly increases until a maximum value is reached. At this point, the ring breaks free from the surface, leaving the solution, and the force drops to zero. Surface tension is directly related to the maximal force required to free the ring from the solution. A similar method in use today employs a Wilhelmy Plate in place of the Du Nouy Ring (Fig. 3.6). By this method, the Plate (usually made of platinum) is slowly lowered to the aqueous interface, whereupon the surface "jumps" onto the plate and the weight of the "jumped" solution is measured. It is remarkable that the simple century old technique to measure surface tension, developed before the concept of a lipid bilayer had even been imagined, has changed very little through the years.

Anything that disrupts the surface, decreases the surface tension. One class of molecules that are very effective at reducing the surface tension of water, are surfactants (surface active reagents). A very important class of surfactant that is involved in membrane studies are detergents (discussed in Chapter 13). Surfactants are amphipathic molecules, having both a polar and nonpolar end (Fig. 3.7). Surfactants accumulate at the air–water interface where their nonpolar tails stick into the air, avoiding unfavorable interaction with water while the polar end of the molecule sits comfortably in the water. The surfactants replace some of the surface interface waters, decreasing the net surface tension. For example, in an early study, the high surface tension of water (72.8 dyn/cm) was reduced to ~7–15 dyn/cm by adding membrane phospholipids to the interface. The implication is that a membrane surface would have a low surface tension.

Surfactants are often detergents (see Chapter 13) that employ reduction in surface tension to function. In one simple demonstration, metal needles are carefully floated on the air–water interface. Although metal is denser than water, the needles can float due to the surface

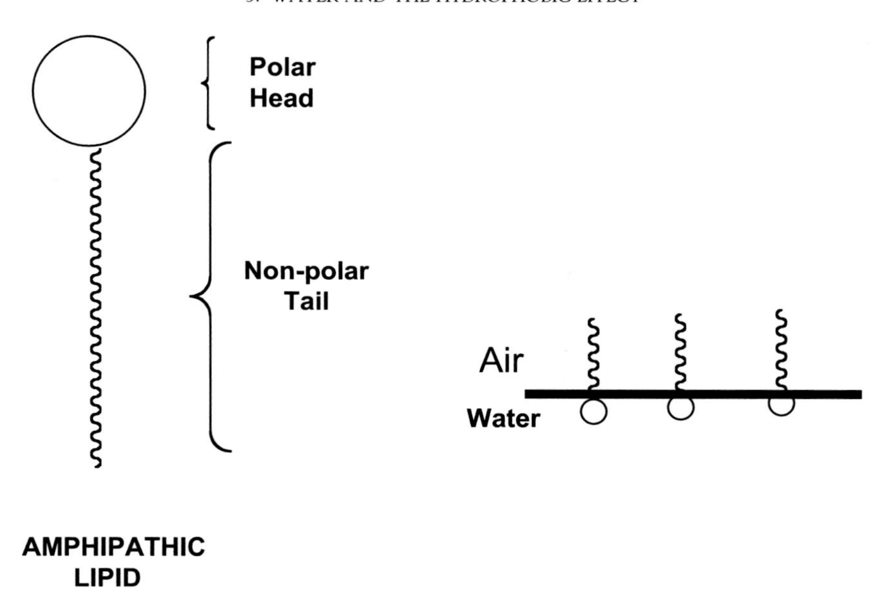

FIGURE 3.7 Structure of an amphipathic lipid and its orientation at the air/water interface.

tension of water. Upon the addition of a small drop of detergent to the aqueous subphase, the needles sink to the bottom. The heavier needles sink before the lighter needles, as the surface tension drops. Detergent surfactants play a critical role in disrupting or even dissolving membranes. They are essential for extracting proteins from membranes in a nondenatured form for subsequent reconstitution studies (Chapter 13). Detergents are a vital tool in membrane studies and are discussed in detail in Chapter 13.

5. THE HYDROPHOBIC EFFECT

It is quite remarkable that membranes are so stable without any covalent bonds holding them together! The subtle forces driving membrane bilayer stability must indeed be strong, if not obvious. These forces are referred to as the Hydrophobic Effect and have been elegantly discussed by Charles Tanford (Fig. 3.8) in his two books, *Hydrophobic Effect: Formation of Micelles and Biological Membranes* in 1973 [14] and *The Hydrophobic Effect* in 1980 [15]. The Hydrophobic Effect is attributed to intramolecular H-bonding between water. Charles Tanford is also the author of *Ben Franklin Stilled the Waves*, an interesting account of the early history of membrane studies [16].

H-bonds are the result of sharing a hydrogen between two electronegative atoms (eg, F, Br, O, N, S etc.). H-bonds cannot form between water and a hydrocarbon, as C and H have nearly equal affinity for electrons. If the hydrocarbon hexane is placed in a box of water in the absence of gravity, hexane will form a perfect sphere, indicating minimal contact surface area with water (Fig. 3.5). Entropy, a measurement of the degree of randomness, would

FIGURE 3.8 Charles Tanford, 1921–2009.

predict that without any other forces, hexane by itself should disperse, decreasing its order. However, observation clearly shows that in water, hexane prefers to exist in what appears to be an unfavorable, highly organized spherical structure. Remove water from the box and hexane would disperse. Therefore the cohesive forces that keep water together (H-bond forces) are much stronger than the hexane entropic forces that would disperse the sphere. Very weak van der Waals forces that would keep the hexane molecules together are also insignificant when compared to the cohesive forces of water. Hexane molecules stay together in water because it is essentially more favorable for the water molecules to H-bond to each other than to engage in very weak interactions with nonpolar hexane molecules. The bulk waters are surrounded by and H-bonded to four other waters, while the waters at the hexane interface are surrounded by and H-bonded to only two other waters (Fig. 3.5). The reduced number of H-bonds available to waters at the hexane interface makes them energetically less favorable. As a result it takes work to drive a hydrocarbon chain (ie, hexane or a fatty acyl chain) into water, creating a new, unfavorable interfacial surface.

The aversion of hydrocarbons for water is known as the Hydrophobic Effect and is the major driving force for all biological structure, including that of membranes. The hydrocarbon tails of membrane phospholipids segregate away from water making the membrane interior dense with hydrocarbon chains, but devoid of water. The energetics is similar to the case of the hexane sphere separating from water.

Fig. 3.9 depicts the five major forces contributing to membrane bilayer stability in water: the Hydrophobic Effect, head group–water interactions, head group–head group interactions, entropy of the caged tails, and van der Waals forces.

The Hydrophobic Effect is by far the most important force in membrane stabilization. In contrast to the hexane sphere discussed previously, membranes are composed of amphipathic lipids that also have a polar, often charged head group that can favorably interact with water, further stabilizing the membrane. Additional stabilizing forces can come from ionic interactions between the head groups of adjacent phospholipids. Another favorable force released upon membrane formation is related to the motion (entropy) of the bilayer

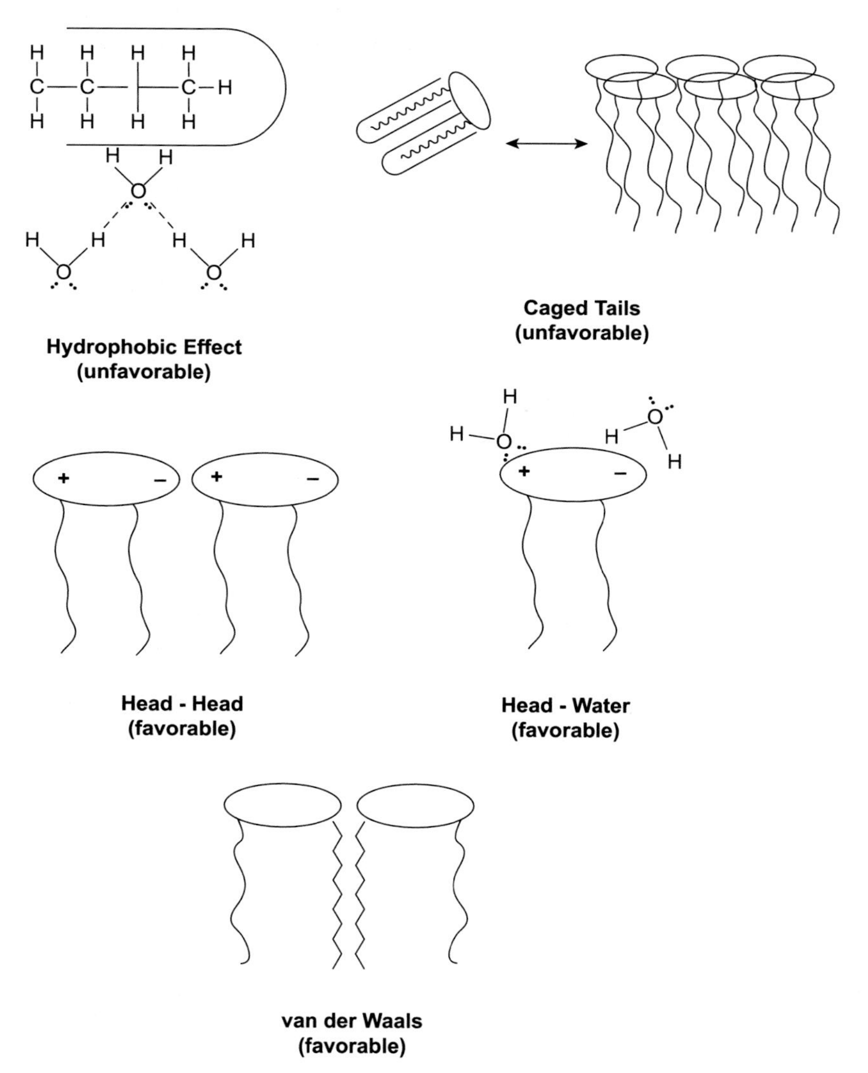

FIGURE 3.9 Five forces that stabilize the membrane lipid bilayer in water.

acyl chains. The hydrocarbon chains of phospholipids exposed to water would be totally surrounded by a stiff water cage where their motion would be highly restricted (low entropy, unfavorable). Upon shedding the water cage, the acyl chains are segregated into the hydrophobic bilayer interior resulting in a large increase in favorable acyl chain motion (increase in entropy).

The membrane interior is also held together very weakly by van der Waals forces. van der Waals forces are a collection of atomic, molecular, and surface interactions discovered by the Dutch scientist Johannes Diderik van der Waals (Fig. 3.10) in the late nineteenth century. For

FIGURE 3.10 Johannes D. van der Waals, 1837–1923.

TABLE 3.3 Bond Energies Expressed as Dissociation Energies for Several Types of Bonds

Bond Energies	
Bond type	**Dissociation energy (kcal)**
Covalent	400
Hydrogen bonds	12–16
Permanent dipole–dipole	0.5–2
van der Waal–London (induced dipole–dipole)	<1

his achievements, van der Waals received the 1910 Nobel Prize in Physics. van der Waals force (for membranes it is actually referred to as the van der Waals–London force) is a very weak force that is neither covalent nor ionic. Instead it is the result of induced dipoles that form instantaneously between two very close molecular surfaces. This force increases with increasing length of the nonpolar part of the adjacent surfaces and so is particularly important to the densely packed acyl chains that comprise the membrane hydrophobic interior. Therefore, there are at least five forces stabilizing the membrane bilayer, the most important being the Hydrophobic Effect.

The relative strengths of several types of bonds are listed in Table 3.3. From this Table it is obvious that covalent bonds are more than an order of magnitude stronger than H-bonds and H-bonds are more than an order of magnitude stronger than van der Waals–London forces.

6. SUMMARY

Water is the most important biochemical in life and is now considered to be a fingerprint of life itself. Despite its apparent simplicity, water houses a plethora of subtle, but extraordinary properties. Liquid water is abundant, small in size, and highly mobile. From a membrane perspective, it is water's propensity to form H-bonds that is responsible for its strong self-attraction (cohesion) and many remarkable properties. Importantly, water is primarily a fluid, but even at physiological temperature, liquid water exists as large extended clusters held together by H-bonds. Water is life's solvent, an essential reagent in life, the major thermal regulator in life, and the driving force for all biological structure. The aversion of hydrocarbons for water, known as the Hydrophobic Effect, is the major stabilizing factor for membranes.

Membranes are composed primarily of lipids and proteins, often with carbohydrates adorning their (outer) surface. Chapter 4 will discuss simple membrane lipids, with an emphasis on fatty acids.

References

[1] Properties of Water. In: *Wikipedia, The Free Encyclopedia*; March 5, 2016.
[2] Hanslmeier A. Water in the Universe. Series: Astrophysics and space science Library. 1st ed., vol. 368. Springer; 2011.
[3] Kotwicki V. Water in the Universe. Hydrolog Sci J 1991;36(1):49–66.
[4] Encrenaz T. Water in the solar system. Ann Rev Astron Astrophys 2008;46:57–87.
[5] Bibring J-P, et al. Perennial water ice identified in the South Polar Cap of Mars. Nature 2004;428:627–30.
[6] Christensen PR. Water at the poles and in permafrost regions of Mars. Geosci World Elem 2006;3(2):151–5.
[7] Didymus JT. Scientists find evidence Mars subsurface could hold life. Digit J Sci January 21, 2013.
[8] NASA/JPL/SSI. Cassini Flyby shows Enceladus venting. Mosaic by Emily Lakdawalla. 2011.
[9] Latimer WM, Rodebush WH. Polarity and ionization from the standpoint of the Lewis theory of valence. J Am Chem Soc 1920;42:1419–33.
[10] Nelson DL, Cox MM. Water. In Lehninger principles of Biochemistry. 5th ed. New York, NY: W.H. Freeman & Co; 2008. p. 43–66.
[11] Matthews R. Wacky water. New Sci 1997;154:40.
[12] Wiggins PM. Role of water in some biological processes. Microb Rev 1990;54:432–49.
[13] Surface tension by the ring method (Du Nouy method) Laboratory Experiments, Physics. Gottingen, Germany: Phywe Systeme GMBH; 2007.
[14] Tanford C. Hydrophobic effect: Formation of Micelles and biological membranes. John Wiley & Sons Inc; 1973. 208 pp.
[15] Tanford C. The hydrophobic effect. John Wiley & Sons Inc.; 1980. 234 pp.
[16] Tanford C. Ben Franklin stilled the waves. Duke University Press; 1989.

CHAPTER

4

Membrane Lipids: Fatty Acids

OUTLINE

1. WHAT ARE LIPIDS?

"Lipid" is derived from the Greek word *lipos* meaning "fat." Therefore, lipids are by definition insoluble in water but soluble in nonpolar organic solvents. Water insolubility is conferred via lipid molecular structure, having large regions of the surface composed of hydrocarbons with very few polar groups. A subset of lipids, those of interest in membrane studies, are structurally schizophrenic, containing segments that are polar and so prefer dissolution into water and more extensive regions that are totally nonpolar and avoid water at all cost. These lipids are referred to as "amphipathic," or containing both hydrophobic and hydrophilic moieties (Fig. 3.7), and are the focus of this book.

An Introduction to Biological Membranes
http://dx.doi.org/10.1016/B978-0-444-63772-7.00004-X

2. WHY ARE THERE SO MANY DIFFERENT LIPIDS?

Lipids have proved to be very difficult to study due to their overwhelming diversity [1]. Although the human body may contain about 1000 major lipids, countless minor lipids also exist. As a result, lipid studies lag far behind protein and nucleic acid studies, and it is often considered that "lipids are the poor relations among wealthier biological macromolecules." At first glance, there may appear to be an almost infinite number of lipid species, and new ones are being discovered each day. It has been estimated that a typical animal cell has about 1000 different major lipid species, which is a large number. But this is a pittance when minor lipids are folded into the calculation. For example, it has been reported that human tears contain more than 30,000 lipid molecular species [2,3]! And this is undoubtedly just the tip of the lipid iceberg. We probably know only a fraction of the total lipids in the biosphere. Comprehending large numbers of very different and often strange molecules makes scientists uncomfortable. The questions have often been raised: Why are there so many different lipids [1]? and can such an array of structures be organized into an understandable and useful classification system?

2.1 Lipid Functions

It is clear that after about 4 billion years of chemical and biological evolution, Nature does not make mistakes. There must be a reason, and hence discoverable function, for each and every lipid species. These functions can be roughly divided into three general categories: storage lipids, structural lipids, and active lipids. Storage lipids are generally nonpolar; they are normally found sequestered away from water in lipid droplets and are not commonly found in membranes. They are storage forms of fatty acids (FAs) (triacyl glycerols, see Chapters 5 and 14) that are densely packed biochemical energy sources. Triacylglycerols contain 9 kcal/g of energy compared with only 4 kcal/g for sugars. Structural lipids form the amphipathic matrix of membranes and are the focus of this book (see Chapters 9—11). Active lipids come in many shapes and sizes and possess potent biochemical activities (see Chapters 20 and 21). Often, they are found in membranes at much lower levels than structural lipids. Membrane lipids make up 5% to 10% of the dry mass of most cells, and storage lipids more than 80% of the dry mass of fat-storing adipocytes. Lipids exhibit a wide diversity of biological functions, which are primarily discussed in the second half of this book (Part II. Membrane Biological Functions). Lipid functions include:

1. Energy source (fats and oils)
2. Membrane bilayer component defining the cell permeability barrier (lipid bilayer)
3. Matrix for assembly and function of many catalytic processes
4. Chaperones for protein folding
5. Light absorber (eg, chlorophyll, retinal)
6. Energy transduction (eg, retinyl component attached to rhodopsin)
7. Electron carrier (eg, coenzyme Q)
8. Hormone (eg, testosterone, progesterone)

9. Vitamins (eg, vitamins A, E, D, and K)
10. Antioxidant (eg, vitamin E)
11. Signaling molecules (eg, ceramide)

3. LIPID CLASSIFICATION SYSTEMS

With so many different lipids, it is not surprising that several different classification systems have been proposed. However, any classification system is artificial and so has advantages and disadvantages. It does not matter what the system is; some lipids will not fit cleanly into a single category and may be placed into several categories or may even fall entirely between categories. One good example of this conundrum is classifying the membrane lipid sphingomyelin (SM; see Chapter 5). SM contains sphingosine and therefore can be classified as a sphingolipid, but it also contains phosphate, making it a phospholipid. Where does SM belong? Each year, countless numbers of previously unknown lipids are isolated and described, and any classification system must be flexible and expandable to accommodate the newcomers. It has become evident that a comprehensive nomenclature system is desperately needed before the task becomes too immense for the scientific community to tackle.

One important theme that runs through all of the proposed membrane lipid classification systems is the fundamental importance of FAs. The simplest division of lipids begins with the process of hydrolysis (discussed in Chapter 5). If the hydrolyzed lipid releases a free FA, it is deemed "complex." If it does not, it is "simple." Ironically, while esterified FAs, found in complex lipids, are an essential component of membrane structure, unesterified free FAs are at most a minor component of membranes and in fact are harmful to membrane stability (see Chapter 5). The two basic lipid types, simple and complex, are then further subdivided. Two classification systems are shown in Tables 4.1 and 4.2. The earlier system published by Hauser and Poupart [4] is presented in Table 4.1.

The newer and more comprehensive system (Table 4.2, from Fahy, E. et al. [5]) is based around "lipidomics," a loosely defined term encompassing several diverse fields of lipid study. Lipidomics is a large-scale study of non–water-soluble molecules (lipids). The term was first introduced in 2001 as an analog to the well-established fields of proteomics (based on proteins) and genomics (based on nucleic acids). The goal of lipidomics is to relate lipid composition to lipid physical properties and biological activities [6]. While the lipidomics-based classification system is comprehensive and is compatible with informatics, from a membrane structure perspective it is basically similar to the simpler system of Hauser and Poupart and is unnecessarily complex for the membrane studies outlined in this book. Only a small fraction of the lipids classified by Fahy et al. have a significant role in membranes. Here, we focus only on the lipids that *significantly* affect membrane structure and function. This book will not generally consider nonmembrane lipids or many other important aspects of lipids, such as nutrition.

The discussion of membrane lipids will begin with a lengthy description of FAs and then will be followed in Chapter 5 by the most important membrane structural lipids, complex FA–containing glycerolipids, glycerophospholipids, sphingolipids, and sterols.

TABLE 4.1 The 1991 Classification System of Hauser and Poupart [4]

1. Nonhydrolyzable (nonsaponifiable) lipids
- Hydrocarbons
  - Simple alkanes
  - Terpenes (isoprenoid compounds)
- Substituted hydrocarbons
  - Long-chain alcohols
  - Long-chain FAs
  - Detergents
  - Steroids
  - Vitamins

2. Simple esters
- Acylglycerols
- Cholesterol esters
- Waxes

3. Complex lipids
- Glycerophospholipids
- Sphingolipids

4. Glycolipids
- Glycoglycerolipids
- Glycosphingolipids
  - Cerebrosides
  - Gangliosides
- Lipopolysaccharides

TABLE 4.2 The 2005 "Comprehensive Classification System" of Fahy et al. [5]

Major categories
FA: Fatty acid
GL: Glycerolipid
GP: Glycerophospholipid
SP: Sphingolipid
ST: Sterol lipid
PR: Prenol lipid
SL: Saccharolipid
PK: Polyketide

This system divides all lipids into eight major categories and further subdivides the categories, assigning a 12-digit identifier to each lipid.

4. FATTY ACIDS

FAs are the major building block of complex membrane lipids, so their properties are fundamental to understanding membrane structure and function. An FA is a monocarboxylic acid usually with a long unbranched aliphatic tail that may be either saturated or unsaturated [7].

4.1 Chain Length

FAs commonly have a chain of 4 to 36 carbons (usually unbranched and even-numbered), which may be saturated or unsaturated. Chains of carbon length 4 to 6 are referred to as short-chain, 8 to 10 medium-chain, and 12 to 24 (or longer) as long-chain. Short-chain FAs are chemically more closely related to sugars than to fats and are not major components of membranes. Most membrane fatty acyl chains are 14, 16, 18, 20, 22, or 24 carbons in length. While present, odd carbon straight-chain FAs are far less abundant. Even carbon numbers reflect FA biosynthesis from two-carbon acetyl units (see Chapter 14). Acyl is a general term meaning an esterified FA of unspecified chain length.

4.2 Saturated Fatty Acids

Acyl chains may contain from zero to six double bonds. FAs with no double bonds belong to the class known as saturated. Saturated FAs, unlike unsaturated FAs, cannot be modified by hydrogenation or halogenations [8]. They are also highly resistant to oxidation because they have no susceptible double bonds. Saturated FAs are named for the saturated hydrocarbon with the same number of carbons. The series of even-numbered carbon saturated FAs from C-2 to C-24 is given in Table 4.3. Presented are the FA melting points (T_ms) that, with the exception of acetic acid, increase with chain length and their water solubility that decreases with chain length. Also note that at physiological temperature (37°C), FAs of C-10 or less are fluid (their T_m is below 37°C) while those of C-12 or more are solid (their T_m is above 37°C). The very short-chain FAs (C-2 and C-4) are completely miscible with water, while the very long-chain FAs (C-20, C-22, and C-24) are essentially insoluble in water. These properties will be important in later discussions of membrane "fluidity" (order), thickness, interdigitation, and permeability (see Chapter 9). The three most common saturated FAs found in membranes are myristic (C-14), palmitic (C-16),and stearic (C-18). These three FAs have been known for a long time as they were discovered before 1850! Stearic acid, for example, was first identified in 1823 by the French biochemist Eugene Chevreul, a full century before Sumner isolated the first enzyme, urease, from jackbean meal in 1925. Chevreul was the earliest pioneer in the isolation of membrane lipids, including stearic acid, oleic acid, and cholesterol (see Chapter 10).

All long-chain FAs, those important in membrane structure, have very limited water solubility but are soluble in organic solvents. The solubility of palmitic acid (16:0) in various organic solvents is presented in Table 4.4. The best solvent, chloroform, has historically been the solvent of choice for efficiently dissolving lipids. In fact, mixtures of chloroform—methanol—water have been the primary solvents used for membrane lipid extractions since

TABLE 4.3 The Series of Even-Numbered Carbon Saturated FAs From C-2 to C-24 With Melting Point and Water Solubility

Systematic name	Trivial name	Structure	Melting point (°C)	Water solubility (g/L, 20°C)
Acetic	Acetic	2:0	16.7	Infinite
Butanoic	Butyric	4:0	−7.9	Infinite
Hexanoic	Caproic	6:0	−3.4	9.7
Octanoic	Caprylic	8:0	16.7	0.7
Decanoic	Capric	10:0	31.6	0.15
Dodecanoic	Lauric	12:0	44.2	0.055
Tetradecanoic	Myristic	14:0	53.9	0.02
Hexadecanoic	Palmitic	16:0	63.1	0.007
Octadecanoic	Stearic	18:0	69.6	0.003
Eicosanoic	Arachidic	20:0	75.3	Insoluble
Docosanoic	Behenic	22:0	79.9	Insoluble
Tetracosanoic	Lignoceric	24:0	84.2	Insoluble

TABLE 4.4 Solubility of Palmitic Acid in Organic Solvents (g/L, 20°C)

Solvent	Solubility
Chloroform	151
Benzene	73
Cyclohexane	65
Acetone	53.8
Ethanol (95%)	49.3
Acetic acid	21.4
Methanol	37
Acetonitrile	4
(water)	0.007

1957 [9]. The large differences in lipid solubility for different organic solvents are an important property in separation of the various FAs for compositional analysis (Chapter 13).

4.3 Double Bonds

Membrane acyl chain double bonds are primarily *cis* (also designated Z) and are almost never conjugated. Normally, the double bonds are separated by intervening methylene groups that result in exceptional chain flexibility especially observed for FAs with multiple double bonds (see Chapter 9). The intervening methylene group is also a major target for lipid peroxidation. Many common membrane FAs are therefore readily oxidized and must be effectively protected by membrane antioxidants (eg, vitamin E; see Chapter 21).

Conjugated double bonds	–C–C=C–C=C–C–
Methylene-interrupted double bonds	–C–C=C–C–C=C–C–

If a single double bond is present, the FA belongs to the monoenoic or monounsaturated class (MUFAs), while FAs with more than one double bond are referred to as polyenoic or polyunsaturated FAs (PUFAs).

4.4 Fatty Acid Nomenclature

In order to define the structure of FAs, it is necessary to describe the number of carbons in the chain as well as the number, type (*cis* or *trans*), and position of each double bond. The nomenclature used in this book will first list the number of carbons in the chain, followed by a colon, and then the number of double bonds. Next, the position (superscript Δ followed by the carbon position of the double bond from the carboxyl terminus) is listed. It is assumed the double bond is *cis*, unless a *t* indicating *trans* is added. For example, elaidic acid, an 18-carbon *trans*-monoeonic acid with the double bond between carbons 9 and 10 is written $18{:}1^{\Delta 9t}$, while the homologous C-18 *cis* FA oleic acid is designated $18{:}1^{\Delta 9}$. The homologous sequence of C-18 FA is shown later (see Fig. 4.3) with the alternate omega designation (discussed next) included.

Melting points (T_ms) are an important general property of FAs that will have significant implications for later discussions of membrane "fluidity" (order), thickness, and permeability (see Chapter 9). Using the C-18 sequence as an example, on adding a first double bond (stearic → oleic) the T_m decreases by a very large −53.4°C (Table 4.5). Addition of a second double bond (oleic → linoleic) substantially further decreases T_m, by −21.2°C. The addition of a third double bond (linoleic → α-linolenic) decreases T_m but only by an additional −6°C. Compare this with the saturated FA series listed in Table 4.3. Saturated FA T_ms *increase* with every additional 2 carbons in chain length but only by a small amount, similar in magnitude to that noted for the addition of a third double bond in the C-18 sequence. T_m increases by +9.2°C in going from C-14 myristic to C-16 palmitic and by +6.5°C in going from palmitic to C-18 stearic. These T_m changes indicate that the effect of double bonds, particularly the first two double bonds, is much greater than that of changing carbon chain length.

TABLE 4.5 T_ms of the 18-Carbon Series of FA

Name	Structure	T_m	Omega
Stearic	18:0	69.6	0
Oleic	$18{:}1^{\Delta 9}$	16.2	9
Linoleic	$18{:}2^{\Delta 9,12}$	−5	6
α-Linolenic	$18{:}3^{\Delta 9,12,15}$	−11	3
γ-Linolenic	$18{:}3^{\Delta 6,9,12}$		6
Elaidic	$18{:}1^{\Delta 9t}$	43.7	9

4.5 Monoenoic Fatty Acids

The MUFA oleic acid was also discovered by Chevreul in his pioneering studies on pork fat in 1823 [10]. Now more than 100 MUFAs are known. The most important MUFAs found in membranes are listed in Table 4.6 with their corresponding melting point (T_m). The omega designation is referred to as n and is discussed later. It is evident in comparing the T_ms that saturated FAs of the same chain length have much higher T_ms than do FAs with one double bond. For example, stearic acid (18:0) has a melting point of 69.6°C, while oleic acid ($18{:}1^{\Delta 9}$) is 16.2°C.

4.6 Polyenoic Fatty Acids

All PUFAs have very low melting points (Table 4.7) and are highly susceptible to oxidative degradation (peroxidation) [11]. UV radiation, high temperature, oxygen, metals, and alkaline conditions are sufficient to alter these molecules by enhancing not only oxidation but

TABLE 4.6 The Most Important MUFAs Found in Membranes [8]

Systematic name	Trivial name	Structure	Melting point (°C)
cis-9-Hexadecenoic	Palmitoleic	16:1(n-7)	0.5
cis-6-Octadecanoic	Petraselinic	18:1(n-12)	30
cis-9-Octadecanoic	Oleic	18:1(n-9)	16.2
trans-9-Octadecanoic	Elaidic	18:1(n-9t)	43.7
cis-11-Octadecanoic	Vaccenic	18:1(n-7)	39
cis-11-Eicosenoic	Gadoleic	20:1(n-9)	25
cis-13-Docosenoic	Erucic	22:1(n-9)	33.4
cis-15-Tetracosenoic	Nervonic	24:1(n-9)	39

TABLE 4.7 Most Important Membrane PUFAs [11]

Systematic name	Trivial name	Structure	Melting point (°C)
9,12-Octadecadienoic	Linoleic	18:2(n-6)	−5
6,9,12-Octadecatrienoic	γ-Linolenic	18:3(n-6)	
9,12,15-Octadecatrienoic	α-Linolenic	18:3(n-3)	−11
5,8,11,14-Eicosatetraenoic	Arachidonic	20:4(n-6)	−50
5,8,11,14,17-Eicosapentaenoic	EPA	20:5(n-3)	−54
7,10,13,16,19-Docosapentaenoic	DPA	22:5(n-3)	
4,7,10,13,16,19-Docosahexaenoic	DHA	22:6(n-3)	−44

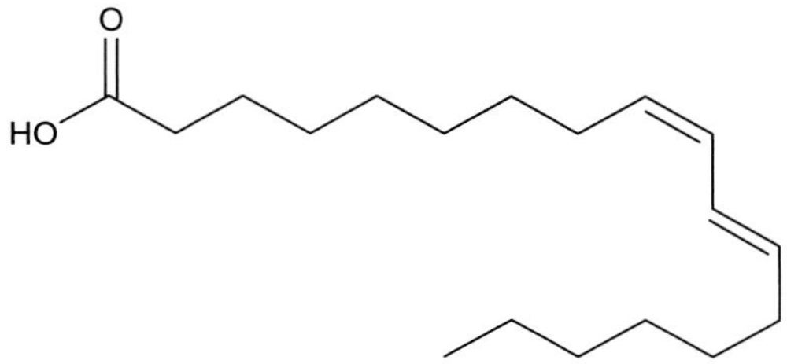

FIGURE 4.1 A conjugated linoleic acid (CLA, $18{:}2^{\Delta 9,11t}$).

also double bond migration and isomerization, resulting in a multitude of complex positional geometric isomers. Positional isomerization can occur in PUFAs yielding structures with double bonds adjacent to one another (conjugated double bonds). Conjugated linoleic acid (CLA, see Fig. 4.1) is a mixture of such isomers (more than 28 have been found to date) and has been used as a measure of free radical activity. CLA also has antibacterial properties and can even inhibit leukemic cell growth.

A wide variety of human health benefits have been reputed for the two long-chain omega-3 PUFAs eicosapentaenoic acid (EPA) and docosahexaenoic acid (DHA; see Fig. 4.2 and discussed in Chapter 23). In fact, it is almost impossible to find a single human affliction that has not been tested with EPA or DHA (see Chapter 23). Of particular importance are various cancers, heart disease, and immunological and neurological disorders [12,13]. Fig. 4.2 depicts structures of the ethyl esters of EPA ($20{:}5^{\Delta 5,8,11,13,14,17}$) and DHA ($22{:}6^{\Delta 4,7,10,13,16,19}$), both of which are used in enteral applications ("tube feeding").

4.7 Omega Designation

The omega nomenclature for FAs was proposed in 1964 by R.T. Holman [14]. This classification divides FAs into metabolic families reflecting how they were synthesized biochemically (discussed in Chapter 14). The system is based on location of the last double bond in

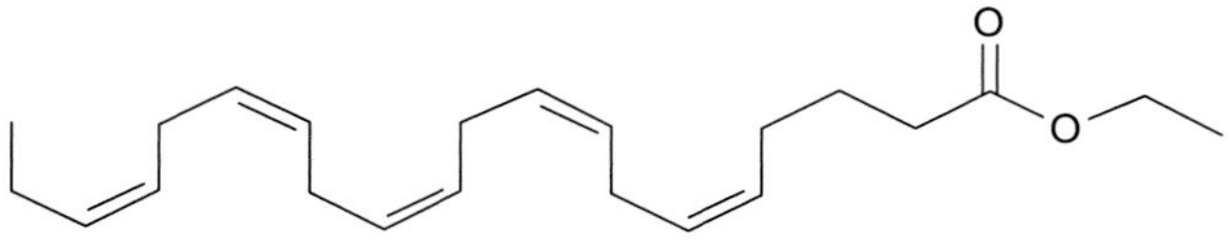

EPA: ethyl ester of eicosapentanoic acid

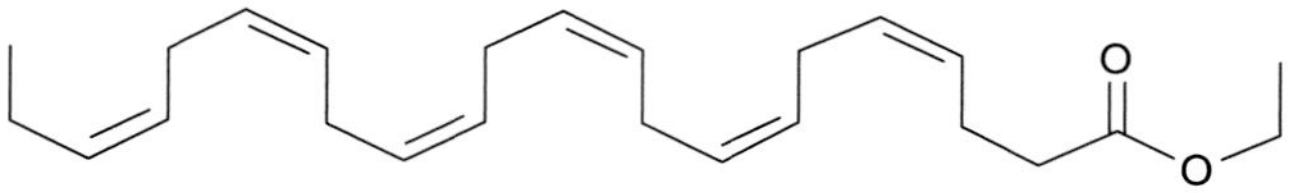

DHA: ethyl ester of docosahexaenoic acid

FIGURE 4.2 Ethyl ester of EPA and DHA.

the chain, specifically how many carbons it is from the terminal or omega methyl group. The terminal carbon is designated omega, *n*, or *ω*. Fig. 4.3 depicts the structures for the 18-carbon membrane FAs: stearic acid, oleic acid, linoleic acid, α-linolenic acid, and γ-linolenic acid. Also listed are the omega designation for each FA. For example, α-linolenic acid ($18{:}3^{\Delta 9,12,15}$) has the last of its three double bonds 3 carbons from the terminal (omega) end and therefore is an omega-3 FA (Table 4.7). Although docosahexaenoic acid ($22{:}6^{\Delta 4,7,10,13,16,19}$) has four more carbons in its chain than α-linolenic acid, it, too, is an omega-3 FA. The omega designation is therefore independent of FA chain length. It is now known that the C-18 omega-3 FA α-linolenic acid is far more similar to the C-22 omega-3 FA DHA in supporting human health than it is to the saturated C-18 stearic acid (18:0) or the monounsaturated C-18 oleic acid ($18{:}1^{\Delta 9}$). By this system, the major classes of FAs are saturated, omega-9, omega-6, and omega-3, although other minor classes of FAs exist.

A strong case has been made for the number and location of double bonds being more important for human health than chain length. The class known as omega-3s is of particular importance. Omega-3 FAs play crucial roles in cardiac, brain, and neurological function, as well as having roles in normal growth and development [12,13]. Because humans cannot make double bonds (desaturate) beyond the Δ9 position, essential omega-3 FAs including α-linolenic, EPA, and DHA must be primarily obtained from the diet (see Chapter 14). The long-chain omega-3 FAs EPA and DHA are often obtained from fish, such as salmon, tuna, and halibut and other marine life including algae and krill [15]. The shorter chain omega-3, α-linolenic acid, is a common component of plants that in humans is only inefficiently converted to the longer chain omega-3s by chain desaturation before the Δ9 position followed by chain elongation (see Chapter 14).

Although rarely found in plants, arachidonic acid (AA, $20{:}4^{\Delta 5,8,11,14}$) is the most abundant of the omega-6 series in animals. In humans, AA is a major component of membranes and is the principal precursor of eicosanoids, including the prostaglandins, isoprostanes, and isofurans (see Chapter 20). It is believed that the omega 6:omega-3 ratio is prognostic of human health. Ideally, this ratio should be between 1:1 and 4:1, but in the American diet, this ratio is currently between 10:1 and 30:1, resulting in high cardiovascular disease.

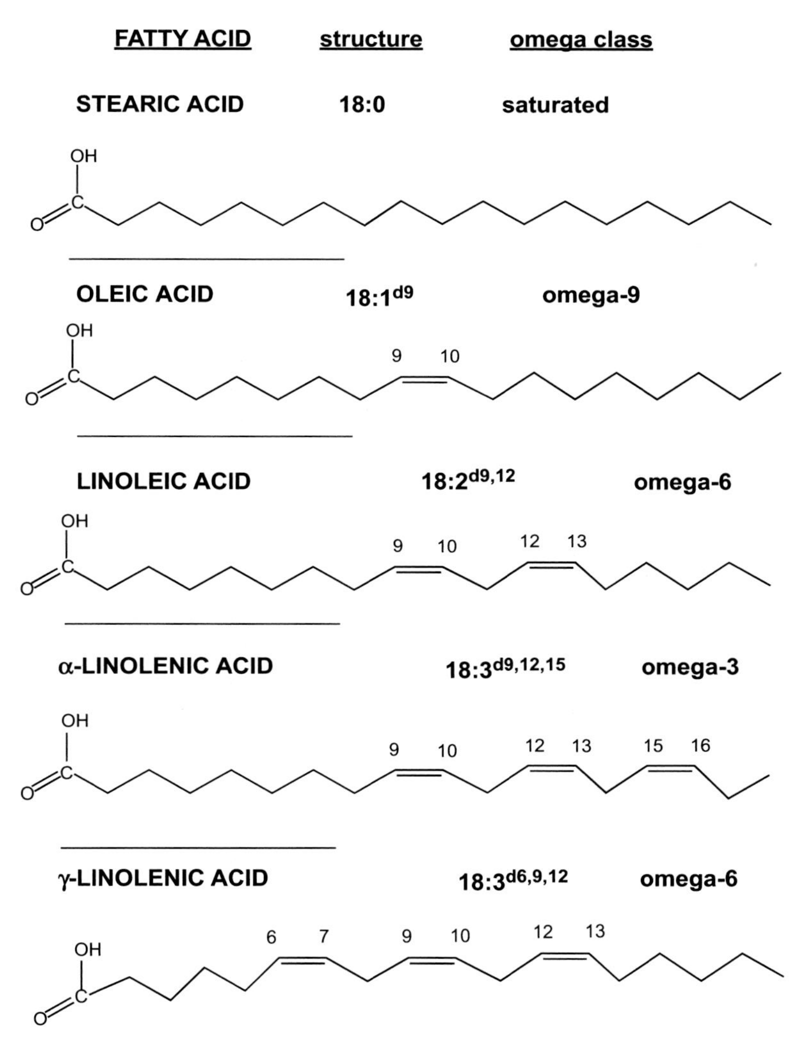

FIGURE 4.3 Structure of the 18-carbon sequence of important membrane FAs. Depicted are stearic acid, oleic acid, linoleic acid, α-linolenic acid, and γ-linolenic acid. Also listed is the omega designation for each of the FAs.

4.8 Most Abundant Membrane Fatty Acids

The FAs most commonly found in mammalian membranes are palmitic (16:0), stearic (18:0), palmitoleic (16:1), oleic (18:1), linoleic (18:2), arachidonic (20:4), and docosahexaenoic (22:6). Table 4.8 presents a typical mammalian membrane FA composition profile [16]. The presented FA compositions are of phosphatidylcholine (PC) isolated from rat liver membranes. FAs are expressed as weight percent.

TABLE 4.8 FA Composition of PC Isolated From Rat Liver Membranes

	FA								
	14:0	16:0	16:1	18:0	18:1	18:2	18:3	20:4	22:6
Rat liver	0.5	29.7	1.0	16.8	10.4	16.8		18.3	3.4
Mitochondria (outer)	0.4	27.0	4.1	21.0	13.5	13.5		15.7	3.5
Mitochondria (inner)	0.3	27.1	3.6	18.0	16.2	15.8		18.5	3.8
Plasma membrane	0.9	36.9		31.2	6.4	12.9	Trace	11.1	
Smooth endoplasmic reticulum	0.4	28.6	3.1	26.5	10.6	14.9		14.0	0.7
Rough endoplasmic reticulum	0.5	22.7	3.6	22.0	11.1	16.1		19.7	2.9
Golgi appartus	0.9	34.7		22.5	8.7	18.1	Trace	14.5	

Data From White DA. The phospholipids composition of mammalian tissues. In: Ansell GB, Hawthorne JN, Dawson RMC, editors. Form and function of phospholipids, 2nd ed., New York: Elsevier Scientific Publishing Company; 1973 pp. 441–482.

4.9 *trans*—Fatty Acids

Although far less common, some *trans* (also designated E) double bonds are routinely found in human membranes [17]. *trans*-FAs (TFAs) are further discussed in Chapter 23. TFAs enter the diet through dairy products, meat, and partially hydrogenated plant oils that are commonly found in processed food. TFAs are suspected to disrupt normal membrane functions, eventually leading to heart disease [18]. The most common TFAs found in humans is elaidic acid ($18{:}1^{\Delta 9t}$), an FA that has probably been in human membranes for millennia. Low levels of elaidic acid are not likely to be the cause of TFA-linked heart disease. Instead, it is probably the bewildering array of unnatural TFAs produced during partial catalytic hydrogenation of plant lipids, a commonly used procedure in processed foods, that cause heart disease. The question remains why *cis*-FAs are essential for human health while TFAs are harmful. This conundrum is further discussed in Chapter 23.

4.10 Branched Chain (Isoprenoid) Fatty Acids

The primary function of double bonds is to disrupt tight lipid packing in membranes, thus increasing the packing free volume ("breathing space") required for normal membrane protein function [19] (see Chapter 9). However, Nature has found other ways to duplicate the function of double bonds. In lower organisms, bulky methyl branches often replace double bonds. Higher organisms also contain repeating methyl branches in long hydrophobic structures called isoprenoids. These long hydrocarbon chains appear to be polymers of a five-carbon branched compound called isoprene (Fig. 4.4). Isoprene is a structural component of natural rubber, and about 95% of industrial isoprene is used to produce *cis*-1,4-polyisoprene, a synthetic rubber used to make diving suits.

In plants, the hydrophobic side chain of chlorophyll is phytol, a 20-carbon (4 isoprene units) isoprenoid. Many other important biochemicals, including the family of retinoids

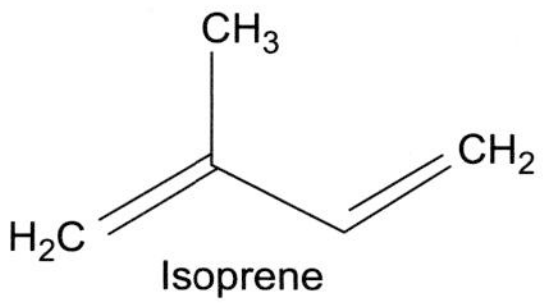

FIGURE 4.4 Isoprene.

(vitamin A) and ubiquinone (coenzyme Q), are anchored to membranes through long-chain isoprenoids (see Chapters 18 and 21).

In 1984, Schmidt reported that many proteins are anchored to membranes via long-chain isoprenoids [20]. This family of proteins has since been termed prenylated proteins and has been shown to facilitate cellular protein–protein interactions and membrane-associated protein trafficking. The largest group of prenylated proteins is the plasma membrane GTP-binding proteins (see Chapter 6) that transduce extracellular signals into intracellular changes via downstream effectors (see Chapter 18). Prenylated proteins are produced by posttranslational modifications where two types of isoprenoids are covalently attached to cysteine residues at or near the carboxyl-terminal (see Chapters 6 and 16). The two isoprenoids are farnesyl (15 carbons, 3 isoprene groups) or geranylgeranyl (20 carbons, 4 isoprene groups).

4.11 Unusual Fatty Acids

The number of frequently occurring FAs is less than 10 in plants and about 20 in animals. However, there is a tireless search for other rare and unusual FAs in various hidden corners of the biosphere. Unlike mammal FAs, plant and bacterial FAs often have odd-numbered carbon chains. Bacteria commonly have methyl branches or even other functional groups including cyclopropyl, cyclobutyl, cyclopentyl, and cyclohexyl rings or epoxy or hydroxyl groups in their FA chains. At one extreme is the most unsaturated FA found in Nature: 28:8 isolated from a dinoflagellate. There have now been thousands of strange FAs identified that are currently considered to be a curiosity in search of a function, and they will not be discussed in this book.

5. SUMMARY

Lipids are insoluble in water and have large regions of their surface composed of hydrocarbons with very few polar groups. Their aversion to water is responsible for the hydrophobic effect that drives membrane stability. Membrane lipids are "amphipathic," with a polar end anchoring the lipid to the aqueous interface and a long hydrophobic segment that forms the oily membrane interior. In human cells, there are more than 1000 major lipid species and countless minor lipids. There have been several attempts to classify lipids, the most recent being a lipidomics-based system by Fahy et al. from 2005. The most important lipid component of membranes is FA. An FA is a monocarboxylic acid with a long (normally 14 to 24 carbons), unbranched, hydrophobic tail that may be either saturated or unsaturated (usually zero to six *cis*, nonconjugated double bonds). FAs are the basic building blocks of complex membrane lipids and are largely responsible for membrane structure, fluidity, and function.

Although FAs comprise the bulk of the membrane hydrophobic interior, they rarely exist in free, nonesterified form. Chapter 5 discusses complex, polar membrane lipids where FAs are normally found esterified to phospholipids, sphingolipids, or some membrane proteins. A major nonesterified class of membrane lipid, sterols (particularly cholesterol), is also discussed.

References

[1] Dowhan W. Molecular basis for membrane phospholipid diversity:Why are there so many lipids? Ann Rev Biochem 1997;66:199–232.
[2] Nicolaides N, Santos EC. The di- and triesters of the lipids of steer and human meibomian glands. Lipids 1985;20:454–67.
[3] Borchman D, Foulks GN, Yappert MC, Tang D, Ho DV. Spectroscopic evaluation of human tear lipids. Chem Phys Lipids 2007;147:87–102.
[4] Hauser H, Poupart G. Lipid structure. In: Yeagle P, editor. The structure of biological membranes. Boca Raton, FL: CRC Press; 1991. p. 3–71 [Chapter 1].
[5] Fahy E, et al. A comprehensive classification system for lipids. J Lipid Res 2005;46:839–62.
[6] Wenk MR. The emerging field of lipidomics. Nat Rev Drug Discov 2005;4:594–610.
[7] Cyberlipid Center. Fatty Acids, online, non-profit scientific organization, www.cyberlipid.org/fa/acid0001.htm.
[8] The AOCS Lipid Library. Fatty acids: straight-chain saturated. Structures, occurrence and biosynthesis. 2011. Lipidlibrary.aocs.org.
[9] Folch J, Lees M, Stanley GHS. A simple method for the isolation and purification of total lipids from animal tissues. J Biol Chem 1957;226:497–509.
[10] The AOCS Lipid Library. Fatty acids: straight-chain monoenoic. Structures, occurrence and biochemistry. 2011. Lipidlibrary.aocs.org.
[11] The AOCS Lipid Library. Fatty acids: methylene interrupted double bonds. Structures, occurrence and biochemistry. 2011. Lipidlibrary.aocs.org.
[12] University of Maryland Medical Center. Contempory medicine. Omega-3 fatty acids. www.umm.edu.
[13] Holman RT. The slow discovery of the importance of omega 3 essential fatty acids in human health. J Nutr 1998;128:427S–33S.
[14] Holman RT. Nutritional and metabolic interrelationships between fatty acids. Fed Proc 1964;23:1062–7.
[15] Holub B. Dietary sources of omega-3 fatty acids. DHA/EPA Omega-3 Institute; 2011. www.dhaomega3.org.
[16] White DA. The phospholipids composition of mammalian tissues. In: Ansell GB, Hawthorne JN, Dawson RMC, editors. Form and function of phospholipids. 2nd ed. New York: Elsevier Scientific Publishing Company; 1973. p. 441–82.
[17] IUFoST Scientific Information Bulletin. Trans fatty acids. 2010.
[18] Mozaffarian D, Katan MB, Ascherio A, Meir J, Stampfer MJ, Walter C, et al. Trans fatty acids and cardiovascular disease. N Engl J Med 2006;354:1601–13.
[19] The AOCS Lipid Library. Fatty acids: branched chain. Structures, occurrence and biosynthesis. 2011. Lipidlibrary.aocs.org.
[20] Schmidt RA, Schneider CJ, Glomset JA. Evidence for post-translational incorporation of a product of mevalonic acid into Swiss 3T3 cell proteins. J Biol Chem 1984;259:10175–80.

CHAPTER

5

Membrane Polar Lipids

OUTLINE

1. INTRODUCTION

Free fatty acids are at best a minor component of most membranes, as large quantities are known to disrupt membranes through a "detergent effect" (see Chapter 13). When the concentration of free fatty acids in solution reaches a threshold value known as the critical micelle concentration (CMC; see Chapter 13), the fatty acids aggregate into mostly spherical structures known as micelles (Fig. 5.1). Other structures, including ellipsoids, are also possible. A typical fatty acid micelle is an aggregate with the hydrophilic carboxylic acid

An Introduction to Biological Membranes
http://dx.doi.org/10.1016/B978-0-444-63772-7.00005-1

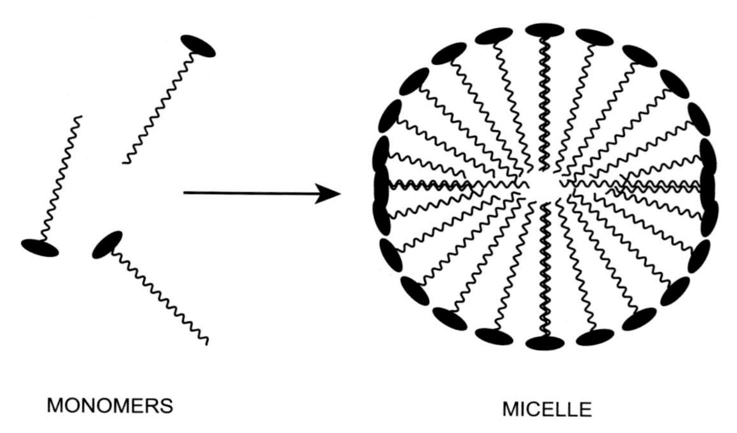

FIGURE 5.1 Amphipathic molecules form micelles when their CMC is exceeded.

heads in contact with the surrounding water and the hydrophobic tails sequestered away from water in the dry micelle center. Micelles can accommodate membrane phospholipids, thus destroying the membrane through the "detergent effect." CMCs for fatty acids decrease with increasing chain length and increase with increasing double bonds.

2. BONDS CONNECTING ACYL CHAINS

By definition, complex lipids are molecules that, when hydrolyzed, release a free fatty acid. The fatty acids are primarily attached to the rest of the lipid via an oxygen ester, although sphingolipids use a nitrogen ester (amide). Even sulfur esters (thioesters) are not uncommon (Fig. 5.2).

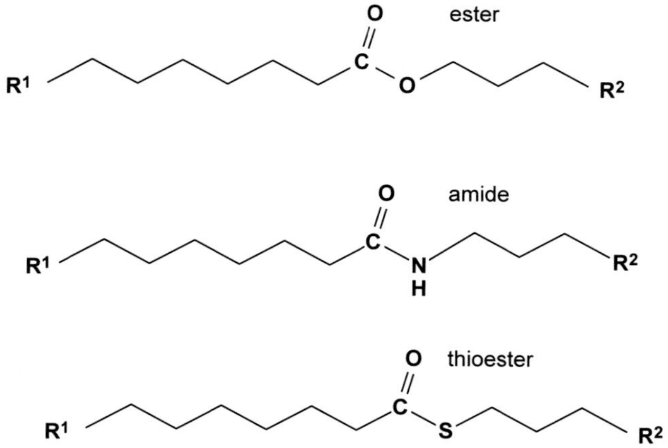

O ester
R1 C O R2

O amide
R1 C N H R2

O thioester
R1 C S R2

FIGURE 5.2 Types of bonds attaching fatty acids in complex lipids.

One important example of an ester-containing lipid is triacylglycerol, a nonpolar lipid called either a fat (if solid at room temperature) or an oil (if liquid at room temperature). A triacylglycerol has three fatty acids attached to a glycerol via ester bonds. A fat has predominantly high T_m saturated fatty acids, while an oil has predominantly low–melting point (T_m) unsaturated fatty acids. Esters can be hydrolyzed by boiling in NaOH (soda ash) or KOH (caustic potash). Boiling animal fat in KOH produces glycerol (also called glycerine) and three K^+ salts of fatty acids (Fig. 5.3). Boiling animal fat with caustic potash is the original method for making soap.

The term used for making soap by hydrolyzing a fat or oil in strong base is "saponification." It is clear that saponification has been known for a very long time [1], at least since the ancient Romans, because a soap-making factory was found in the ruins at Pompeii. Also, the word *sapo*, Latin for "soap," was first reported by Pliny the Elder in his great work *Historia Naturalis* [2]. Pliny described the making of soap from tallow and ashes. The origin of the word "saponification" is most interesting. It has been suggested to come from Sapo Mountain near Rome. According to Roman legend from about 1000 BC, women washing clothes in the Tiber River noticed that their clothes were cleaner if they washed them in a particular area. On the mountain above the special spot in the river was found a place where animals were sacrificed on an altar. The animal fat was inadvertently mixed with heated ashes and lye, producing a crude soap that washed into the Tiber. Hence the term saponification! Although this makes for a great story that for many years has been propagated by soap manufacturers, it is probably only a myth as Sapo Mountain is fictitious. Also, the first archeological evidence of soap manufacturing came from ancient Babylon (2800 BC), and a formula for soap consisting of water, alkali, and cassia oil was written on a Babylonian clay tablet around 2200 BC [3].

While many fatty acyl chains are attached to lipids and proteins by hydrolyzable ester or amide bonds, others are attached by nonhydrolyzable ether linkages. Ether bonds are particularly prevalent in very primitive organisms including Archaea. Archaea live in extreme conditions where membrane ester linkages would be highly susceptible to hydrolysis. They characteristically have long-chain (32 carbons), branched hydrocarbons linked at each end via ether linkages to glycerols that have either phosphate or sugar residues attached. These unusual lipids are twice the length of regular membrane phospholipids and so can span the bilayer while being anchored in both internal and external aqueous spaces. Another

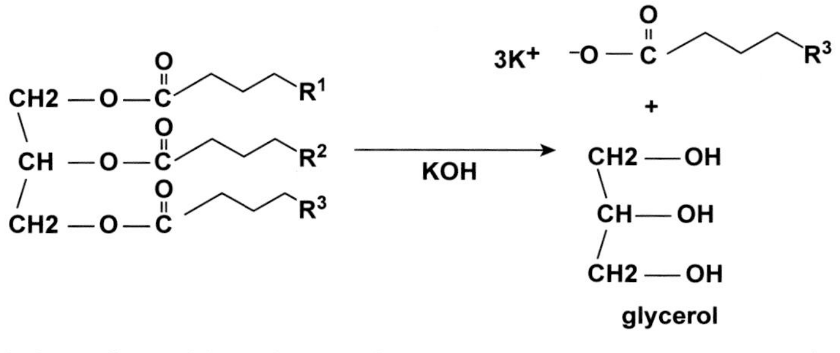

FIGURE 5.3 Hydrolysis of triacylglycerol (saponification) produces glycerol and three K^+ salts of fatty acids.

curiosity of these membrane lipids is that the glycerol central carbon isomer is opposite of that found in phospholipids (discussed and shown later in Fig. 5.7). It is interesting to note that while primitive organisms have stable ether linkages, higher organisms have emphasized less-stable ester linkages. Esters can be cleaved via saponification (chemical hydrolysis) or via the action of phospholipases (enzymatic cleavage, see Fig. 5.18), while ethers are resistant to both processes.

In humans, the members of one unusual phospholipid class called plasmalogens [4] have both an ester-linked acyl chain and an ether-linked chain (for the structure of plasmalogens, see Fig. 5.4). This raises the intriguing possibility that the ether linkage found in plasmalogens may be a vestige held over from an ancient past. For many years, plasmalogens and other ether lipids were considered little more than biological curiosities. However, their large abundance in Nature indicates that ether–lipid bonds must play significant roles in life processes, particularly those related to membranes. Vertebrate heart is particularly enriched in ether-linked lipids, where about half of the phosphatidylcholine (PC) is plasmalogen. Elevated levels of ether-linked lipids in cancer tissues and their role as a platelet-activating factor (important in molecular signaling) have greatly stimulated interest in these compounds. Although their molecular function remains unknown, it seems possible that it is related in some way to its resistance to hydrolysis or its ability to protect cells against the damaging effects of singlet oxygen. Two main types of ether bonds exist in natural lipids: alkyl and alkenyl (vinyl) ethers (Fig. 5.4). The double bond adjacent to the oxygen atom in the alkenyl ether has a *cis* or Z configuration.

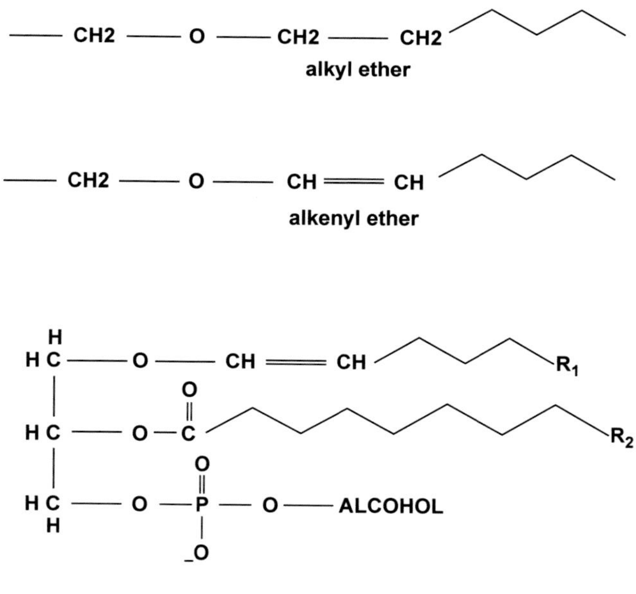

— CH2 — O — CH2 — CH2

alkyl ether

— CH2 — O — CH = CH

alkenyl ether

H
HC — O — CH = CH — R_1
HC — O — C(=O) — R_2
HC — O — P(=O)(O⁻) — O — ALCOHOL
H

PLASMALOGEN

FIGURE 5.4 Alkyl ether and alkenyl ether linkages and a plasmalogen. Plasmalogens have an ether link at the *sn*-1 position and an ester link at the *sn*-2 position.

3. PHOSPHOLIPIDS

The members of the major class of animal membrane polar lipids are commonly referred to as phospholipids (Comprehensive Classification System: glycerophospholipids) (for a general review, see [5]). Phospholipids [5,6,7,8,9] are composed of one glycerol, one phosphate, two fatty acids, and an alcohol. The molecule is constructed by removing four waters, creating four esters. The general structure of a phospholipid is depicted in Fig. 5.5. The name of the phospholipid class is derived from the alcohol. In animals, there are seven major classes of phospholipids (see Fig. 5.9 later).

A phospholipid is amphipathic, having both a polar head group and two very hydrophobic apolar acyl tails. Often, a phospholipid is represented by a simple cartoon drawing, as shown in Fig. 5.6. The oval head contains glycerol, phosphate, the alcohol, and the ester bonds attaching the two acyl chains to the glycerol.

3.1 Fatty Acids

The fatty acids are esterified to the glycerol at carbon positions C-1 (called the *sn*-1 position) and C-2 (the *sn*-2 position). As a general rule, in mammals, the *sn*-1 chain is saturated

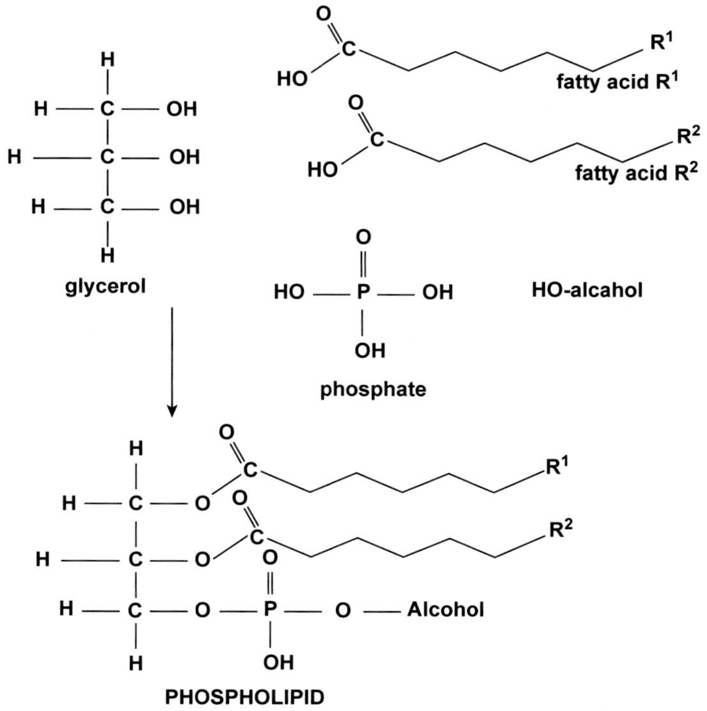

FIGURE 5.5 A phospholipid is formed from the condensation (dehydration) of one glycerol, two fatty acids, one phosphate, and one alcohol.

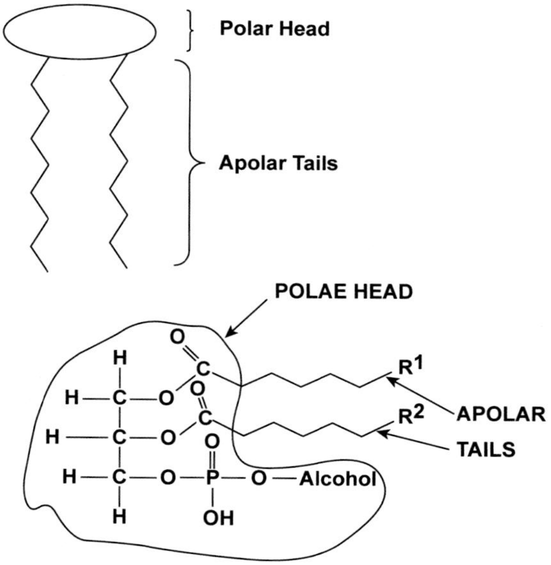

FIGURE 5.6 Cartoon depicting the amphipathic nature of a phospholipid.

while the *sn*-2 chain is unsaturated. The phosphate is esterified to the glycerol C-3 position. The types and structures of membrane fatty acids were discussed in Chapter 4.

3.2 Glycerol

Although the molecule glycerol is not optically active (it does not have an asymmetric carbon with four different groups attached), once the other phospholipid components are attached, the *sn*-2 position of glycerol becomes asymmetric (it now has four different groups attached) and hence is optically active. Consider the case for glycerol-3-phosphate. It comes in two optical isomers, D and L, differing only by the orientation of the same groups about the *sn*-2 carbon (Fig. 5.7). The biologically correct form is the D-isomer. While generally unappreciated, biological phospholipids are similar to amino acids and sugars in being optically active.

3.3 Phosphate

Phosphoric acid is a tri-acid with three different dissociable groups having pK_a values of 12.8, 7.2, and 2.1 (Fig. 5.8). As a first alcohol (glycerol) is esterified to phosphate, the highest pK_a, 12.8, is eliminated. Attachment of a second alcohol removes the next highest pK_a, 7.2, leaving only the lowest pK_a of 2.1. Because six of the seven common phospholipids have two phosphate oxygens esterified, the only dissociable phosphate group left has a pK_a of 2.1. At physiological pH, this group is almost totally dissociated (approximately 99.999% dissociated) resulting in a full negative charge on the phosphate.

L glycerol-3-phosphate D glycerol-3-phosphate

FIGURE 5.7 Optical isomers of glycerol-3-phosphate.

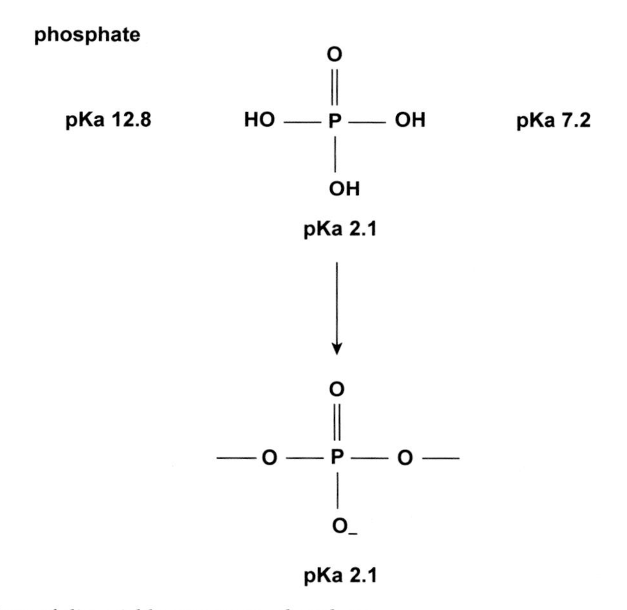

FIGURE 5.8 pK_a values of dissociable groups on phosphate.

All seven classes of phospholipids are amphipathic molecules that contribute to lipid bilayer structure. They vary from no net charge (they are zwitterions) to anions. None have a net positive charge at physiological pH. Therefore, one function of most phospholipids is to provide negative charge density to membranes. In addition, each phospholipid class undoubtedly provides other, perhaps unique functions to membranes. However, it is clear that phospholipids have no inherent catalytic activity. Catalysis is provided by resident membrane proteins. Currently, only a few of the more obvious unique phospholipid functions have been discovered. Unraveling these functions is complex, requiring a combination of in vivo and in vitro studies. Genetic mutations do not directly affect the phospholipids, but rather their biosynthetic pathways may indirectly alter many membrane events. Other more subtle functions for each class of phospholipids will undoubtedly be found in the future.

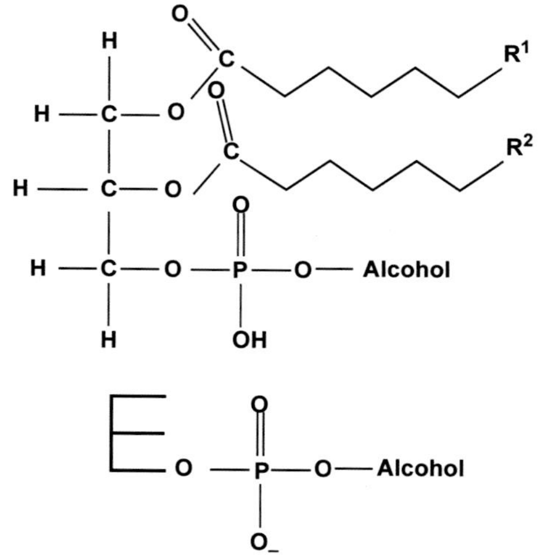

Phospholipid Name	Abbreviation	Alcohol
Phosphatidic Acid	PA	None
Phosphatidylethanolamine	PE	Ethanolamine
Phosphatidylcholine	PC	Choline
Phosphatidylserine	PS	Serine
Phosphatidylinositol	PI	Inositol
Phosphatidylglycerol	PG	Glycerol
Cardiolipin (Diphosphatidylglycerol)	CL	Phosphatidylglycerol

FIGURE 5.9 Simplified structure for phospholipids. Also included are the names and associated alcohol for each of the seven major phospholipid classes in mammalian membranes.

Fig. 5.9 depicts a simplified chemical structure that will be used in the discussion of the individual phospholipid classes. Also included are the names and associated alcohol for each of the seven major phospholipids.

3.3.1 *Phosphatidic Acid (Fig. 5.10)*

Phosphatidic acid (PA) is the smallest and simplest phospholipid and so is the precursor for other more-complex, alcohol-containing phospholipids [10]. As with all other phospholipids, PA composes part of the membrane lipid bilayer and contributes to the membrane's physical properties. At physiological pH, PA has a net charge of about −1.5 and so is a strong anion that contributes negative charge density to the membrane. PA has a unique shape, possessing a small, highly charged head group whose charge lies very close to the glycerol

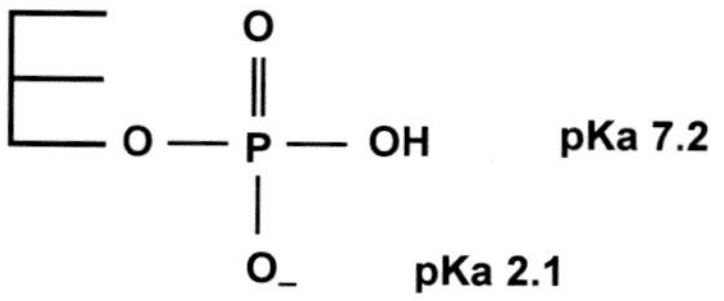

FIGURE 5.10 Phosphatidic acid.

backbone. PA's shape allows this phospholipid to induce high membrane curvature stress and is probably related to its significant role in membrane vesicle fission and fusion events. Membrane PAs can be aggregated by Ca^{2+}, resulting in the formation of PA-rich microdomains that are highly sensitive to pH, temperature, and cation concentration (see Chapter 11). PA also induces negative curvature stress on adjacent proteins, affecting their activity. The activity of more than 10 proteins including protein kinase C and phospholipase C have been shown to be altered by PA (for example, see [11]).

PA is a vital cell lipid that is maintained at extremely low levels in the cell by the activity of potent lipid phosphate phosphohydrolases. Lipid phosphate phosphohydrolases rapidly convert PA to diacylglycerol (DAG), a metabolic precursor for many lipids. PA has emerged as a new class of lipid mediators involved in diverse cellular functions in plants, animals, and microorganisms (see Chapter 20) [12].

3.3.2 *Phosphatidylethanolamine (Fig. 5.11)*

Phosphatidylethanolamine (PE), infrequently referred to as cephalin from the word *cephalic* meaning "pertaining to the head," is the second most prevalent phospholipid in humans but is the principal phospholipid in bacteria. This is due to the ability of humans to convert PE to PC, whereas bacteria cannot and so bacteria are relegated to having PE as their major membrane phospholipid. PE, PC, and cholesterol are also the major components of egg yolk, a ready and inexpensive source of these lipids. In humans, PE is found particularly in nervous tissue including the white matter of the brain, nerves, and in the spinal cord. PE is a primary amine-containing phospholipid and therefore has a highly reactive chemical handle that can be easily derivatized. This property will be exploited in Chapter 9 for various membrane biochemical studies. Because the pK_a of the PE amine is about 8.5, at physiological pH the amine is mostly, but not completely, protonated. With the fully dissociated phosphate, PEs have a slightly negative net charge. PEs are therefore not quite a full zwitterion. PE's primary role in membranes is as a major structural lipid. PE's head group is not only chemically reactive but also small and poorly hydrated. With unsaturated chains, PEs are pyramid shaped, with a wide base and narrow head. Phospholipid shapes are discussed in Chapter 10. Unlike PCs, in the absence of other phospholipids, PEs prefer nonlamellar phase including reverse hexagonal H_{II} phase, an unusual membrane structure that is discussed in Chapter 10. PEs can also form hydrogen bonds with neighboring polar groups, greatly influencing membrane structure.

3.3.3 *Phosphatidylcholine (Fig. 5.12)*

Natural PCs are often referred to by an old term, lecithin. Actually, lecithin is a complex mixture of many lipids, the predominant one being mixed acyl chain PCs. PC is the major

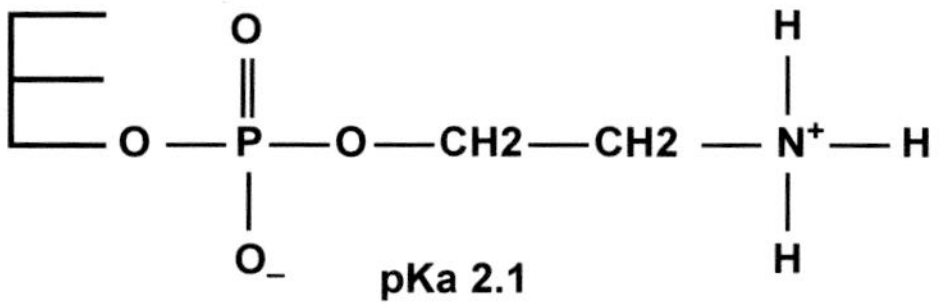

O H
O—P—O—CH2—CH2—N^+—H
O^- pKa 2.1 H

FIGURE 5.11 Phosphatidylethanolamine.

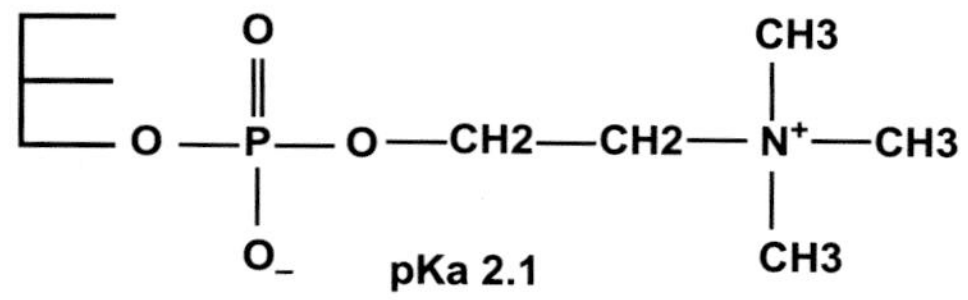

FIGURE 5.12 Phosphatidylcholine.

phospholipid found in animal membranes and therefore is the most important membrane structural lipid component. PC is ideally suited to fit comfortably into membranes. The molecule is can-shaped, with its polar head group being about the same width as the sum of its apolar tails (Chapter 10). In addition, PC is a true zwitterion possessing a formal positive and negative charge but no net charge. Therefore, PC does not exhibit negative charge repulsion, a potential problem with most other phospholipids. In humans, PC can be made by methylating the nitrogen of PE three times, a process missing in prokaryotes. In addition, dietary lecithin is the major source of the essential biochemical choline.

The major structural phospholipids in animal membranes are PC more than PE (Table 5.1). These lipids will predominantly dictate many membrane properties including "fluidity." Although "fluidity" is to a large extent dictated by the length and degree of unsaturation of the acyl chains, the nature of the polar head group is also important. As demonstrated in Table 5.2, PEs with identical chains generally have higher T_m values than PCs by about 20–40°C and therefore at a fixed temperature are less "fluid" than PCs.

3.3.4 *Phosphatidylserine (Fig. 5.13)*

While PC and PE are predominantly structural lipids responsible for maintaining the lipid bilayer matrix of membranes, phosphatidylserine (PS) has many additional functions [13,14]. In fact, PSs, along with phosphatidylinositols (PIs), are the most functionally diverse and dynamic of all of the phospholipids. Although PS is an essential lipid in all human cells,

TABLE 5.1 Average Lipid Composition of a Mammalian Liver Cell

Lipid	Mol (%)
PC	45–55
PE	15–25
PI	10–15
PS	5–10
PA	1–2
CL	2–5
SM	5–10
Cholesterol	10–20

TABLE 5.2 Melting Point (T_m) Values for PCs and PEs Where Both the *sn*-1 and *sn*-2 Acyl Chains are Identical

Acyl chain	diPC (°C)	diPE (°C)
Myristic	23.6	51
Palmitic	41.3	63
Stearic	58	82
Oleic	−22	15
Elaidic	5	41

PEs have higher T_ms than do homologous PCs by more than 20°C.

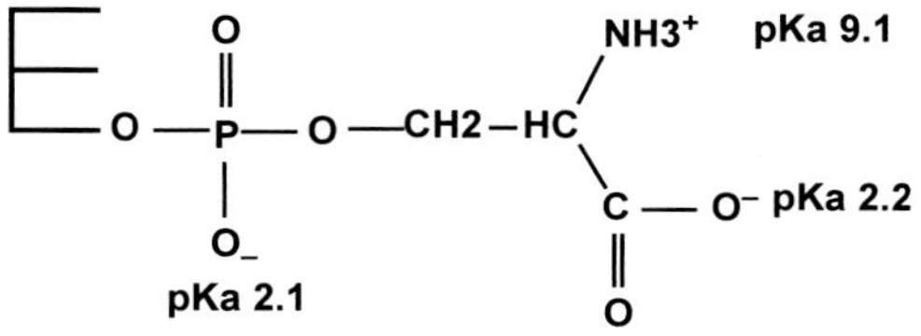

FIGURE 5.13 Phosphatidylserine.

its major functions are associated with the brain, where it is far more abundant than in other organs. Many clinical trials have suggested that PS supports a variety of brain functions that diminish with age [15,16]. Dietary supplementation with PS can alleviate, ameliorate, or sometimes even reverse age-related decline of memory, learning, concentration, word skills, and mood and may also improve capacities to cope with stress and maintain internal circadian rhythms. From this very brief list of general physiological processes, it is clear that PS must serve a number of distinct molecular functions in addition to just being a component of the lipid bilayer.

PS is involved in both activating and anchoring a variety of proteins to membranes. Included in this long list of proteins are several ATPases, kinases, receptors, and the crucial signal transducing proteins, protein kinase C (PKC) and adenylate cyclase [17,18]. PS is also involved in cell–cell recognition and communication processes.

Normally, PS is located almost exclusively in the membrane inner leaflet. It is unusual for any other phospholipid to be about 100% asymmetrically distributed across a membrane (see Chapter 9). This condition is maintained by ATP-dependent amino-PL translocases (also known as flipases). If PS accumulates in the outer membrane leaflet, as it does with age, it signals that the cell should be recycled. In addition, under physiological conditions, PS is an anion and so, similar to PA, interacts strongly with divalent metals like Ca^{2+}.

If we examine the alcohol structures of PE, PC, and PS, it is clear how closely related the three phospholipids are and how easy it might be to interconvert one to the other (Fig. 5.14). PE can be converted to PC by simply methylating the ethanolamine nitrogen three times. This step is not possible in bacteria (the enzyme is missing) and so bacteria lack PC. PS can be converted to PE via simple decarcoxylation.

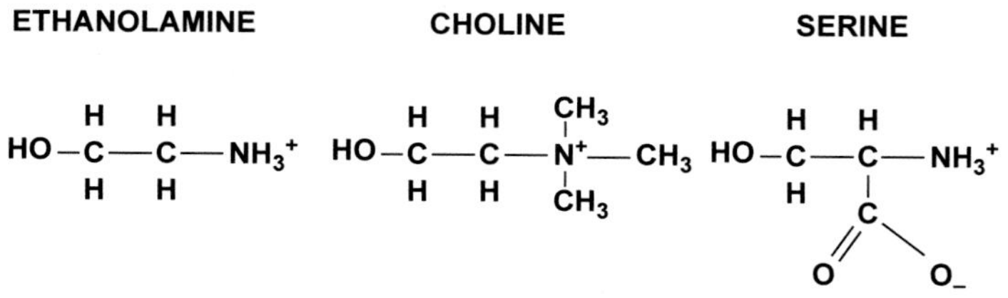

ETHANOLAMINE CHOLINE SERINE

FIGURE 5.14 Close relationship of the head group alcohols of PE, PC, and PS.

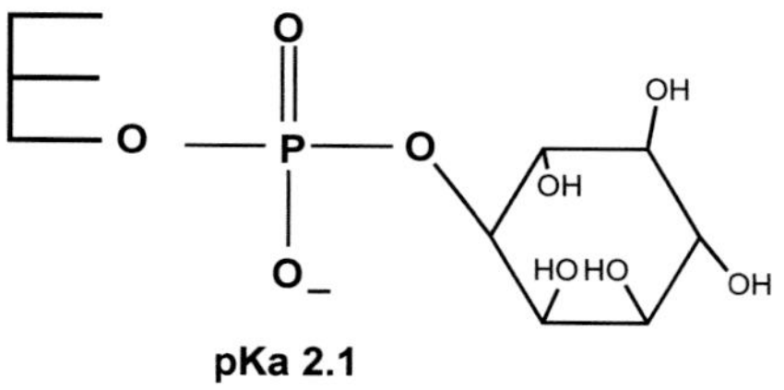

pKa 2.1

FIGURE 5.15 Phosphatidylinositol.

3.3.5 Phosphatidylinositol (Fig. 5.15)

PI is another anionic lipid at physiological pH and so contributes to the lipid bilayer and to the surface negative charge density of membranes. In fact, PIs are considered to be one of the most acidic of the phospholipids. Its low concentration (PI is generally less abundant than even PS, Table 5.1) indicates that this phospholipid plays only a small structural role in membranes. However, it has been appreciated for some time that PI does have several crucial roles in cell function [19]. Although PI is present in all tissues and cell types, it is especially abundant in brain, where it can account for 10% of the total phospholipids. Consistent with the other phospholipids, PI has a saturated fatty acid in the *sn*-1 chain and an unsaturated fatty acid in the *sn*-2 chain. What is unusual about PI is the highly specific nature of its chains. Most PIs are the single molecular species, 18:0,20:4 PI.

PIs hold a central role in cell signaling and regulation [20,21]. Often, PI is phosphorylated at various positions on the inositol chain, most commonly positions 4 and 5, although position 3 can also be phosphorylated (Fig. 5.16). Positions 2 and 6 are not phosphorylated due to steric hindrance. Phosphorylated PIs are referred to as phosphatidylinositol phosphates or polyphosphoinositides. They are usually present at low levels, typically at about 0.5–1% of the total lipids, and are associated with the plasma membrane inner leaflet. The major phosphoinositide is PI 4,5-bisphosphate (PIP_2).

Because the phosphoinositides are highly anionic, they are effective at enhancing nonspecific electrostatic interactions with cellular proteins, particularly to protein "pH domains," at inner leaflet plasma membrane interfaces. PIs are, in essence, an inert storage form of the potent biological signaling second messenger molecules, membrane-bound DAG and water-soluble inositol phosphates. These molecules are released on hydrolysis of, for example, PIP_2 by phospholipase (PL)C (Fig. 5.16). In fact, PIs are the major source of DAG in cells. DAG, in turn, affects the activity of a large, important family of signaling enzymes

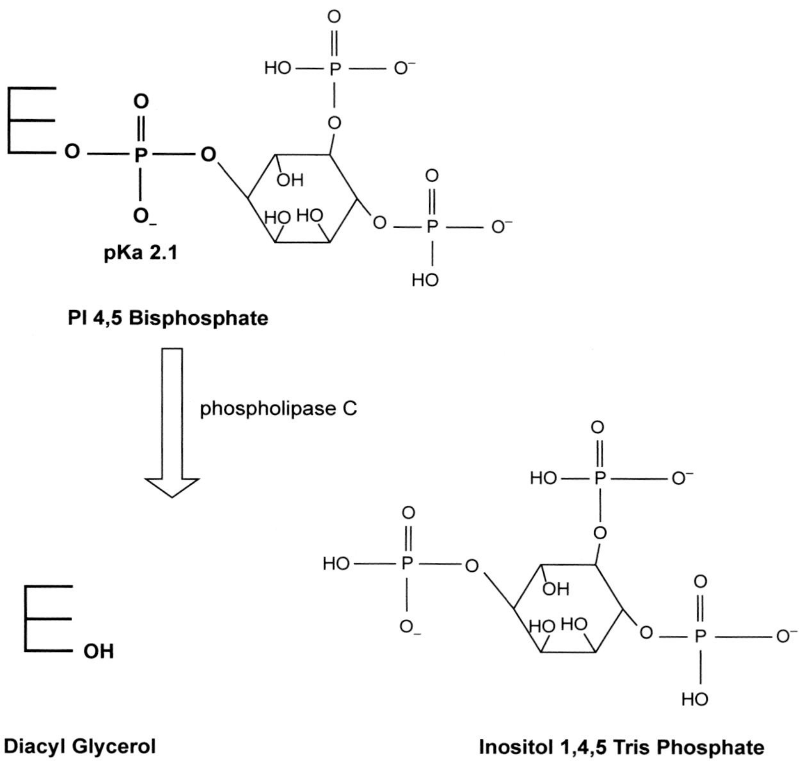

FIGURE 5.16 Hydrolysis of PIP2 by PLC.

known as PKCs. The water-soluble inositol phosphates affect the opening of channels for K^+, Na^+, Ca^{2+}, and other ions. Stimulation of Ca^{2+} release from the endoplasmic reticulum is an important signaling event. PIP_2 has also been shown to affect cell shape, motility, and other processes by interacting with actin on the cytoskeleton. PIP_2 is also a cofactor for PLD that produces PA, yet another signaling molecule (see Chapter 20). In the nucleus, the phosphoinositides affect DNA repair, transcription, and RNA dynamics. PI is also involved in binding a variety of proteins to the membrane surface via a glycosyl bridge [22]. Proteins bound in this way are referred to as being glycosyl-phosphatidylinositol–anchored proteins (see Chapter 6). It is clear that the phospholipid PI plays important roles in many aspects of cellular biochemistry in addition to being a component of the membrane lipid bilayer.

3.3.6 Phosphatidylglycerol (Fig. 5.17)

Phosphatidylglycerol (PG) [23] is another anionic lipid found in mammalian membranes in low amounts (1–2% of the total phospholipids). PG, therefore, is also not a major membrane structural lipid. However, PG is the second most abundant lipid in lung surfactant, comprising up to 11% of the surfactant lipids, and its level is used to determine the maturity of a baby's lung. PG differs from the other phospholipids already discussed in possessing two

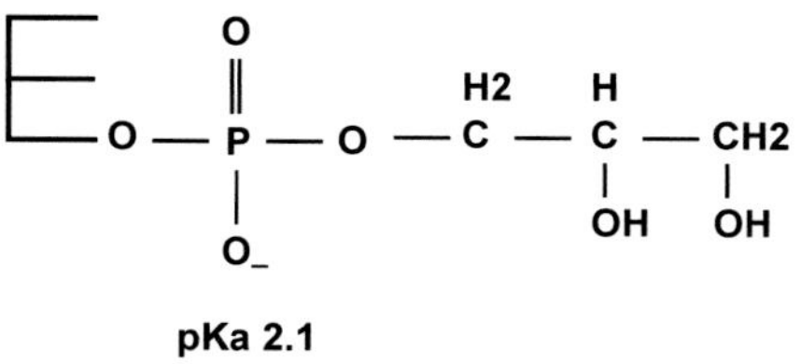

pKa 2.1

FIGURE 5.17 Phosphatidylglycerol.

optically active carbon centers (the *sn*-2 position in the phosphatidyl group and the central carbon of the alcohol glycerol) and in often having a more unsaturated chain occupying the *sn*-1 position. It appears that the major membrane role for PG in animals is as a precursor for mitochondrial cardiolipin (CL). CL (discussed in Chapter 18) is essential for normal electron transport and oxidative phosphorylation. While only a minor component of animal membranes, PG can be the main component of some bacterial membranes and is abundant in plant membranes. In plants, PG comprises about 10% of the thylakoid (photosynthetic) membrane lipids. In higher plants, PG is unique in that it contains a high proportion of *trans*-3-hexadecenoic acid. This *trans*—fatty acid is located exclusively in the *sn*-2 position. However, its role in photosynthesis is not known.

3.3.7 CL – Cardiolipin (Diphosphatidylglycerol) (Fig. 5.18)

CL is the most unusual of the seven common mammalian phospholipids. It is essentially two PAs glued together by a glycerol. It has an oversized head containing two negative charges due to the two dissociated phosphates and four acyl chains (Fig. 5.19). CL has three potential optically active carbons, one on each of the two *sn*-2 phosphatidyl groups and the third on the central carbon of the linking glycerol. Saturated chain CLs form normal lamellar phase (the structure is can-shaped), while the more biologically relevant CLs containing highly unsaturated chains prefer nonlamellar phases (the structure is a truncated pyramid with a wider base than head). Under biological conditions where CL is dispersed with other phospholipids (primarily PCs and PEs), the mixture prefers the lamellar (normal bilayer) phase. Nonlamellar phases are discussed in Chapter 10.

CL is found primarily in the mitochondrial inner membrane where it constitutes about 20% of the total lipid and is closely linked with electron transport and oxidative phosphorylation (discussed in Chapter 18) [24]. Particularly effective in supporting the bioenergetic processes are mitochondrial CL with a high content of linoleic acid ($18{:}2^{\Delta 9,12}$). CL aggregates and

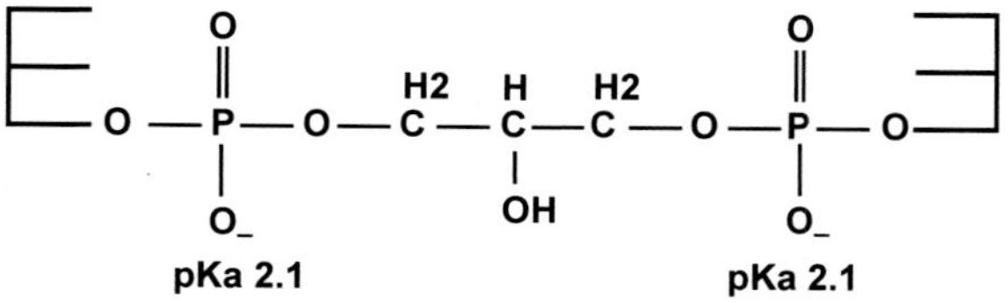

pKa 2.1 pKa 2.1

FIGURE 5.18 CL (diphosphatidylglycerol).

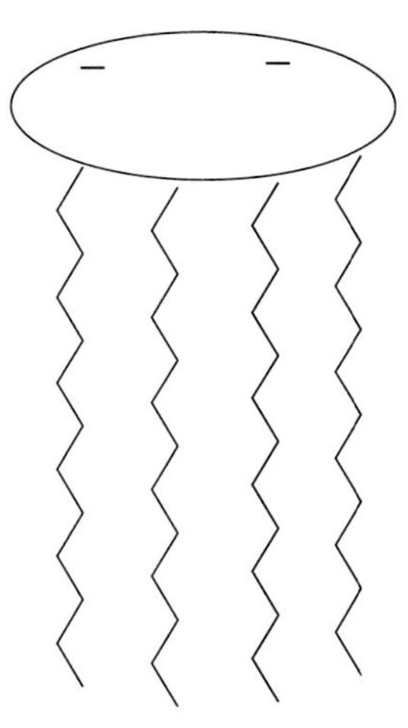

FIGURE 5.19 Cartoon depicting the structure of CL.

stabilizes important electron transport protein complexes including cytochrome c oxidase, NADH dehydrogenase, ATP synthase, and various mitochondrial carrier proteins (discussed in Chapter 18). While most lipids are made in the endoplasmic reticulum, CL is synthesized on the matrix side of the inner mitochondrial membrane (see Chapter 14).

CL was first isolated by M. Pangborn [25] and since has been linked to a bewildering array of human metabolic disorders, including Tangier's disease, Alzheimer's disease, Parkinson's disease, and problems associated with apoptosis. CL from a cow heart is even used as an antigen in the Wassermann test for syphilis. A rare genetic disorder, Barth syndrome, is associated with abnormal CL biosynthesis. Sufferers of this condition are always male, have mitochondria that are abnormal, and are incapable of sustaining adequate production of ATP. CL touches many aspects of membrane function and human health.

4. SPHINGOLIPIDS

All members of the family of sphingolipids [26,27,28] contain the parent C-18 amino alcohol sphingosone (Fig. 5.20). More than 60 different sphingolipids have been found in human membranes, primarily differing by the components attached to the C-1 alcohol. A variety of fatty acids are also esterified to the nitrogen attached at C-2, forming a saponifiable amide linkage. The first three carbons in sphingosine are structurally analogous to the glycerol in phospholipids. Therefore, like phospholipids, sphingolipids have a polar head group that interacts with water and two hydrophobic tails. However, while both phospholipid acyl chains are variable and can be hydrolyzed, one hydrophobic chain in sphingolipids is neither variable nor hydrolizable (it is sphingosine), and the other, the variable chain, is attached via a nitrogen ester (amide), not an oxygen ester. The sphingosine chain also contains a permanent *trans* double bond between carbons 4 and 5, while the variable acyl chain can be either saturated or contain *cis* double bonds.

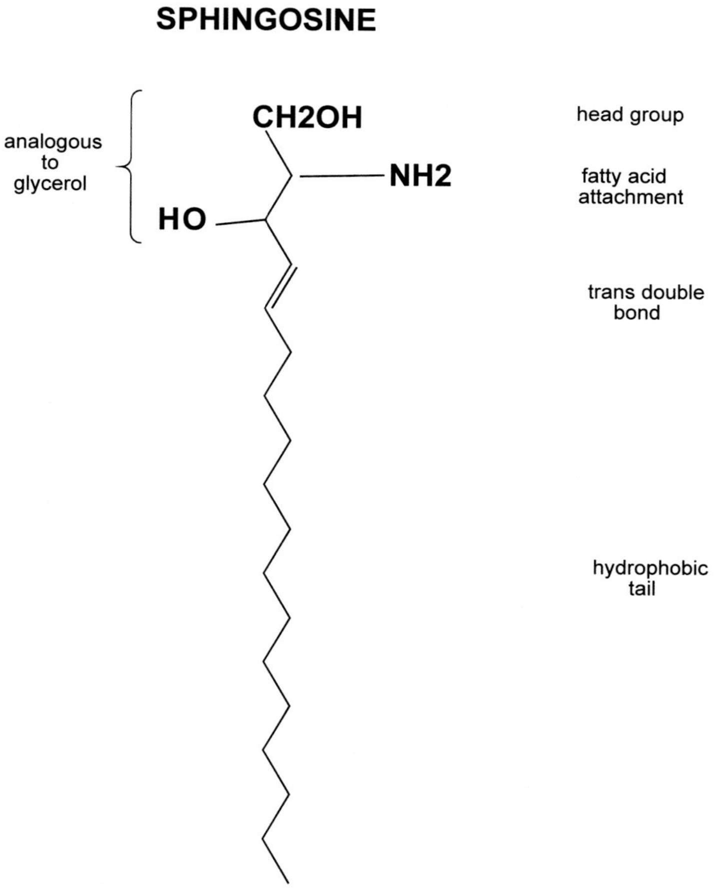

FIGURE 5.20 Sphingosine.

A wide variety of polar head groups may replace the simple proton attached to the –OH at C-1 resulting in the different sphingolipid families listed in Table 5.3.

Most abundant of the human sphingolipids is sphingomyelin (SM), where the head group is phosphocholine (or, less commonly, phosphoethanolamine). SM is rarely found in plants and bacteria. The variable chain in most SMs is longer and more saturated than most phospholipid acyl chains. For example, the most common acyl chains in myelin SM are lignoceric acid (24:0) and nervonic acid (24:1). The general structure of SM, therefore, resembles PC (or PE). Ceramide with a simple –OH polar head is structurally similar to diacylglycerol. Cerebrosides contain a single sugar as its head group. Cerebrosides with galactose are found in the plasma membrane of neuronal cells, while those with glucose are in the plasma membrane of non-neuronal cells. Globosides have two or more sugars, usually D-glucose, D-galactose, or *N*-acetyl-D-galactosamine. Gangliosides have several sugars (they are oligosaccharides) and have at least one unusual, characteristic, nine-carbon

TABLE 5.3 Head Groups Attached to the C-1 Position of Sphingosine

Name	Attached group
Ceramide	−H
Sphingomyelin	Phosphocholine, phosphoethanolamine
Cerebroside	Glucose or galactose
Globoside	Di-, tri-, or tetra-saccharide
Ganglioside	Complex oligosaccharide (always with at least one sialic acid)

anionic sugar, sialic acid (*N*-acetylneuraminic acid). Sialic acid is discussed with the other membrane sugars in Chapter 7. Gangliosides are further subdivided into families or series that are characterized by the number of sialic acid residues on the sphingolipid:

One sialic acid	GM series
Two sialic acids	GD series
Three sialic acids	GT series
Four sialic acids	GQ series

The specific function of only a few sphingolipids is known. Carbohydrates of some sphingolipids define human blood groups and determine blood type. Gangliosides are found on the outer surface of the plasma membrane, where they present points of recognition for extracellular molecules or surfaces of neighboring cells. The absence of specific sphingolipid degrading enzymes are known to be the cause of a number of deadly, incurable genetic lysosomal storage diseases, including Tay-Sachs disease, Gaucher's disease, and Niemann-Pick disease. Some of these diseases are discussed in Chapter 7.

5. STEROLS

Sterols [29,30] are major components of many animal, plant, and fungal membranes but are absent in prokaryotes. In bacterial membranes, sterol function is replaced by hopanoids, sterol-like lipids. In animals, the major sterol is the much maligned cholesterol; in fungi, ergosterol; and in plants, β-sitosterol. It is evident from their structures that these three sterols are very similar, differing only by slight modifications to their floppy tails (Fig. 5.21).

Sterols are very water insoluble and so readily partition into membranes. Membrane sterols can be thought of as having three distinct parts, each with a different function (Fig. 5.22). Sterols are anchored to the aqueous interface via a polar −OH group. The rest of the molecule is hydrophobic. The four rigid sterol rings are responsible for the sterols' major function, controlling membrane fluidity and packing free volume (breathing space for protein function [see Chapter 11]). The function of the floppy tail is far less obvious and is uncertain. Overall,

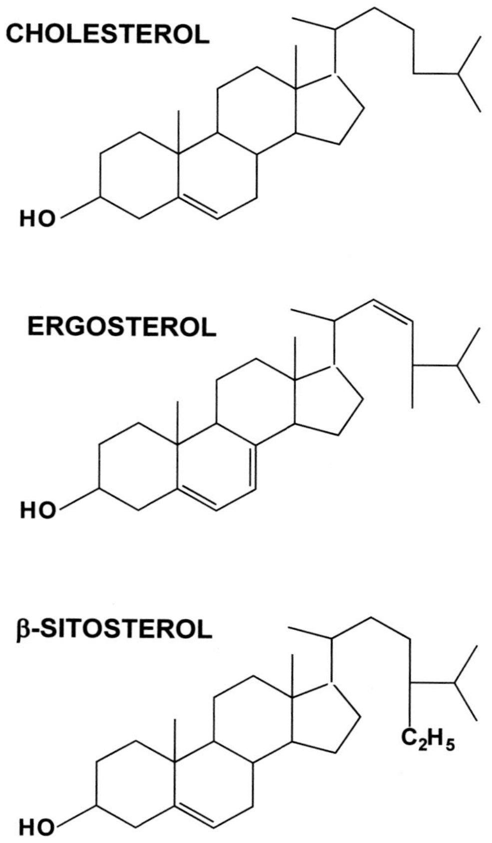

CHOLESTEROL

HO

ERGOSTEROL

HO

β-SITOSTEROL

C_2H_5

HO

FIGURE 5.21 Structure of the sterols cholesterol (animal), ergosterol (fungi), and β-sitosterol (plant).

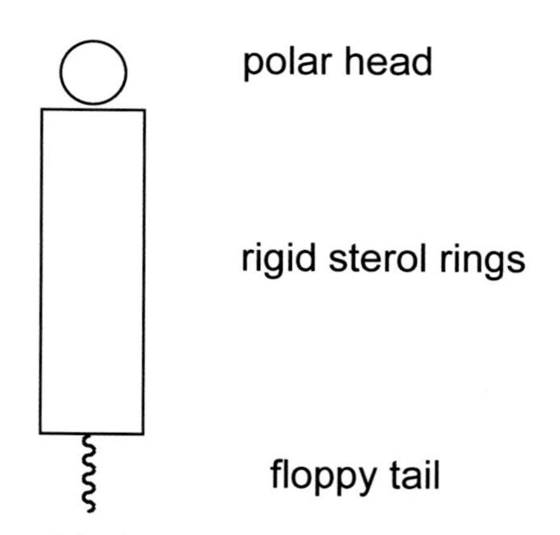

FIGURE 5.22 Cartoon depiction of the three sections of a sterol.

the molecule is flat. The molecular mode of action of the best-understood sterol, cholesterol [31,32,33], will be discussed in detail in Chapters 9 and 10.

Interest in cholesterol is primarily due to its strong link to high blood pressure and heart disease in humans. However, most aspects of cholesterol are not evil. Cholesterol was first isolated from gallstones around 1758 by de La Salle. In 1815, Chevreul isolated cholesterol from the unsaponifiable fraction of animal fat and coined the term "cholesterine" (from the Greek word *khole* for "bile"). The major location of cholesterol in vertebrates is the brain, which accounts for one quarter of the body's total cholesterol. In the plasma membrane of animal cells, cholesterol is usually the major polar lipid, often composing more than 50% of the total membrane lipids. It has even been reported that the ocular lens membrane is about 80 mol% cholesterol! Cholesterol is, therefore, a major structural component of at least the plasma membrane, where it is involved in providing mechanical strength, controlling phase behavior, and supporting membrane lateral organization and stability. Cholesterol has been proposed to be the molecular "glue" that holds together essential cell signaling "lipid rafts (see Chapter 8). In addition, cholesterol is a precursor for all steroid hormones (including progesterone, testosterone, and estradiol), bile salts, and vitamin D (see Chapters 20 and 21). Therefore, cholesterol has two distinct roles: as a structural component of membranes and as a precursor for other essential biochemicals.

In addition to free cholesterol, the sterol can be found esterified to various fatty acids, allowing it to be harmlessly stored as cytosolic droplets or in lipoprotein particles (low-density lipoproteins; see Chapter 14). Cholesterol can also be found esterified to a type of secreted polypeptide signaling molecule encoded by the hedgehog gene family. In addition, cholesterol can be sulfated (cholesterol sulfate) or glycosylated (cholesterol glucoside). Indeed, it seems cholesterol is one of those molecules that Nature really likes!

The cholesterol story presents an interesting conundrum. Cholesterol provides such a fundamentally essential role in the plasma membrane of animals that it would be easy to conclude that cholesterol or a structurally similar molecule (e.g., β-sitosterol or ergosterol) should be essential for life, yet sterols are absent in prokaryotes and these organisms are perfectly functional. Also, in an animal cell, while the plasma membrane is often essentially saturated in cholesterol, the mitochondrial inner membrane is almost completely devoid of cholesterol! Without any cholesterol, how does the mitochondrial inner membrane accomplish the same basic functions as the cholesterol-rich plasma membrane? Cholesterol plays an important role in preventing leakiness to many solutes across the plasma membrane, yet the cholesterol-free mitochondrial inner membrane must be impermeable to the smallest solute, a proton, to function.

For many years, it was believed that cholesterol was nearly absent from plants. However, its presence in higher plants is now universally accepted. Still, plants produce a staggering number of plant sterols collectively referred to as phytosterols. More than 250 different phytosterols have been discovered to date. With such a large number of sterols commonly found in plants, it is obvious that large quantities of phytosterols must enter the human diet [34]. Yet photosterols are systematically excluded from the human body, primarily at the level of the intestinal epithelium. Interestingly, it has been well documented that plant sterols, particularly β-sitosterol, do compete with cholesterol uptake, thus providing a potentially beneficial role of reducing human cholesterol levels.

6. MEMBRANE LIPID DISTRIBUTION

Table 5.1 lists an "average" mammalian cell polar lipid composition. These numbers are representative of many published examples (eg, see [35]). Note the values are reported as ranges and not precise values. Unlike protein and nucleic acid compositions that are defined, lipid compositions can vary considerably with different organisms, different tissues in the same organism, different organelles within a cell, different membrane domains within a single membrane, and even different phospholipid molecular species in the same cell. Lipid composition can vary with diet, environmental conditions, and age of the organism. However, there are some general conclusions that are usually observed for mammalian cells:

- The major structural lipids are PC > PE.
- There are no cationic lipids.
- The anionic lipids, PA, PS, PI, PG, and CL, are all found at lower levels than the structural lipids, PC and PE.
- PC is a true zwitterion and so has no net charge, while PE is almost a full zwitterions that has a slight net negative charge.
- Cholesterol can vary from trace amounts (inner mitochondrial membrane) to being the major polar lipid (over 50 mol% in the plasma membrane) depending on the particular membrane.
- CL is found almost exclusively in the mitochondrial inner membrane.
- Free, unesterified fatty acids and lysophospholipids (Fig. 5.23) are found in very low amounts where they adversely affect membrane structure. High enough amounts of free fatty acids or lysophospholipids will kill a cell through the "detergent effect."
- Fatty acids are distributed unevenly between the two phospholipid acyl chains. The *sn*-1 chains are predominantly, but not exclusively, saturated (often 16:0 or 18:0), while the *sn*-2 chains are mostly unsaturated—18, 20, or 22 carbons long.
- The large number of unsaturated acyl chains means that on average at physiological temperature, most, if not all, of the membrane is in the melted, fluid state.

7. PLANT LIPIDS

The most characteristic feature of plants is photosynthesis, and photosynthetic organisms have evolved unique membrane lipids [36] to support this process. Photosynthetic organisms, including plants, blue green algae, and some bacteria, contain a large amount of glycerolipids in their thylakoids that must somehow be indispensable to photosynthesis. These lipids often contain the sugar galactose (and so are galactolipids) or sulfur (and so may be sulfolipids) [37]. Because photosynthesis is so prevalent, the major glycerolipid, monogalactosyldiacylglycerol (MGDG, Fig. 5.24), is the most abundant membrane lipid on Earth. The glycerolipids are totally devoid of phosphate and, as a result, most have no net charge. An exception is sulfoquinovosyldiacylglycerol (SQDG, Fig. 5.24), which is an anion at physiological pH. Phosphate-free plant lipids may have evolved to cope with a soil environment that often has very low phosphate levels; in fact, phosphate is usually the limiting factor in plant growth.

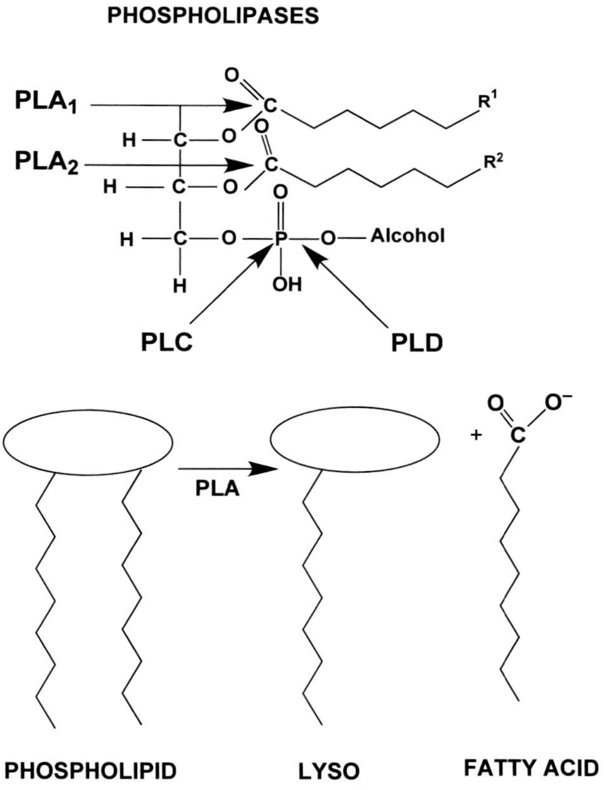

PHOSPHOLIPASES

PLA1 PLA2 PLC PLD

PLA

PHOSPHOLIPID LYSO FATTY ACID

FIGURE 5.23 Hydrolysis of phospholipids by phospholipases. Sufficient quantities of lysophospholipids or free fatty acids can destroy membranes by a "detergent effect."

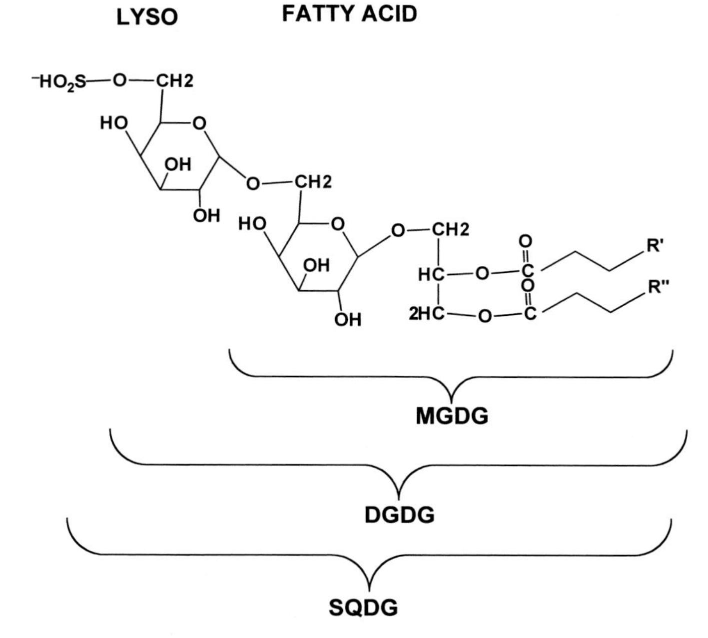

FIGURE 5.24 Structure of three closely related galactosyl diacylglycerols commonly found in plant thylakoid (photosynthetic) membranes.

TABLE 5.4 Polar Lipid Composition of the Plant Thylakoid (Photosynthetic) Membrane

Lipid	Mol (%)
MGDG	48
DGDG	25
PG	13
SQDG	8
PC	2
Other	4

Despite their unwieldy names, there is a very simple structural relationship between MGDG, DGDG and SQDG (Fig. 5.24). MGDG has a relatively small polar head and, with its normal contingent of unsaturated acyl chains, a wide hydrophobic base. Therefore, like PEs with unsaturated acyl chains, MGDG is pyramid shaped and prefers nonlamellar phase (reverse H_{II} phase; see Chapter 10). The addition of a second galactose in DGDG widens the head sufficiently that the molecule becomes can-shaped and so, like PC, prefers lamellar phase. SQDG is anionic and, like PA, PS, PI, PG, and CL in animals, is found in much lower amounts than MGDG and DGDG, the thylakoid membrane structural lipids. The major thylakoid lipids are listed in Table 5.4. Note that the major animal membrane phospholipid, PC, is found at only 2 membrane mol% in plant thylakoids.

8. MEMBRANE LIPIDS FOUND IN LOW ABUNDANCE

There are countless numbers of unusual lipids found in membranes in almost undetectable amounts. For example, the facile oxidation of polyunsaturated fatty acids produces countless byproducts whose structures and biological functions are largely unknown. A few of the more common lipids whose structures and function are appreciated are discussed next and in subsequent chapters.

8.1 Free Fatty Acids and Lysolipids

Under physiological conditions, phospholipids are degraded (hydrolyzed) by a variety of phospholipases (Fig. 5.23) [38]. The phospholipases A_1 and A_2 cleave the hydrophobic acyl chains located at positions *sn*-1 and *sn*-2, respectively. Both products of the hydrolysis, free fatty acids and lyso-phospholipids, are harmful to membranes and so are normally present in trace amounts. Significant quantities of a free fatty acid produce a detergent effect that can result in membrane phospholipids dissolving into fatty acid micelles (Fig. 5.1). The remaining portion of the phospholipid is named a lyso-phospholipid. "Lyso" refers to the molecule's ability to disrupt or lyse a membrane. Lyso-phospholipids are cone-shaped with a wide polar head and a narrow apolar tail. Molecules with this shape do not readily fit into regular lamellar phase packing, thus distorting the membrane (see Chapter 10). High levels of free fatty acids and lyso-phospholipids are usually the result of artifacts arising from the membrane isolation procedure or from membranes isolated from an aged or diseased cell. An exception to this is found in chromaffin granule membranes, which appear to be naturally high in free fatty acids. Chromaffin granules are organelles in chromaffin cells of the adrenal medulla, where epinephrine and norepinephrine are synthesized, stored, and released.

The action of PLC on phospholipids produces DAG and the water-soluble phospho-alcohol characteristic of the phospholipid class. As discussed earlier and in Chapter 20, DAG is a potent activator of several membrane proteins.

8.2 Other Important Membrane Lipids

This book will discuss three categories of essential membrane lipids in chapters found in Part II. Membrane Biological Functions. These categories include electron carriers (Chapter 18 – ubiquinone), bioactive lipids (Chapter 20 – ceramides, diacylglycerol, eicosanoids, steroid hormones, and phosphatidic acid), and lipid-soluble vitamins (Chapter 21 – vitamins A, D, E, and K).

9. SUMMARY

A major component of biological membranes is complex polar lipids, lipids that, when hydrolyzed, release free fatty acids. Free fatty acids themselves are a minor component of most membranes, as they can disrupt membrane structure. As a result, fatty acids are usually esterified. Major animal structural lipids can be classified into three basic types: phospholipids, sphingolipids, and sterols. Member of the major type of membrane lipid, phospholipids, are divided into seven structural classes. Each phospholipid has two esterified fatty acyl chains, one saturated (*sn*-1 chain) and the other unsaturated (*sn*-2 chain), that control properties of the membrane hydrophobic interior. The second membrane lipid type is sphingolipid, so designated because they all contain the parent C-18 amino alcohol sphingosine. More than 60 different sphingolipids have been found in human membranes. The third major lipid type is sterol, particularly cholesterol. In mammalian plasma membranes, cholesterol usually composes more than 50 mol% of all lipids, where it controls membrane "fluidity," lipid packing, and permeability. Cholesterol has been proposed to be the molecular "glue" that holds together membrane lipid rafts. Lipid composition can vary with diet, environmental conditions, and the age of the organism.

By weight, the major component of most membranes is protein, and proteins are responsible for all membrane biochemical activity. Chapter 6 will discuss the many types of proteins that are found in membranes.

References

[1] Cavitch SM. The natural soap book. Storey Publishing; 1994.

[2] Pliny the Elder. Historia Naturalis, XXVIII, 191; 1469.

[3] Good Scents Candles & Soap. History and use of soap. www.goodscentscandles.us/soaphistory.php.

[4] Kanno S, Nakagawa K, Eitsuka T, Miyazawa T. In: Yanagita T, Knapp H, editors. Chapter 14. Plasmalogen: a short review and newly-discovered functions. Dietary in fats and risk of chronic diseases. AOCS Press. Lipidat; 2006.
[5] Fahy E, Subramaniam S, Brown HA, Glass CK, Merrill Jr AH, Murphy RC, et al. A comprehensive classification system for lipids. J Lipid Res 2005;46:839–62.
[6] Cevc G, editor. Phospholipids handbook. New York: Marcel Dekker; 1997.
[7] Hanahan DJ. A guide to phospholipid chemistry. Oxford University Press; 1997. 214 pp.
[8] Hawthorn JN, Ansell GB, editors. Phospholipids, vol. 4. Elsevier; 1982. 484 pp.
[9] AOCS The Lipid Library. Complex glycerolipids. 2011. Lipidlibrary.aocs.org.
[10] Leray C, Cyber Lipid Center. Resource site for lipid Studies. 2011. www.cyberlipid.org/phlip/pgly02.htm.
[11] Moritz A, De Graan PN, Gispen WH, Wirtz KW. Phosphatidic acid is a specific activator of phosphatidylinositol-4-phosphate kinase. J Biol Chem April 1992;267:7207–10.
[12] Moolenaar WH, Kruijer W, Tilly BC, Veerlaan I, Bierman AJ, de Laat SW. Growth factor-like action phosphatidic acid. Nature 1986;323:171–3.
[13] AOCS The Lipid Library. Phosphatidylserine and related lipids structure, occurrence, biochemistry and analysis. Phosphatidylserine – structure and occurrence.
[14] Vance JE, Steenbergen R. Metabolism and functions os phosphatidylserine. Prog Lipid Res 2005;44:207–34.
[15] Almada A. Phosphatidylserine boosts brain function. Nutrition Science News; 2001.
[16] Kidd, PM. Phosphatidyl Serine and Aging-Can This Remarkable Brain Nutrient Slow Mental Decline? ImmuneSupport.com.
[17] Abe T, Lu X, Jiang Y, Boccone CE, Quian S, Vattem KM, et al. Site-directed mutagenesis of the active site of diacylglycerol kinase α: calcium and phosphatidylserine stimulate enzyme activity via distinct mechanisms. Biochem J 2003;375:673–80.
[18] Newton AC. Protein kinase C: structure, function. Regul J Biol Chem 1995;270:28495–8.
[19] AOCS. The Lipid Library. Phosphatidylinositol and related lipids structure, occurrence, composition and analysis1. Phosphatidylinositol.
[20] Vivanco I, Sawyers CL. The phosphatidylinositol 3-Kinase–AKT pathway in human cancer. Nat Rev Cancer 2002;2:489–501.
[21] Kuksis A. Inositol phospholipid metabolism and phosphatidyl inositol kinases. Elsevier; 2003.
[22] Ferguson MAJ, Williams AF. Cell-surface anchoring of proteins via glycosyl-phosphatidylinositol structures. Annu Rev Biochem 1988;57:285–320.
[23] AOCS. The Lipid Library. Phosphatidylglycerol and related lipids. Structure, occurrence, composition and analysis.
[24] Houtkooper RH, Vaz FM. Cardiolipin, the heart of mitochondrial metabolism. Cell Mol Life Sci 2008;65: 2493–506.
[25] Pangborn M. Isolation and purification of a serologically active phospholipid from beef heart. J Biol Chem 1942;143:247–56.
[26] AOCS. The Lipid Library. Sphingolipids. An Introduction to sphingolipids and membrane rafts.
[27] Dickson RC. Sphingolipid functions in *Sacromyces cerevisiae*. Ann Rev Biochem 1998;67:27–48.
[28] Gunstone F. Fatty acid and lipid chemistry. Blackie Academic and Professional; 1996. p. 43–4.
[29] Parish EJ, Nes WD, editors. Biochemistry and function of sterols. CRC Press; 1997. 278 pp.
[30] Patterson GW, Nes WD, editors. Physiology and biochemistry of sterols; 1991. Am Oil Chem Soc. Champaign, IL. 223 pp.
[31] Yeagle P. Chapter 7, the roles of cholesterol in biology of cells. In: Yeagle P, editor. The structure of biological membranes. Boca Raton (FL): CRC Press; 1992.
[32] Yeagle PL. Chapter 7. The roles of cholesterol in the biology of cells. In: The structure of biological membranes. 2nd ed. Boca Raton (FL): CRC Press; 2005.
[33] Finegold LX, editor. Cholesterol in membrane models. Boca Raton (FL): CRC Press; 1993. 265 pp.
[34] Dutta PC, editor. Phytosterols as functional food components and nutraceuticals nutraceutical science and technology. New York: Marcel Dekker; 2004. 406 pp.
[35] White DA. The phospholipids composition of mammalian tissues. In: Ansell GB, Hawthorne JN, Dawson RMC, editors. Form and function of phospholipids. 2nd ed. New York: Elsevier Scientific Publishing Company; 1973. p. 441–82.

[36] Williams JP, Khan MU, Lem-Kluwer NW. Physiology, biochemistry, and molecular biology of plant lipids. New York: Academic Press; 1997. 418 pp.
[37] AOCS. The Lipid Library. Mono- and digalactosyldiacylglycerols and related lipids from plants and Microorganisms.Structure. Occurrence, biosynthesis and analysis.
[38] Gelb MH, Lambeau G. PLA2: a short phospholipase review. Cayman Chemical; 2003. Issue 14.

CHAPTER

6

Membrane Proteins

OUTLINE

1. INTRODUCTION

Membranes are composed primarily of lipids (Chapters 4 and 5) and proteins (this chapter) with a variable amount of carbohydrates (Chapter 7) attached to the surface [1]. By number, lipids are the major component of all membranes where they exceed proteins by approximately 40 to 1 up to 200 to 1 and are responsible for the basic membrane structure and environment. Lipids however do not directly express any biochemical catalytic activity. The amount of proteins in a membrane varies over a wide range, depending on how biochemically

An Introduction to Biological Membranes
http://dx.doi.org/10.1016/B978-0-444-63772-7.00006-3

active the membrane is (Chapter 1). It has been estimated that more than half of all proteins have some association with a membrane. At the low extreme is the myelin sheath whose primary function is to insulate nerve cells. Approximately 20% of the myelin sheath by weight is protein. Myelin lipids outnumber the proteins by approximately 200 to 1. At the other extreme is the mitochondrial inner membrane that supports many complex functions, including electron transport, oxidative phosphorylation (Chapter 18), and numerous trans-membrane transport systems. By weight this membrane is approximately 75% protein. But even here, lipids outnumber the proteins by more than 40 to 1. Since different types of membranes have very different functions but have similar basic physical properties, there is far more functional diversity observed in membrane proteins than is found in membrane lipids.

2. THE AMINO ACIDS

The first basic question is why some proteins are water-soluble while others partition into the oily interior of a membrane. Since all proteins are a linear string of the same 20 amino acids, it must be that the types and arrangement of the amino acids ultimately determine the eventual location of the protein. The term amino acid [2,3] pertains to any small molecule that possesses both an amine and carboxylic acid group. There are about 500 known types of amino acids of biological origin. The first amino acid was discovered in 1806 by the French Chemists Louis Nicolas Vauquelin and Pierre Jean Robiquet from asparagus juice, and so was named asparagine [4]. Vauquelin (Fig. 6.1) was prolific, authoring some 376 papers covering many topics. He is best known for his discovery of chromium (1797) and beryllium (1798), and for producing liquid ammonia at atmospheric pressure.

FIGURE 6.1 Louis N. Vauquelin, 1763–1829.

There are 20 common amino acids, so-called because they are coded for by the genetic code. It is a natural tendency of scientists to categorize everything and this includes the amino acids. Although there have been several proposed amino acid classifications, they are all artificial and so have advantages and disadvantages. Since this book is primarily interested in membrane proteins, a first approach would be to simply divide the amino acids into water-soluble (polar) and water-insoluble (nonpolar) classes based on the structure of the amino acids' side chains. While this is easy to do with most amino acids, some present a problem. For example what do you do with glycine that has no real side chain (only a −H) and tyrosine that has a polar −OH attached to a nonpolar benzene ring?

Table 6.1 presents a simple classification system that was first used to compare water-soluble and membrane proteins. By this analysis, an initial comparison was made between two mitochondrial proteins, the very water-soluble cytochrome C and the very water-insoluble mitochondrial "structural protein". Surprisingly, both proteins had almost identical ratios of polar to nonpolar amino acids. Additional analysis of many proteins indicated that almost all proteins fall within the same range of having 45−55% polar and 45−55% nonpolar amino acids.

The two basic types of classifiers are "lumpers" and "splitters". Dividing the 20 common amino acids into just two classes, nonpolar and polar amino acids, is an example of lumping (Table 6.1). There have been other systems that further subdivide (split) the amino acids into multiple classes. Table 6.2 presents a commonly used (splitter) system that is based on the chemical structure of the side chains.

It is now clear that neither a protein's net amino acid composition nor its distribution into totally artificial classifications dictate whether the protein will reside in a membrane or not. If

TABLE 6.1 List of Nonpolar and Polar Amino Acids

Nonpolar	Polar
Alanine (ala, A)	Glycine (gly, G)
Valine (val, V)	Serine (ser, S)
Leucine (leu, L)	Threonine (the, T)
Isoleucine (ile, I)	Cystine (cys, C)
Proline (pro, P)	Glutamine (gln, Q)
Phenylalanine (phe, F)	Tyrosine (tyr, Y)
Tryptophan (trp, W)	Asparagine (asn, N)
Methionine (met, M)	Lysine (lys, K)
	Arginine (arg, R)
	Histidine (his, H)
	Glutamic acid (glu, E)
	Aspartic acid (asp, D)

TABLE 6.2 A Classification System Based on the Chemical Structure of the Amino Acid Side Chains

Aliphatic	Aromatic	Polar uncharged	Sulfur-containing	Cation	Anion	Imino
Gly	Phe	Ser	Cys	Lys	Asp	Pro
Ala	Tyr	Thr	Met	His	Glu	
Val	Trp	Asn		Arg		
Leu		Gln				
Ile						

the net amino acid composition has little, if anything to do with locating a protein into a membrane, the driving force must reside within the amino acid sequence. In a protein's three-dimensional structure, the more hydrophobic amino acids should reside together forcing that portion of the protein into the membrane while the more hydrophilic amino acids should locate into protein domains that will interact favorably with water. Therefore, membrane proteins, analogous to membrane lipids (Figure 5.7), must also be amphipathic (Fig. 6.2), having both hydrophobic and hydrophilic regions.

The outstanding question is how the amino acid sequence relates to a protein's location either into or out of the membrane. One direct possibility would be to determine the precise location of each amino acid in a protein by X-ray crystallography. Unfortunately this has proven to be a very difficult task for most membrane proteins. There are now more than 30,000 known water-soluble globular proteins and more than 20,000 membrane proteins. However, as it is much easier to obtain crystals from globular proteins, more than 6000 crystallography structures of these proteins have been determined compared to a few dozen structures of integral membrane proteins. So, are there other methods to identify protein domains that prefer a location in the hydrophobic core of a membrane? To answer this question, the amino acids cannot simply be lumped into classes, but instead must be characterized individually. This approach involves hydropathy plots. By this method each

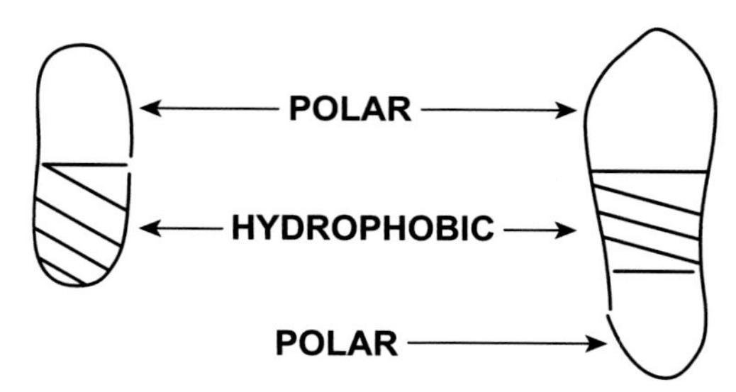

FIGURE 6.2 Amphipathic nature of membrane integral proteins. The protein on the left would sit on one side of the membrane (it would be either an ecto- or endo-protein) while the protein on the right would span the membrane (it would be a trans-membrane protein).

amino acid is assigned a unique Hydropathy Index value that expresses its lipid/water partition. These values represent a measure of the free energy change accompanying the movement of an amino acid from a nonpolar solvent to water. Some amino acids are very hydrophobic while others are very hydrophilic. Still others are in-between. Since the assigned Hydropathy Index values are entirely empirical, a number of very different Hydropathy Index tables have been suggested. Probably the most used approach for this type of membrane study came from Kyte and Doolittle in 1982 [5]. Their method is based on a hydropathy scale that measures the relative hydrophilicity/hydrophobicity of each of the 20 common amino acids. The values are an amalgam of experimentally derived observations taken out of the literature. Table 6.3 lists the amino acids and their Kyte-Doolittle Hydropathy Index values from the most hydrophobic (isoleucine) to the most hydrophilic amino acid (arginine).

TABLE 6.3 Kyte-Doolittle Hydropathy Index Values for the 20 Common Amino Acids [5]

Amino acid	Hydropathy index
Isoleucine	4.5
Valine	4.2
Leucine	3.8
Phenylalanine	2.8
Cysteine	2.5
Methionine	1.9
Alanine	1.8
Glycine	−0.4
Threonine	−0.7
Tryptophan	−0.9
Serine	−0.8
Tyrosine	−1.3
Proline	−1.6
Histidine	−3.2
Glutamic acid	−3.5
Glutamine	−3.5
Aspartic acid	−3.5
Asparagine	−3.5
Lysine	−3.9
Arginine	−4.5

It is obvious that the Kyte-Doolittle Hydropathy Index values are for the most part consistent with the first crude classification of the amino acids into polar and nonpolar. By Kyte-Doolittle, most of the nonpolar amino acids have a positive Index while most of the polar amino acids have a negative Index. A wide variety of hydropathy value tables have now been proposed, and all are empirical amalgamations of experimentally derived values. Although all of the hydrothapy methods have similar amino acid rankings, all have slight variations in some of the relative amino acid rankings. Their empirical numbers also vary over different ranges, than do the Kyte-Doolittle Hydropathy Index of +4.5 to −4.5.

Once the Hydropathy Index for the amino acids is established, the next step is to plot the Index value for each amino acid in the protein's sequence from N-terminal to C-terminal. Also, a midpoint line that is the Hydropathy Index average of all the amino acids is drawn. Unfortunately, the first pass through the amino acid sequence assigning the actual Hydropathy Index value for each amino acid in sequence is extremely "noisy" and cannot be interpreted. Therefore a second and even a third pass are required where the Hydropathy Index of a window of amino acids surrounding a central amino acid is averaged and this value is assigned to the central amino acid in the window. This is referred to as a "rolling average". For example using a window of 7 amino acids, amino acid 15 would be assigned the average Hydropathy Index of amino acids 12 through 18. The window would then be shifted to amino acid 16 and the average Hydropathy Index for amino acids 13 through 19 would be assigned to this central amino acid (amino acid 16). This continues down the entire amino acid chain. A third smoothing pass ("rolling average") may be required to further diminish noise. Readily available computer programs allow the user to easily change the window size from 5 to 25 amino acids and does all of the calculations automatically. Kyte and Doolittle applied this methodology to several proteins whose structure had been determined by crystallography. They found for water-soluble globular proteins, a "remarkable correspondence between the interior positions of their sequence and the regions on the hydrophobic side of the midpoint line, as well as the exterior portions and the regions on the hydrophilic side" [5]. With membrane proteins whose crystallography structures are far less available, possible trans-membrane amino acid stretches can be readily determined. For example, a stretch of 21 hydrophobic amino acids (approximately 33 Å) is enough to cross the lipid bilayer. By Kyte-Doolittle, the hydrophobic stretch of amino acids that may cross the membrane must have averaged Index values of greater than 1.25. Fig. 6.3 is a Kyte-Doolittle Hydropathy Index plot of the single membrane span protein glycophorin [5]. This plot was made from a single pass with a window set at seven amino acids. The membrane spanning α-helix can be readily picked out between amino acids 75 to 94.

3. HOW MANY MEMBRANE PROTEIN TYPES ARE THERE?

How many membrane protein types exist, is a very difficult question to answer with any certainty, as it is subject to interpretation. One membrane type gradually blends into another. A first crude approximation could classify proteins as either peripheral (extrinsic) or integral (intrinsic). Peripheral proteins are essentially water-soluble globular proteins that are attached to the membrane surface through electrostatic interactions. They do not

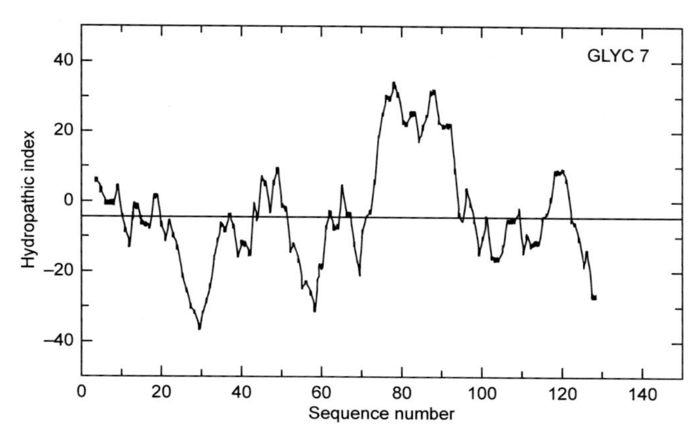

FIGURE 6.3 Hydropathy plot for the single membrane span protein glycophorin. The membrane spanning α-helix can be found between amino acids 75 to 94 [5].

significantly penetrate the membrane hydrophobic interior although they may exhibit some weak hydrophobic interaction. Peripheral proteins can be readily removed from the membrane by simply altering the ionic strength of the media or adjusting the pH. The removed protein is freely soluble in water. Integral proteins penetrate the membrane hydrophobic interior to varying extents and so can only be removed from the membrane by far more drastic measures. Integral proteins can only be removed by destroying the membrane and the extracted protein is isolated as water-insoluble aggregates. While classifying proteins as either peripheral or integral is easy and not subject to much controversy, it is not very satisfying, as most membrane proteins fit into the integral class. In this book, membrane proteins will be classified primarily by the extent and nature of their hydrophobic interactions with membranes. As more is discovered about membrane proteins, devising a meaningful classification system becomes ever more challenging. Overlaps between classes are all too common. Although certainly not perfect, the membrane classification system outlined in Table 6.4 will be employed. This system is an expansion of that proposed for membrane proteins by Robert Gennis in his 1989 book *Biomembranes: Molecular Structure and Function* [1].

3.1 Peripheral Proteins

Peripheral proteins are held to the membrane surface primarily through electrostatic or hydrogen bonds [6]. They can be divided into two basic types: those that are attached through charges on the surface of integral proteins, and those that are attached to the anionic head groups of phospholipids (Fig. 6.4).

Peripheral proteins can be removed easily and cleanly from the surface of membranes by either altering the pH or the media salt concentration. One commonly used method that in fact often defines peripheral proteins employs release with alkaline carbonate. If the peripheral protein is attached via Ca^{2+}, a chelating agent will release the protein from the membrane. By definition, a true peripheral protein exhibits little or no interaction with the

TABLE 6.4 Membrane Protein Classification System

I. Peripheral
  A. Bound to protein
  B. Bound to lipid
II. Amphitropic
III. Integral
  A. Endo and Ecto
  B. Trans-membrane
    Type I: Single trans-membrane span by α-helix
    Type II: Multiple trans-membrane spans by α-helices
    Type III: Membrane domains of several different polypeptides forming a channel through the membrane – β-barrels
IV. Lipid-linked
  A. Myristoylated
  B. Palmitoylated
  C. Prenylated
  D. Glycosylphosphatidylinositol-linked

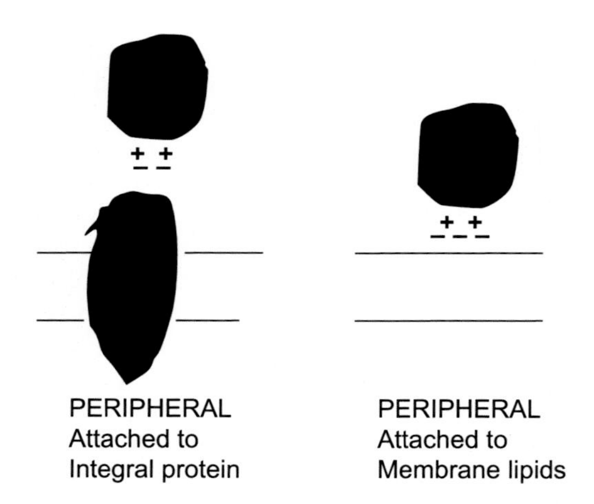

FIGURE 6.4 Peripheral protein attached to an integral protein (left) or to anionic phospholipids (right).

membrane hydrocarbon interior and is removed from the membrane without attached lipids. There are numerous well-known examples of peripheral proteins, many of which are equally at home classified as amphitropic proteins. Classic examples of peripheral proteins include cytochrome c (binds to an integral protein [7,8]) and the myelin basic protein (MBP) (binds to membrane phospholipids [9]).

Cytochrome c is probably the best-studied peripheral protein. Its function as an essential component in the mitochondrial electron transport system where it links Complex III to Complex IV has been known for many decades (Chapter 18). More recently, this very small, primitive protein has been found to be an example of a "moonlighting" protein, where it performs an entirely different function, that of promoting apoptosis (Chapter 24) [7].

Cytochrome c is weakly bound to cytochrome c oxidase on the outer side (intermembrane space) of the inner mitochondrial membrane [8]. Cytochrome c can be detached from the mitochondrial membrane by simply washing in 0.15 M KCl. In fact, the protein is often inadvertently lost from the mitochondria during routine isolation and must be replaced. The facile loss of cytochrome c is what makes this peripheral protein such a good trigger for apoptosis (see Chapter 24).

The best example of a peripheral protein that attaches primarily through electrostatic forces to anionic phospholipids on the surface of a membrane is MBP [9]. The myelin sheath is considered to be the least dynamic of all membranes and so has very few proteins. Function of the major myelin proteins, myelin proteolipid protein, MBPs, and myelin-associated glycoprotein, is probably structural. MBP is involved in myelination of nerves in the central nervous system, and defects in this protein are believed to be important in multiple sclerosis (Chapter 22).

3.2 Amphitropic Proteins

Amphitropic proteins [10] are a relatively new class whose importance is rapidly becoming more apparent. Unfortunately there is a lot of overlap between the peripheral proteins that are bound to membrane surface phospholipids, integral ecto- and endo-proteins, and amphitropic proteins. Amphitropic proteins can alternate between conformations that are either water-soluble or weakly bound to the membrane surface through a combination of electrostatic and hydrophobic forces. Since the membrane interaction is reversible, an amphitropic protein coexists as both a free globular protein and a membrane-bound protein. The water-soluble globular form is converted into the membrane-bound form following a conformation change induced by phosphorylation, acylation, or ligand binding that exposes a previously inaccessible membrane binding site on the protein. Membrane interaction is often based on production of a hydrophobic loop of amino acids, sometimes an α-helix, that can penetrate into the lipid bilayer.

There are many important amphitropic proteins that are integrally involved in cell signaling events. Examples include Src kinase, protein kinase C, and phospholipase C. Upon binding with the appropriate substrate, conformation of the globular form of the amphitropic exzyme is altered, exposing a hydrophobic segment on the protein that inserts into the membrane where it comes into close association with its membrane target. For example, in *Escherichia coli*, globular pyruvate oxidase binds to its substrate pyruvate and cofactor thiamine pyrophosphate, exposing a hydrophobic helix in the protein. The modified enzyme then binds to the membrane, transferring electrons from the cytoplasm to the electrons transport chain in the membrane. The category of amphitropic proteins also includes some water-soluble channel-forming polypeptide toxins (eg, colicin A and α-hemolysin).

3.3 Integral Proteins

It is estimated that 20–30% of all cellular proteins are integral and these proteins are tightly and permanently bound to the membrane [11]. While peripheral and amphitropic proteins are weakly bound to the surface of membranes and can be readily dissociated producing lipid-free globular proteins, integral proteins are integrated into the membrane hydrophobic interior and can only be removed by more drastic means. Historically, integral proteins have been separated from membranes through acetone and other nonpolar organic solvent extractions, chaotropic agents or detergents (Chapter 13).

The earliest attempts to isolate integral proteins involved the brute force method of dissolving away the membrane lipids by use of cold acetone. Proteins precipitate into a white powder called an "acetone powder". Of course the proteins are completely denatured and virtually worthless for biochemical analysis. Chaotropic agents have been successfully employed to isolate some still functional integral proteins. Chaotropic agents function by disrupting water structure, thus eliminating the Hydrophobic Effect that is the major force stabilizing membranes. Although not as harmful as acetone, chaotropic agents may also denature membrane integral proteins. Chaotropic agents are highly water-soluble solutes that essentially pull all of the water away from a membrane, destabilizing it. The best example of a chaotropic agent is 6–8 M urea. Other examples include high concentrations of guanidinium chloride, lithium perchlorate, and thiocyanate. At present the best method for isolating integral proteins involves the use of detergents. This large topic is discussed in Chapter 13, describing protein reconstitution into membranes.

Although thousands of globular proteins have been crystallized and their 3-D structures accurately resolved by crystallography, only a few dozen integral membrane proteins have been resolved [12]. The problem of obtaining 3-D structures for the integral proteins is in obtaining pure crystals in the first place [13]. While peripheral proteins are obtained as monomers that are free of membrane lipids, integral proteins are isolated as aggregates with lipids (and/or detergents) still attached. The aggregates are almost impossible to crystallize. Therefore structural information about integral proteins is often obtained by indirect methods like enzyme degradation studies and hydropathy plots.

3.3.1 *Endo- and Ecto-Proteins*

Endo- and ecto-proteins do not traverse the entire membrane. Instead they are inserted into the membrane's hydrophobic interior from either the cytoplasmic side (endo-protein) or the extracellular side (ecto-protein). An example of such a protein is cytochrome b_5. Cytochrome b_5 is an electron transport protein associated with the endoplasmic reticulum (ER). It is a component of the mixed function oxidase system involved in the desaturation of fatty acids. Intact cytochrome b_5 can only be removed from the ER membrane by detergents. Therefore, by definition it is an integral protein but since it has none of its 152 amino acids exposed on one side of the membrane it is not a trans-membrane protein. When exposed to trypsin, cytochrome b_5 is cleaved at the aqueous interface, producing a 104 amino acid water-soluble N-terminal peptide and a 48 amino acid membrane-soluble C-terminal peptide (Fig. 6.5) [14]. The C-terminal membrane-bound segment has many hydrophobic amino acids and is depleted in hydrophilic amino acids.

3.3.2 *Trans-Membrane Proteins*

A trans-membrane protein must span the entire membrane with segments exposed on both the outside and inside aqueous spaces. The membrane that must be spanned is composed of a lipid bilayer that can be divided into three sections (Fig. 6.6). The inner hydrocarbon region is approximately 27 to 32 Å thick. The very narrow boundary region between the hydrophobic inner core and the hydrophilic interfacial regions is approximately

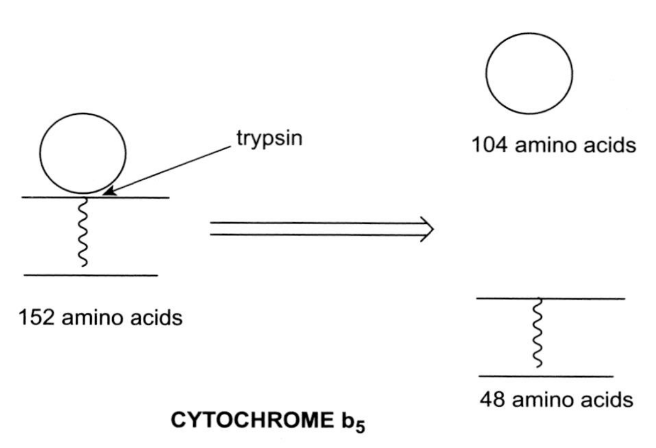

FIGURE 6.5 Cleavage of cytochrome b_5 by trypsin producing a 104 amino acid, N-terminal, water-soluble protein and a membrane-bound 48 amino acid, C-terminal peptide.

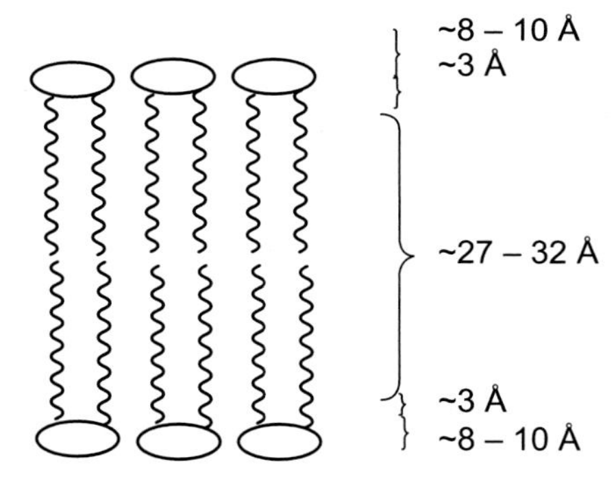

FIGURE 6.6 Three sections of the membrane lipid bilayer. The inner hydrocarbon region is ~27 to 32 Å thick. The very narrow boundary region between the hydrophobic inner core and the hydrophilic interfacial regions is ~3 Å. Finally the polar head group region is ~8–10 Å thick.

3 Å. Finally the polar head group region is approximately 8–10 Å, although this may be wider in membranes that include large amounts of carbohydrate-rich components.

All membrane proteins are 100% asymmetrically distributed with respect to the membrane and do not flip-flop across the membrane. Therefore, trans-membrane proteins are locked in place. A characteristic of many trans-membrane proteins is the presence of tyrosines and tryptophans at the aqueous interface [15]. These amino acids serve as interfacial anchors that can interact simultaneously with the membrane hydrophobic interior and the aqueous exterior. Also, for reasons not entirely clear, the cationic amino acids lysine, histidine, and arginine occur more commonly on the cytoplasmic side of membranes. This is referred to as the "positive-inside rule" [16]. The inner (cytoplasmic) leaflet is also the location of the majority of membrane primary amine lipids, phosphatidylethanolamine and phosphatidylserine [17]. Virtually all integral proteins have at least one stretch of approximately 20 hydrophobic amino acids that implies the existence of a trans-membrane α-helix. α-Helices are found in all types of biological membranes and so represent a common membrane structural motif [18]. Since each amino acid in an α-helix spans approximately 1.5 Å, it takes approximately 20 amino acids to cross the hydrophobic interior (approximately 30 Å). As expected, the trans-membrane α-helices are generally devoid of polar amino acids while nonmembrane spanning portions of the protein are usually enriched in these amino acids. α-Helices can cross the membrane a single time or multiple times.

4. TYPE I. SINGLE TRANS-MEMBRANE α-HELIX: GLYCOPHORIN

Since erythrocytes, unlike other human cells, do not have any complicating internal membranes, and are easily obtained in pure form from blood, they have served for decades as the "laboratory for membrane studies". Many of the classic membrane biochemical and biophysical studies discussed in later chapters were first worked out on erythrocytes, and many of the membrane protein studies were first done on resident glycophorin. However, despite the fact that glycophorin is one of the most studied of all membrane proteins, its biological function remains uncertain. Glycophorins a and b are the major glycoproteins in the erythrocyte plasma membrane and bear the antigenic determinants for the MN and Ss blood groups. By weight, 60% of glycophorin is carbohydrate that is entirely attached to the protein's outer leaflet surface. Abundant sialic acids (Chapter 7) impart a negative charge to the erythrocyte outer surface. The negative charges on glycophorin cause the erythrocytes to repel one another, preventing them from clumping in the blood. With age, the loss of sugars triggers destruction of old erythrocytes.

Glycophorin a is a single span integral protein of 131 amino acids whose cartoon structure is depicted in Fig. 6.7 [19]. All of glycophorin's carbohydrates are attached to amino acids on the N-terminal, exterior segment. These carbohydrates are attached to only a few of the first 50 amino acids of glycophorin's 80 amino acid exterior sequence. As with cytochrome b_5, this

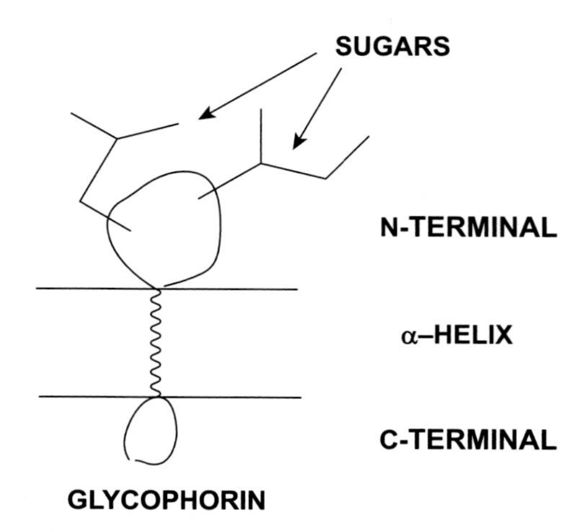

FIGURE 6.7 Cartoon structure of glycophorin showing extra-cellular (N-terminal), trans-membrane (α-helix) and intracellular (C-terminal) domains [19].

extracellular segment can be cleaved by trypsin. The attached sugar tetrasaccharides are O-linked to serine or threonine while the larger oligosaccharides are N-linked to asparagines. O-linked serine and threonine and N-linked asparagine are the usual way sugars are linked to membrane proteins (Chapter 7). Glycophorin a is held in the membrane via a 22-amino acid single span trans-membrane α-helix [20]. Finally, glycophorin a has a 29 amino acid C-terminal segment that extends into the aqueous space in the erythrocyte interior. This segment is totally devoid of sugars. Both the N-terminal and C-terminal nonmembrane domains have many hydrophilic amino acids that are not found in the trans-membrane α-helix segment. Glycophorin a is locked in place by an α-helix that is flanked by the positively charged amino acids, lysine and arginine. These cationic amino acids interact with the negatively charged phospholipid head groups. In the erythrocyte membrane, glycophorin a probably exists as dimers that form a coiled structure involving the α-helices.

5. TYPE II. MULTIPLE TRANS-MEMBRANE SPAN BY α-HELICES: BACTERIORHODOPSIN

Proteins span membranes via two major structures, α-helices or β-barrels. The first of these structures that was shown to be located inside membranes is the α-helix, a secondary structure first formulated in 1948 by Linus Pauling while lying in bed, sick from a cold. The legend states that boredom led Pauling to start playing with paper and scissors and voila, the α-helix was born! A single span α-helix integral protein, glycophorin a, was discussed previously. More common than the single span proteins are the multiple span α-helix proteins. Of particular importance is a very large family of proteins that have seven membrane-spanning α-helices. These proteins have been referred to as the "magnificent seven" due to their biochemical importance [21]. Examples of proteins containing the seven α-helix motif include the large family of G-protein-coupled receptors (GPCR) (Chapter 18), including: the visual receptor rhodopsin, the olfactory receptor, and a variety of channels including voltage gated potassium channels, mechanosensitive channels, aquaporin, chloride channels, and polysaccharide transporters. Included in the G-protein coupled seven α-helix family are receptors for most of the important hormones and signaling molecules in man including: γ-aminobutyric acid, adenosine, bradykinin, opioid peptides, somatostatin, vasopressin, dopamine, epinephrine, histamine, glucagons, acetylcholine, serotonin, prostaglandins, platelet activating factor, leukotrienes, calcitonin, and follicle stimulating hormone. The seven α-helix motif is so important that it has been estimated that more than half of all commercial pharmaceuticals are modulators of GPCRs. It is no wonder that they are known as the "magnificent seven"!

The best studied of the seven α-helix motif proteins is bacteriorhodopsin, a product of the salt loving archaea *Halobacterium* [22]. Under anaerobic conditions, this bacteria switches its metabolism to produce enormous quantities of bacteriorhodopsin, a seven α-helix trans-membrane protein that accumulates in two-dimensional crystalline patches known as a "purple membrane". The "purple membrane" may occupy up to 50% of the cell surface. This highly unusual feature allows for the easy isolation and crystallization of bacteriorhodopsin. Bacteriorhodopsin's function is bioenergetics. Under anaerobic conditions, it captures light to generate a trans-membrane proton gradient that is used to drive adenosine triphosphate synthesis. The light absorbing entity is a retinal (vitamin A aldehyde,

Chapter 21) connected through a Schiff base to a bacteriorhodopsin lysine buried deep in the membrane hydrophobic interior.

The crystallographic structure of bacteriorhodopsin has been determined to 1.55 Å resolution [23]. The protein was shown to have seven trans-membrane α-helices accounting for 80% of the protein's mass. The α-helices are connected by short, extra-membrane, nonhelical loops (Fig. 6.8, Bottom) [25]. In agreement with the Kyte-Doolittle hydropathy plot analysis

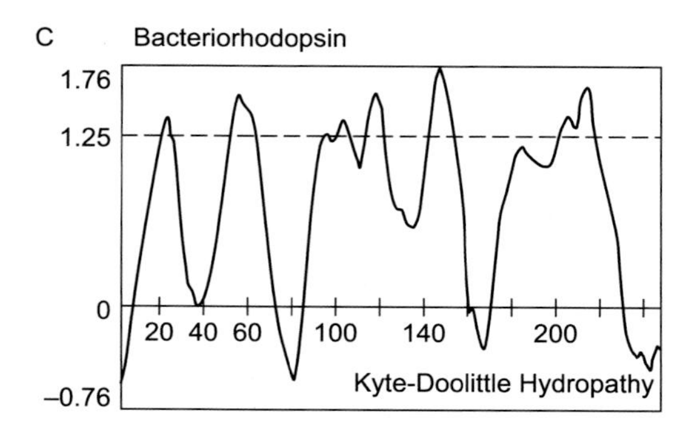

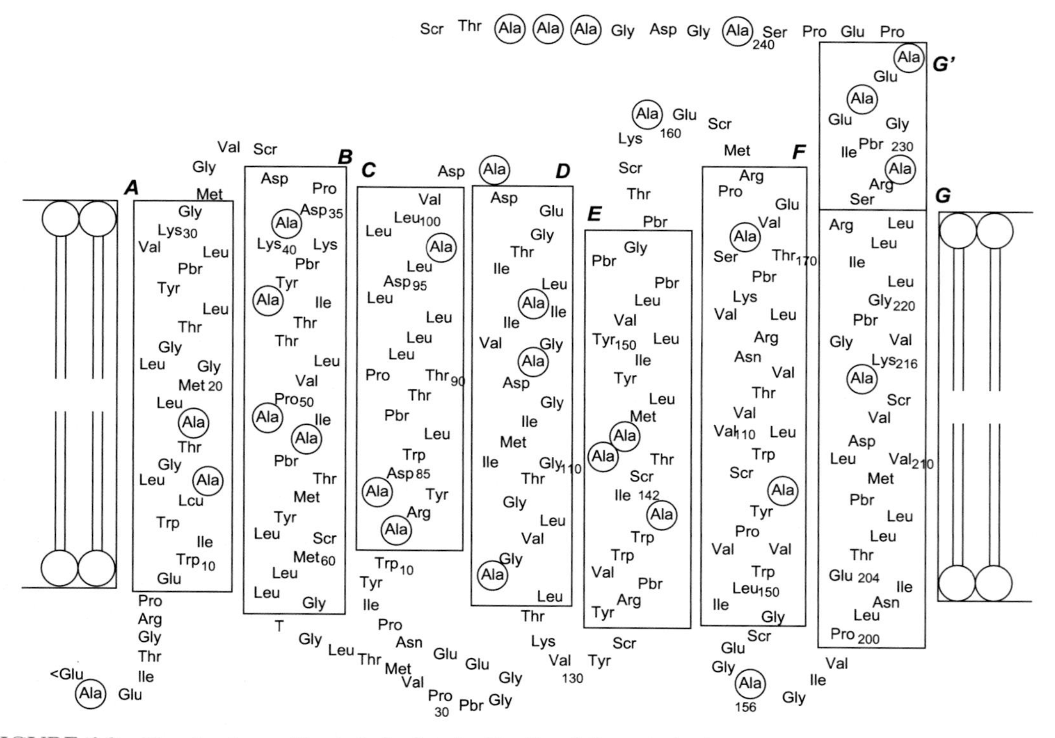

FIGURE 6.8 **The structure of bacteriorhodopsin.** Top Panel: Bacteriorhodopsin Kyte-Doolittle Hydropathy plot. Bottom Panel: trans-membrane sequence demonstrating the seven trans-membrane α-helices [24,25].

(Fig. 6.8, Top) [24], each α-helix is composed of about 20 amino acids, the vast majority of which are hydrophobic. However there are several amino acids found in the α-helices that would not be expected based on their Hydropathy Index values listed in Table 6.3. These include lysine-216 that binds to the retinal, two prolines that normally terminate α-helices, as well as the charged amino acids lysine, arginine, aspartic acid, and glutamic acid. Since each of the seven α-helices has at least one of these unfavorable amino acids, application of a Kyte-Doolittle hydropathy plot without employing a rolling average to smooth the curves would not detect any 20 or more amino acid α-helix spans in the protein. The seven α-helices are clustered together and are oriented a little off perpendicular to the membrane plane. Since this structural motif is typical of the entire family of GPCRs, bacteriorhodopsin was then used to model the other seven α-helix family members until some additional crystallography data finally became available in 2007.

6. TYPE III. MULTIPLE TRANS-MEMBRANE SPANS BY β-BARRELS

The second major type of trans-membrane structure is the β-barrel [26]. This structure is stabilized inside the membrane by the same hydrophobic forces that stabilize α-helix integral proteins and, furthermore, the protein's folding is facilitated by water-soluble chaperones. β-Barrels are very common in the outer membrane of gram-negative bacteria (they account for 2–3% of the bacterial genes), but are also found in the leaky outer membranes of chloroplasts, and mitochondria. β-Barrels are composed of 20 or more β-sheet segments that line a cylinder that spans the membrane. It is through this cylinder that various polar solutes diffuse. Fig. 6.9 shows side views of two typical bacterial outer membrane β-barrel proteins [27]. In β-barrels the trans-membrane segments are comprised of only 7–9 amino acids, compared to the 20 plus amino acids in a trans-membrane α-helix. Also, every other amino acid in a β-barrel is hydrophobic, alternating with hydrophilic amino acids. This makes detecting trans-membrane β-barrel segments by hydropathy plots much more difficult than α-helix segments. In a β-barrel, hydrophobic amino acids face the acyl chains of the membrane hydrophobic interior while the hydrophilic (mostly charged) amino acids line the inside of the aqueous trans-membrane channel. The β-barrel strands are arranged in an antiparallel fashion. Currently there are more than 50 known β-barrel structures that fall into 3 categories: up-and-down, greek key, and jelly roll. Bacterial porins [27] are composed of 16–18 strands that are connected by beta turns on the cytoplasmic side and long loops of amino acids on the extracellular side. The porin channel is partially obstructed by an amino acid loop called the eyelet that gives the channel solute selectivity [28].

7. TYPE IV. LIPID-ANCHORED PROTEINS

Another classification involves proteins that have lipid-anchors. Many of these proteins however are actually trans-membrane proteins that have additional lipid-anchors. They could be classified as either a type of integral protein or as a lipid-anchored protein. For example, rhodopsin is the prototypical example of a seven α-helix membrane-spanning

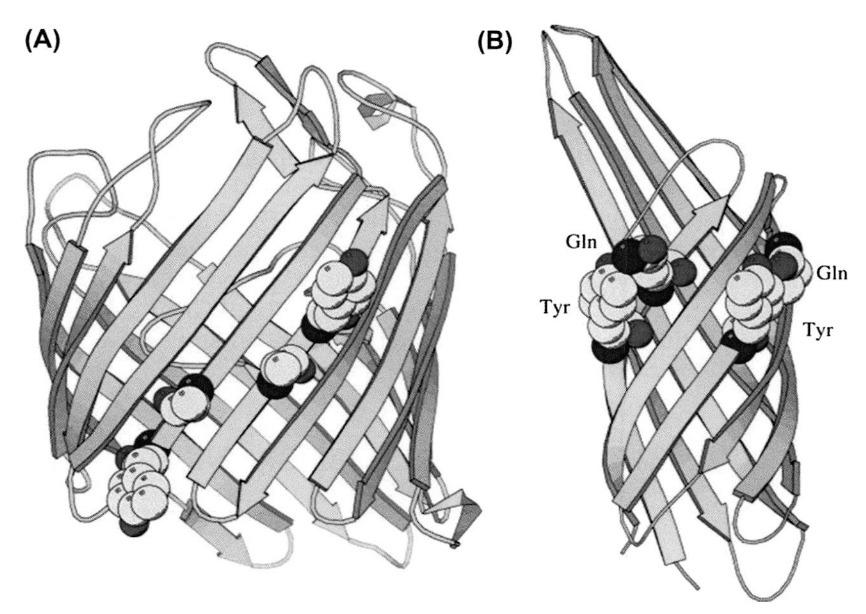

FIGURE 6.9 Two "typical" examples of trans-membrane β-barrel proteins, (A) a AY2 motif in OmpF and (B) a YQ2 motif in OmpX. Omp stands for outer membrane protein of which there are several distinct types. *Reproduced with permission Wikipedia Commons, Wikipedia, the free encyclopedia. March 11, 2007. File: Sucrose porin 1a0s.pn[g].*

protein but also is anchored to the membrane through covalently attached myristic acid. The choice of classification for this protein is arbitrary. Although there are additional examples of lipid-anchors, the most important involve myristic acid (Fig. 6.10), palmitic acid (Fig. 6.11), prenylated hydrophobic acyl chains (Fig. 6.12), and proteins attached through glycosylphosphatidylinositols (GPIs) (Fig. 6.13). The GPI-anchored proteins are always attached to the exo-cytoplasmic (outer leaflet) surface, while the others are lipid-anchored to the endo-cytoplasmic (inner leaflet) surface of cells. Since lipid-anchors by themselves are very weak, these proteins are often additionally attached by other, usual interactions (electrostatic and hydrophobic).

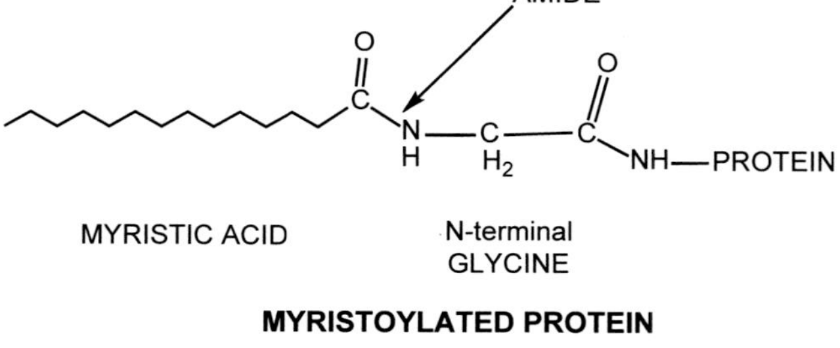

AMIDE
O
O
C
N
H
C
H_2
C
NH—PROTEIN
MYRISTIC ACID
N-terminal
GLYCINE
MYRISTOYLATED PROTEIN

FIGURE 6.10 Myristoylated protein. Protein is N-myristoylated through an amide to its N-terminal glycine.

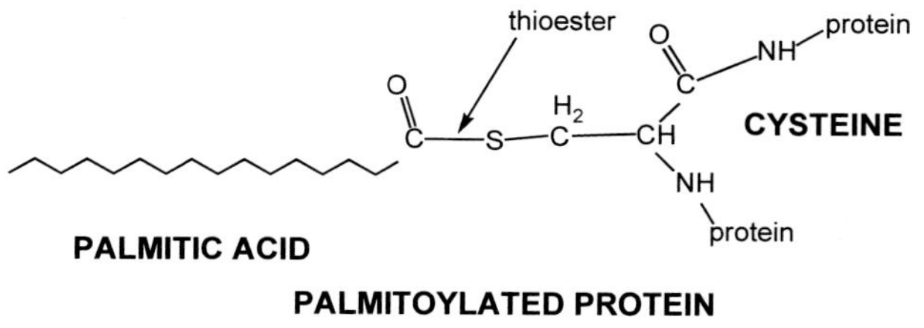

FIGURE 6.11 Palmitoylated protein. Protein is S-palmitoylated through a thioester to a cysteine near the protein's N-terminal.

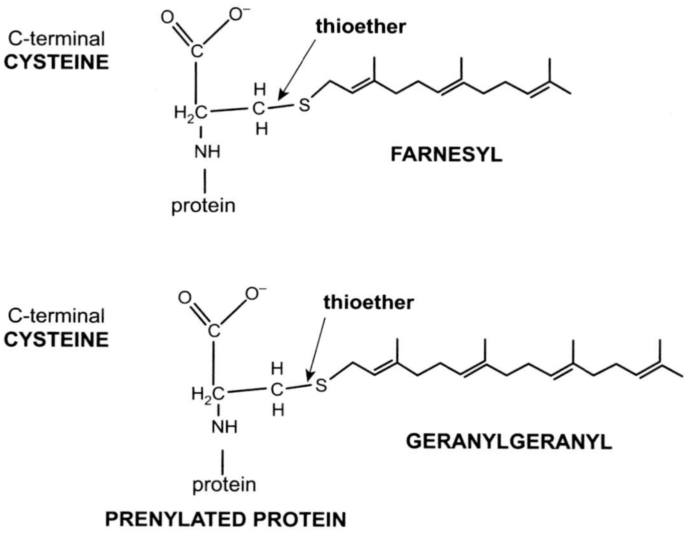

FIGURE 6.12 Prenylated proteins are attached through a thio-ether to either farnesyl (15-carbons, 3-isoprene units) or geranylgeranyl (20-carbons, 4-isoprene units) at a C-terminal cysteine on the protein.

7.1 Myristoylated Lipid-Anchored Proteins

The 14-carbon saturated fatty acid, myristic acid (14:0), is often employed as a lipid-anchor for proteins [29]. This may seem an odd choice, since myristic acid is found in membranes only at very low levels compared to the other commonly employed fatty acid lipid membrane anchor, palmitic acid (16:0) (Chapter 4). Myristoylation was first reported in 1982 for the proteins calcineurin B [30] and the catalytic subunit of the cyclic AMP-dependent protein kinase [31]. Myristic acid was found to be acylated to the anchored protein via an amide linkage to an N-terminal glycine (Fig. 6.10). This linkage is found only in eukaryotes and viral proteins and is established early in translation (Chapter 16 for posttranslational modifications). Generally, it is an irreversible protein modification.

Additional anchoring for myristoylated proteins is provided by electrostatic interactions between positively charged protein side chains and negatively charged membrane phospholipids. Some N-myristoylated proteins undergo additional fatty acyl modifications by

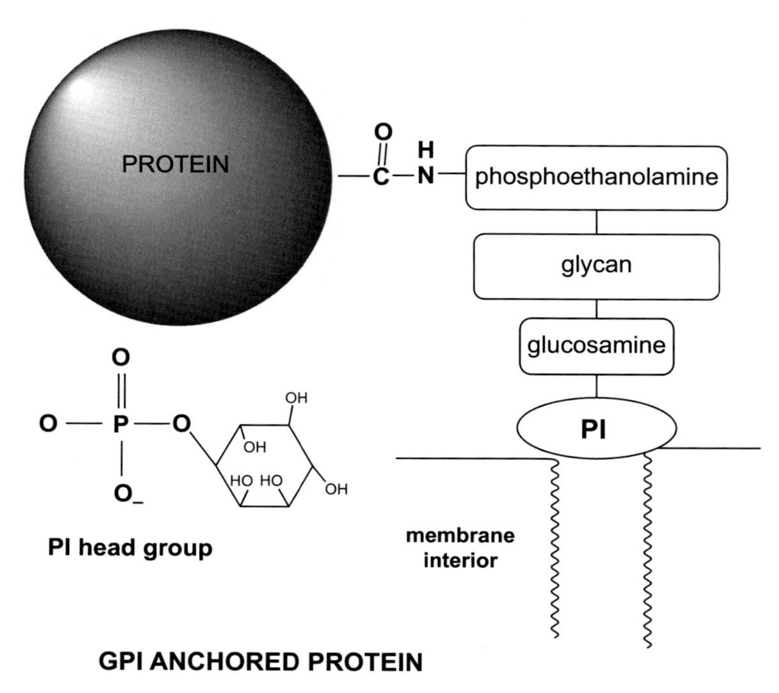

FIGURE 6.13 Cartoon drawing depicting the structure of a typical GPI-anchored protein.

attachment of palmitoyl groups to cysteines via reversible thioesters. An example of these dually acylated proteins include members of the Src family of tyrosine kinases (eg, Fyn, Lck). N-Myristoylated proteins have a variety of important functions including roles in several signal transduction cascades. Under unusual circumstances where myristic acid is limiting, some other fatty acids including shorter-chain and unsaturated, can be attached to the N-terminal glycine.

7.2 Palmitoylated Lipid-Anchored Proteins

Palmitoylated (16:0) proteins are the most extensively studied of the lipid-anchored proteins [32]. The first member of the palmitoylated class was identified by Braun and Rohn in 1969 in the *E. coli* cell wall [33]. Later, palmitoylation was found in the myelin proteolipid protein (PLP). PLP is the major protein in the myelin sheath, accounting for approximately 40% of membrane proteins. PLP is a four α-helix span trans-membrane protein that contains six cysteine residues that are thought to be acylated, primarily with palmitic acid. Other less common acylations involve stearic acid (18:0) and oleic acid ($18{:}1^{\Delta 9}$). Acylation of a fatty acid to the sulfur side chain of cysteine involves reversible formation of a thioester (Fig. 6.11). In fact being a reversible fatty acid attachment is perhaps the most important feature of this lipid linkage. Reversibility is in sharp contrast to the myristoylation (Section 7.1) and prenylation (Section 7.3).

Facile palmitoylation and depalmitoylation has the potential to regulate enzyme activity using acyltransferases and thioesterases, respectively. This is highly reminiscent of the better appreciated enzyme control mechanisms employing phosphorylation by kinases and dephosphorylation by phosphatases. Among the large and ever-growing family of palmitoylated proteins are rhodopsin, band three of erythrocytes, and the lung surfactant proteolipid. Although acylation through a thioester to cysteine is the most important palmitoylation, palmitic acid has also been reported to be O-acylated to serines and threonines, and N-acylated to the ε-amine side chain of lysines. In the signaling pathway protein Sonic Hedgehog, palmitic acid is reported to be attached to an N-terminal cysteine via an amide linkage.

7.3 Prenylated Lipid-Anchored Proteins

Membrane integral proteins can also be lipid-anchored through irreversible thioether linkages to very hydrophobic isoprene groups. These proteins are thus termed prenylated proteins (Fig. 6.12) [34]. In 1978 Kamiya et al. [35] discovered the first prenylated protein (actually a simple 11-amino acid polypeptide), a fungal mating factor. As discussed in Chapter 5, prenylated groups are repeating 5-carbon branched units of 3-methyl-2-buten-1-yl and derive their name from isoprene. Although it appears that the major function of these prenylated chains is to anchor the attached protein to the hydrophobic core of membranes, these groups have an additional, very different and vital role in membranes. Prenyl groups have been shown to be important for protein—protein binding through specialized prenyl-binding domains [36]. Therefore prenylated groups are both lipid-anchors and are intimately involved in localizing certain proteins to specific domains or to specific proteins within the membrane.

Membrane prenylation involves either 15-carbon (3 isoprene units) farnesyl or 20-carbon (4 isoprene units) geranylgeranyl (see Chapter 4), covalently attached through a thioether to cysteine residues at or near the C-terminus of proteins (Fig. 6.12). Once formed, thioethers cannot be readily cleaved and so are considered to be permanent attachments. As with myristoylated and palmitoylated proteins, prenylation anchors the protein to the cytoplasmic side (inner leaflet) of the membrane. Known prenylated proteins include nuclear lamins, Ras and Ras-associated G proteins, and protein kinases. Since prenylation is essential for the function of many proteins involved in cellular signaling and trafficking pathways, the process is becoming a major therapeutic target for multiple diseases. Already there are more than 20,000 known prenylated proteins, and this number increases daily. Therefore, it is reasonable to assume that many more prenylated proteins have yet to be discovered.

7.4 Glycosylphosphatidylinositol-Anchored Proteins

While myristoylation, palmitoylation, and prenylation all anchor proteins to the cytoplasmic side (inside) of membranes, some extracellular proteins are anchored to the outer surface of cells through the acyl chains of GPI (Fig. 6.13) [37,38]. GPI-anchored proteins were discovered at about the same time (1976) by Ikazawa [39] and by Low [40].

GPI-anchored proteins are not trans-membrane but also cannot be removed by alkaline carbonate. Therefore, by definition, they are not peripheral proteins. Their major membrane anchor seems to be through the acyl chains of phosphatidylinositol (PI) that without additional electrostatic or hydrophobic interactions would be very weak. GPI-anchored proteins only face outside and are normally clustered. The anchored protein is indirectly linked to the membrane from its C-terminal through a phosphoethanolamine, to a glycan (three mannose sugars), to a glucosamine, and finally to the inositol sugar of PI (Fig. 6.13).

GPI-anchored proteins are found in most eukaryotes but not bacteria. Although GPI-anchors are not considered to be a common motif of membrane protein structure, there are some 45 GPI-anchored proteins in man. In yeast, where all GPI proteins are known, only 15 of the 5790 proteins (about one in 400 proteins) have GPI-anchors. GPI-anchored proteins have a wide variety of essential functions including enzymatic, antigenic, adhesion, membrane organization, and roles in several receptor-mediated signal transduction pathways. Among the better known GPI-anchored proteins are: carbonic anhydrase, alkaline phosphatase, Thy-1, and BP-3. The GPI-anchored protein can be released from the membrane upon hydrolysis by phospholipase C, and in fact this is how they were discovered [39,40]. There are several, mostly rare human genetic disorders that have been linked to GPI proteins. Perhaps the best known is Marfan Syndrome, whose most famous carrier was reputed in a 1962 theory to be Abraham Lincoln. Although it has received a great deal of attention, this interesting story is no longer believed. It is more likely that Lincoln had a different disorder, multiple endocrine neoplasia type 2B, that causes skeletal features almost identical to Marfan Syndrome.

8. SUMMARY

Soon after Gorter and Grendel proved the existence of lipid bilayers, it became obvious that proteins were also a component of membranes. It is a protein's amino acid sequence, not its net content, that locates it to membranes. Hydropathy Plots are used to predict which segments of a protein cross the membrane. It is not clear how many distinct types of membrane proteins exist. A first crude approach identifies peripheral (loosely attached to the membrane surface by electrostatics) and integral (penetrates into the hydrophobic interior). Integral proteins span membranes via two major structures, α-helices and β-barrels. α-Helices can cross the membrane single or multiple times. A very large family of proteins, including GPCRs, has seven membrane-spanning α-helices. β-Barrels are composed of 20 or more β-sheet segments that line a membrane-spanning cylinder. Many integral proteins are further anchored to the membrane cytoplasmic surface by lipids, primarily myristyl, palmityl, or prenyl groups. Some outer surface proteins are anchored through the acyl chains of GPI. Most membrane surface proteins are heavily glycosylated.

The final major membrane component is carbohydrates (sugars) that, if present, are attached to the membrane outer leaflet. Chapter 7 will discuss the carbohydrates that are commonly found attached to membrane lipids (glycolipids) and proteins (glycoproteins).

References

[1] Gennis RB. Biomembranes: molecular structure and function. New York: Springer-Verlag; 1989.
[2] Barrett GC, Elmore DT. Amino acids and peptides. London: Cambridge University Press; 1998. 240 pp.
[3] Nelson DL, Cox MM. Lehninger Principles of Biochemistry. 5th ed. New York, NY: W.H. Freeman & Co; 2008. 1262 pp.
[4] Bradford H, Vickery C, Schmidt LA. The history of the discovery of the amino acids. Chem Rev 1931;9:169–318.
[5] Kyte J, Doolittle RF. A simple method for displaying the hydropathic character of a protein. J Mol Biol 1982;157:105–32.
[6] Marsh D, Horvath LI, Swamy MJ, Mantripragada S, Kleinschmidt JH. Interaction of membrane-spanning proteins with peripheral and lipid-anchored membrane proteins: perspectives from protein-lipid interactions (Review). Mol Membr Biol 2002;19:247–55.
[7] The New World Encyclopedia. Cytochrome c. www.newworldencyclopedia.org/entry/Cytochrome_c.
[8] Garber EA, Margoliash E. Interaction of cytochrome c with cytochrome c oxidase: an understanding of the high – to low-affinity transition. Biochim Biophys Acta 1990;1015:279–87.
[9] Boggs JM. Myelin basic protein. Hauppauge, NY: Nova Scientific Publishers; 2008. 249 pp.
[10] Johnson JE, Cornell RB. Amphitropic proteins: regulation by reversible membrane interactions (review). Mol Memb Biol 1999;16:217–35.
[11] Molecular Cell Biology. NCBI (National Center for Biotechnology Information). Section 3.4 Membrane Proteins. www.ncbi.nlm.nih.gov.
[12] Membrane Proteins of Known Structure. From the Stephen White laboratory. Irvine: University of California; 2016.
[13] Caffrey M. Membrane protein crystallization. J Struct Biol 2003;142:108–32.
[14] Ozols J, Gerard C, Nobrega FG. Proteolytic cleavage of horse liver cytochrome b5. Primary structure of the heme-containing moiety. J Biol Chem 1976;251:6767–74.
[15] De Planque MR, Bonev BB, Demmers JA, Greathouse DV, Koeppe 2nd RE, Separovic F, et al. Interfacial anchor properties of tryptophan residues in transmembrane peptides can dominate over hydrophobic matching effects in peptide-lipid interactions. Biochemistry 2003;42:5341–8.
[16] Von Heijne G, Gavel Y. Topogenic signals in integral membrane proteins. Eur J Biochem 1988;174:671–8.
[17] Op den Kamp JAF. Lipid asymmetry in membranes. Ann Rev Biochem 1979;48:47–71.
[18] General Principles of Membrane Protein Folding and Stability. From the Stephen White laboratory. Irvine: University of California; 2014.
[19] Tomita M, Marchesi VT. Amino-acid sequence and oligosaccharide attachment sites of human erythrocyte glycophorin. Proc Nat Acad Sci USA 1975;72:2964–8.
[20] Furthmayr H. Structural composition of glycophorins and immunochemical analysis of genetic variants. Nature 1978;271:519–24.
[21] Sakmar TP. Twenty years of the magnificent seven. The Scientist 2005;19:22–5.
[22] Lanyi JK. Bacteriorhodopsin. Ann Rev Physiol 2004;66:665–88.
[23] Luecke H, Schobert B, Richter H-T, Cartailler J-P, Lanyi JK. Structure of bacteriorhodopsin at 1.55 Å resolution. J Mol Biol 1999;291:899–911.
[24] Engelman DM, Goldman A, Steitz TA. The identification of helical segments in the polypeptide chain of bacteriorhodopsin. Methods Enzym 1982;88:81–9.
[25] Krystek SR, Metzler WJ, Novotny J. Hydrophobic profiles for protein sequence analysis. Curr Protoc Protein Sci 2001. Unit 2.2.
[26] Wimley WC. The versatile β-barrel membrane protein. Curr Opin Struct Biol 2003;13:404–11.
[27] Wikipedia Commons. Wikipedia, the free encyclopedia. March 11, 2007. File: Sucrose porin 1a0s.pn[g].
[28] Achouak W, Heulin T, Pages JM. Multiple facets of bacterial porins. FEMS Microbiol Lett 2001;199:1–7.
[29] Farazi TA, Waksman G, Gordon JI. The biology and enzymology of protein N-myristoylation. J Biol Chem 2001;276:39501–4.
[30] Aitken A, Cohen P, Santikarn S, Williams DH, Calder AG, Smith A, et al. Identification of the NH_2-terminal blocking group of calcineurin B as myristic acid. FEBS Lett 1982;150:314–8.
[31] Carr SA, Biemann K, Shoji S, Parmelee DC, Titani K. n-Tetradecanoyl is the NH_2-terminal blocking group of the catalytic subunit of cyclic AMP-dependent protein kinase from bovine cardiac muscle. Proc Natl Acad Sci USA 1982;79:6128–31.

[32] Bijlmakers M-J, Marsh M. The on–off story of protein palmitoylation. Trends Cell Biol 2003;13:32–42.
[33] Braun V, Rohn K. Chemical characterization, special distribution and function of lipoprotein (murein lipoprotein) of the *E. coli* cell wall. The specific effect of trypsin on the membrane structure. Eur J Biochem 1969;10:426–38.
[34] Gelb MH, Scholten JD, Sebolt-Leopold JS. Protein prenylation: from discovery to prospects for cancer treatment. Curr Opin Chem Biol 1998;2:40–8.
[35] Kamiya Y, Sakurai A, Tamura S, Takahashi N. Structure of rhodotorucine 4 a novel lipopeptide, inducing mating tube formation in *Rhodosporidium toruloides*. Biochem Biophys Res Commun 1978;83:1077–83.
[36] Kloog Y, Cox AD. Prenyl-binding domains: potential targets for Ras inhibitors and anti-cancer drugs. Seminars Cancer Biol 2004;14:253–61.
[37] Ferguson MAJ, Williams AF. Cell-surface anchoring of proteins via glycosyl-phosphatidylinositol structures. Annu Rev Biochem 1988;57:285–320.
[38] Ikezawa H. Glycosylphosphatidylinositol (GPI)-anchored proteins. Biol Pharm Bull 2002;25:409.
[39] Ikezawa H, Yamanegi M, Taguchi R, Miyashita T, Ohyabu T. Studies on phosphatidylinositol phosphodiesterase (phospholipase C type) of *Bacillus cereus*. I. purification, properties and phosphatase-releasing activity. Biochim Biophys Acta 1976;450:154–64.
[40] Low MG, Finean JB. Non-lytic release of acetylcholinesterase from erythrocytes by a phosphatidylinositol-specific phospholipase C. FEBS Lett 1977;82:143–6.

CHAPTER

7

Membrane Sugars

OUTLINE

1. MEMBRANE SUGARS – INTRODUCTION

In addition to lipids and proteins a third major component exists primarily on the outer surface of the cell plasma membrane. This component is a collection of very water-soluble carbohydrates (sugars) [1–4]. The word carbohydrate is a contraction of "hydrated carbon". $C_n(H_2O)_n$ is the empirical formulation of most sugars. For example, the most common sugar, glucose (Fig. 7.1), is $C_6H_{12}O_6$ or $C_6(H_2O)_6$, and is thus a "hydrated carbon".

Unlike membrane lipids and proteins, carbohydrates have no hydrophobic segments and so are never found integrated into the nonpolar interior of membranes. Instead they sit on the membrane surface where they are favorably surrounded by water. Their location, almost exclusively on the outer surface of the plasma membrane, strongly indicates that their main function is to interact with the cell's external environment. It is well established that, like lipids, carbohydrates have no enzymatic functions and so must influence membrane biochemistry through indirect methods. General roles for membrane carbohydrates include: biosynthetic sorting, protecting membrane proteins from proteolysis, assisting in protein folding (serve as chaperones), enhancing protein stability, involvement in cell identification, recognition and cell–cell adhesion, providing immunological properties (they are antigens), and serving as receptors. Carbohydrates clearly have a large impact on plasma membrane function.

An Introduction to Biological Membranes
http://dx.doi.org/10.1016/B978-0-444-63772-7.00007-5

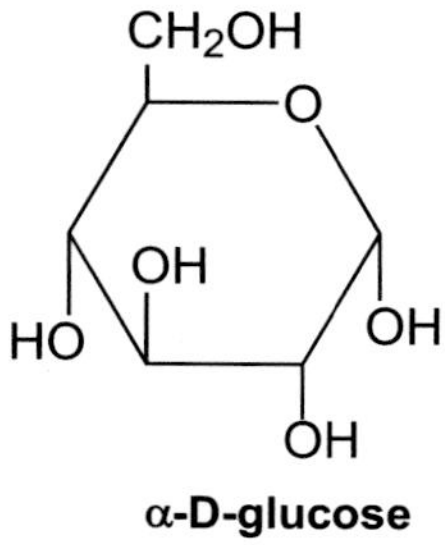

α-D-glucose

FIGURE 7.1 α-D-Glucose.

The amount of carbohydrate on a membrane protein varies tremendously from 0% in multiple span trans-membrane proteins to 60% by weight for glycophorin and greater than 85% for some blood group substances. In fact some plasma membranes have so many surface carbohydrates that in cross-section they resemble a hairbrush, with a lot of sugars (bristles) on the outside and none on the inside. The term used to describe these membranes is glycocalyx or "sugar coat" [5]. A glycocalyx is commonly found on the outer surface of bacteria and mammalian epithelial cells (Fig. 7.2). Membrane sugars are attached to either lipids (glycolipids [6–8]) or proteins (glycoproteins [7–10]).

There are nine basic sugars commonly found attached to membranes: α-D-glucose, α-D-galactose, α-D-mannose, α-L-fucose, α-L-arabinose, α-D-xylose, N-acetyl glucosamine, N-acetyl galactosamine, and N-acetyl neuraminic acid (sialic acid) (Fig. 7.3). The carbohydrates can be either single sugars or short (less than 15 unit) polysaccharides and are always located on the extracellular membrane leaflet.

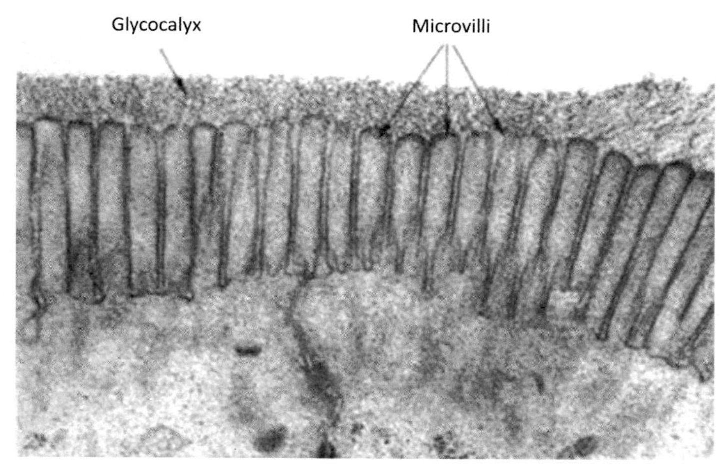

FIGURE 7.2 Glycocalyx. Electron micrograph of a cross section of microvilli in the human digestive track. The glycocalyx can be seen at the top of the columnar microvilli. *Reproduced with permission Neu J, Hilton B. Update on host defense and immunonutrients. Clin Periinatol 2002;29(1):41–64.*

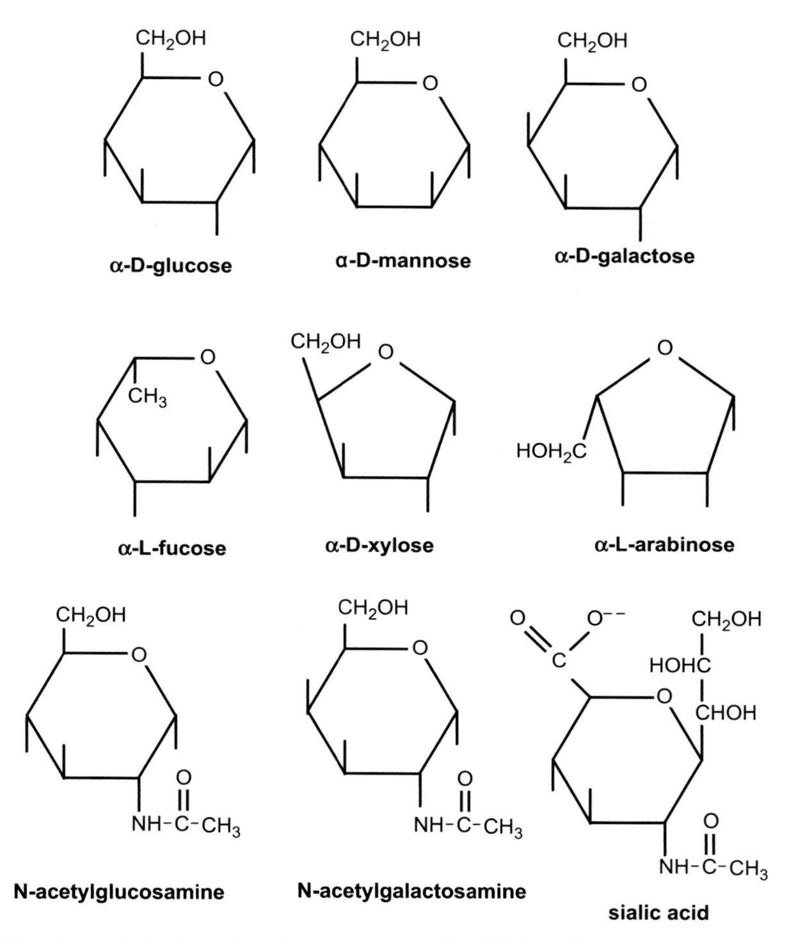

FIGURE 7.3 The nine carbohydrates found on membrane glycolipids and glycoproteins.

One underappreciated aspect of carbohydrates is their enormous potential for information storage. Carbohydrates exist in either an aldehyde or a ketone series. In addition, each carbon location that has a carbon with four different groups on it (positions 2 through 5 for glucose) comes in two orientations or isomers. An aldehyde 6-carbon (hexose) sugar therefore has four asymmetric carbons, each with two isomers or 2^4 or 16 different possible isomers. The potential for information storage is greatly enhanced when two or more sugars are connected through what is known as a glycosidic bond. For example, linking two glucoses together by removing water between carbon number 1 of one glucose (referred to as the anomeric carbon) to the alcohol at position number 4 of the second glucose, produces the disaccharide maltose (Fig. 7.4).

It is generally assumed that biological information storage resides within the primary sequence of proteins and nucleic acids. However, it is becoming more evident that carbohydrates have far more information storage capacity than these other biopolymers. This has led

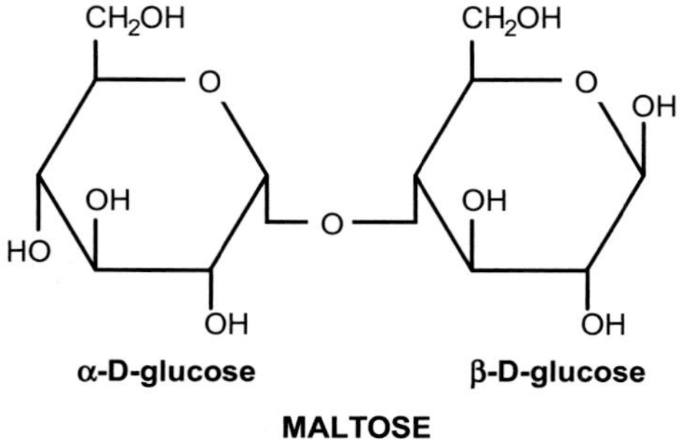

FIGURE 7.4 The disaccharide maltose. Two α-D-glucose sugars are attached through a glycosidic bond from the anomeric carbon (C-1) of the left glucose to C-4 of the right glucose.

to a search for the elusive "carbohydrate code". Many different sugar links are possible for even disaccharides and this number increases logarithmically upon each additional linked sugar in the polysaccharide. In a 1997 paper titled "Information Capacity of the Carbohydrate Code", R.A. Laine calculated the possible number of ways sugars could be linked [11]. For a linear tri-saccharide, there are 12 possible linkage sites (4 on each sugar) that can be arranged into 6,000,000 different structures. This is in sharp contrast to a tri-peptide composed of the 20 different common amino acids that can be arranged in only 8000 (20^3) different ways. Therefore carbohydrates have about a three order of magnitude greater information storage capacity than do proteins.

Of the nine sugars commonly found in membranes (Fig. 7.3), one is highly unusual. Sialic acid [12,13] is a 9-carbon carboxylic acid that is an anion under physiological conditions (Fig. 7.3). Actually sialic acid is a family of similar sugars, the most common being N-acetylneuraminic acid. Sialic acid was named by Gunnar Blix in 1952 after the Greek word for saliva. By the time Blix gave the name to sialic acid, he had been working on this family of compounds for more than 15 years. Sialic acid is widely found in gangliosides and glycoproteins of animal plasma membranes, but is also found in most other organisms including plants, fungi, and bacteria. Glycoproteins of cancer cells that can metastasize are particularly rich in sialic acid. The negative charge on sialic acid has many diverse functions. some advantageous, some not. For example, anionic sialic acid helps keep erythrocytes from clumping in the blood stream but it is also the binding site and entry port for the Human Influenza Virus (HIV). Sialic acid-rich oligosaccharides help retain water close to the cell surface and thus are involved in water uptake and retention. Sialic acid also "hides" mannose units from incorrectly reacting with components of complement. Erythrocytes have a 120-day life span after which they are targeted for destruction. Sialic acid plays an important role in this process. Aged erythrocytes are known to lose sialic acid from their membranes, exposing the remaining neutral sugars to host antibodies. After antibody binding, the senescent cell is removed from circulation by the reticuloendothelial system (RES), primarily within the spleen. This mechanism is responsible for removing one million old erythrocytes per second from the blood. The RES also presents a major problem in prematurely removing drug-containing targeted liposomes from the circulatoratory system (Chapter 23).

2. GLYCOLIPIDS

In animals, the major glycolipids are sphingolipids (Chapter 5 [14–16]). The glycolipids include cerebrosides that have only one sugar attached (either glucose or galactose), globosides that have either a di-, tri-, or tetra-saccharide attached, but do not contain sialic acid and gangliosides that have complex oligoaccharides attached that includes at least one sialic acid residue.

The name ganglioside was coined by E. Klenk in 1942 for lipids isolated from ganglion cells of the brain. In fact gangliosides can amount to 6% of brain lipid weight and constitute 10–12% of the total lipid content of neuronal membranes (20–25% of the membrane outer leaflet lipids). More than 60 gangliosides that differ in the number and arrangement of the sugars, particularly sialic acid, have now been identified. In the plasma membrane outer leaflet, glycolipids accumulate into lipid rafts (Chapter 8) and are therefore involved in cell signaling. They serve as receptors of signaling proteins including: interferon, epidermal growth factor, nerve growth factor, and insulin. In addition, gangliosides bind specifically to viruses and to various bacterial toxins, such as those from botulism, tetanus, and cholera. It has been proposed that toxins utilize the gangliosides to hijack an existing retrograde transport pathway from the plasma membrane to the endoplasmic reticulum. Interestingly, binding of cholera toxin to gangliosides is the primary experimental marker used to identify lipid rafts on the surface of plasma membranes. Due to information stored within the "carbohydrate code" and important functions of sialic acid, gangliosides are also intimately involved in complex cell–cell recognition and serve as essential antigens in immunology. Gangliosides are not found outside of the animal kingdom.

Malfunctions in the enzymatic degradation of sphingosine-containing glycolipids, particularly the gangliosides, have been linked to a family of devastating, incurable genetic diseases that are lumped together as lysosomal storage diseases (LSDs) [17]. These diseases all result when a specific lytic enzyme is defective or missing, resulting in abnormal accumulation of the enzyme's substrate in the lysosomal membrane. One series of these diseases results from the failure to degrade the ganglioside: $GM_1 \rightarrow GM_2 \rightarrow GM_3$ (Fig. 7.5). In GM_1, G stands for glycoside, M for mono (or one) sialic acid and 1 indicates it is the first mono-sialo ganglioside ever characterized.

The best known disease involving failure to degrade a ganglioside (generally known as gangliosidoses or sphingolipidoses) is Tay-Sachs Disease [18]. Tay-Sachs was the first gangliosidosis identified (in 1881) and results from a genetic mutation that fails to produce the enzyme hexoseaminidase A that normally would convert $GM_2 \rightarrow GM_3$ + galactosamine (Fig. 7.5). As a result GM_2 accumulates in the lysosomal membrane producing a plethora of eventually fatal complications. Accumulation of GM_2 occurs primarily in the lysosomes of tissues that are normally enriched in gangliosides (neurons and brain) resulting in severe neurologic symptoms and death before age 5. Symptoms include: seizures, mental retardation, paralysis, blindness, extreme sensitivity to noise, and the appearance of an unusual cherry-red macular spot. The trait is carried by 1/27 of normal adults of Eastern European (Ashkenazi) Jewish origin and is 100 times more prevalent in this population than other populations.

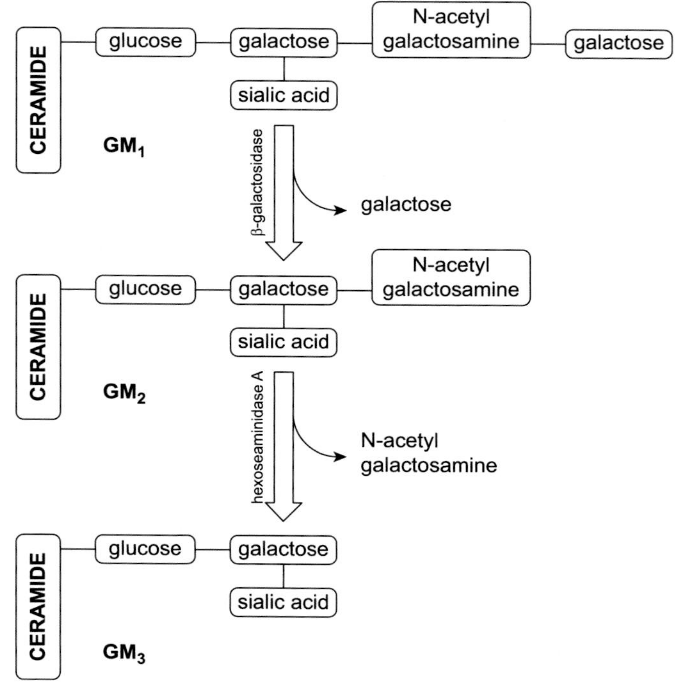

FIGURE 7.5 Enzymatic degradation of the ganglioside $GM_1 \rightarrow GM_2 \rightarrow GM_3$.

Approximately 50 lysosomal storage diseases have now been identified, all resulting in the accumulation of unhydrolyzed membrane sphingolipids. These relatively rare, incurable genetic diseases occur in less than 1 in 100,000 births. They primarily affect children, resulting in premature deaths, often within a few months or years of birth. Table 7.1 lists a few of the better known LSDs.

3. GLYCOPROTEINS

Most plasma membrane proteins have attached sugars, all of which face the outside. On the plasma membrane outer surface there is a much higher percentage of proteins with attached sugars (glycoproteins) than lipids with attached sugars (glycolipids). Glycoproteins [9,10] are ubiquitous in nature, although they are relatively rare in bacteria. The sugars are attached cotranslationally or posttranslationally (Chapter 16) to a protein through either the nitrogen of asparagine, or the oxygen of serine, threonine, hydroxylysine, or hydroxyproline via a process known as glycosylation (Fig. 7.6). Attachment to asparagine is known as N-glycosylation and to serine, threonine, hydroxylysine, or hydroxyproline as O-glycosylation. N- and O-glycosylations are very different. The sugar directly bound to asparagine is N-acetylglucosamine. O-linked sugars tend to be shorter (mono-, di-, or trisaccharides) and more diverse than are the N-linked sugars that contain much longer

TABLE 7.1 Some Better Known Lysosomal Storage Diseases With the Missing Enzyme and Accumulated Sphingolipid

Disease	Missing enzyme	Accumulating sphingolipid
Fabrey	α-Galactosidase A	Gb3
Gauchers	β-Glucosidase	Glucoceramide
GM_1 Gangliosidoses	β-Galactosidase	GM_1
Hunter	Iduronate-2-sulfatase	Mucopolysaccharides
Krabbe	Galactosylceramidase	Galactoceramide
Niemann-Pick	Sphingomyelinase	Sphingomyelin
Pompe	α-Glucosidase A	Gb3
Sandhoff	Hexoseaminidase A & B	GM_2
Tay-Sachs	Hexoseaminidase A	GM_2

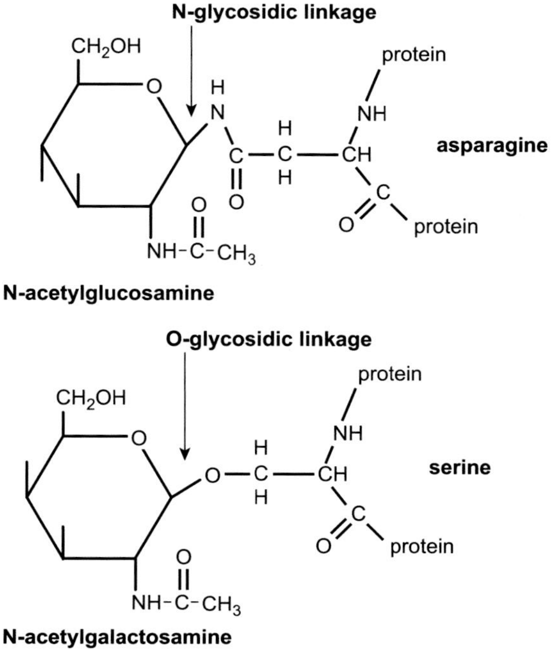

FIGURE 7.6 Attachment of sugars to proteins occurs at either asparagines (the sugar is N-linked) or to serine, threonine, hydroxylysine or hydroxyproline (the sugar is O-linked).

polysaccharides. For most membrane glycoproteins, N-linked sugars are more common in dynamic enzymatic proteins, while O-linked sugars are predominant on structural proteins. Therefore, there are a number of different locations (amino acids) where sugars can be attached. In addition there may be a variety of different sugars that can be attached at a single location. This complexity leads to what is known as microheterogeneity. For example, ovalbumin has only one site of glycosylation yet more than a dozen different attached sugars have been identified at this site.

In the late 1990s computer-based sequencing searches were applied to membrane glycosylation by Nathan Sharon and colleagues [19,20]. In one study of known membrane proteins, the questions addressed were, how common are glycoproteins and how are the sugars attached to the proteins. Complete sequences of 1823 integral membrane proteins with extracellular features were identified. Of these, 1676 glycoproteins (92% of the total integral proteins) were identified, and of these, 1630 or 97.3% were N-glycosylated. This leaves only 46 (2.8%) possible O-glycosylated proteins. 116 multiple span proteins were not glycosylated, leaving only 14 possible single span proteins at most that were not glycosylated. From this study, two major conclusions can be drawn. The vast majority of plasma membrane integral proteins are glycosylated and of these there are many times more N-glycosylations than O-glycosylations.

4. SUMMARY

In addition to lipids and proteins, a third major component, carbohydrates, exists primarily on the outer surface of the cell plasma membrane where they can be anchored to lipids (glycolipids) or proteins (glycoproteins). Carbohydrates are highly water-soluble and so are never found in a membrane's hydrophobic interior. Their main function is to interact with the cell's external environment. The amount of carbohydrate on a membrane protein varies from 0% to greater than 85% by weight. There are nine basic sugars commonly found attached to membranes, as either single sugars or short (less than 15 unit) polysaccharides, and are always located on the extracellular membrane leaflet. One sugar, sialic acid, is highly unusual, being a 9-carbon carboxylic acid that is an anion under physiological conditions. In animals the major glycolipids are sphingolipids. Sphingolipid (ganglioside) sugar metabolism disorders constitute the approximately 50 fatal LSDs.

So far we have discussed the major components of membranes, lipids, proteins, and carbohydrates. Chapter 8 will assimilate these components into working models for membrane structure. The Chapter covers lipid bilayers to lipid rafts.

References

[1] Nelson DL, Cox MM. Lehninger principles of biochemistry. 5th ed. New York (NY): W.H. Freeman & Co; 2008. 1262 pp.

[2] Stick RV, Williams S. Carbohydrates: the essential molecules of life. 2nd ed. Elsevier Science; 2009. 496 pp.

[3] Wang PG, Bertozzi CR, editors. Glycochemistry - principles, synthesis and applications. Marcel Dekker; 2001. 696 pp.

[4] Fraser-Reid BO, Tatsuta K, Thiem J, Coté GL, Flitsch S, Ito Y, et al., editors. Glycoscience - chemistry and chemical biology. 2nd ed. Springer; 2008. 2875 pp.

[5] Reitsma S. The endothelial glycocalyx: composition, functions, and visualization. Europ J Physiol 2007;454: 345–59.
[6] The AOCS Lipid Library. Complex glycerolipids. 2011. lipidlibrary.aocs.org/lipids/complex.html.
[7] Hirabayashi Y, Ichikawa S. Roles of glycolipids and sphingolipids in biological membrane. In: Fukuda M, Hindsgaul O, editors. The frontiers in molecular biology series. IRL Press at Oxford Press; 1999. p. 220–48.
[8] Moss GP. IUPAC-IUB Joint Commission on Biochemical Nomenclature (JCBN). Nomenclature of glycolipids. 1997. http://www.chem.qmul.ac.uk/iupac/misc/glylp.html.
[9] Kornfeld R, Kornfeld S. Comparative aspects of glycoprotein structure. Ann Rev Biochem 1976;45:217–38.
[10] Tauber R, Reutter W, Gerok W. Role of membrane glycoproteins in mediating trophic responses. Gut 1987;28(Suppl.):71–7.
[11] Laine RA. Information capacity of the carbohydrate code. Pure Appl Chem 1997;69:1867–73.
[12] Schauer R. Chemistry, metabolism, and biological functions of sialic acids. Adv Carbohydr Chem Biochem 1982;40:131–234.
[13] Schauer R. Sialic acids and their role as biological masks. Trends Biochem Sci 1985;10:357–60.
[14] AOCS. The lipid library. Sphingolipids. An Introd Sphingolipids and Membrane Rafts.
[15] Dickson RC. Sphingolipid functions in *Sacromyces cerevisiae*. Ann Rev Biochem 1998;67:27–48.
[16] Gunstone F. Fatty acid and lipid chemistry. Blackie Academic and Professional; 1996. p. 43–4.
[17] Futerman AH, van Meer G. The cell biology of lysosomal storage disorders. Nat Rev Mol Cell Biol 2004;5:554–65.
[18] National Institutes of Neurological Disorders and Stroke (NINDS). National Institutes of Health. Ninds Tay-Sachs disease information page.
[19] Sharon N, Lis H. Glycoproteins: structure and function. In: Gabius H-J, Gabius S, editors. Glycosciences-status and perspectives. Weinheim, Germany: Chapman and Hall; 1997. p. 123.
[20] Apweiler R, Hermjakob H, Sharon N. On the frequency of protein glycosylation, as deduced from analysis of the SWISS-PROT database. Biochim Biophys Acta 1999;1473:4–8.

CHAPTER

8

From Lipid Bilayers to Lipid Rafts

OUTLINE

1. DEVELOPMENT OF MEMBRANE MODELS

By 1935 all of the basic elements were in place for someone to propose a realistic model of a membrane (Chapter 2, [1]). Comparing cell permeability measurements with solute partition coefficients, Overton had suggested the lipid nature of a membrane, while Gorter and Grendel had proposed that there was the precise amount of lipid in the erythrocyte membrane to exactly cover the erythrocyte cell surface twice. Thus, at least the erythrocyte membrane was a lipid bilayer. In fact, the then obscure Gorter-Grendel paper implied that the membrane was entirely lipid! Using trans-membrane electrical measurements, in 1925 Fricke correctly determined that a membrane was ~33 Å thick. Also in 1925, Sumner purified the first protein, urease. The prevailing opinion at that time was that biological catalysts were small, hard to isolate organic molecules that were hidden in all the "cellular junk". Although it took almost a full decade for the importance of protein catalysts (enzymes) to be universally accepted, by 1935 it was obvious that proteins were to be a major player in understanding life. Proteins were soon found to be a major component of all membrane compositional analysis. One readily measurable membrane property, surface tension, was severely impacted by

An Introduction to Biological Membranes
http://dx.doi.org/10.1016/B978-0-444-63772-7.00008-7

both membrane lipids and proteins. The surface tension of pure water is 72.8 dyn/cm at 25°C. Phospholipids reduce this to ~7–15 dyn/cm. Proteins further reduce the surface tension down to that of a biological membrane (~0.1 dyn/cm). The conclusion was that proteins must play an important role in membrane structure. However, by 1935 the only proteins that were understood to any extent were water-soluble globular proteins. Nothing was known about membrane proteins.

Questions about overlap between the related fields of biochemistry, protein structure and function, and membranes were starting to emerge by 1935. Although it was clear that different membranes supported different biochemical functions, it was not known whether the structures of different membranes were completely different or if they had important features in common. In other words, are the various membranes more similar or different? It was clear that membranes were "semipermeable", being permeable to some things while impermeable to others. How was semipermeability accomplished? Any proposed model for membrane structure would have a lot of things to consider [2].

The first realistic model of a biological membrane was that of James Danielli and Hugh Davson, initially proposed in 1935 [3]. Their "Pauci-Molecular" model was subsequently tweaked and altered as membrane science progressed [2]. The history of membrane models will be discussed around some major mile posts, starting with the Danielli-Davson model and progressing through the Robertson "Unit Membrane" to the Singer "Fluid Mosaic" model and finally to Simons' "Lipid Raft" model. Two misdirected models that do not employ a lipid bilayer and so clearly fall off this major track, will also be discussed.

2. THE DANIELLI-DAVSON "PAUCI-MOLECULAR" MODEL

In 1935 Danielli and Davson proposed their model for a biological membrane [3]. Their "Pauci-Molecular" model was the first to employ a lipid bilayer. While the seminal experiment of Gorter and Grendel [4] was the first to demonstrate the possibility of a lipid bilayer barrier (membrane) surrounding a cell, these authors did not actually propose a membrane model. Instead, 10 years later Danielli and Davson used the lipid bilayer as the foundation for a new membrane model. Ironically the original 1935 Danielli-Davson paper did not even mention the Gorter-Grendel work! They realized that a membrane was not just lipid, as suggested by Gorter-Grendel, and somehow proteins had to be incorporated. Since at that time only globular proteins were known, Danielli and Davson simply attached globular proteins loosely to the lipid bilayer surface in the form of a protein–lipid–protein sandwich (Fig. 8.1A). Note their original model had no membrane-penetrating protein. Later Danielli, realizing the membrane had to be semipermeable, added thin, peptide-lined trans-membrane channels (Fig. 8.1B).

One important feature of this simple model is that it was proposed to be the basic foundation of all biological membranes, and at the heart of their model was the lipid bilayer. These concepts have remained intact through all subsequent revisions of the "Pauci-Molecular" model. The major problems with this first model were the failure to incorporate proteins into the membrane bilayer interior and the total lack of dynamics in the model. The "Pauci-Molecular" model is static. The model predicts that all membranes would be identical and fails to indicate how variety would be achieved, nor does the model take into account

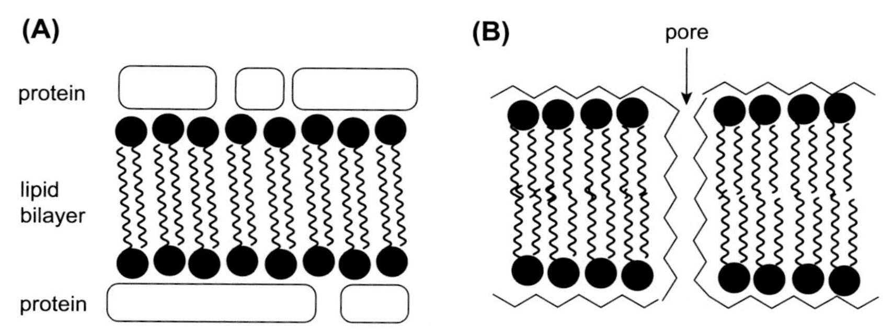

FIGURE 8.1 The Danielli-Davson "Pauci Molecular" model. Part (A) shows the original model consisting of a lipid bilayer with globular proteins covering the surface. Part (B) shows a later revision of part (A) with a trans-membrane peptide pore.

membrane asymmetry. Also the globular protein coat would prevent favorable interactions between the phospholipid head groups and water. While the Danielli-Davson "Pauci-Molecular" membrane model was a good start, and crucial parts of the model, particularly the lipid bilayer, are still invoked today, significant alterations were needed.

3. THE ROBERTSON "UNIT MEMBRANE" MODEL

In a 1957 meeting of the American Physiological Society, J. David Robertson proposed a slight modification of the Danielli-Davson model membrane [5,6]. He called his model the "Unit Membrane". Robertson was generally regarded as the premier electron microscopist of cell membranes, and electron micrographs (EMs) formed the basis of his model. Upon staining membranes with $KMnO_4$, Robertson always observed what is often referred to as "railroad tracks", two dark bands separated by a light band (Fig. 8.2, [7]). Robertson defined the "Unit Membrane" as "measuring ~75 Å across, consisting of two parallel dense lines 25 Å wide separated by a light zone ~25 Å wide". By the "Unit Membrane" model, all membranes have the same basic structure. They are composed of a lipid bilayer with protein monolayers attached covalently to both sides, very similar to the Danielli-Davson model. Whereas the Danielli-Davson model was noncommittal concerning the thickness of a membrane, Robertson's model was not.

As with the Danielli-Davson model, Robertson's "Unit Membrane" model had its advantages and disadvantages. Robertson's EM pictures allowed for a precise membrane measurement. Indeed, Robertson's EM images were the first direct observations of membrane structure. Robertson found the same tri-laminar structure ("railroad tracks") in the plasma membranes from a wide variety of cell types as well as from internal cellular membranes including those of the mitochondria, endoplasmic reticulum, and the nuclear envelope. Unfortunately the same flaws that eventually doomed the Danielli-Davson model also doomed the Robertson model. The "Unit Membrane" is a static model and fails to appreciate either the dynamic aspects of membranes or the asymmetric nature of membrane structure. Membrane dynamics would have to await further advances in technology to address these problems (Chapters 9, 10, and 11).

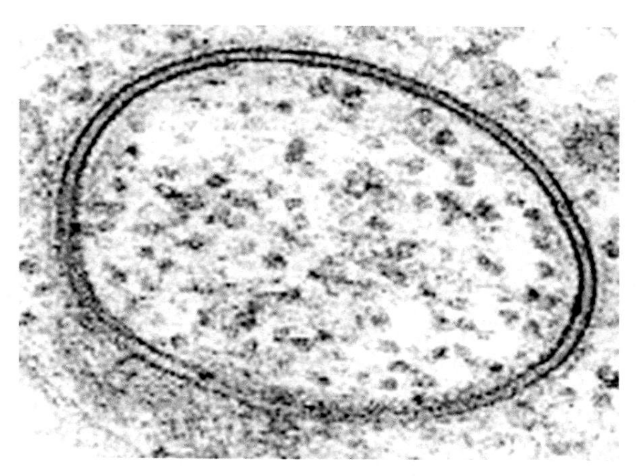

FIGURE 8.2 A typical electron micrograph of a plasma membrane demonstrating the "railroad tracks" on which Robertson based his "Unit Membrane" model [7]. The plasma membrane is observed to be two dark lines surrounding a light interior.

The eventual replacement of Robertson's use of $KMnO_4$ by gluteraldehyde and OsO_4 indicated that the intracellular membranes are not directly connected to one another as Robertson predicted, but instead are connected through tiny vesicles. The newer methodology also showed that the plasma membrane has ties to the internal cytoskeleton. The "Unit Membrane" model was so similar to the "Pauci-Molecular" model, the two are often blended into one Danielli-Davson–Robertson model membrane.

4. BENSON AND GREEN LIPOPROTEIN SUBUNIT MODELS

Through the 1960s, membrane science was still in its infancy and the most important missing aspects involved membrane asymmetry and dynamics. It should not be surprising that even the venerable lipid bilayer was in question by some. Two well-disseminated membrane models replaced the lipid bilayer with lipoprotein subunits. The Lipoprotein Subunit models were proposed by Andrew Benson (for photosynthetic thylakoids, [8]) and David Green (for mammalian mitochondria, [9]). These models are depicted in Fig. 8.3.

These models received considerable attention due to the stature of their inventors. Andrew Benson (Fig. 8.4) was a world-renowned plant physiologist from his early work on the Calvin–Benson cycle of photosynthetic dark reactions. David Green (Fig. 8.5) was a leading mitochondriac whose reputation was based on fractionating the mitochondrial electron transport–oxidative phosohorylation system into five isolatable complexes (Chapter 18). Partial or complete mitochondrial oxidative phosphorylation could be reconstituted by simply mixing the appropriate complexes. Therefore both Benson and Green were experts on highly specific and sophisticated membranes, Benson the photosynthetic thylakoid and Green the bioenergetic inner mitochondrial membrane. Both of these membranes are not "typical". They are at the upper range of membrane protein–lipid ratios. Therefore, in these complex membranes, the lipid bilayer would likely be reduced. Both the Benson and Green

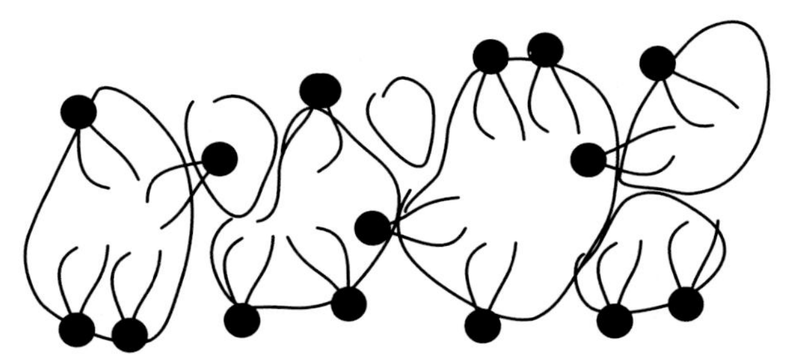

FIGURE 8.3 The Benson – Green Lipoprotein Subunit model membrane. Proteins are depicted as large, open sack-like structures while phospholipids are the small, black spheres with attached acyl chains. Note the complete absence of a lipid bilayer.

FIGURE 8.4 Andrew A. Benson, 1917–2015.

FIGURE 8.5 David E. Green, 1910–1983.

models were essentially a lipoprotein monolayer devoid of a lipid bilayer (Fig. 8.3). The Green model (also called the "Protein Crystal" model) was particularly well-described. In his 1970 Preceedings of the National Academy of Sciences paper [9], Green states "The essential principle underlying this model is that when membrane proteins polymerize, the points of contact between proteins are few, and cavities lined with predominantly nonpolar amino acids are formed. Phospholipid molecules become oriented with the fatty chains inserted into the cavities while the polar heads remain on the surface of the membrane. This orientation applies to both faces of the membrane continuum. All the lipid known to be present in membranes can be accommodated in this manner." Therefore, no lipid bilayer was required. While this proposal was quite interesting, it failed to explain membrane structure for most membranes, ignored the rapidly accumulating data on the lipid bilayer-based models (for example lateral diffusion), and ignored the likelihood that a single mutation could destroy an entire membrane. These lipoprotein-based model membranes disappeared with the advent of Singer's "Fluid Mosaic" model. Before they vanished however, they did clearly indicate a major problem with all membrane models, the lipid bilayer "sea" must be very limited in scope and must be very crowded (Chapter 11, [10]).

5. SINGER-NICOLSON "FLUID MOSAIC" MODEL

By 1972, information concerning membrane structure was rapidly accumulating from a wide variety of new approaches. It was becoming clear that the Danielli-Davson—Robertson model for membranes was inadequate. Of particular concern was the failure to properly account for membrane structure and topology, membrane protein diversity, and the lack of appreciation for membrane dynamics. The major positive aspect of the Danielli-Davson—Robertson model was having a lipid bilayer as its centerpiece.

In what was to become an iconic paper on membranes, in 1971 Seymour J. Singer (Fig. 8.6) and Garth L. Nicolson (Fig. 8.7) proposed a new membrane model they called the "Fluid

FIGURE 8.6 Seymour J. Singer, 1924-.

FIGURE 8.7 Garth L. Nicolso., 1943—http://www.cancercontrolsociety.com/bio2000/nicolson.html.

Mosaic" model [11]. With some later modifications, this model is the one still *in vogue* today. In this chapter, basic components of the "Fluid Mosaic" model will be presented, while experiments on membrane physical properties supporting the model will be discussed in Chapter 9. In their paper, Singer and Nicolson defined a membrane as "an oriented, two-dimensional viscous solution of amphipathic proteins (or lipoproteins) and lipids in instantaneous thermodynamic equilibrium". Their model retained the lipid bilayer, but substantially altered many other aspects of the static Danielli-Davson—Robertson model.

A major problem with Robertson's "Unit Membrane" was assigning the two dark bands he noted in all of his EM pictures to continuous protein coats on both sides of a white lipid bilayer core. We now know that shielding the polar lipid headgroups from water would substantially diminish bilayer stability. Also, these protein coats were symmetrically distributed on each side of the membrane and did not span the hydrophobic core. Therefore, the Unit Membrane could not possibly account for trans-membrane vectorial (directional) biochemical activity. The "Fluid Mosaic" model took into account the reality of membrane protein structure (Chapter 6). Both peripheral and integral proteins were proposed to be distributed asymmetrically across the membrane (Fig. 8.8). Carbohydrates were attached to proteins (glycoproteins) and lipids (glycolipids), but were found only on the outer membrane surface. Integral proteins were described as being "globular and amphipathic" and "having a significant fraction of its volume embedded in the membrane". This early description of protein structure was later modified to include trans-membrane α-helical and β-barrel spans (Chapter 6). Membrane components were distributed in a "mosaic structure" where both proteins and lipids were dispersed inhomogeneously, resulting in lateral and trans-membrane patches called domains [12].

A second important element Singer and Nicolson added to their model can be loosely described as "membrane dynamics". The Danielli-Davson—Robertson model was essentially a frozen picture. The "Fluid Mosaic" model recognized for the first time that a biological membrane is dynamic, fluid, and ever changing. Although identifying the vast diversity

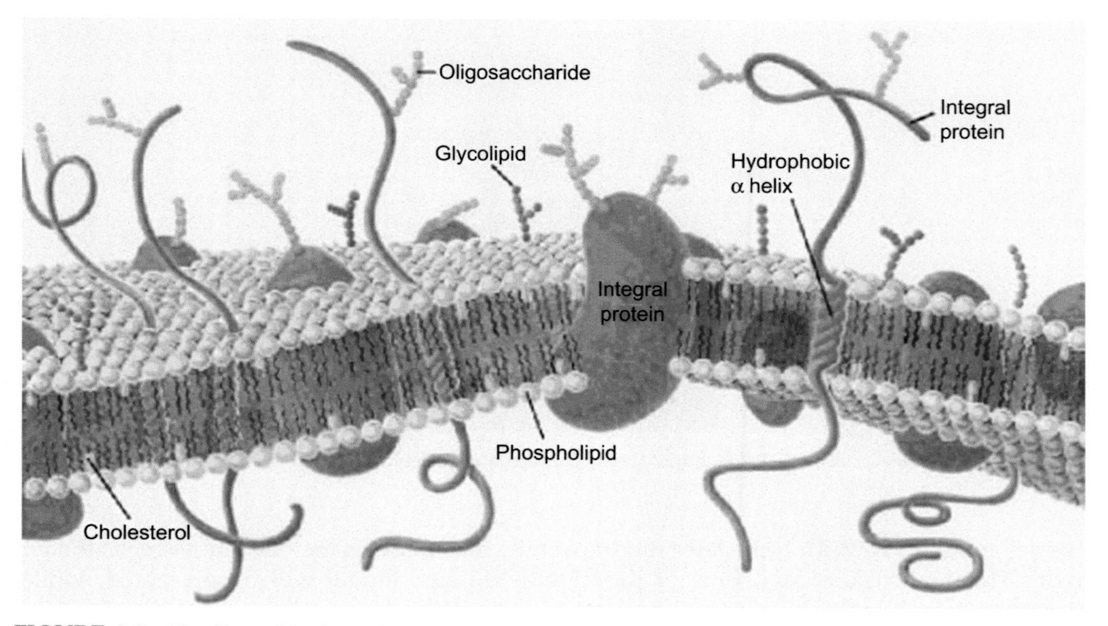

FIGURE 8.8 The Singer-Nicolson Fluid Mosaic Model for membrane structure.

of membrane lipids, proteins, and carbohydrates is by itself a difficult task, it is membrane dynamics that presents the most challenges. Dynamics involves how the various membrane components interact with one another. A variety of techniques are required to cover the vast size and time domains encountered in membrane studies. Various aspects of membrane dynamics are discussed in Chapter 9.

The original "Fluid Mosaic" model predicted totally free lateral diffusion of both lipids and proteins. This concept was later modified by proposing limited diffusion for some proteins due to their interactions with the essentially immobile cytoplasmic proteinaceous mesh known as the cytoskeleton [13]. The "Fluid Mosaic" model was the first to recognize both membrane fluidity and lateral mobility. The existence of an extensive lipid bilayer matrix and free lateral diffusion in the bilayer presented a dilemma. A bilayer should not be able to support long-range (few tenths of a micron or greater) order, although short-range order should be feasible. However, some long-range order, like that observed in the synapse (Chapter 18), is known to exist. Also, a number of agents extrinsic to the membrane can result in clustering of membrane proteins (see the following section on lipid rafts). Detailed mechanisms explaining long-range order were not ready to be included in the Singer-Nicolson 1972 *Science* paper. Possible discussions on structural coupling between distant lateral sites on the membrane (*cis* coupling) and between sites on opposite sides of the membrane (*trans* coupling) were also premature.

Table 8.1 lists the salient features of the "Fluid Mosaic" model. Some of these features (eg, β-barrels and lipid-linked proteins) were discovered after the "Fluid Mosaic" model was first formulated in 1972.

TABLE 8.1 Salient features of the "Fluid Mosaic" model

I. Membrane lipids:
- **A.** Molecular structures are amphipathic (membrane lipids are polar).
- **B.** Most membrane lipids exist in a lipid bilayer matrix.
- **C.** Some lipids are bound to protein (lipoproteins).
- **D.** Trans-membrane lipid distribution is partially asymmetric.
- **E.** Most membrane lipids exist in a fluid state.
- **F.** Membrane lipids exhibit rapid, heterogeneous lateral diffusion in a membrane.
- **G.** Membrane lipids exhibit slow trans-membrane diffusion (flip-flop).
- **H.** Lipids have no biochemical activity.

II. Membrane proteins:
- **A.** Membrane protein molecular structures are amphipathic.
- **B.** Trans-membrane protein distribution is totally asymmetric.
- **C.** Membrane protein lateral diffusion rate varies from rapid to immobile.
- **D.** Absolutely no trans-membrane protein diffusion (flip-flop) occurs.
- **E.** Membrane proteins exhibit substantial diversity (biochemical activities).
- **F.** Protein types:
  - **1.** Peripheral (essentially globular)
    - **a.** bound to lipid (electrostatic)
    - **b.** bound to protein (electrostatic)
  - **2.** Amphitropic
  - **3.** Integral: Endo-, ecto- and trans-membrane
    - **a.** α-Helix: Single span and multiple Span
    - **b.** β-Barrel
    - **c.** Lipid-linked
      - Myristoylated, palmitoylated, and prenylated
      - Glycosylphosphatidylinositol-anchored (GPI-linked)
  - **4.** Glycoproteins (general type of protein that has attached sugars)

III. Membrane carbohydrates:
- **A.** Totally water-soluble (no interaction with membrane interior).
- **B.** Glycolipids
- **C.** Glycoproteins
- **D.** Membrane distribution is asymmetrical.
- **E.** Always found on membrane outer surface.
- **F.** Lateral diffusion is determined by the attached lipid or protein.
- **G.** No trans-membrane diffusion (flip-flop) occurs.
- **H.** Carbohydrates have no biochemical activity.

6. SIMONS LIPID RAFT MODEL

In the original 1972 "Fluid Mosaic" model, Singer and Nicolson did not postulate lateral heterogeneity into functional membrane domains although it was generally accepted that membrane phospholipids and proteins were probably heterogeneously distributed. However, in the 1970s, reports were starting to emerge that membrane "lipid clusters" may be common features of biological membranes. In a 1982 paper in the *Journal of Cell Biology*, Karnovsky and Klausner formalized the concept of lipid domains in membranes [14]. Interestingly, one type of lipid cluster that kept arising was that of cholesterol and sphingolipids [15,16]. Indeed protein-free lipid model membranes showed that cholesterol has a higher affinity for sphingomyelin (SM) than it does for other phospholipids [16]. Although lipid

FIGURE 8.9 Kai Simons, 1938-.

phase separations in membranes were starting to achieve acceptance, their biological relevance awaited a link to a biological function. The link proved to be spectacular when cholesterol- and sphingolipid-rich microdomains were closely associated with essential cellular signaling events. However, the real interest in these microdomains did not come until after they received the very catchy and descriptive name "Lipid Raft". This critical link was made by Kai Simons (Fig. 8.9) in a 1997 paper in *Nature* [17]. Membrane studies would be changed forever. Since 1997, Lipid Rafts have generated countless primary research papers and many excellent reviews (eg, [18–25]. However, one wonders if these microdomains were correctly called "phase separated cholesterol- and sphingolipid-rich, liquid ordered microdomains" instead of "Lipid Rafts" whether they would have achieved such notoriety!

Lipid rafts are cholesterol- and sphingolipid-rich, liquid ordered (l_o) state platforms (microdomains) that float in a nonraft liquid disordered matrix (Fig. 8.10). The l_o state has properties midway between those of a liquid crystalline (disordered, fluid) and gel (ordered, solid) state. These states are discussed in detail in Chapter 9. Briefly, the acyl chains of lipid rafts are more saturated than are the acyl chains of the surrounding membrane matrix and so pack more tightly, producing the l_o state. It is believed that lipid rafts are held together by cholesterol, reflecting the sterol's high affinity for sphingolipids. Typically raft cholesterol levels are twice that found in membrane nonraft regions, while the SM concentration is ~50% higher and the phosphatidylcholine level is reduced in proportion to the increase in SM. As a result of the tight packing, lipid raft portions of the membrane are thicker than the surrounding, more loosely packed membrane and so protrude above the surrounding membrane surface (Chapter 11). Lipid rafts are believed to house a variety of raft-characteristic proteins that are intimately involved in cell signaling events. Often these proteins affect the phosphorylation state that can be modified by local kinases and phosphatases to affect downstream signaling.

An original hallmark of lipid rafts is their ability to be isolated from the remaining membrane matrix by extraction in cold (4°C) nonionic detergents (eg, Triton X-100 or Brij-98).

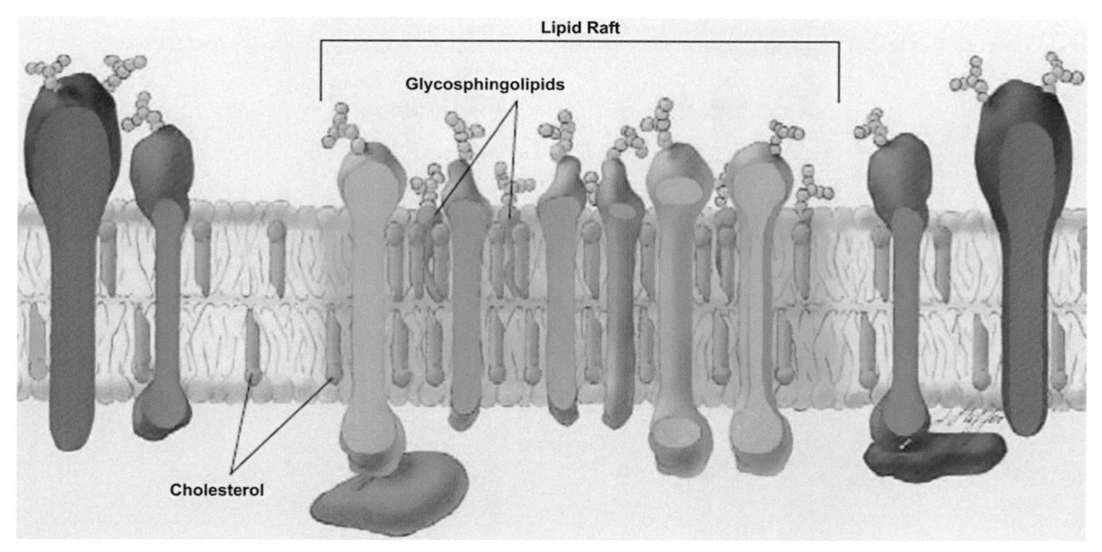

FIGURE 8.10 Simons' lipid raft model.

Under cold, detergent conditions, the membrane matrix is solubilized while the lipid rafts are insoluble (they are detergent resistant). Because of their composition and detergent resistance, lipid rafts are also referred to as detergent-insoluble glycolipid-enriched complexes (GEMs or DIGs) or Detergent Resistant Membranes (DRMs). It is believed, although by no means proven, lipid rafts are normally extremely small (~10–200 nm) but their size can be greatly enhanced by various natural and artificial cross-linking agents.

In a 2006 Keystone Symposium on Lipid Rafts and Cell Function [26], lipid rafts were defined as "small (10–200 nm), heterogeneous, highly dynamic, sterol- and sphingolipid-enriched domains that compartmentalize cellular processes. Small rafts can sometimes be stabilized to form larger platforms through protein–protein interactions".

While the lipid raft story is compelling, it is wrought with potential problems. In fact rafts have occasionally been referred to as "unidentified floating objects" or UFOs. The cold temperature detergent procedure is crude and very slow compared to the likely timescale of lipid raft stability, and is full of potential artifacts. The older raft isolation procedures are being improved and nondetergent methodologies have been developed (Chapter 13) [27]. The very small raft size, being below the classical diffraction limit of a light microscope, has made direct observation of rafts very difficult. Raft size also brings into question their stability. Even if lipid rafts exist, they may only occur on a timescale that is irrelevant to biological, diffusion-controlled processes.

One question that is not close to resolution is how many types of microdomains constitute a functional membrane. While lipid rafts have garnered almost all of the attention, it seems likely that there probably exists many types of less obvious, and hence as yet undiscovered, nonraft microdomains. Even traditional lipid rafts may be composed of countless subraft domains. Currently two basic types of rafts have been identified, planar lipid rafts (also referred to as noncaveolar, or glycolipid rafts) and caveolae. Planar rafts are

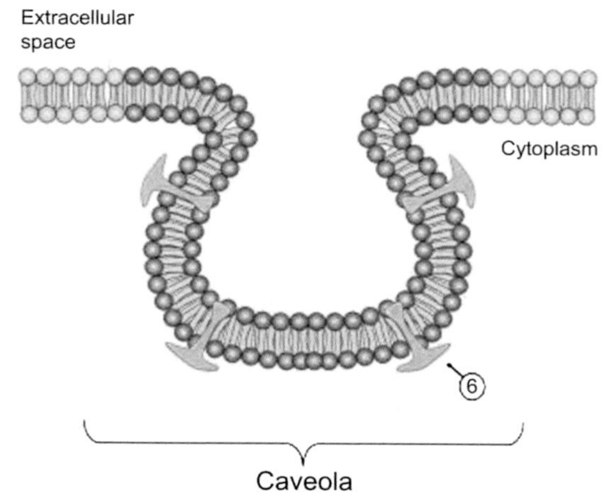

FIGURE 8.11 Caveolae, a type of lipid raft.

continuous with the membrane surface while caveolae are flask-shaped invaginations (Fig. 8.11). Caveolae are easily distinguished from planar rafts by their unique flask-shape and the presence of a characteristic protein, caveolin. In contrast, planar rafts contain flotillin proteins. The function of both caveolin and flotillin is to attract and retain signaling molecules into rafts.

While planar lipid rafts and caveolae have been fairly well-defined, membranes are composed of such a vast array of lipids and proteins that other distinct types of raft and non-raft microdomains will undoubtedly be discovered. Two examples of these nontraditional membrane domains have been recently proposed, "ceramide-rich platforms" [28], and "poly-unsaturated fatty acid (PUFA)-rich micro-domains".

Ceramides are an example of a type of "bioactive lipid" (Chapter 20) that in relatively low concentration have a significant influence on membrane structure and therefore function [21,29]. Like lipid rafts, ceramide-rich platforms are liquid ordered (l_o) microdomains, but unlike rafts they are enriched in C-24 fatty acids and have greatly reduced levels of cholesterol. They are believed to cluster as a unique set of stress-induced signaling proteins attached to ceramides, forming the platform.

While lipid rafts are tightly packed, cholesterol- and saturated acyl chain-rich, l_o microdomains, many membrane phospholipids have highly unsaturated PUFA chains that should have trouble accumulating in rafts. The very extreme example of a PUFA is docosahexaenoic acid (DHA), which at 22-carbons and 6 double bonds is the longest and most unsaturated fatty acid commonly found in membranes (Chapter 23, [30]). DHA has a repeating series of ($=CH-CH_{(2)}-CH=$) units generating an extremely flexible structure that rapidly isomerizes through many conformational states (Chapter 10, [31]). This makes DHA's accumulation into tightly packed lipid rafts problematic. In addition, it is well established that DHA has a strong aversion for cholesterol (Chapter 10), further diminishing its accumulation into rafts [32]. The existence of DHA-rich membrane microdomains coexisting with rafts has been hypothesized. Such liquid disordered (l_o) domains would be compositionally and organizationally the antithesis of lipid rafts and so would support nonraft cellular functions.

7. SUMMARY

By 1935, basic elements were in place to propose a realistic membrane model. This was accomplished by Danielli and Davson in their "Pauci-Molecular" model, based on a lipid bilayer that had proteins uncomfortably attached in the form of a protein-lipid bilayer-protein sandwich. A slight improvement on this model was proposed by Robertson in his 1957 Unit Membrane model. His model was based on EMs, the first direct observations of membrane structure, and featured images that resembled 'railroad tracks'. In the 1960s, Benson (for the photosynthetic thylakoid) and Green (for the mitochondrial inner membrane) proposed lipid bilayer-free models, called the Lipoprotein Subunit model. These early membrane models did not appreciate membrane asymmetry or dynamics. In 1972 Singer and Nicholson proposed their "Fluid Mosaic" model that retained the lipid bilayer, but accounted for previous model shortcomings by recognizing membrane fluidity and lateral mobility. An extension of this model was provided by Kai Simons' 1997 "Lipid Raft" model that proposed sphingolipid- and cholesterol-rich, liquid ordered, signaling microdomains.

The next three chapters will discuss various membrane physical properties. These properties are essential to understanding how membranes work. Chapter 9 will discuss membrane thickness, bilayer stability, membrane protein, carbohydrate and lipid asymmetry, transmembrane movement (flip-flop), membrane lateral diffusion, membrane–cytoskeleton interaction, lipid melting behavior, and membrane fluidity.

References

[1] Eichman P. From the lipid bilayer to the fluid mosaic: a brief history of model membranes. SHiPS Resource Center for Sociology, History and Philosophy in Science Teaching.
[2] De Weer PA. Century of thinking about cell membranes. Ann Rev Physiol 2000;62:919–26.
[3] Danielli JF, Davson H. A contribution to the theory of permeability of thin films. J Cell Comp Physiol 1935;5:495–508.
[4] Gorter E, Grendel F. On bimolecular layers of lipids on the chromocytes of the blood. J Exptl Med 1925;41:439–43.
[5] Robertson JD. The cell membrane concept. J Physiol 1957;140:58P–9P.
[6] Robertson JD. The molecular structure and contact relationships of cell membranes. Prog Biophys Biophys Chem 1960;10:344–418.
[7] Wikipedia. History of cell membranes. www.thefullwiki.org.
[8] Benson AA. On the orientation of lipids in chloroplasts and cell membranes. J Am Oil Chem Soc 1966;43:265–70.
[9] Green DE, Vanderkooi G. Biological membrane structure, I. The protein Crystal model for membranes. Proc Natl Acad Sci USA 1970;66:615–21.
[10] Ellis RJ. Macromolecular crowding: obvious but unappreciated. Trends Biochem Sci 2001;26:597–604.
[11] Singer SJ, Nicolson GL. The fluid mosaic model of the structure of cell membranes. Science 1972;175:720–31.
[12] Devaux PF, Morris R. Transmembrane asymmetry and lateral domains in biological membranes. Traffic 2004;5:241–6.
[13] Doherty GJ, McMahon HT. Mediation, modulation, and consequences of membrane-cytoskeleton interactions. Ann Rev Biophys 2008;37:65–95.
[14] Karnovsky MJ, Kleinfeld AM, Hoover RL, Klausner RD. The concept of lipid domains in membranes. J Cell Biol 1982;94:1–6.
[15] Estep TN, Mountcastle DB, Barenholz Y, Biltonen RL, Thompson TE. Thermal behavior of synthetic sphingomyeline cholesterol dispersions. Biochemistry 1979;18:2112–7.
[16] Demel RA, Jansen JW, van Dijck PW, van Deenen LL. The preferential interaction of cholesterol with different classes of phospholipids. Biochim Biophys Acta 1977;465:1–10.

[17] Simons K, Ikonen E. Functional rafts in cell membranes. Nature 1997;387:569–72.
[18] Brown DA, London E. Functions of lipid rafts in biological membranes. Ann Rev Cell Dev Biol 1998;14:111–36.
[19] Simons K, Toomre D. Lipid rafts and signal transduction. Nat Rev Mol Cell Biol 2000;1:31–9.
[20] Edidin M. Shrinking patches and slippery rafts: scales of domains in the plasma membrane. Trends Cell Biol 2001;11:492–6.
[21] Edidin M. The state of lipid rafts: from model membranes to cells. Ann Rev Biophys Biomol Struct 2003;32:257–83.
[22] Pike LJ. Lipid rafts: heterogeneity on the high seas. Biochem J 2004;378:281–92.
[23] Simons K, Vaz W. Model systems, lipid rafts, and cell membranes. Ann Rev Biophys Biomol Struct 2004;33:269–95.
[24] Pike LJ. The challenge of lipid rafts. J Lipid Res 2009;50:S323–8.
[25] Lingwood D, Simons K. Lipid rafts as a membrane-organizing principle. Science 2010;327:46–50.
[26] Pike LJ. Rafts defined: a report on the Keystone symposium on lipid rafts and cell function. J Lipid Res 2006;47:1597–8.
[27] Smart EJ, Ying Y-S, Mineo C, Anderson RGW. A detergent-free method for purifying caveolae membrane from tissue culture cells. Proc Natl Acad Sci USA 1995;92:10104–8.
[28] Schenck M, Carpintessu A, Grassme H, Lang F, Gulbins F. Ceramide physiological and pathological aspects. Arch Biochem Biophys 2007;462:171–5.
[29] Yu C, Alterman M, Dobrowsky RT. Ceramide displaces cholesterol from lipid rafts and decreases the association of the cholesterol binding protein caveolin-1. J Lipid Res 2005;46(8):1678–91.
[30] Stillwell W, Wassall SR. Docosahexaenoic acid: membrane properties of a unique fatty acid. Chem Phys Lipids 2003;126:1–27.
[31] Wassall SR, Stillwell W. Docosahexaenoic acid domains: the ultimate non-raft membrane domain. Chem Phys Lipids 2008;153(1):57–63.
[32] Wassall SR, Stillwell W. Polyunsaturated fatty acid-cholesterol interactions: domain formation in membranes [Review]. Biochim Biophys Acta 2009;1788(1):24–32.

CHAPTER

9

Basic Membrane Properties of the Fluid Mosaic Model

OUTLINE

An Introduction to Biological Membranes
http://dx.doi.org/10.1016/B978-0-444-63772-7.00009-9

Every aspect of membrane structural studies involves parameters that are very small and very fast. As a result a variety of highly specialized, usually esoteric biophysical methodologies must be employed, often simultaneously. It is the nature of the highly sophisticated techniques used to investigate membrane structure that unfortunately forms an almost insurmountable impediment between physical and biological approaches to membrane studies. This chapter will investigate several fundamental membrane properties using specific examples taken from the research literature. Emphasis will be placed on how the technique was employed to address a particular question. It is not the objective of this chapter, or this book, to discuss the intricate details of each technique, but rather to understand what questions the technique can address. In other words, the emphasis will be placed on application and not detailed methodology.

1. SIZE AND TIME DOMAINS

Generally, parameters of interest to the study of biological membranes will range in size from microns (10^{-6} m) to Angstroms (Å) (10^{-10} m), and in time from milliseconds (10^{-3} s) to picoseconds (10^{-12} s). Both size and time ranges are so vast that multiple instrumentations must be employed. Size domains are demonstrated by use of the electromagnetic spectrum shown in Fig. 9.1. The electromagnetic spectrum is the range of all possible electromagnetic radiation frequencies. It has no distinct beginning nor does it have a distinct end. It covers wavelengths from the size of the universe down to a fraction of the size of an atom. The spectrum is continuous and infinite. However, for convenience, the spectrum is classified into several, overlapping ranges. The wavelengths, from longest to shortest are: radiowaves, infrared region, visible region, ultraviolet region, X-rays, and gamma rays. The shorter the wavelength, the higher is its frequency and energy. While short wavelength radiation has better resolving power, it is also more destructive. Long wavelength radiowaves can be used for some crude imaging, such as ultrasound to safely image a fetus still in the womb, but would be of little use in imaging a membrane. Electromagnetic radiation between 380–760 nm can be detected by the human eye and so is referred to as visible light. Shorter wavelength blue light has better resolving power than does longer wavelength red light. However, the wavelength of even blue light is much longer (380 nm) than the thickness of a biological membrane (a little less than 10 nm or 100 Å). Even including proteins, membrane thickness is generally less than 10 nm. Therefore, a biological membrane cannot be directly imaged with a light microscope, but must employ much shorter wavelength radiation, primarily destructive X-rays. It is unfortunate that nonharmful, long wavelength radiation has poor resolving power while good resolution can only be achieved with biologically harmful short wavelength radiation.

Also complicating studies of biological membranes are the very fast times associated with many biological events (microseconds to picoseconds). Fig. 9.2 shows the time ranges over which several important biophysical techniques commonly employed in membrane studies can be used. Of these techniques the fastest is Raman = infrared (IR) > electron spin resonance (ESR) = fluorescence polarization (FP) > nuclear magnetic resonance (NMR). There are however modifications of each of these techniques that can substantially expand the normal range to much faster times. The technique must match the question

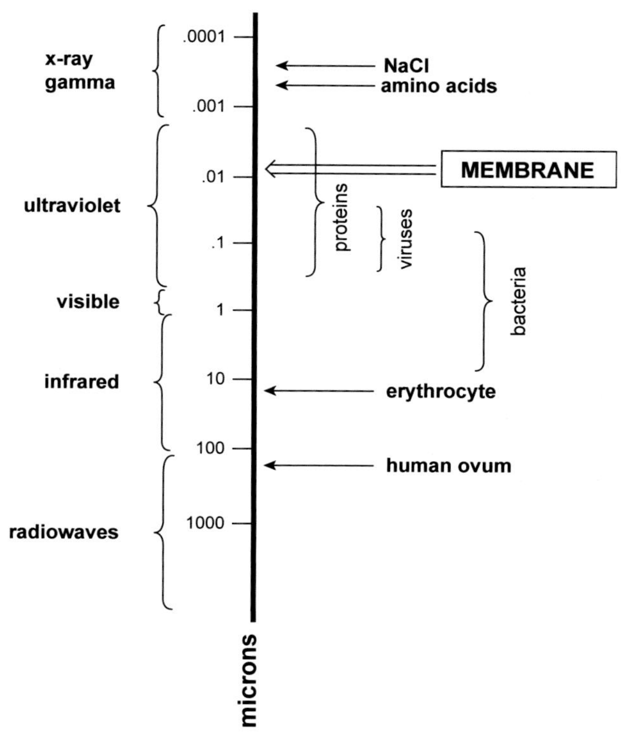

FIGURE 9.1 The electromagnetic spectrum and the size of various important biological components. The average thickness of a biological membrane is a little less than 10 nm or 100 Å.

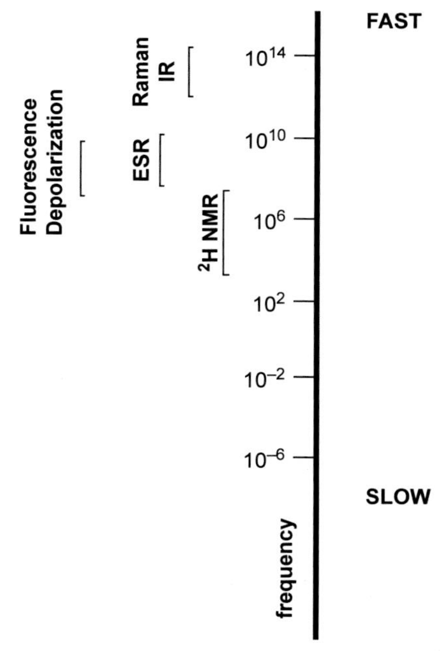

FIGURE 9.2 Time domains for several biophysical techniques. Frequencies are per second. *ESR*, Electron spin resonance; *IR*, infrared; *NMR*, nuclear magnetic resonance.

being addressed. For example, early ESR studies indicated that lipids in immediate contact with membrane proteins (termed annular lipids) have different properties than the bulk bilayer lipids (Chapter 10). However, by NMR all lipids are the same and therefore annular lipids do not exist. The reason for this discrepancy is that ESR detects events that are 10 times faster than can be observed by slower NMR.

2. MEMBRANE THICKNESS

Although not fully appreciated at the time, the thickness of a biological membrane was accurately determined by Hugo Fricke [1]. Using trans-membrane electrical measurements, Fricke correctly determined that the hydrophobic center of a membrane was ~33 Å thick. This was in close agreement with the early pioneers working on lipid monolayers, Rayleigh, Pockels, and Langmuir, who determined that the thickness of a triolein lipid monolayer was ~13–16 Å (Chapter 2). Doubling this value to account for membrane lipids existing in bilayers, reported in 1925 by Gorter and Grendel, yields a correct hydrophobic membrane interior of ~26–32 Å. Fricke's failure to appreciate the importance of his own measurements stemmed from his lack of understanding the bilayer. It was not until 1957 when Robertson imaged the membrane using electron microscopy (EM) (Chapter 8) that the full membrane span of about 75 Å, including lipid bilayer and protein, was established.

Two other important methods that have been used to determine membrane thickness (planar bimolecular lipid membrane analysis and X-ray diffraction) are considered in Sections 2.1 and 2.2.

2.1 Planar Bimolecular Lipid Membranes

For the decades between 1925 and 1960, the lipid bilayer was believed to be an essential feature of membranes. However, it was not clear if the lipid bilayer was stable enough to actually exist. Several investigators, including Irwin Langmuir (in approximately 1937), tried but failed to make a lipid bilayer using the reverse process involved in making soap bubbles. Whereas a lipid bilayer is water–lipid bilayer–water, a soap bubble is air–inverse lipid bilayer–air (Fig. 9.3). Fascination with soap bubbles is centuries old and its early scientific investigation is summarized in a classic book by C.V. Boys [2].

In 1961, Mueller and co-workers reported the first preparation of a large, stable lipid bilayer they termed the planar bimolecular lipid membrane (BLM) [3]. They found that BLMs could be made from a variety of polar lipids dissolved in organic solvents. Originally these investigators used a chloroform–methanol–brain lipid-extract since they were interested in the nerve process of excitability (Chapter 18). The solution was spread over a submerged 1–2 mm hole in a Teflon cup (Fig. 9.4, top) by use of a small artists brush or later by a syringe. The hole was monitored by 90° reflected light through a small telescope. After application of the lipid solution, the hole appeared gray, but soon developed many bands of color, the result of partial destructive interference (Fig. 9.4, bottom). Eventually, a black spot indicative of total destructive interference appeared. The black spot rapidly expanded to occupy the entire hole. Using Bragg's law it was a simple calculation to determine that the BLM was ~40–80 Å thick. This experiment proved that the lipid bilayer was indeed stable!

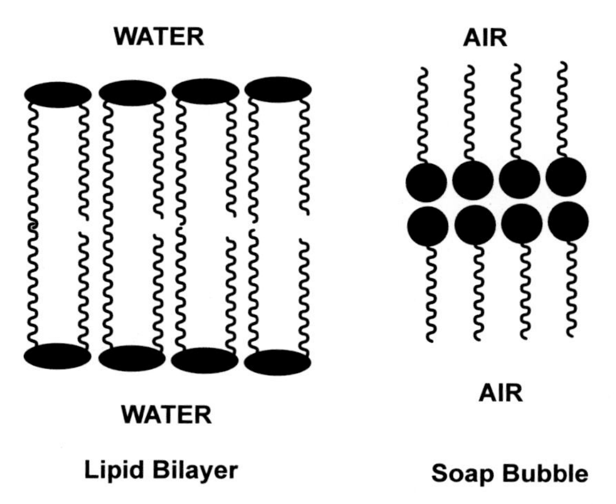

FIGURE 9.3 The lipid bilayer compared to a soap bubble. Both are made from amphipathic molecules, only their orientation is reversed.

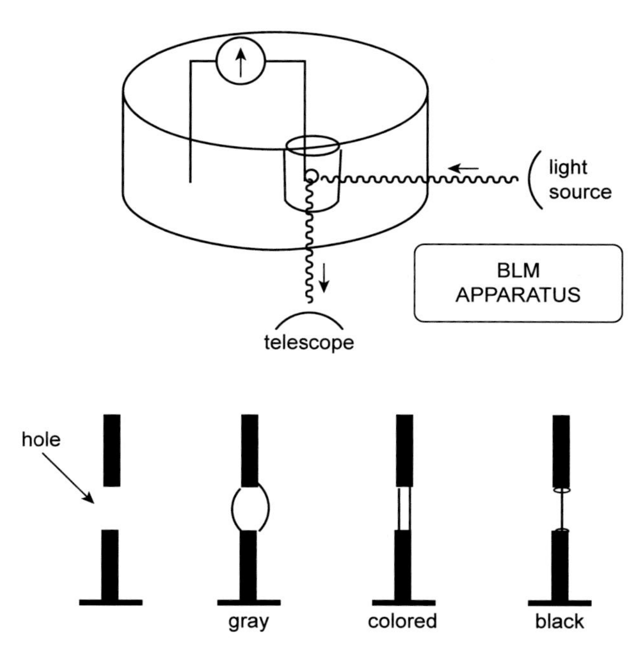

FIGURE 9.4 Apparatus for making the original planar bimolecular lipid membrane (top). Appropriate polar lipids, dissolved in an organic solvent, were added across a small aperture in a Teflon cup. The aperture was observed using a small telescope by 90° reflected light. Initially the lipid solution appeared gray. As the membrane thinned, partial destructive interference of the reflected light made the developing membrane appear multi-colored (bottom). Eventually a black spot appeared in the thick, colored membrane identifying a small area that had become bimolecular. The spot rapidly filled almost the entire aperture indicating the entire membrane had become a lipid bilayer. Various membrane electrical properties were measured using electrodes on either side of the aperture.

2.2 X-ray Diffraction

X-ray diffraction can generate limited high resolution structural information, but only for membranes that have highly ordered crystalline repeating units. A few biological membranes (eg, myelin sheath and the rod outer segment) are naturally stacked and so are ideally suited for X-ray diffraction studies. In fact as early as 1935, X-ray studies on myelin were consistent with the presence of a lipid bilayer. In addition, mitochondria and erythrocyte membrane vesicle preparations and phospholipid vesicles (liposomes) can be collapsed by centrifugation, also making them suitable for X-ray diffraction. The electron density profiles of all membranes look very similar by X-ray diffraction. They consist of a low electron density hydrocarbon interior flanked by a high electron density polar group on either side. Fig. 9.5 shows an electron density profile of dimyristoyl-*sn*-glycero-3-phosphocholine (DMPC) and DMPC–cholesterol liposomes [4]. This bilayer has a 43 Å separation between polar head groups. The addition of more lipids or even proteins only slightly modify the profile. Details of membrane protein structure, for example, cannot be obtained by X-ray diffraction. Therefore, X-ray diffraction does not yield molecular detail of membrane structure, but does support the concept of a lipid bilayer.

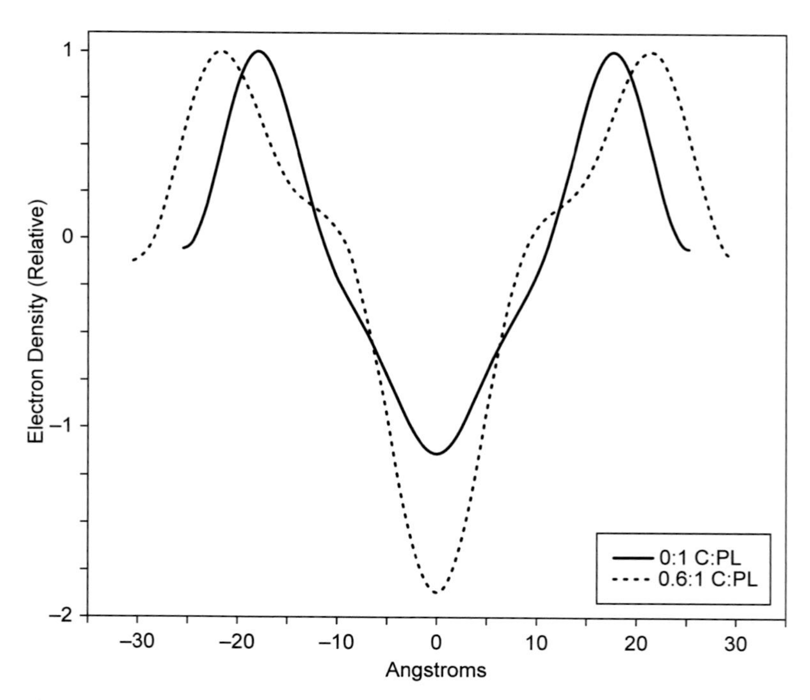

FIGURE 9.5 Electron density profiles obtained by X-ray diffraction for lipid vesicles (liposomes) made from DMPC (solid line) or DMPC–cholesterol (0.6:1, mol:mol, dotted line). The most electron-dense part of the spectrum corresponds to the polar head group regions. The least electron-dense region is the bilayer center found between the two bilayer leaflets [4].

3. MEMBRANE ASYMMETRY

Even before membrane asymmetry had been experimentally proven, it was universally agreed that proteins and probably carbohydrates should be 100% asymmetrically distributed across the membrane for functional reasons. The situation for membrane lipids, however, was not as certain.

3.1 Protein Asymmetry

It seems logical that all functional membrane proteins should be oriented in only a single trans-membrane direction to prevent futile cycles. For example, if the function of a protein was to pump a solute into a cell, it would not make any sense if some of the same solute transport proteins were oriented in the opposite direction, thus pumping the same solute back out of the cell. But logic is not proof. A classic 1964 EM paper by Humberto Fernandez-Moran is often cited as the first definitive evidence that membrane proteins are 100% asymmetrically distributed across membranes. Fernandez-Moran (Fig. 9.6) was a pioneer in EM and has been credited for developing the diamond knife and ultra microtome for EM use. Most amazing is the fact that Fernandez-Moran was able to obtain his PhD from the University of Munich in 1944!

In 1964 Fernandez-Moran published a paper with David Green who later proposed the lipoprotein model for membrane structure (Chapter 8). In this paper Fernandez-Moran used negative staining with phosphotungstate to obtain EM images of what was then known as a mitochondrial elementary particle (EP) [5]. We now know EP is actually the

FIGURE 9.6 Humberto Fernandez-Moran, 1924–1999.

F1 ATPase (Chapter 18). It was the unique shape of this particle when imaged by EM that facilitated its use as a membrane directional marker. To quote Fernandez-Moran "...the elementary particle (EP), consists of three parts: (1) a spherical or polyhedral head piece (80 to 100 Å in diameter); (2) a cylindrical stalk (about 50 Å long and 30 to 40 Å wide); and (3) a base piece (40 × 110 Å)." All of the stalked EPs faced the same direction (towards the mitochondrial matrix) (Fig. 9.7). Later it was shown that all three bioenergetic membranes, the mitochondrial inner membrane, the bacterial plasma membrane, and the thylakoid, have similar shaped ATPases. For each membrane, all of the ATPases face in the same direction (inward for the mitochondrial inner membrane and the bacterial plasma membrane, and outward for the thylakoid). No ATPase has been found to simultaneously face in both directions.

In addition to the classic Fernandez-Moran experiment, there have been countless other indications that membrane proteins are always 100% asymmetrically distributed across membranes. Using patch clamp techniques, all ion channels have been shown to orient in only one direction. The substrate binding site of many membrane enzymes have been shown to be only associated with one side of the membrane. For example, the Na^+/K^+ ATPase hydrolyzes adenosine triphosphate (ATP) only from the inside and pumps Na^+ out of the cell while pumping K^+ into the cell (Chapter 19). Hormones and membrane–protein specific antibodies will bind to only one membrane surface.

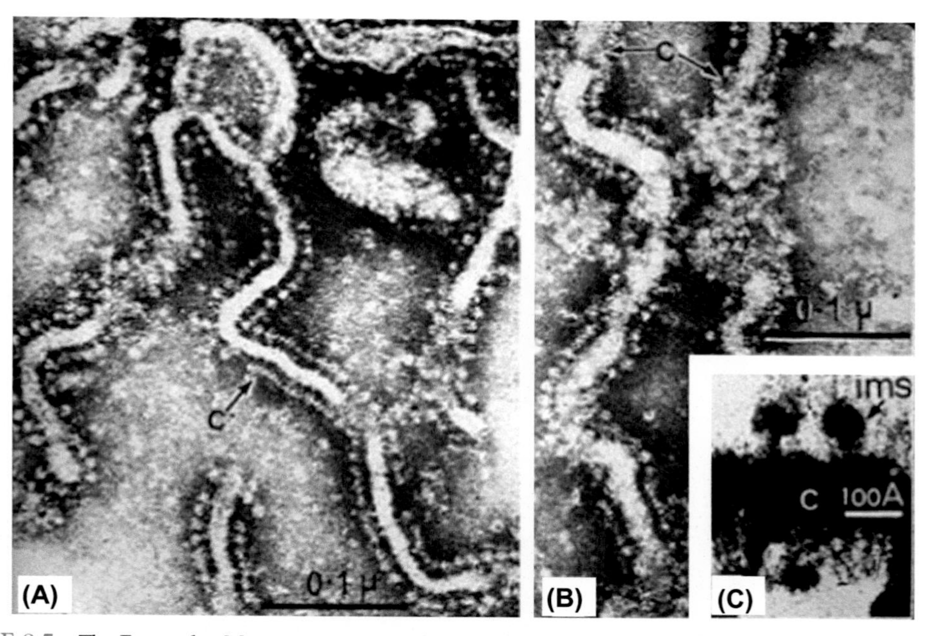

FIGURE 9.7 The Fernandez-Moran experiment. The mitochondrial inner membrane (the cristae) was imaged by electron microscopy using negative staining with phosphotungstate. The mitochondrial elementary particles, now known, to be the F1 ATPase, are the white lollipop-shaped objects. They are all attached to the same membrane leaflet (the inner leaflet of the mitochondrial inner membrane) and face into the matrix. The thick dark tube-like space is the matrix where the elementary particles protrude.

3.2 Carbohydrate Asymmetry

The primary function of membrane-bound carbohydrates is to interact with the extracellular environment by providing the cell surface with a unique sugar identity (Chapter 7). Function strongly implies, but does not prove, that carbohydrates should face outside the cell. Fortunately, carbohydrates are relatively easy to access and label, and so a variety of methodologies are available. The best and most employed methodology involves lectins.

3.3 Lectins

Lectins are small, sugar-binding proteins that are highly specific for their sugar ligand [6]. In fact, the term "lectin" is derived from the Latin word *legere* meaning "to select." Lectins are ubiquitous in living organisms and are often referred to as "agglutins" due to their ability to aggregate animal cells. Lectins are nonenzymic in action and nonimmune in origin. Their most common biological role is assisting in various kinds of recognition events, particularly those involving cells and proteins. In addition, lectins can target mannose-6-phosphate-containing hydrolytic enzymes to the lysosome, play important roles in the immune system by recognizing carbohydrates that are found exclusively on pathogens, clear certain glycoproteins from the circulatory system, and are routinely used for blood typing.

The lectin, ricin, isolated from the seeds of the castor bean, is currently employed as a highly toxic biochemical warfare agent. A famous incident involving ricin occurred in 1978 when the Bulgarian dissident writer Georgi Markov was stabbed in his calf by an umbrella while waiting for a bus on the Waterloo Bridge in London. The umbrella, wielded by a member of the Soviet KBG, injected a fatal dose of ricin into Markov's calf [7].

In addition to its many important biological functions, lectins have been employed as biochemical tools to study membranes. Lectins are routinely used in affinity chromatography, affinity electrophoresis, and affinity immunoelectrophoresis to isolate glycoproteins and plasma membrane vesicles. This aspect of using lectins as biochemical tools in isolating plasma membranes is discussed in Chapter 12. There are now hundreds of commercially available lectins, the most commonly used being concanavalin A or Con A. A brief list of commonly used lectins and their sugar ligands can be found in Table 12.1. Unmodified lectins are, of course, not detectable and so would be useless for membrane imaging studies. Therefore a variety of tags have been successfully attached to lectins and many of these are now commercially available. Included in the modified lectins are fluorescent-labeled (visible light), radiolabeled (autoradiography), and heavy metal-labeled (EM) lectins. All lectin-based imaging studies confirm that carbohydrates are 100% asymmetrically distributed across the membrane with sugars always facing the outside.

3.4 Lipid Asymmetry

It probably surprised no one that proteins and carbohydrates were shown to be 100% asymmetrically distributed across membranes. Their functions would strongly imply this. However, it was not as evident what to expect of lipid asymmetry. The major function of membrane lipids is to provide the fundamental bilayer that serves as the barrier controlling leakiness as well as the environment to support membrane biochemical (protein) activity.

There is no reason to think that these functions would require lipid asymmetry. But this would have to be tested [8,9].

There are now three major and very distinct methods to measure lipid asymmetry in membranes: chemical modification, enzymatic modification, and lipid exchange. Any successful technique has to cause a measurable alteration of a membrane lipid, modifying the lipid without destroying the membrane or making the membrane leaky to the reagents involved in the measurement. The chemical, enzymatic, or lipid exchange agents must only alter the external leaflet lipid without being exposed to the inner leaflet lipid and the procedure must be fast compared to the rate of trans-membrane lipid diffusion (flip-flop).

3.5 Chemical Modification

The first report of lipid asymmetry in membranes came from Mark Bretscher's lab at Oxford [10]. Bretcher (Fig. 9.8) developed the primary amine binding reagent, formylmethionylsulphone methyl phosphate (FMMP, Bretscher's Reagent, Fig. 9.9) for his studies. He reported that with intact erythrocytes, most phosphatidylethanolamine (PE) and phosphatidylserine (PS) (primary amine lipids) could not be labeled by FMMP. However, almost all PE and PS could be labeled in leaky erythrocyte ghosts. Therefore, he concluded that most PE and PS reside on the inner leaflet of erythrocytes. Later FMMP was replaced by the more available trinitrobenzenesulfonate (TNBS, Fig. 9.9). Both compounds are employed identically. The intact cells are exposed to the reagent and after a short incubation the cells are

FIGURE 9.8 Mark Bretscher, 1940–.

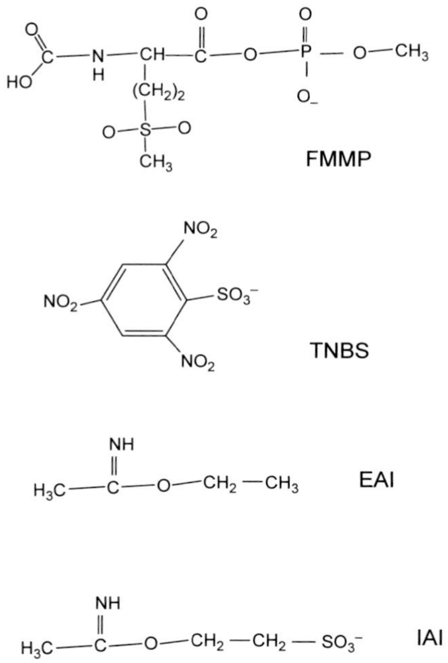

FIGURE 9.9 Types of primary amine tags used to label PE and PS in membrane lipid asymmetry studies. *FMMP*, formylmethionylsulphone methyl phosphate, Bretscher's Reagent; *TNBS*, trinitrobenzene sulfonate; *EAI*, ethylacetimidate; *IAI*, isethionylacetimidate.

removed from the solution and washed free of the reagent. The membrane lipids are then extracted and separated, usually by thin layer chromatography. These methods are discussed in Chapter 13. The amounts of labeled (outer leaflet) and nonlabeled (inner leaflet) PE and PS are quantified and the outside to inside ratio for both lipids determined.

Another interesting pair of primary amine-detecting reagents used in lipid asymmetry determinations is ethylacetimidate (EAI) and isethionylacetimidate (IAI). Both compounds, shown in Fig. 9.9, attach to PE and PS by the same mechanism. The difference between the two reagents is that IAI is charged and cannot penetrate the membrane and so only labels outer leaflet lipids. In contrast, EAI is uncharged and so can readily cross membranes, labeling both inner and outer leaflet lipids.

3.6 Enzymatic Modification

PE and PS both have a highly reactive primary amine in their polar head group. Therefore they lend themselves well to covalent attachment by a number of reagents. However, the remaining membrane lipids do not possess such a handle and so other methods had to be sought. Chemical methods do exist that can modify the nonprimary amine head groups, but these methods are much harsher and usually result in membrane destruction. Enzymatic alteration of a polar head group can replace chemical modification. Enzymes have a tremendous advantage in being highly specific for the substrate lipid. The large size of an enzyme also makes it impossible to cross an intact membrane. Therefore, the enzyme will only alter the outer leaflet lipid. While a variety of enzymes have been employed, only three will be discussed here, phospholipase D (PLD), sphingomyelinase (SMase), and cholesterol oxidase.

PLD [specific for Phosphatidylcholine (PC), $PLD_{(PC)}$] can be used to measure the asymmetry of PC. The enzyme catalyzes the reaction:

$$\text{PC} \underset{\text{PLD}_{(\text{PC})}}{\rightarrow} \text{PA} + \text{Choline}$$

Since the enzyme cannot cross the membrane, it only converts outer leaflet PC to phosphatidic acid (PA). It does not affect the inner leaflet PC. The PC and PA content is determined before and after hydrolysis by $PLD_{(PC)}$. The decrease in PC equals the increase in PA, and this represents the outer leaflet PC content. From this measurement, the membrane asymmetry of PC can be determined.

In a similar fashion, SMase can be used to measure the membrane asymmetry of sphingomyelin (SM). The enzyme catalyzes the following reaction:

$$\text{SM} \underset{\text{SMase}}{\rightarrow} \text{Ceramide} + \text{phosphorylcholine}$$

As with $PLD_{(PC)}$, only the outer leaflet SM is hydrolyzed and the SM and ceramide content is determined before and after enzyme treatment. The decrease in SM equals the increase in ceramide, and this represents the outer leaflet SM content. From this measurement, the membrane asymmetry of SM can be determined.

Some other types of phospholipases are less useful since they produce products that destabilize the membrane. For example, phospholipase A_1 or A_2 produce lysophospholipids and free fatty acids, both of which adversely affect membrane structure. Phospholipase C produces the nonlamellar phase-preferring compound diacylglycerol (DAG) (Chapter 5).

Cholesterol is not hydrolysable and so its asymmetry must be determined using a very different method. Cholesterol also presents an additional problem since its inherent rate of trans-membrane diffusion (flip-flop) is so much faster than phospholipids. Accurate measurement of cholesterol asymmetry is difficult and no method is without flaws. One common method employs use of cyclodextrin (Chapter 10) [11]. A less commonly used method is outlined in Fig. 9.10. It presents a theoretically easy UV method that would be available to most laboratories. As the C-3 hydroxyl group of cholesterol is oxidized to a ketone by the enzyme cholesterol oxidase, its UV absorbance increases. Since the enzyme is too large to cross a membrane, it can only oxidize cholesterol exposed on the outer surface of a cell. Therefore if cells are mixed with cholesterol oxidase in a quartz cuvette, the UV absorbance will increase rapidly until all of the outer leaflet cholesterol is oxidized. Subsequently, slower oxidation can be measured as cholesterol flips from the inner to the outer leaflet. From these measurements cholesterol's membrane asymmetry and flip-flop rate can be determined.

3.7 Erythrocyte Lipid Asymmetry

Since the 1925 Gorter and Grendel experiment establishing the lipid bilayer, erythrocytes have been "the laboratory for membrane studies." It was in the erythrocyte that lipid

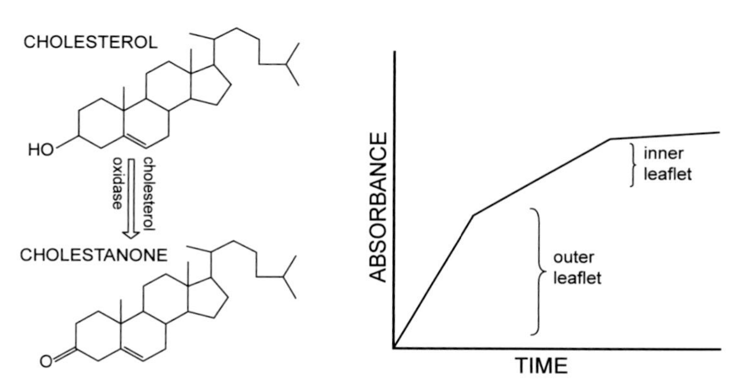

FIGURE 9.10 Conversion of cholesterol to cholestanone by cholesterol oxidase. The initial rapid increase in absorbance is due to oxidation of outer leaflet cholesterol to UV light-absorbing cholestanone. The slower oxidation occurs after inner leaflet cholesterol has flipped to the outer leaflet. Cholesterol asymmetry can be determined from the ratio of inner to outer leaflet cholesterol.

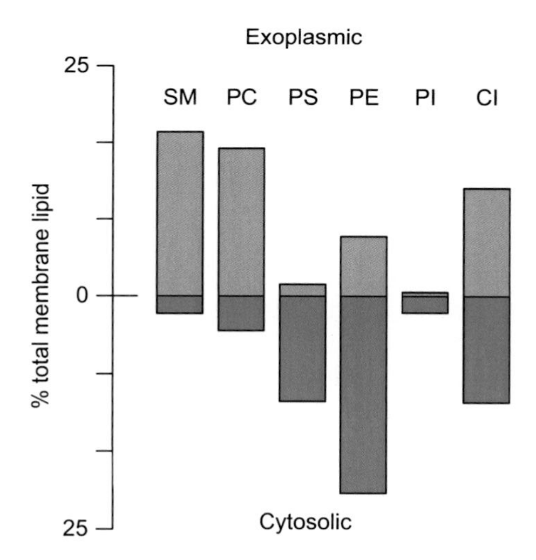

FIGURE 9.11 Distribution of common membrane lipids between the inner and outer leaflets of erythrocytes. The choline-containing lipids [sphingomyelin (SM) and phosporylcholine (PC)] are predominantly found in the outer (exoplasmic) leaflet, while the primary amine lipids (PE and PS) are associated with the inner (cytoplasmic) leaflet. Also, phosphatidylinositol (PI) is found primarily on the inner leaflet, while cholesterol (CL) is equally distributed between the leaflets.

asymmetry was first established. Fig. 9.11 depicts asymmetry of the basic erythrocyte lipids. The choline-containing lipids (PC and SM) are predominantly in the membrane outer leaflet while the primary amine lipids (PE and PS) are found predominantly in the membrane inner leaflet. PS is unusual in being 100% asymmetrically distributed to the inner leaflet. This is attributed to the presence of an ATP-dependent flipase [12–14] whose function it is to take any outer leaflet PS and immediately flip it back to the inner leaflet, preventing accumulation of outer leaflet PS. PA, phosphatidylinositol (PI), and PI 4,5-bisphosphate are predominantly located in the inner leaflet while cholesterol is mainly associated with the outer leaflet, although its rapid flip-flop rate would indicate it moves back and forth across the membrane spending more time in the outer leaflet. The sugar-containing gangliosides are always in the outer leaflet (100% asymmetric). Cholesterol is evenly distributed across the membrane.

It is not an easy task to understand the variety of forces driving and maintaining partial lipid asymmetry [15–17]. For example, lipid asymmetry must be severely altered at membrane segments that are tightly curved. In order to fit the enormous quantity of internal membranes into the tiny intracellular space (Chapter 1), there must be countless places in internal membranes that are under tremendous curvature strain. These sections are difficult to study since the strain is instantly released upon breaking open the cell. Therefore membrane curvature is best studied by making spherical lipid vesicles of known diameter. The tightest vesicle that is achievable with phospholipids is ~200 Å in diameter. These vesicles, referred to as small unilamellar vesicles (SUVs) (Chapter 13), are characterized by having two-thirds of their lipids in the outer leaflet and one-third crunched into the inner leaflet. In mixed lipid SUVs, the larger head group lipids (PC, SM, glycolipids) tend to accumulate in the less densely packed outer leaflet, while lipids with smaller head groups (eg, PE and cholesterol) accumulate in the more compact inner leaflet. SUVs undergo a series of fusions driven by the release of curvature strain. With fusion, the membrane becomes essentially planar in the new large unilamellar vesicles (LUVs) (Chapter 13). In the LUVs, lipid asymmetry is lost.

For example, if a mixed lipid PC–PS SUV is made at pH 7, PC is predominantly found in the outer leaflet, while PS is more prevalent in the inner leaflet. At pH 7, PC is larger than PS. However, if the pH is increased, PS flips to the outside and PC moves to the inside. At high pH, PS becomes larger than PC due to negative charge repulsion of the PS head group. At pH 7, PS has a net charge of −1, while at high pH the charge of PS is −2. The problem with size is far more complex than this as the size of a phospholipid is also dependent on the nature of the acyl chains. The phospholipid base diameter increases with the number of double bonds (Chapter 11). One would therefore expect that lipids with more double bonds would be forced into the outer leaflet in membrane segments with high curvature. But there is an additional complication as lipids with more double bonds are more "compressible" than lipids with saturated chains (Chapter 11). In sharp contrast, cholesterol is almost totally noncompressible. The net impact of all of these forces is unknown and at present unpredictable.

Lipid partial asymmetry was first worked out in the erythrocyte and the results are depicted in Fig. 9.11. If the erythrocyte is transformed into a spherical "ghost" upon emulsion in hypotonic media (the erythrocyte is "hemolyzed"), much of the intracellular cytoskeleton is lost. The ghost still has lipid asymmetry, but it is much reduced compared to the intact erythrocyte. If the erythrocyte lipids are extracted and used to make an LUV, no lipid asymmetry is observed. If the LUVs are sonicated to produce SUVs, lipid asymmetry is once again established. However, this lipid asymmetry does not resemble either that observed for the intact erythrocyte or the erythrocyte ghost, but instead reflects curvature strain.

4. LATERAL DIFFUSION

A key aspect of the Fluid Mosaic model is "fluidity." The model predicts that the membrane is in constant flux (it is fluid) with components always in the process of exchanging lateral positions with other components. The first proof of membrane lateral mobility was the iconic 1970 immunofluorescence paper by Frye and Edidin [18]. Subsequent, improved methodologies to follow lateral diffusion include fluorescence recovery after photobleaching (FRAP) and single particle tracking (SPT), discussed in Sections 4.2 and 4.3 respectively.

4.1 The Frye-Edidin Experiment

By 1970 it was evident that cells were not rigid structures but changed shape as the cell moved or developed pseudopodia. The objective of the Frye-Edidin experiment (Fig. 9.12) was to prove that plasma membrane (surface) proteins could move laterally in the membrane. These investigators used two very different cell lines, one a human line, the other a mouse line. The experiment is depicted in Fig. 9.12. Each cell type had different surface antigens and so could be readily distinguished by fluorescent-labeled antibodies. The technique is referred to as immunofluorescence. The antibody for the mouse antigen was labeled with fluorescein and so fluoresced green, while the antibody against the human antigen was labeled with rhodamine and appeared red. The cells were fused at 37°C with a Sendai virus producing a heterokaryon that initially had one half of the fused cell green and one half red. After 40 min, the red and green colors were totally mixed. One interpretation of this observation was that the plasma membrane proteins were free to diffuse laterally in the plane of the membrane. However, Frye and Edidin did realize that other explanations were possible and they tested these. Protein synthesis inhibitors (puromycin, cycloheximide, and chloramphenicol) and the uncoupler 2,4-dinitrophenol had no effect on the process. Therefore, intermixing of the surface antigens was not the result of newly synthesized antigens at various locations in the plasma membrane, nor was the intermixing due to an energy-dependent translocation. However, if the temperature was reduced from 37°C to 15°C, the process was greatly slowed. The dependence on temperature was consistent with a change in membrane viscosity that would affect diffusion rates. This paper concluded that a membrane "is not a rigid structure, but is fluid enough to allow free diffusion of surface antigens resulting in their intermingling within minutes after the initiation of fusion."

This seminal paper in membrane studies was one of the very first in Michael Edidin's long and distinguished career at Johns Hopkins (Fig. 9.13). He is also well respected in an area far removed from membranes. He is a world expert on the history of watch making and repair!

4.2 Fluorescence Recovery After Photobleaching

While the Frye-Edidin experiment clearly established that membrane proteins can diffuse laterally, it proved to be a difficult method to obtain accurate rates of diffusion. The Sendai virus is not a reliable fusogen. Its fusion agent is likely a component picked up from the host cell as the virus escapes and so varies tremendously in effectiveness from preparation to preparation. In addition, while the time that fusion begins ($T = 0$) can be accurately

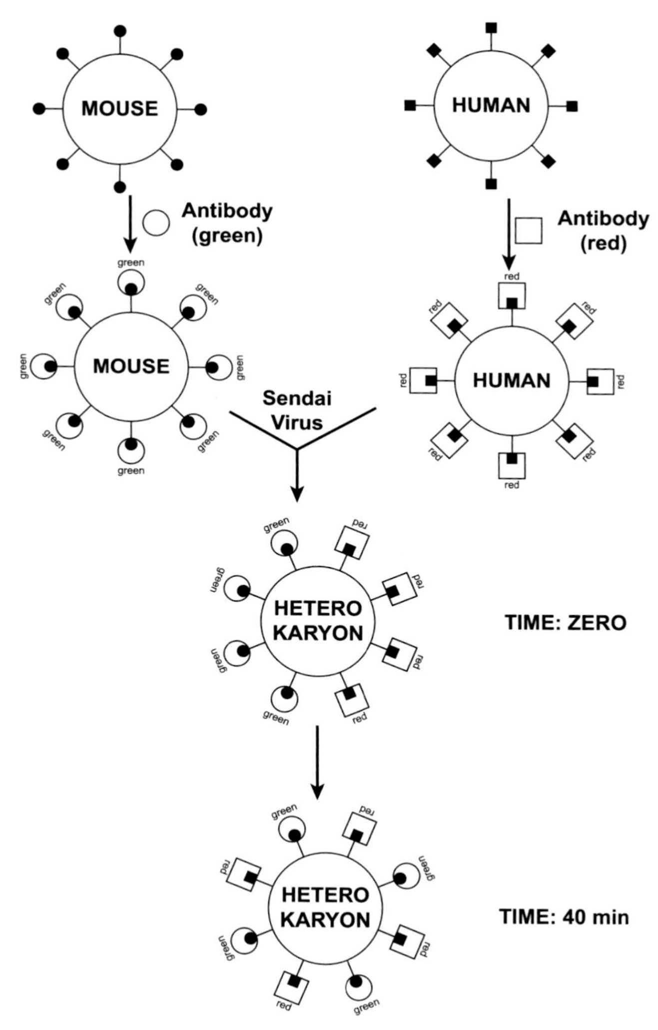

FIGURE 9.12 The Frye-Edidin experiment. Cell cultures were obtained from a mouse and a human cell line. Each cell type had different surface antigens. Fluorescent-labeled antibodies were obtained for each cell line. The antibody for the mouse antigen was labeled with fluorescein and so appeared green while the antibody against the human antigen was labeled with rhodamine and so appeared red. The cells were fused with a Sendai virus producing a heterokaryon that initially had one half of the mega-cell green and one half red. After 40 min the red and green colors were totally mixed indicating membrane protein lateral diffusion.

determined, the time at which complete mixing is achieved is imprecise. Therefore other, more reliable and accurate methods to measure lateral diffusion in membranes were sought. Advances in laser technology led to the accelerated development of what is now the major technique in determining lateral diffusion of lipids and proteins in membranes. This technique is referred to as FRAP.

FIGURE 9.13 Michael Edidin.

FRAP had its beginnings in a 1976 paper by Axelrod et al. [19]. The technique requires a high quality light microscope, a general low intensity light source, and a highly focused, high intensity light (normally a laser). FRAP begins by uniformly labeling the surface of a cell with a fluorescent-labeled antibody or other lipid- or protein-specific fluorescent tag. For example, a fluorescent lectin could be bound to a specific glycolipid or glycoprotein. The next step involves acquiring a basic fluorescence background image of the cell surface that will be photobleached. A small diameter, intense light pulse is then used to rapidly photobleach a small spot on the cell surface. Originally the light source was a broad spectrum mercury or xenon light and was used in conjunction with a color filter to assure the spot would receive light that matches the absorbance of the fluorescent probe. Tunable lasers have now replaced the older light sources. The intense laser pulse photobleaches a spot of predetermined size and shape. This spot appears dark on a highly fluorescent, bright, unbleached background. The laser destroys the fluorescence of the probe but has no effect on the membrane component to which the tag is associated. With time, the still fluorescent-labeled membrane components diffuse into the photobleached spot, while the nonfluorescent (bleached) components diffuse out of the spot. Eventually, the once bleached spot recovers most, but not all its fluorescence.

A plot of data obtained from a typical FRAP experiment is shown in Fig. 9.14 [20]. The initial fluorescence background for the unbleached cell is indicated by line (1). Photobleaching (arrow) decreases fluorescence in the spot (2). The net loss in fluorescence is X. Brownian motion allows for recovery of fluorescence into the spot (3). Eventually fluorescence returns in the spot to a new baseline (4) that is lower than the prebleached fluorescence (1). The fluorescence recovery is Y ($Y < X$). The % recovery is therefore $Y/X \times 100$ and represents the percent of the fluorescent-tagged membrane component that is diffusible. The % of the

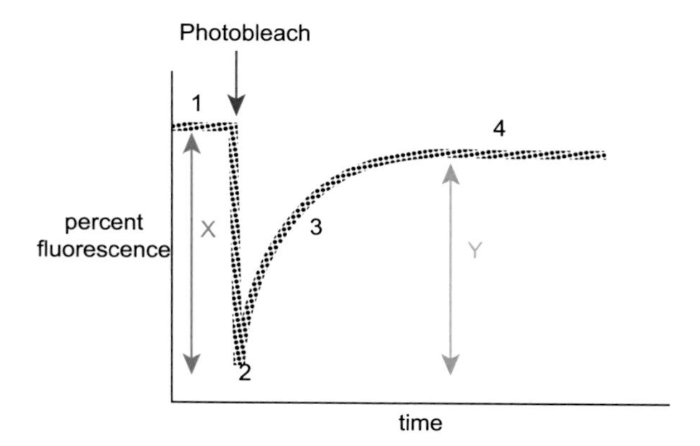

FIGURE 9.14 Data obtained for a typical fluorescence recovery after photobleaching experiment. The initial fluorescence background for the unbleached cell is indicated by line (1). Photobleaching (arrow) decreases fluorescence in the spot (2). The net loss in fluorescence is X. Brownian motion allows for recovery of fluorescence in the spot (3). Eventually fluorescence returns in the spot to a new baseline (4) that is lower than the pre-bleached fluorescence (1). The fluorescence recovery is Y ($Y < X$). The % recovery is therefore $Y/X \times 100$ and represents the percent of the membrane component that is diffusible. The percent of the membrane component that is non-diffusible is $100-(Y/X \times 100)$. Lateral mobility is directly related to the slope of the recovery process. A steeper slope indicates a faster recovery and higher diffusion rate [20].

membrane component that is nondiffusible is $100-(Y/X \times 100)$. Lateral mobility is directly related to the slope of the recovery process. A steeper slope indicates a faster recovery and higher diffusion rate. The diffusion rate D can be obtained from the following equation:

$$D = w^2/4t_{1/2}$$

where w is the width of the photobleaching beam and $t_{1/2}$ is the time required for the bleach spot to recover half of its initial intensity.

4.3 Single Particle Tracking

While FRAP clearly has advantages over the original Frye-Edidin method, it still just gives an averaged diffusion rate for an ensemble of particles. Single Particle Tracking (SPT), as the name implies, can follow diffusion of one molecule at a time [21]. Akiharo Kusumi (Fig. 9.15) of the Kyoto University School of Medicine is recognized as the world leader in this emerging field [22]. In his laboratory, the temporal resolution of SPT has been pushed down to 25 μs and spatial resolution down to nanometers. As might be expected, this complex technology has resulted in a completely new paradigm in our understanding of membrane structure. SPT development was driven by a basic observation that confused scientists for decades. If movement in a membrane is the result of free Brownian motion, why do membrane components diffuse 10–100-fold slower in a cell (plasma) membrane than they do in an artificial lipid bilayer membrane? Michael Sheetz [23] proposed that trans-membrane proteins might be contained by the cytoskeleton ("fence" theory). However, membrane technology was not sufficiently advanced to confirm this hypothesis. Conformation had to await the development of SPT.

FIGURE 9.15 Akiharo Kusumi, 1953–.

SPT had its beginnings in a paper by De Brabander et al. [24], but was, and continues to be primarily advanced by Kusumi and colleagues. In what is now a classic 1993 Biophysical Journal paper, Kusumi reported on lateral membrane diffusion of the cell–cell recognition adhesion receptor E-cadherin in cultured mouse karatinocytes [25]. The paper compared SPT to the then more established technique of FRAP. Receptor movement was observed by SPT after E-cadherin was bound to an anticadherin monoclonal antibody coupled to a 40 nm gold particle. Motion was followed by video-enhanced differential interference contrast microscopy at a temporal resolution of 30 ms and at a nanometer special precision. After plotting the results as mean square displacement of the gold particle against time, four characteristic types of motion were observed (Fig. 9.16):

A. Stationary mode where diffusion is less than 4.6×10^{-12} cm^2/s (6%).
B. Simple Brownian diffusion mode (28%).
C. Directed diffusion mode indicated by unidirectional motion (2%).
D. Confined diffusion mode where Brownian diffusion is between 4.6×10^{-12} and 1×10^{-9} cm^2/s and diffusion is confined within a limited area by the cytoskeleton network (64%). Diffusion was confined into many small domains 300–600 nm in diameter.

Confined diffusion represented almost two-thirds of the total diffusion patterns. Later Kusumi noticed that in rat kidney cells, diffusion appeared to be a connected series of confined diffusion mode domains. In other words the overall, low-resolution diffusion measured by FRAP was composed of two parts, Brownian motion that is confined within corrals and "hop diffusion" that connects the confined zones. Unfortunately, these observations

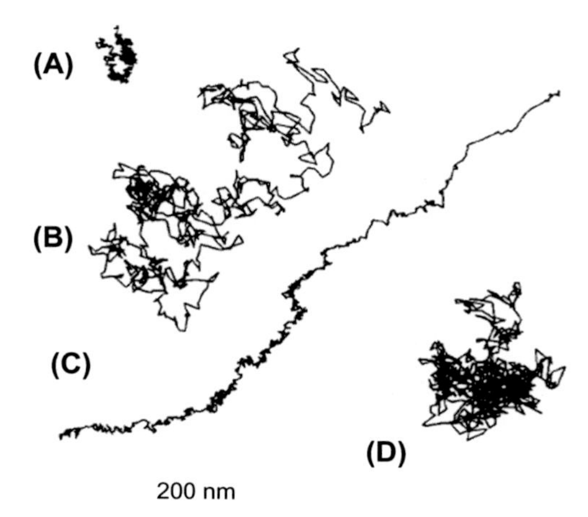

FIGURE 9.16 Four characteristic motions observed by single particle tracking (SPT) for E-cadherin in the plasma membrane of a cultured mouse keratinocyte [25]. The four motions are: (A) Stationary mode; (B) Simple Brownian diffusion mode; (C) Directed diffusion mode; and (D) Confined diffusion mode. SPT trajectories of E-cadherins were recorded for 16.7 s (500 video frames) [25].

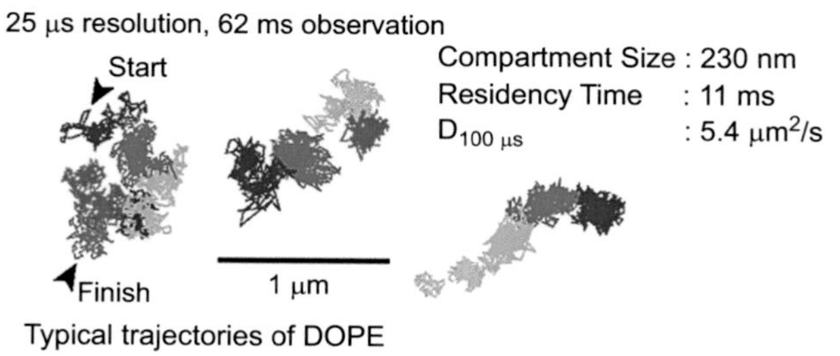

FIGURE 9.17 Hop diffusion of the membrane phospholipid dioleoylphosphatidylethanolamine (DOPE) (18:1,18:1 PE). DOPE spends most of its time trapped in a compartment (~11 ms) before hopping out and diffusing to another compartment where it is again trapped. DOPE diffuses as fast within the confined domains as it does in a protein-free lipid bilayer.

could not be confirmed in erythrocytes or other cells. Kusumi realized that a faster methodology was needed. He obtained ultrafast filming technology from the field of explosives research [22]. This technology employed a camera that obtained 40,000 frames per second with a temporal resolution of 25 μs. This produced a 1000 times faster resolution than before (30 ms for the E-cadherin experiment), allowing Kusumi to even measure hop diffusion for membrane lipids, which hop at very fast rates (Fig. 9.17).

The "fence" (corral) model for membrane structure poses some major questions that must be addressed. For example, what affect does the relatively enormous attached gold particle have on the measurements? Since it is likely that only the plasma membrane has a close association with the cytoskeleton, do the other intracellular membranes display corrals? How does "raft" theory fit in with "fence" theory? Does one negate the existence of the other or are they compatible?

4.4 Conclusions for Membrane Lateral Diffusion Rates

The Frye-Edidin experiment proved that membrane proteins were free to diffuse laterally in the cell plasma membrane at a rate sufficient to completely encircle a bacterial cell in 1 s and a eukaryotic liver cell in 1 min. The diffusion rate was 5×10^{-11} cm^2/s. For a similar sized, but water-soluble protein (hemoglobin), the diffusion rate was much faster, 7×10^{-7} cm^2/s. The effective membrane viscosity was therefore 7×10^{-7} cm^2/s/ 5×10^{-11} cm^2/s or $\sim 10^3 - 10^4$ times more viscous than water. Viscosity of the membrane bilayer interior therefore resembles that of light machine oil.

The diffusion rate for membrane proteins was shown to span a large range, from that approaching free phospholipids ($\sim 10^{-8}$ cm^2/s) through to essentially being immobile (bound to the cytoskeleton, $\sim 10^{-12}$ cm^2/s). Interactions of many trans-membrane proteins with the cytoskeleton account for protein immobility and are now well established. A first indication of this occurred in the erythrocyte where removal of the cytoskeleton increased band 3 lateral diffusion 40-fold. Phospholipid diffusion was shown to be independent of the head group (ie, all phospholipids have about the same fast diffusion rate of $\sim 10^{-8}$ cm^2/s), but diffusion is impacted by membrane phase (discussed below and in Chapter 10). For example, lipid diffusion in gel state bilayers is $\sim$100–1000 times slower than in fluid, liquid crystalline state bilayers. Lipid diffusion is also known to be impacted by the presence of cholesterol, generating the liquid ordered (l_o) state. This must be considered with regard to the structure, stability, and function of l_o state lipid rafts. In general, membrane components diffuse 10–100 times slower in biological membranes than in protein-free lipid bilayer membranes, implying hindered diffusion due to crowding (Chapter 11).

It is generally accepted that biological membrane components, lipids and proteins, are heterogeneously distributed into countless numbers of bewildering and short-lived domains. An early experiment of Michael Edidin employing lateral diffusion measurements supported the concept of membrane heterogeneity [26]. Edidin's experiment used FRAP to follow the diffusion of two carbocyanine dyes, one with two short lipid chains (C_{10}, C_{10} DiI) and one with two long lipid chains (C_{22}, C_{22} DiI). The esterified fatty acyl chains have very different melting points (T_ms). C-10 (decanoic acid) has its T_m at 31.6°C while the T_m of C-22 (behenic acid) is 79.9°C (see Table 4.3). At 37°C, C-10 would be in the melted, liquid crystalline state while C-22 would be in the solid, gel state. Edidin predicted that the short chain dye (C_{10}, C_{10} DiI) would partition into a more fluid (disordered) phase while the longer chain dye (C_{22}, C_{22} DiI) would prefer a more ordered phase. If the membrane lipid bilayer was homogeneous he reasoned, both dyes should have similar diffusion rates, but if the membrane bilayer was hetergeneous the two fluorescent probes should exhibit different diffusion rates. When he tested this hypothesis on sea urchin and mouse eggs he found large differences between the diffusion rates for each dye. His conclusion was that membrane bilayers have considerable patchiness that must exist for at least a few minutes. The nature of this patchiness is at the heart of membrane domain studies.

5. LIPID TRANS-MEMBRANE DIFFUSION (FLIP-FLOP)

Earlier in this chapter it was discussed that trans-membrane asymmetry is always absolute for proteins and carbohydrates. Therefore their rate of trans-membrane diffusion or flip-flop is zero. It never happens. However, lipids do exhibit partial lipid asymmetry and so one would assume that lipid flip-flop might be possible. However, if flip-flop was very fast, lipid asymmetry could not be maintained. If flip-flop was very slow, a lipid's asymmetry would reflect the membrane sidedness during its synthesis and it would be hard to explain how each membrane has a different asymmetry for each lipid. The well-established fact that all membranes exhibit partial lipid asymmetry indicates that the process of flip-flop might not be easy, but it should be possible (Fig. 9.18).

As a lipid undergoes flip-flop, two thermodynamically unfavorable events must occur simultaneously. The lipid hydrophobic tails must be exposed to water and the polar headgroup must be exposed to the hydrophobic bilayer interior (Fig. 9.18). From this, a rough estimate of the energy of flip-flop can be made. It requires ~2.6 kcal of free energy to transfer a mole of methane from a nonpolar medium to water at 25°C. Free energy required to transfer a mole of zwetterionic glycine from water to acetone is about 6 kcal at 25°C. Therefore it would very roughly require approximately 97 kcal to flip a phospholipid across a membrane (assuming an average chain length of 17.5 carbons per chain). This means that only 1 phospholipid in $\sim 10^{15}$ would have enough energy to flip and so the inherent rate of flip-flop would be quite slow, but not impossible. Indeed this has been shown to be the case [27]. This slow rate can, however, be substantially increased (10^3–10^5-fold) by incorporating any transmembrane protein (Fig. 9.18, examples include glycophorin [28] and cytochrome b5 [29]). Although lateral exchange of nearest lipid neighbors in the same membrane leaflet and exchange across the membrane from one leaflet to another, traverse approximately the same distance (~4 nm), the lateral exchange rate is $\sim 10^{10}\times$ faster than is flip-flop. Nevertheless, lipid flip-flop is an essential component of membrane dynamics. The importance of flip-flop begins with the biogenesis of the phospholipid and ends with cell death [30].

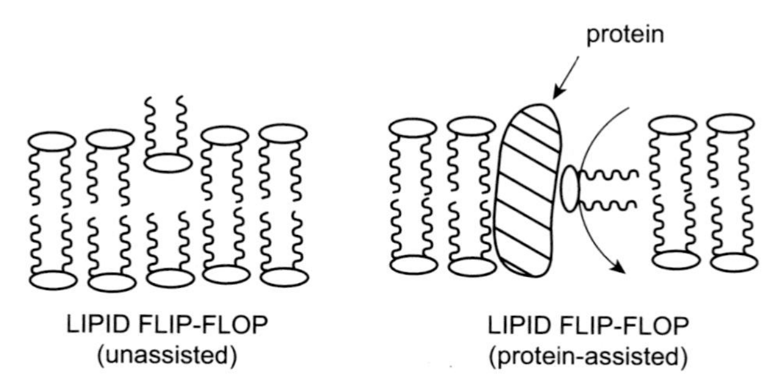

FIGURE 9.18 Lipid flip-flop, unassisted (left) and protein-assisted (right). A protein's surface can increase phospholipid flip-flop rate 10^3–10^5-fold.

5.1 Chemical Method to Determine Flip-Flop Rates

Since flip-flop is essentially the kinetics of asymmetry, any method that can be used to measure lipid asymmetry should, in principle, also be applicable to measuring flip-flop. For example, the rate of flip-flop of the primary amine phospholipids PE and PS can be determined by combining the methods discussed previously using trinitrobenzene sulfonate (TNBS) and IAI for determining lipid asymmetry. The structure of these compounds is shown in Fig. 9.8 and their application for determining flip-flop in Fig. 9.19.

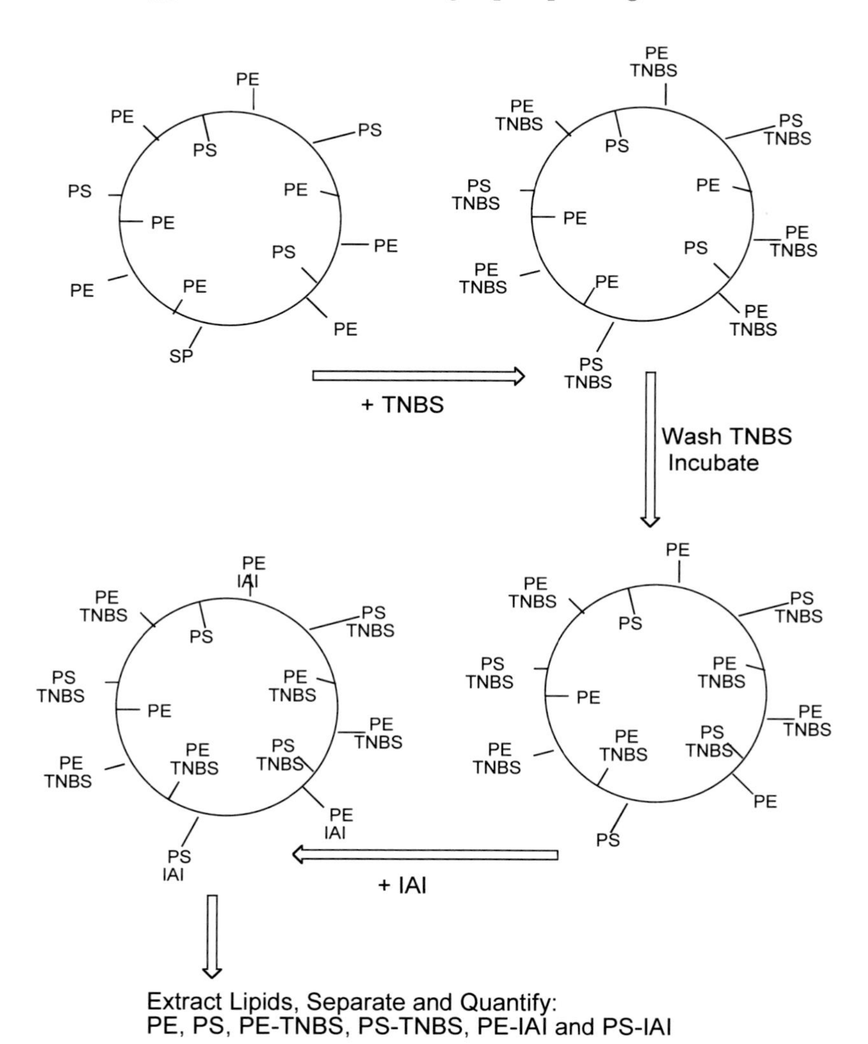

FIGURE 9.19 Determining flip-flop rates for the primary amine phospholipids PE and PS by use of the primary amine labels, TNBS and IAI.

If a sealed membrane vesicle (LUV) is exposed to TNBS, only the outer leaflet PE and PS will be labeled since the reagent does not cross the membrane. The unreacted TNBS is then washed away from the TNBS-modified vesicles and the vesicles are allowed to incubate for a time ($t = X$). During this time, some unmodified (inner leaflet) PE and PS can flip to the outer leaflet while roughly an equivalent amount of PE–TNBS and PS–TNBS can flip in. At $t = X$ the vesicles are then exposed to another nonpermeable primary amine reactive reagent, IAI. IAI then reacts with the PE and PS that had flipped from the inner leaflet to the outer leaflet during the time interval X. After this, the nonreacted IAI is washed away from the vesicles and the lipids are extracted, separated, and quantified (Chapter 13). From the measured amount of PE, PS, PE–TNBS, PS–TNBS, PE–IAI, and PS–IAI, the rate of flip-flop for PE and PS can be estimated.

5.2 Phospholipid Exchange Protein Method to Determine Flip-Flop Rates

Phospholipid exchange proteins (PLEPs, also called lipid transfer proteins, or TPs), offer a versatile approach to measure lipid asymmetry and flip-flop. PLEPs were first discovered by Wirtz and Zilversmit at Cornell University [31]. These are a ubiquitous family of low-molecular weight proteins whose function is to shuttle various lipids from membrane to membrane [32]. Since many PLEPs exhibit absolute specificity for a particular phospholipid, they do not have to be totally purified but can be employed as a crude cellular mixture. The experiments can be constructed in almost limitless combinations to address lipid asymmetry and flip-flop. A simple example to determine the inherent flip-flop rate for PC in a protein-free lipid vesicle (liposome) is depicted in Fig. 9.20.

LUVs are made from ^{32}P–PC (the method is discussed in more detail in Chapter 13). To the LUVs are added mitochondria that provide a large excess of unlabeled (mitochondrial) PC. Mitochondria are chosen since they are easily obtained and are much bigger and heavier than the ^{32}P–PC LUVs. A simple low-speed centrifugation is sufficient to completely separate the LUVs and mitochondria. To start the experiment, a source of crude $PLEP_{(PC)}$ specific for PC is added. The mixture is incubated until all of the radiolabeled outer leaflet ^{32}P–PC (black head group in Fig. 9.20) is exchanged for the nonlabeled mitochondrial PC (white head group in Fig. 9.20). The large excess of mitochondrial PC assures that all of the initial LUV exoleaflet ^{32}P–PC is replaced by nonradiolabeled mitochondrial PC. The now radiolabeled mitochondria and $PLEP_{(PC)}$ are then removed, leaving the LUVs with a ^{32}P–PC inner leaflet and a nonradiolabeled mitochondrial PC outer leaflet. The vesicles are allowed to incubate for a period of time $t = X$ during which flip-flop can occur. At time $t = X$, the vesicles are exposed to new mitochondria and $PLEP_{(PC)}$. Any ^{32}P–PC that has flipped from the inner leaflet to the outer leaflet during the time $t = X$ will now appear in the mitochondria that can be centrifuged and counted. The method is sensitive enough to detect only a few flipped ^{32}P–PC lipids. This number (PCs flipped as followed by counts over time $t = X$) is compared to the total counts in the vesicles where only the inner leaflet was radiolabeled. This experiment is then repeated for several longer incubations. From this an extrapolated half-life for PC flip-flop can be obtained.

Inherent phospholipid flip-flop rates, as determined in protein-free model lipid bilayer membranes, are very slow, often with half-lives in the days to weeks range [27]. In sharp contrast, James Hamilton [33] has shown that the flip-flop rates for free fatty acids are

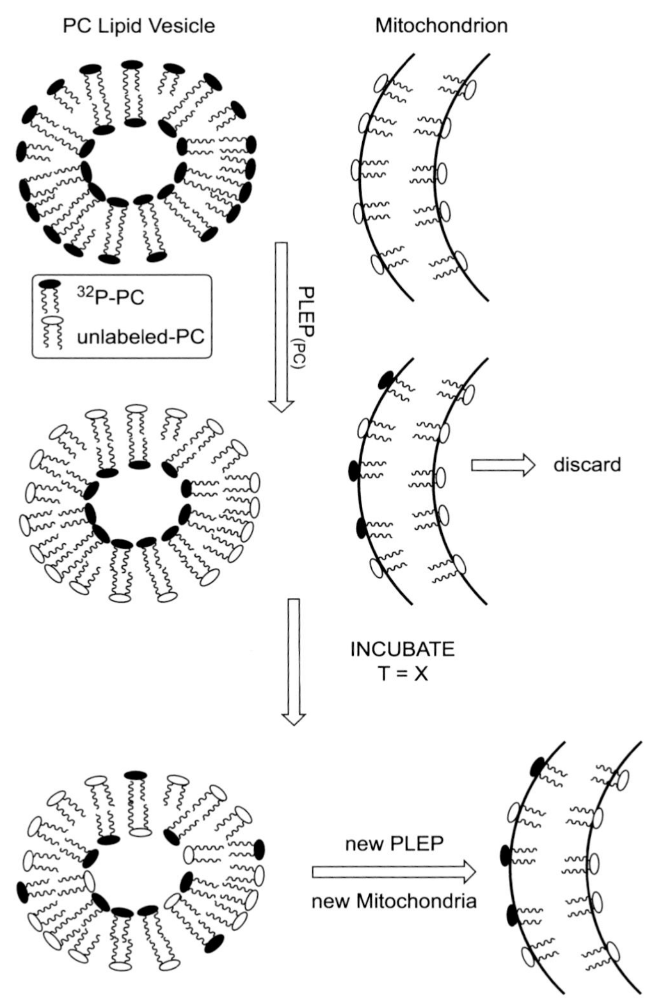

FIGURE 9.20 Use of a $PLEP_{(PC)}$ to determine the inherent rate of flip-flop for PC in a protein-free lipid bilayer (liposome). The PCs are either ^{32}P—PC (black head group) or non-labeled (^{31}P—PC) mitochondrial PC (white head group).

very fast, in the seconds to milliseconds range. Therefore it appears that moving the phospholipid polar head group into the hydrocarbon interior must be the major impediment to flip-flop. Supporting this is the very rapid flip-flop rate for DAG whose polar head group is only a simple uncharged alcohol. Indeed, Horman and Pownall [34] showed that at pH 7.4, flip-flop rates increased in the order PC < PG (phosphatidylglycerol) < PA < PE, where the rate for PE was at least 10 times greater than for the homologous PC phospholipid. They also reported that the flip-flop rate for PA increased 500-fold when the solution pH was decreased from 7.4 to 4. At pH 4, the net negative charge on PA is much reduced compared to pH 7.4. Finally, Middelkoop et al. [35] reported the affect of acyl chain composition on the flip-flop rate of various PCs (16:0,16:0 PC < 18:1,18:1 PC < 16:0,18:2 PC < 16:0,20:4 PC). They reported that for erythrocytes at 37°C, flip-flop half-lives decreased with the increasing number of double bonds.

Another way of looking at lipid flip-flop is to determine the affect of the bilayer hydrophobic environment (acyl chain components) on the flip-flop of a single lipid species. Armstrong et al. [36] measured flip-flop rates for the fluorescent phospholipid, nitrobenzoxadiazole (NBD)–PE by quenching NBD fluorescence with dithionite. As the number of double bonds comprising the bilayer increased, so did the flip-flop rate (Table 9.1). For example, flip-flop of NBD–PE in 22:6,22:6 PC with 12 double bonds was at least 10^3 times greater than for 16:0,16:0 PC with no double bonds. These investigators demonstrated that flip-flop is related to the acyl chain double bond content in the lipid bilayer interior.

5.3 Flippase, Floppase, and Scramblase

From the very first determinations of lipid asymmetry in erythrocytes, it was evident that lipid asymmetry was complex. Particularly troublesome was the observation that the distribution of PS, unlike the other phospholipids, was almost totally asymmetric. Bretscher realized the importance of PS being found exclusively on the inner membrane leaflet. He proposed the existence of an energy-dependent membrane protein whose function it was to flip PS from the outer membrane leaflet to the inner leaflet, thus preserving the total asymmetric distribution of PS [10]. He even coined the term "flippase" for this protein. An ATP-dependent flippase was later discovered by Seigneuret and Devaux [12]. This protein proved to be only one of a family of related membrane proteins called "flippases," "floppases," and "scramblases" whose functions are to correctly distribute the various phospholipids across membranes [15,16].

TABLE 9.1 Flip-Flop Rates (Expressed as Half-Lives) for Nitrobenzoxadiazole (NBD)–PE as Followed By Quenching of NBD Fluorescence With Dithionite. Large Unilamellar Vesicles Were Made From the Listed Phosphorylcholines (PC) With 1% NBD–PE Incorporated (25°C)

Phosphorylcholine	Flip-flop half life ($t_{1/2}$) (h)
16:0,16:0	>70
18:0,18:1	11.5 ± 1.6
18:1,18:1	8.1 ± 0.6
18:0,18:2	4.9 ± 0.9
18:2,18:2	1.9 ± 0.4
18:3,18:3	1.3 ± 0.3
18:0,22:6	0.29 ± 0.02
22:6,22:6	0.087 ± 0.006

Newly synthesized phospholipids are initially incorporated into the inner membrane leaflet after which they are subsequently distributed by a variety of lipid translocators [15,16]. Although the term flippase initially referred in general to any lipid translocator, it now refers to proteins that move lipids to the inner leaflet (cytofacial) membrane surface. They are ATP-dependent transporters. The flippase is highly selective for PS and functions to keep this lipid from appearing on the outer surface of the cell. Upon aging, PS starts to accumulate on the cell exterior, triggering apoptosis (cell death). Floppases are the opposite of flippases and are ATP-dependent transporters that move lipids to the outer (exofacial) leaflet membrane surface. They have been associated with the ATP-binding cassette class of transporters. The third family of lipid translocators is the scramblases whose function is to move lipids bi-directionally across the membrane without the use of ATP. Scramblases are inherently nonspecific and function to randomize the distribution of newly synthesized phospholipids and may be involved in membrane disruption events associated with cell death (apoptosis) (Chapter 24). In the case of the well-studied erythrocyte, net flip-flop rates are PS > PE > PC > SM.

5.4 Some General Conclusions From Flip-Flop Studies

Flip-flop of proteins and carbohydrates (in the form of glycoproteins and glycolipids) is absolutely prohibited. Inherent rates of membrane phospholipid flip-flop are very slow, with half-lives varying from days to weeks. This slow rate can be substantially enhanced (10^3–10^5-fold) by the simple inclusion of a trans-membrane protein. Cholesterol, DAG, and free fatty acids have high inherent rates of flip-flop. Cell membranes have families of lipid translocators known as flippases, floppases, and scramblases whose functions are to accurately maintain proper lipid asymmetry.

6. LIPID MELTING BEHAVIOR

The simple cartoon depiction of a lipid bilayer shown in Chapter 5 is deceptive. The lipid bilayer, existing in what is more correctly termed the "lamellar phase," is highly dynamic and can actually exist in many related phases in addition to the lamellar state. Although the lamellar phase dominates biological membrane structure, many and perhaps even most membrane polar lipids, when isolated and added to water, would prefer to exist in exotic nonlamellar phases. It is the lipid mixtures that stabilize lamellar phase. A variety of nonlamellar phases are discussed in Chapter 10. However, even the seemingly simple lamellar phase is complex and can exist in several variations. We will first discuss the simple melting of the lamellar phase from the solid-like gel (L_β' for PCs) to the fluid-like liquid crystalline (L_α) state. Lipid bilayers should not be thought of as existing as either a hard solid or a water-like fluid but more like what would be observed upon melting bacon grease. The gel state is a soft solid while the liquid crystalline state is a viscous fluid.

A major theme of Chapter 4 was the affect of fatty acyl chain length and degree of unsaturation on "melting" temperature (expressed as T_ms). But why do fatty acyl chains melt at all? Fig. 9.21 depicts the steps involved in melting of a saturated fatty acyl chain. A section of an unmelted, *trans* state, saturated fatty acid is shown in Fig. 9.21, part A. Newman

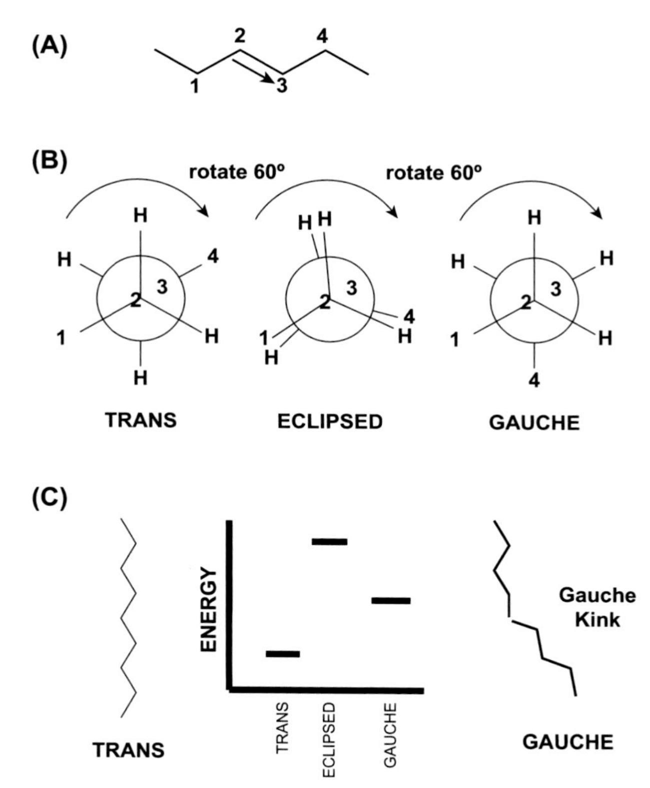

FIGURE 9.21 Melting of a saturated fatty acyl chain. Part A depicts a 4-carbon section somewhere along a saturated acyl chain existing in the unmelted *trans* state. Part B demonstrates Newman projections for the 4 sequential carbons depicted in A. Upon heating the chain, rotation occurs around C-2 and C-3. The *trans* configuration has the lowest possible energy (part C) as the largest groups, C-1 and C-4, are as far apart as possible. As the large C-4 passes a H, steric hindrance is encountered as depicted in the *eclipsed* state. The *eclipsed* state is an unfavorable, high-energy transient state (part C). Further rotation relieves some of the steric stress, resulting in another low energy state referred to as the *gauche* state. The *gauche* state is of lower energy than the *eclipsed* state, but higher energy than the *trans* state, since there is more steric interference between the closer, bulky C-1 and C-4 groups. This lower energy state creates a '*gauche* kink' in the chain (part C). It is the accumulation of '*gauche* kinks' in the chain that results in chain melting.

projections of a short segment of the chain are shown in Fig. 9.21, part B. Upon heating the chain, rotation occurs around all C–C bonds, for example the bond connecting C-2 and C-3. The *trans* configuration (Fig. 9.21, part B, left) has the lowest possible energy (Fig. 9.21, part C) as the largest groups, C-1 and C-4, are as far apart as possible. Upon clockwise rotation, as the large C-4 passes a H, steric hindrance is encountered as depicted in the *eclipsed* state (Fig. 9.21, part B, middle). The *eclipsed* state is an unfavorable, high-energy state (Fig. 9.21, part C). Further rotation relieves some of the steric stress resulting in another low energy state referred to as the *gauche* state (Fig. 9.21, part B, right). The *gauche* state is lower energy than the *eclipsed* state, but higher energy than the *trans* state since there is more steric interference between the closer, bulky C-1 and C-4 groups. This lower energy state creates a "*gauche* kink" in the chain (Fig. 9.21, part C, far right). It is the accumulation of "*gauche* kinks" in the chain that results in melting. At 37°C there are ~2 kinks per acyl chain and the chain exists in the melted, liquid crystalline state. An acyl chain is essentially tethered to the bulky head group. The kinks then rapidly (~10^{-9} s) work their way down the chain and are released at the bilayer interior. For this reason, acyl chains exhibit minimal motion near the polar head and maximal motion at the methylene chain terminus (the omega end). The chain is more fluid at its omega end than at its alpha end.

Due to the lack of "*gauche* kinks," saturated acyl chains below their melting point pack tightly while the same chains above their melting point pack loosely (Fig. 9.21). Lipid packing as related to melting affects several important, basic membrane properties. Compared to gel-state membranes, liquid crystalline state membranes:

1. Are more poorly packed.
2. Occupy a larger area per lipid.
3. Are more fluid.
4. Are more permeable.
5. Are thinner.
6. Are less stable.
7. Are more "dynamic."

6.1 Differential Scanning Calorimetry

The most direct method for monitoring melting in lipid bilayers is by differential scanning calorimetry (DSC). The technique was developed by E.S. Watson and M.J. O'Neill in 1960 and was rapidly moved into commercial production in 1963. DSC is a nonperturbing, analytical technique that can accurately monitor thermotropic phase behavior in lipid bilayers [37–42]. From DSC, one can obtain T_ms, transition enthalpies and information on phase. In general, this technique involves heating a reference solution devoid of the membrane of interest and a sample containing a dilute aqueous membrane suspension. The technique measures the difference in the amount of heat required to increase the temperature of the sample and reference at a chosen rate. During the scan, temperature of the sample and reference is very accurately kept the same. It is the heat required to maintain equal temperature that is important.

The phospholipid most studied by DSC has been dipalmitoylphosphatidylcholine (DPPC) (16:0,16:0 PC), the workhorse of DSC membrane studies. This lipid is inexpensive and readily available in highly pure form from a number of commercial sources. The lipid is resistant to oxidation (it has two saturated acyl chains) and has a T_m (41.3°C) that is not near the troublesome ice transition (0°C). Also, the T_m is not so high as to render the lipid susceptible to massive heat-induced hydrolysis. A typical DSC scan for DPPC is shown in Fig. 9.22. The main transition occurs at 41.3°C (Table 5.1). A much smaller transition, called the pretransition, is observed at ~35.6°C.

Since the pretransition is not found in all phospholipid classes, it probably is related to the polar head group. For example, (16:0,16:0 PE, DPPE) has no pretransition and has a T_m of 63°C. Profound differences in the DSC scans between DPPC and DPPE have been attributed to differences in head group size and hydration (Fig. 9.23). PC has a much larger and more

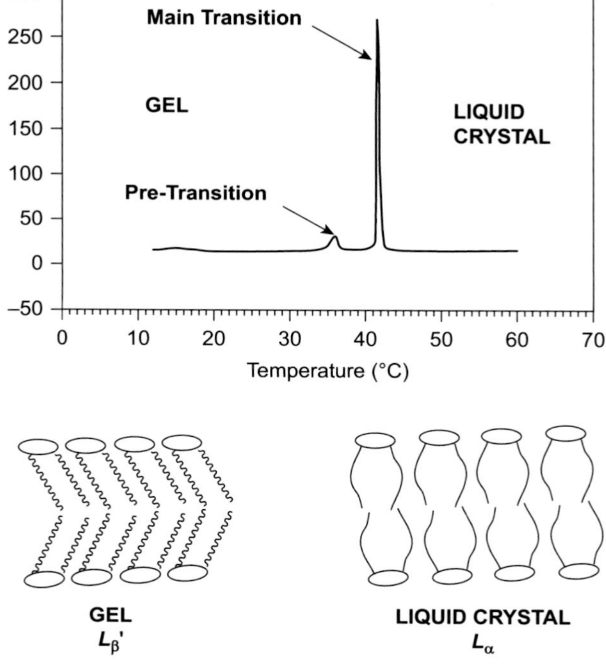

FIGURE 9.22 Differential scanning calorimetry scan of the gel (L_{β}') to liquid crystal (L_{α}) transition for dipalmitoylphosphatidylcholine (DPPC) (top panel). For disaturated chain phosphorylcholines in the gel state, the chains are tilted in relation to the polar head group. Upon melting to the liquid crystal state, the tilt disappears (bottom panel).

hydrated head group than does PE. As a result, at temperatures below T_m (in the gel state), the diameter of the PC head group (S) is wider than the combined diameters of the two all *trans* acyl chains (2Σ). Therefore $S > 2\Sigma$. The chains must then tilt relative to the head group until their combined diameters are the same size, $S = 2\Sigma$. Without tilting there would be a gap beneath the head that would have to be filled with water, a thermodynamic disaster. The larger the head group, the larger is the tilt angle. For example, in the gel state, DMPC (14:0,14:0 PC) has a tilt angle of 12° while the much larger cerebroside (with a sugar head group) has a tilt angle of 41°. For DPPE which has a small, poorly hydrated head group, in the gel state $S = 2\Sigma$ and the tilt angle is zero. Upon melting, the *gauche* kinks substantially increase the diameter of the acyl chains whereupon their sum becomes equal to or greater than the head diameter ($S \leq 2\Sigma$), eliminating the tilt angle.

As depicted in Fig. 9.23, gel state DPPC exhibits chain tilt, while DPPE does not. This may explain why PCs have a lower T_m than PEs. The chain tilt puts DPPC in a configuration more amenable to melting. The typical DSC scan shown in Fig. 9.22 for DPPC shows both the main chain melting transition and the minor pretransition. Both transitions are believed to be part of the same melting process. In the gel state (L_{β}'), the chains of DPPC are tilted. The tilt is lost upon transitioning into the melted liquid crystalline (L_{α}) state. Something unusual happens to DPPC in the temperature range spanning the pretransition and main transition ripple

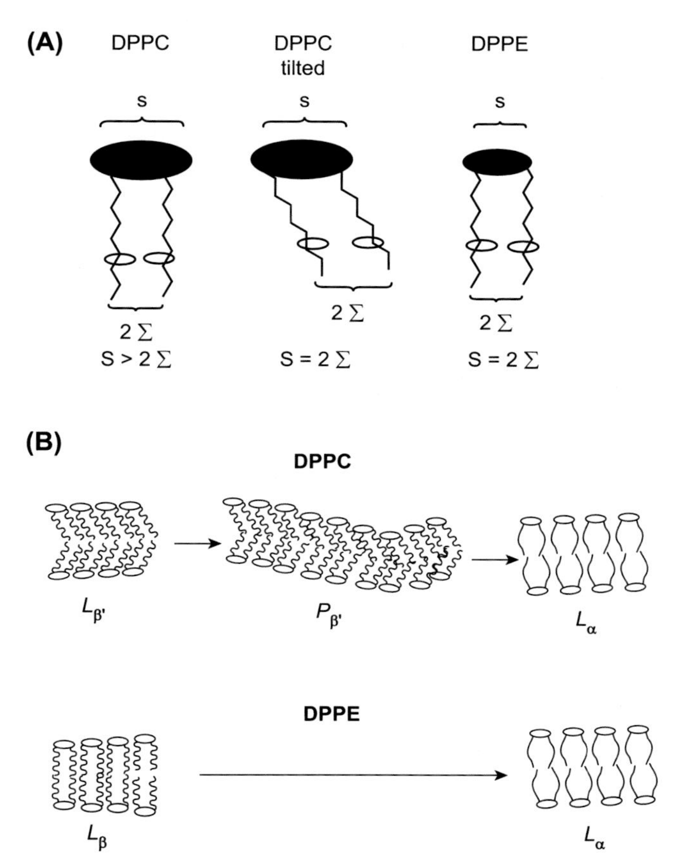

FIGURE 9.23 Gel to liquid crystal phase transitions for dipalmitoylphosphatidylcholine (DPPC) and dipalmitoylphosphatidylethanolamine (DPPE). Chain tilt noted for DPPC in the gel state (L_{β}') is not seen with DPPE (L_{β}). Neither DPPC nor DPPE exhibit chain tilt in the liquid crystal (L_{α}) state. Another unusual feature of the DPPC differential scanning calorimetry scan that is not observed for DPPE membranes is the existence of 'Ripple Structure' (P_{β}') between the pretransition and the main transition. PEs exhibit neither a pretransition nor ripple structure.

structure occurs. Fluorescence and ESR (Chapter 10) methodologies indicate a coexistence of other membrane lipids, eliminates the pretransition and hence the ripple phase. More unusual, however, is the shape of the membrane over this temperature range. At the pretransition, the flat membrane in the gel phase transforms into a periodically undulated bilayer, named the ripple or P_{β}' phase. The ripple phase has been shown by electron density profiles obtained from small angle X-ray scattering to be an asymmetric undulation pattern resembling a saw-tooth with a repeating length of 120–160 Å, depending on acyl chain length. It is not at all clear if ripple structure has any biological importance as very small amounts of lipid 'contaminants', including other membrane lipids, greatly diminish the likelihood of ripple phase. Ripple phase is also not observed for heteroacid molecular species that are predominant in membranes. Heteroacid phospholipids have different acyl chains attached to the *sn*-1 and *sn*-2 locations.

The DSC curve for DPPC in Fig. 9.22 shows a narrow main transition. The narrower the transition, the more lipids melt as a single unit. Almost any lipid contaminant or a fast temperature ramp (scan) will reduce and broaden the curve. In fact, an absolutely pure-phospholipid bilayer scanned at an infinitely slow temperature ramp would theoretically result in a vertical line where all of the gel state lipid would melt at once. Lipid contaminants broaden and reduce the main transition. Fig. 9.24 shows the affect of increasing amounts of cholesterol on the DPPC main transition. Even 1 mol% cholesterol noticeably affects the transition, and by 10 mol% cholesterol, most of the transition has disappeared. By 30 mol% cholesterol, the DPPC main transition is completely obliterated.

From the shape of the curve, one can estimate the number of lipids melting as a unit. This empirical number, referred to as the "cooperativity unit," can be simply calculated from the DSC curve, as first described by Susan Mabrey and Julian Sturtevant [37] and nicely summarized by Donald Small [39]. The cooperativity unit is defined as:

$$\text{Cooperativity Unit} = \Delta H^{\circ}_{\text{vH}} / \Delta H^{\circ}_{\text{cal}}$$

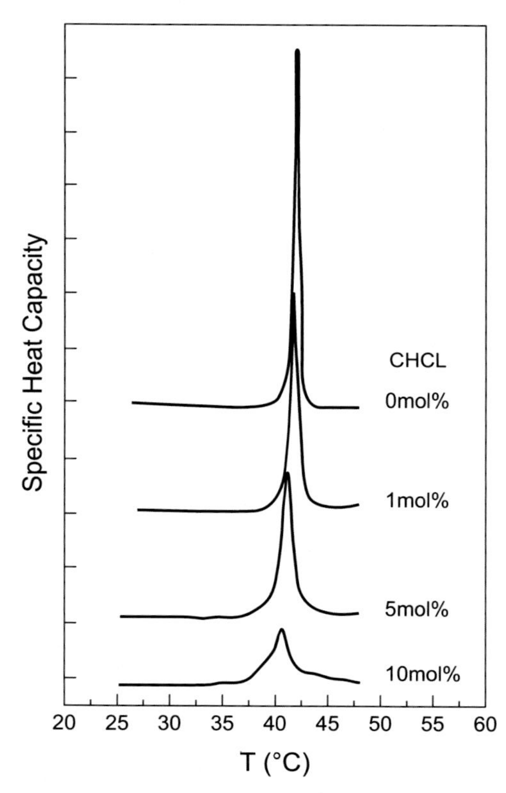

FIGURE 9.24 The affect of increasing levels of cholesterol on the main transition of dipalmitoylphosphatidylcholine as followed by differential scanning calorimetry. The main transition shifts to lower temperature and exhibits a decrease in enthalpy (decrease in cooperativity) with increasing cholesterol.

Where ΔH°_{vH} is the van't Hoff enthalpy and ΔH°_{cal} is the effective enthalpy which is the calculated area beneath the curve of the thermal transition expressed in cal/mol/gram. The van't Hoff enthalpy is defined by Julian Sturtevant as:

$$\Delta H^{\circ}_{vH} = 6.9T^2_m/[(T_2 - T_1)/2]$$

where T_m is the midpoint of the transition (°K) and the expression $[(T_2{-}T_1)/2]$ is the peak width at one half the peak height. T_1 is the onset temperature and T_2 is the final (offset) temperature of the transition, both expressed in °K.

Julian M. Sturtevant (Fig. 9.25) was a pioneer in the field of biophysical chemistry with a world reputation on the application of thermochemistry to biological systems. He published into his 90s, and died at the age of 97.

Since the DSC scan shape, and hence calculated cooperativity units, can be altered by changing the scan rate (a faster rate produces a broader curve and hence apparent lower cooperativity), the cooperativity unit calculations only have meaning compared to another lipid run under identical conditions. DSCs have the ability to scan over a wide range of temperature ramps. A 'normal' scan rate would be ~5°C/h. This rate is sufficiently fast to allow for a complete scan in a reasonable amount of time, but not so fast as to obliterate curve structure. Also, lipids may become hydrolyzed or oxidized if the sample is left in the DSC crucible for too long. However, some very slow scan rates (~0.1°C/h) may be employed to examine events occurring over a very narrow temperature range.

6.2 Fourier Transform-Infrared Spectroscopy

Although DSC is the best method to analyze lipid phase transitions, many other methodologies have successfully been employed. A very different method involves Fourier Transform Infrared (FT-IR) spectroscopy. FT-IR is a versatile, nonperturbing, powerful technique that can be used to obtain information on all regions of the phospholipid molecule simultaneously. For a general discussion of FT-IR see [43,44]. Instead of (slowly) scanning through

FIGURE 9.25 Julian M. Sturtevant, 1908–2005.

the entire spectrum as is done in most UV—visible applications, FT-IR irradiates the sample with all IR wavelengths at once and the entire signal is captured and subjected to Fourier analysis. This allows for multiple spectra to be rapidly obtained and the signals averaged resulting in vastly improved sensitivity. Nanogram samples can be accurately analyzed by FT-IR. Modern FT-IR spectrometers also have the advantage of being able to eliminate the large, troublesome water absorbance, allowing for the ability to measure membranes dispersed in aqueous solution [45,46].

FT-IR spectroscopy detects vibrations during which the electrical dipole moment changes. Of course this time is very short ($\sim 10^{11}$ s^{-1}). Most of the measured absorbance bands are associated with stretching and bending vibrations that occur at discrete positions in the mid-IR spectral range (2500—25,000 nm). Traditionally, IR studies do not report wavelengths but rather wave numbers. The spectral range for FT-IR is therefore 4000—400 cm^{-1}. Of particular importance to membrane studies are the absorbance bands at about 2900 (C—H stretching), 1740 (C=O stretching in esters and carboxylic acids), 1400 (C—O stretching in carboxylates), 1235 (P=O stretching in phosphate esters), and 1000 (C—O stretching in carbohydrates) cm^{-1}.

For phospholipids, the acyl tails (including *trans/gauche* ratios) are followed by C—H vibrations (2800—3100 cm^{-1}), the glycerol-acyl chain interface by ester stretching (1700—1750 cm^{-1}), and the head group phosphate stretching (1220—1240 cm^{-1}) [45,46].

Therefore, FT-IR spectra provide a complete 'molecular snapshot' yielding information on phospholipid conformation, and dynamics from the tip of the polar head group to the omega end of the acyl tail. Fig. 9.26 shows the gel to liquid crystal phase transition as determined by the absorbance maximum of the symmetric stretching band of CH_2 groups in liposomes composed of: (a) 16:1,16:1 PC; (b) 18:1,18:1 PC; (c) 14:0,14:0 PC; and (d) 16:0,16:0 PC [46]. T_ms are in close agreement with values obtained by other methods including DSC [46].

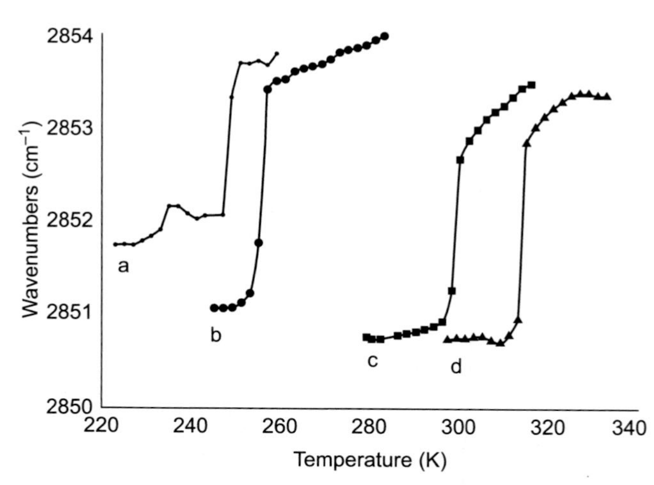

FIGURE 9.26 Gel to liquid crystal phase transition as determined by the absorbance maximum of the symmetric stretching band of CH_2 groups in liposomes composed of: (a) 16:1,16:1 PC; (b) 18:1,18:1 PC; (c)14:0,14:0 and (d) 16:0,16:0 PC. T_ms are in close agreement with values obtained by other methods including differential scanning calorimetry [46].

6.3 Fluorescence Polarization

Another method that is commonly used to follow lipid phase transitions and fluidity in membranes is fluorescence polarization (FP). Fluorescence was first reported, and the term coined, by George Gabriel Stokes [46,47]. Stokes (Fig. 9.27) noted that the mineral fluorite (calcium fluoride) emitted visible light when illuminated with 'invisible radiation' (later identified as UV light). It was from this simple, but seemingly unrelated observation, that the most powerful and versatile technique in modern membrane studies, fluorescence, had its humble beginnings.

The technique of fluorescence is reviewed in two comprehensive books by Joseph Lakowitz [48,49]. Fluorescence occurs after a molecule (a fluorophore) absorbs light (is excited) at one wavelength and radiates (emits) light at a longer wavelength. Basic fluorescence is depicted in Fig. 9.28. The specific frequency of excitation and emission are dependent on the characteristics of the fluorophore. The unexcited molecule has its electrons in the lowest energy or ground state, S0. Upon absorbing light, a ground state electron is boosted to its first (electronically) excited state, S1. A fluorophore in its excited S1 state can relax by two basic radiative pathways, fluorescence and phosphorescence. The excited electron can undergo nonradiative relaxation (internal conversion) by losing heat to its surroundings. Finally, the still excited electron returns to the ground state (S1 $\rightarrow$ S0) in a process referred to as fluorescence. The electron spin direction does not change during fluorescence. However, if the electron spin changes direction during the excited state, a triplet state, *T*, results. Emitted radiation during return to the ground state ($T \rightarrow$ S0) is called phosphorescence and is much slower than fluorescence. Since most membrane studies involve rapid-time events, only fluorescence, and not phosphorescence, will be discussed.

The time domains involved in fluorescence are very fast. The process of light absorption and production of the excited state (S0 $\rightarrow$ S1) takes $\sim 10^{-11}$ s. It takes $\sim 10^{-8}$ s to return from the excited to the ground state (S1 $\rightarrow$ S0). The average time in which the fluorophore stays in its excited state before emitting a photon is referred to as the fluorescence lifetime.

FIGURE 9.27 George Gabriel Stokes, 1819–1903.

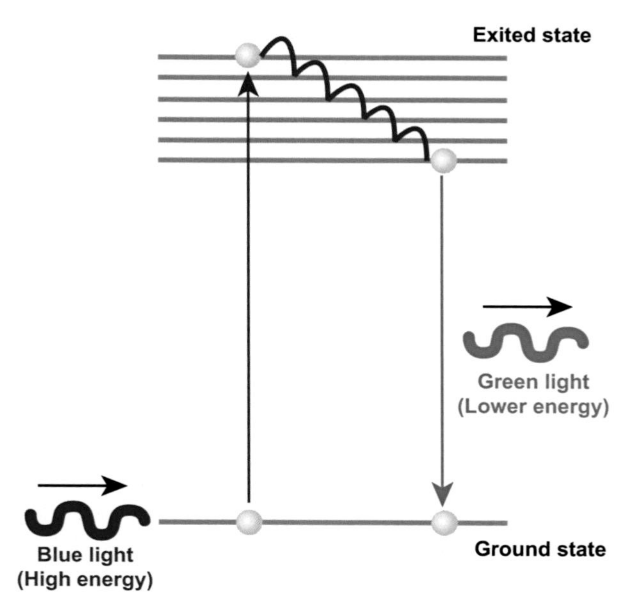

FIGURE 9.28 Basic fluorescence. Light is absorbed by the fluorophore and a ground state electron is boosted to an excited state (10^{-11} s). The electron then loses some of its energy by nonradiative relaxation (internal conversion) to a lower energy excited state before returning to the ground state in the process called fluorescence ($\sim 10^{-8}$ s). The emitted fluorescent light is of a longer wavelength (lower energy) than the light that was initially absorbed.

It is the time between these two events, light absorbance and fluorescence, that is important for membrane structure studies. In contrast, phosphorescence is relatively slow ($>10^{-3}$ s) and is of little use in membrane studies.

Fluorescence and ESR are 'fast' techniques (ie they can be used to monitor fast processes). Since NMR is greater than an order of magnitude slower, events that can be seen with fluorescence and ESR may not be detectable by NMR. Detectable frequencies are $\sim 10^5$ s^{-1} for ^2H NMR and $\sim 10^8$ s^{-1} for ESR and fluorescence. The fluorescence technique most often used to investigate membrane structure, FP, was first described by the French physicist Jean Baptiste Perrin in 1926 and is depicted in Fig. 9.29. Perrin was a diverse and prolific physicist in the first quarter of the 20th century, winning many awards including the 1926 Nobel Prize in Physics.

Plane-polarized light, produced by passing light through a polarizer, is used to excite a fluorophore [48–50]. FP is based on the principle that if the fluorophore is immobile, fluoresced light is emitted in the same plane as the plane-polarized excited light. If the exciting light is polarized in the vertical direction, then the fluoresced light will also be polarized in the vertical direction. This, however, is dependent upon a lack of movement of the fluorophore during the excited state. If the fluorophore remains stationary from the time it absorbs light until the time it emits light, the fluorescent light will remain completely polarized. However, if the molecule rotates or tumbles during this time, some of the light will be polarized in different directions (the emitted light becomes depolarized). The extent of depolarization is monitored by comparing the intensity of vertically polarized fluorescent light (observed

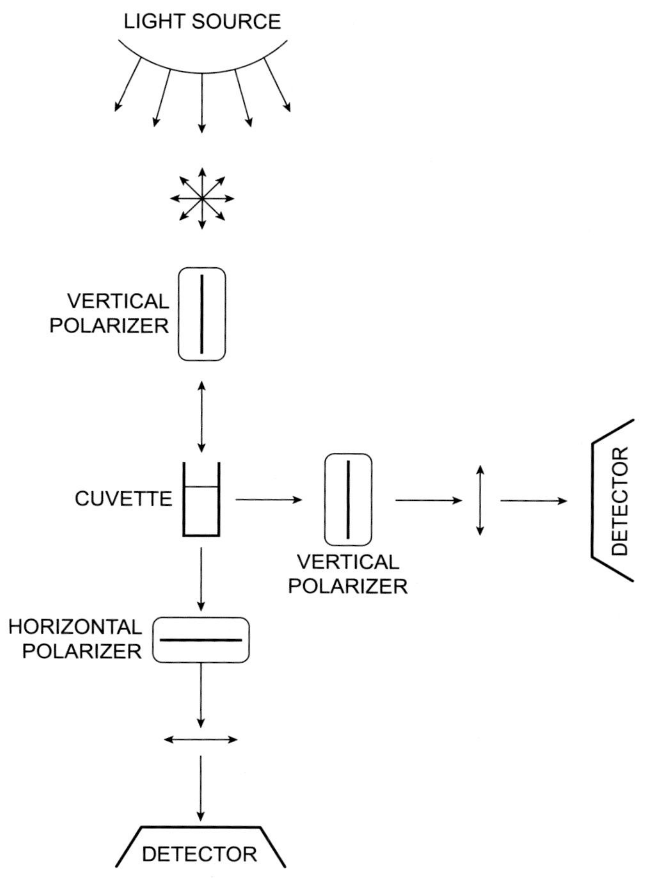

FIGURE 9.29 Method to measure fluorescence polarization. Light that is initially polarized in all planes is first passed through a polarizer selecting for light that will be polarized in only one plane. This light passes through a sample containing a fluorophore in a 4-sided cuvette. Upon emerging from the sample, the intensity of light polarized at planes 90° to one another (vertical and horizontal) is measured. The Anisotropy can then be calculated from these intensities as described in the text.

through a vertical polarizer) to depolarized light (observed through a horizontal polarizer). If the fluorescing molecule is very large or immobile during the fluorescent excited state, all of the light will be seen through the vertical polarized light detector. The horizontal detector, attached behind the horizontal polarizer, will detect no light. If the fluorescing molecule is very small or rapidly tumbling, then equal amounts of emitted light will be observed through the vertical and horizontal polarizer detectors. The degree of depolarization can be expressed as either a Polarization or an Anisotropy value. Although the two values can be readily interconverted, Anisotropy is more frequently encountered.

$$\text{Anisotropy} = I_{(\mathrm{V})} - I_{(\mathrm{H})}/I_{(\mathrm{V})} + 2I_{(\mathrm{H})}$$

Where I is the emitted light intensity as measured through the vertical (V) or horizontal (H) polarizer detectors. If no molecular motion occurs between the time of light absorbance

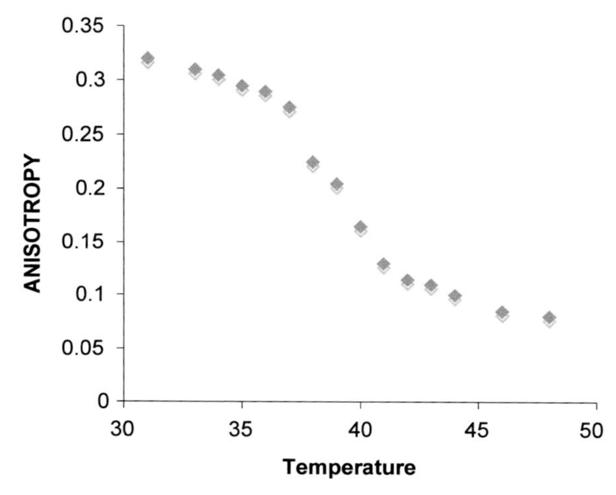

FIGURE 9.30 Gel to liquid crystal transition of dipalmitoylphosphatidylcholine (DPPC) as followed by the steady state anisotropy of 1,6-diphenylhexatriene.

and the time of emission, Anisotropy has a limiting value of 0.4. If rapid motion occurs, the limiting value for Anisotropy is 0. From the decay of the Anisotropy it is possible to obtain an order parameter that has the same definition as for ESR and NMR.

FP can also be used to monitor lipid phase transitions [51]. Fig. 9.30 shows the gel-to-liquid crystal transition for DPPC as followed by the fluorescent probe 1,6-diphenylhexatriene (DPH) (Fig. 9.31). DPH has historically been the most used of all FP probes for membrane studies [52].

DPH possesses several important properties justifying its use in FP investigations. This fluorophore partitions strongly into membranes, is highly fluorescent, does not bind to proteins, and is sensitive to membrane physical state. The transition temperature, T_m, of DPPC as determined by FP (Fig. 9.30) agrees with those reported by DSC (Fig. 9.22) and FT-IR (Fig. 9.26).

7. MEMBRANE FLUIDITY

The term "fluidity" is one of the most used, yet one of the most poorly understood in the life sciences [53]. While everyone has a general feel for what fluidity means, the devil is in the details. A dictionary definition of a fluid is something that is 'capable of flowing, not a solid'. In biology, fluidity refers to the viscosity of the lipid bilayer component of a cell membrane. For water, fluidity is defined as the inverse of viscosity. Historically, viscosity was easily estimated by simply measuring how fast a spherical marble falls through a solution. The concept of viscosity was therefore developed around isotropic motion with equal forces on all sides of a perfect sphere. But how much of this applies to a complex, highly anisotropic biological membrane?

In a biological membrane there are three complex types of motion to consider. First there is nonhomogeneous lateral movement in the plane of the lipid bilayer. This is complicated by

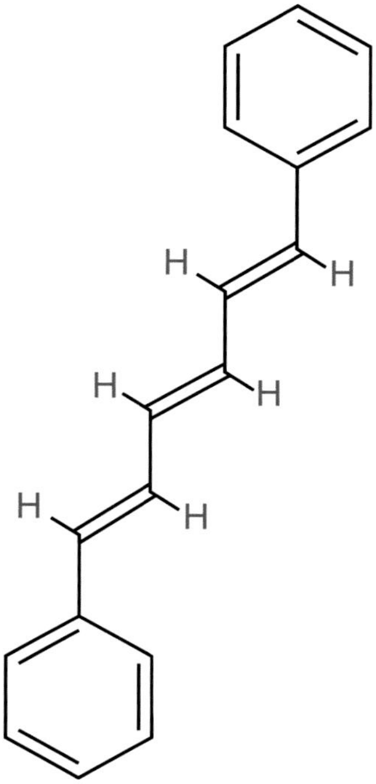

FIGURE 9.31 1,6-Diphenylhexatriene (DPH).

lipid domains of various compositions and properties, crowding by multiple proteins and protein complexes, and interactions with the submembrane cytoskeleton. Next there is rotational movement around the longitudinal axis of the molecule. Finally there is transmembrane movement or flip-flop.

In addition, regardless of how membrane fluidity is defined, there will be a fluidity gradient from the aqueous interface through to the bilayer interior. The extreme heterogeneity of biological membranes makes analysis of fluidity far more complicated than simply dropping a marble through a solution. Usually with membranes, molecular motion is monitored with either spin probes or fluorescent probes [54]. However, unlike marbles, the probes are not perfect spheres and so do not behave isotropically. Instead, they exhibit preferred orientations (they are anisotropic) in the bilayer and may preferentially partition into proteins or different membrane domains. As a result, membrane probes are sensitive to both rates of motion and to constraints on motion, and so report on both dynamics and order. So, in biological membranes the term fluidity is complex, technique-dependent, and quantifies a variety of parameters including:

1. Rotational correlation times.
2. Order parameters.
3. Steady state anisotropy.
4. Partitioning of probes between the membrane and water.

Although a number representing fluidity can be obtained, its precise definition may be a matter of contention. Further complicating the problem is the enormous number of membrane probes that are commercially available. The major provider of membrane probes has for many years been Molecular Probes in Eugene, Oregon (now owned by Invitrogen, Carlsbad, CA). Chapter 13 of the Molecular Probes catalog, entitled 'Probes for Lipids and Membranes', provides an excellent review of the topic.

It is clear from several different types of measurements that fluidity gradients exist from the aqueous interface through to the bilayer interior. This has often been demonstrated using FP as well as ESR. Both of these techniques require adding an anisotropic bulky, and hence perturbing, probe at different locations down the acyl chain. Both FP and ESR show a gradual, continuous increase in fluidity from the carbons near the alpha or carboxylate end to the omega or terminal carbon [54]. Unfortunately, the bulkiness of the fluoro- or spin-probes disturbs the chains to such an extent that important detail of the fluidity gradient can be lost.

NMR has been used to monitor the fluidity gradient without perturbing the system [55]. Without getting bogged down in experimental details, two types of NMR techniques (order parameter and T1 relaxation measurements) have demonstrated that fluidity increases at a steady but slow rate for about the first 8–10 carbons (called the plateau region), after which fluidity progressively increases through the rest of the chain. This pattern is characteristic of all bilayers. However, details concerning the size and location of the plateau and the fluidity of the most fluid terminal region vary depending on the lipid composition of the bilayer. In fact the shape of the fluidity gradient can provide a crude fingerprint of the bilayer.

The NMR-derived order parameter (SCD) can be used to estimate how isotropic the motion is at a particular carbon. A low SCD indicates a large range of motion and hence high fluidity. The range of values can vary depending on what is being measured, but often range from ~0 (no motion) to ~1.0 (isotropic motion). Fig. 9.32 presents SCD values for the *sn*-1 palmitoyl chain of DPPC and 1-palmitoyl-2-oleoyl-*sn*-glycero-3-phosphocholine

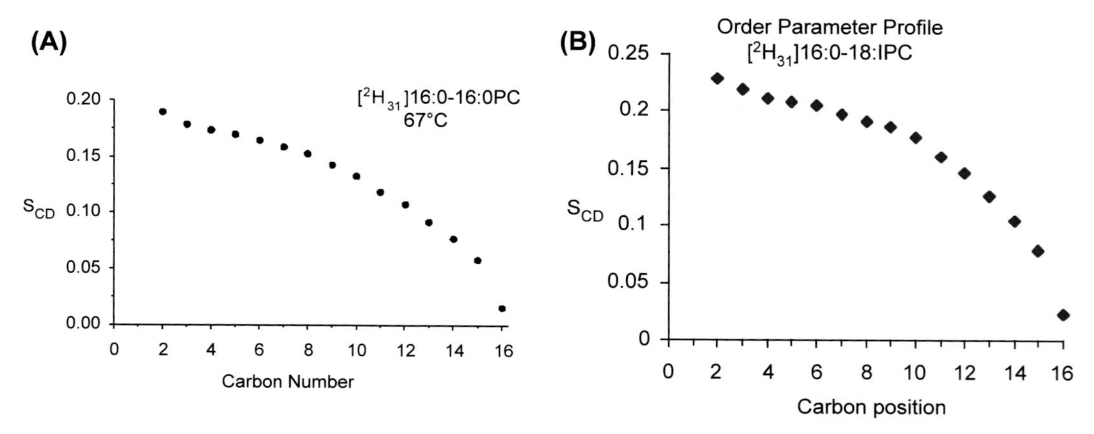

FIGURE 9.32 Nuclear magnetic resonance-derived order parameters (SCD) for the palmitoyl group on the *sn*-1 chain of dipalmitoylphosphatidylcholine (DPPC) (16:0,16:0 PC) (A) and 1-palmitoyl-2-oleoyl-*sn*-glycero-3-phosphocholine (POPC) (16:0,18:1 PC) (B).

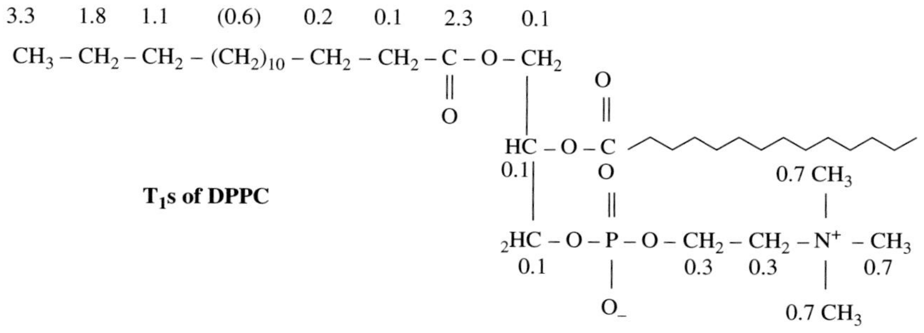

```
3.3    1.8    1.1     (0.6)       0.2     0.1    2.3     0.1
CH3 – CH2 – CH2 – (CH2)10 – CH2 – CH2 – C – O – CH2
                                        ||      |      O
                                        O       |      ||
                                              HC – O – C ∧∧∧∧∧∧∧∧∧∧∧∧∧∧
                                              0.1|     O              0.7 CH3
        T1s of DPPC                              |     ||                |
                                             2HC – O – P – O – CH2 – CH2 – N+ – CH3
                                              0.1      |      0.3    0.3    |   0.7
                                                       O–                 0.7 CH3
```

FIGURE 9.33 ^{13}C-nuclear magnetic resonance-derived T1s for dipalmitoylphosphatidylcholine. A high T1 means the carbon exhibits a lot of motion. A lower T1 indicates less motion. 'Micro-viscosity' of the carbons of DPPC varies 70-fold from the most restricted carbons near the polar head group to the least restricted carbon at the omega terminus of the acyl chain.

(POPC) (16:0,18:1 PC). The typical trans-membrane fluidity gradient pattern with a plateau region can be easily seen, but it is also clear that the two profiles, while having a similar general shape, do exhibit differences. For example, the plateau region of POPC is more pronounced than for DPPC. The profile shown for DPPC was obtained at a relatively high temperature of 67°C. At a reduced temperature, the plateau region would be longer and more pronounced than in Fig. 9.32.

Fast molecular motion can also be followed by T1 relaxation using NMR. A large T1 indicates more rapid motion. Fig. 9.33 shows T1s for various carbons on DPPC. Again a substantial plateau region is evident. The viscosity varies about 70-fold from the highly fluid omega methyl terminus to carbons on the less flui polar head group.

The biophysical techniques used to study membrane fluidity are best suited to compare changes in membrane physical properties brought about by environmental and membrane-soluble agents. It is predicted that anything that will increase membrane viscosity will decrease fluidity. Examples include, decreasing temperature, increasing pressure, increasing cholesterol content, and altering environmental pH and ionic strength. In addition, the nature of the phospholipid head groups and acyl chains, as well as favorable and unfavorable lipid–lipid and lipid–protein interactions, will also affect fluidity. These concepts are discussed in Chapter 10 under "Homeoviscous Adaptation."

8. MEMBRANE DYNAMICS: HARDEN M. MCCONNELL AND DENNIS CHAPMAN

The history of membrane studies is not reflected by a continuous progression, but instead, advances are made in spurts. Examples of particularly productive periods of membrane research include: the period around 1820 (discovery of many membrane lipids), the late 19th century (advances in lipid monolayer studies and understanding the membrane permeability barrier), the early 1920s (discovery of lipid-soluble vitamins), 1925 (discovery of proteins and the lipid bilayer), the late 1960s to early 1970s (profound discoveries concerning dynamic membrane properties and proposal of the Fluid Mosaic model), and the late

1990s (proposal of signaling Lipid Raft domains). Perhaps the most prolific period for membrane studies spanned the late 1960s and early 1970s and was led by Harden M. McConnell (Stanford University) (Fig. 9.34) and Dennis Chapman (several affiliations in London) (Fig. 9.35).

By about 1970, the static components of membranes (size, composition, and arrangement) (Chapter 1) were, for the most part, well established. But things were about to dramatically change. New biophysical techniques were developed and applied to membranes, probing for the first time membrane dynamics. These studies had a profound effect on our understanding

FIGURE 9.34 Harden M. McConnell, 1927–2014. *Reproduced from http://www.chemistry.msu.edu/www/assets/Image/photos/portraits/mcconnellc.jpg.*

FIGURE 9.35 Dennis Chapman, 1927–1999. *Reproduced from http://www.pcbiomaterials.com/history.html.*

of membranes. Harden M. McConnell and Dennis Chapman pioneered many of these novel investigations.

Harden M. McConnell has had a very long, diverse, and productive career, often focusing on membrane properties. His early contributions to membrane studies, emanating from the 1960s and 1970s, include: developing the technique of spin-labels for use in membranes, elucidating the electronic structure of unsaturated hydrocarbons, and measuring the rates of translational diffusion of lipids in bilayer membranes as well as the rates of trans-membrane phospholipid flip-flop. McConnell provided some of the earliest evidence for the fluidity of biological membranes, discovered fluidity gradient profiles, and lipid phase separation. Two examples of McConnell's early membrane work follow.

In 1971, McConnell realized that certain nitroxide free radicals could be attached to fatty acids and report on molecular motion (fluidity) at different locations throughout the membrane interior [56]. McConnell synthesized a series of α-lecithins having a paramagnetic N-oxyl-4′,4′-dimethylozazoleidine at selected positions on a fatty acid chain. Quantitative analysis of the paramagnetic resonance spectra of the spin-labeled lecithins were compared for liposomes and the walking leg nerve fiber of *Homarus americanus*. Analysis provided an estimate of the probabilities of *gauche* and *trans* conformations about C—C single bonds at various positions in liposomes and in a biological membrane. The fluidity gradient profiles for both membranes appeared strikingly similar. He observed that fluidity gradients are all relatively rigid for the first 8—10 carbons with rapidly increasing motions continuing down the chain. The initial rigid portion of the profile is referred to as the "plateau." Each gradient profile is slightly different, creating a unique "fingerprint" for each membrane.

In 1973, McConnell turned his spin-label technique to the emerging problem of lipid phase separation in membranes [57]. He monitored temperature-induced phase separations in aqueous dispersions (liposomes) of binary mixtures of DMPC, DPPC, or 1,2-distearoyl-*sn*-glycero-3-phosphocholine with DPPE. His method was based on partitioning of the spin-label 2,2,6,6-tetramethylpiperidine-1-oxyl (Tempo) between the fluid hydrocarbon bilayer region and the surrounding aqueous phase. As membranes "freeze" with decreasing temperature, Tempo is excluded from the membrane. Tempo spectra were shown to change abruptly at temperatures corresponding to the onset and completion of lateral phase separations. From this data, McConnell was able to construct lipid phase diagrams (Chapter 10). Subsequently, lipid phase separations were also demonstrated in biological membranes.

McConnell's recent research is still concerned with the physical chemistry of biological membranes. These studies range all the way from lipid monolayers at the air—water interface to the regions of membrane—membrane contact that are essential in immunology. A further contribution was his introduction of supported lipid bilayers to mimic cell surfaces, including mimicking antigen presentation. In 1983 McConnell founded Molecular Devices Corporation, a company that produces instrumentation for biochemical analysis and drug discovery. The company has grown to over 1000 employees.

At the same time Harden M. McConnell was beginning his iconic studies on membrane dynamics at Stanford, in London, Dennis Chapman was initiating his parallel and equally important investigations, also focusing on membrane dynamics [58]. The contributions of these two investigators cannot be overstated. Our current concept of membrane structure and function is heavily dependent on their insights.

Chapman's entrance into the field of membranes began with model studies on soaps and progressed to phospholipids. He soon realized that properties of amphipathic molecules (Chapter 3) were very different in anhydrous form compared to when hydrated. Water properties were responsible for these altered properties in membranes (see the Hydrophobic Effect discussed in Chapter 3). For example, in water the hydrocarbon chains undergo a partial melting process some 200° below the anhydrous capillary melting point. Many of Chapman's studies monitored membrane properties as a function of temperature. In an early report from 1964, he employed a temperature-controlled IR spectrometer to demonstrate the liquid crystalline nature of water-dispersed phospholipids [59]. Chapman developed the application of many biophysical instrumentations (eg, IR, DSC, ESR, FT-IR, and proton magnetic resonance spectroscopy) to probe membrane dynamics. A major focus was in precisely defining the heretofore vague concept of membrane fluidity [60]. These studies included temperature-, phospholipid-, and cholesterol-dependent changes in phase behavior [61] that decades later led to the contemporary concept of lipid rafts. The various phases he observed were shown to affect the lateral diffusion, rotation [62], and hence activity, of membrane proteins. Chapman also pioneered chemical (catalytic hydrogenation) ways to alter the degree of unsaturation of a membrane bilayer interior and simultaneously monitor fluidity and enzymatic activity.

Harden M. McConnell and Dennis Chapman transformed the notion that living cell membranes were static structures into our present view of fluid and dynamic molecular assemblies. The contributions of McConnell and Chapman have led to many of the ideas incorporated into this book.

9. SUMMARY

Every aspect of membrane structural studies involves parameters that are very small (micron to angstrom) and very fast (microseconds to picoseconds), requiring an array of esoteric biophysical methodologies. Membrane parameters discussed in this chapter included: membrane thickness, bilayer stability, membrane protein-, carbohydrate- and lipid-asymmetry, lipid trans-membrane movement (flip-flop), membrane lateral diffusion, membrane cytoskeleton interaction, lipid melting behavior, membrane fluidity, and the contributions of Harden M. McConnell and Dennis Chapman to membrane dynamics. These basic membrane properties were discovered by countless investigators over approximately seven decades and formed the basis of the Singer-Nicholson Fluid Mosaic model of membrane structure that is currently in vogue.

Chapter 10 will continue the discussion of membrane physical properties including: lipid affinities, lipid phases, lipid–protein interactions (hydrophobic match), and lipid interdigitation.

References

[1] Fricke H. The electrical capacity of suspensions with special reference to blood. J Gen Physiol 1925;9:137–52.

[2] Boys CV. Soap bubbles: their colors and forces which mold them. Dover Publications; 1958 (This is a reprint of the revised 1911 publication by C.V. Boys).

[3] Mueller P, Rudin DO, Tien HT, Wescott WC. Reconstitution of cell membrane structure in vitro and its transformation into an excitable system. Nature 1962;194:979–80.

[4] Tulenko TN, Chen M, Mason PE, Mason RP. Physical effects of cholesterol on arterial smooth muscle membranes: evidence of immiscible cholesterol domains and alterations in bilayer width during atherogenesis. J Lipid Res 1998;39:947–56.
[5] Fernandez-Moran H, Oda T, Blair PV, Green DE. A macromolecular repeating unit of mitochondrial structure and function. Correlated electron microscopic and biochemical studies of isolated mitochondria and submitochondrial particles of beef heart muscle. J Cell Biol 1964;22:63–100.
[6] Sharon N, Lis H. Lectins. 2nd ed. Dordrecht, The Netherlands: Springer; 2007.
[7] Edwards R. Poison-tipumbrella assassination of Georgi Markov reinvestigated. The Telegraph; June 19, 2008.
[8] Op den Kamp JAF. Lipid asymmetry in membranes. Ann Rev Biochem 1979;48:47–71.
[9] Devaux PF. Static and dynamic lipid asymmetry in cell membranes. Biochemistry 1991;30:1163–73.
[10] Bretscher MSJ. Phosphatidyl-ethanolamine: differential labeling in intact cells and cell ghosts of human erythrocytes by a membrane-impermeable reagent. Mol Biol 1972;71:523–8.
[11] Steck TL, Ye J, Lang Y. Probing red cell membrane cholesterol movement with cyclodextrin. Biophys J 2002;83:2118–25.
[12] Seigneuret M, Devaux PF. ATP-dependent asymmetric distribution of spin-labeled phospholipids in the erythrocyte membrane: relation to shape changes. Proc Natl Acad Sci USA 1984;81:3751–5.
[13] Devaux PF. Phospholipid flippases. FEBS Lett 1988;234:8–12.
[14] Daleke DL. Phospholipid flippases. J Biol Chem 2007;282:821–5.
[15] Daleke DL. Regulation of transbilayer plasma membrane phospholipid asymmetry. J Lipid Res 2003;44:233–42.
[16] Boon JM, Smith BD. Chemical control of phospholipid distribution across bilayer membranes. Med Res Rev 2002;22:251–81.
[17] Pomorski T, Menon AK. Lipid flippases and their biological functions. Cell Mol Life Sci 2007;63:2908–21.
[18] Frye LD, Edidin M. The rapid intermixing of cell surface antigens after formation of mouse-human heterokaryons. J Cell Sci 1970;7:319–35.
[19] Axelrod D, Koppel DE, Schlessinger J, Elson EL, Webb WW. Mobility measurements by analysis of fluorescence photobleaching recovery kinetics. Biophys J 1976;16:1055–69.
[20] University of California Irvine, Developmental Biology Center; 2006.
[21] Saxton MJ, Jacobson K. Single particle tracking: applications to membrane dynamics. Ann Rev Biophys Biomol Struct 1997;26:373–99.
[22] Abbott A. Cell biology: hopping fences. Nature 2005;433:680–3.
[23] Sheetz MP. Membrane skeletal dynamics: role in modulation of red cell deformability, mobility of transmembrane proteins, and shape. Semin Hematol 1983;20:175–88.
[24] De Brabander M, Geuens G, Nuydens R, Moeremans M, de Mey J. Probing microtubule-dependent intracellular motility with nanometre particle video ultramicroscopy (nanovid ultramicroscopy). Cytobios 1985;43:273–83.
[25] Kusumi A, Sako Y, Yamamoto M. Confined lateral diffusion of membrane receptors as studied by single particle tracking (nanovid microscopy) of calcium-induced differentiation in cultured epithelial cells. Biophys J 1993;65:2021–40.
[26] Wolf DE, Kinsey W, Lennarz W, Edidin M. Changes in the organization of the sea urchin egg plasma membrane upon fertilization: indications from the lateral diffusion rates of lipid-soluble fluorescent dyes. Dev Biol 1981;81:133–8.
[27] Thompson TE, Huang C. In: Andreoli TE, Hoffman JF, Fanestil DD, editors. Physiology of membrane disorders. New York: Plenum Press; 1978 [chapter 2].
[28] Van der Steen AT, Taraschi TF, Voorhout WF, De Kruijff B. Barrier properties of glycophorin-phospholipid systems prepared by different methods. Biochim Biophys Acta 1983;733:51–64.
[29] Greenhut SF, Roseman MA. Cytochrome b5 induced flip-flop of phospholipids in sonicated vesicles. Biochemistry 1985;24:1252–60.
[30] Just W. Trans bilayer (flip-flop) lipid motion and lipid scrambling in membranes. FEBS Lett 2010;584:1779–86.
[31] Wirtz KWA, Zilversmit DB. Exchange of phospholipids between liver mitochondria and microsomes in vitro. J Biol Chem 1968;243:3596–602.
[32] Wirtz KWA. Phospholipid transfer proteins. Ann Rev Biochem 1991;60:73–99.
[33] Hamilton JA. Transport of fatty acids across membranes by the diffusion mechanism. Prostag Leuko Essen Fat Acids 1999;60:291–7.

[34] Homan R, Pownal HJ. Transbilayer diffusion of phospholipids: dependence on headgroup structure and acyl chain length. Biochim Biophys Acta 1988;938:155–66.
[35] Middelkoop E, Lubin BH, Op den Kamp JA, Roelofsen B. Flip-flop rates of individual molecular species of phosphatidylcholine in the human red cell membrane. Biochim Biophys Acta 1986;855:421–4.
[36] Armstrong VT, Brzustowicz MR, Wassall SR, Jenski LJ, Stillwell W. Docosahexaenoic acid and phospholipid flip-flop. Arch Biochem Biophys 2003;414:74–82.
[37] Mabrey S, Sturtevant JM. Investigation of phase transitions of lipids and lipid mixtures by high sensitivity differential scanning calorimetry. Proc Natl Acad Sci USA 1976;73:3862–6.
[38] Sturtevant JM. Some applications of calorimetry in biochemistry and biology. Ann Rev Biophys Bioeng 1974;3:35–51.
[39] Small DM. In: Handbook of lipid research. The physical chemistry of lipids. From Alkanes to phospholipids. New York: Plenum Press; 1986. p. 33–6.
[40] Lewis RN, Mannock DA, McElhaney RN. Differential scanning calorimetry in the study of lipid phase transitions in model and biological membranes: practical considerations. Methods Mol Biol 2007;400:171–95.
[41] Sugar IP. Cooperativity and classification of phase transitions. Application to one- and two-component phospholipid membranes. J Phys Chem 1987;91:95–101.
[42] Heimburg T. Thermal biophysics of membranes tutorials in biophysics. Wiley-VCH; 2007.
[43] Griffiths PR, De Haseth JA. Fourier-transform infrared spectroscopy. In: A series of monographs on analytical chemistry and its applications. 2nd ed. Wiley-Interscience; 2007.
[44] Newport Corporation. Introduction to FT-IR spectroscopy. 2011 [Irving, CA], www.newport.com.
[45] Mantsch HH, McElhaney RN. Phospholipid phase transitions in model and biological membranes as studied by infrared spectroscopy. Chem Phys Lipids 1991;57:213–26.
[46] Pastiorius AMA. Biochemical applications of FT-IR spectroscopy. Spectrosc Eur 1995;7:9–15.
[47] Stokes GG. On the change of refrangibility of light. Philos Trans R Soc Lond 1852;142:463–562.
[48] Lakowicz JR. Principles of fluorescence spectroscopy. New York: Plenum Press; 1983.
[49] Lakowicz JR. Principles of fluorescence spectroscopy. 2nd ed. New York, NY: Kluwer Academic/Plenum Publishers; 1999.
[50] Albani JR. Fluorescence polarization. In: Principles and applications of fluorescence spectroscopy. Wiley Online Library; 2008 [chapter 11].
[51] Papahadjopoulos D, Jacobson K, Nir S, Isac I. Phase transitions in phospholipid vesicles fluorescence polarization and permeability measurements concerning the effect of temperature and cholesterol. Biochim Biophys Acta 1973;311:330–48.
[52] Lentz B. Membrane 'fluidity' as detected by diphenylhexatriene probes. Chem Phys Lipids 1989;50:171–90.
[53] Kates M, Kuksis A, editors. Membrane fluidity: biophysical techniques and cellular regulation. Clifton, New Jersey: The Humana Press, Inc; 1982.
[54] Anderson HC. Probes of membrane structure. Ann Rev Biochem 1978;47:359–83.
[55] Collins JM, Dominey RN, Grogan WM. Shape of the fluidity gradient in the plasma membrane of living HeLa cells. J Lipid Res 1990;31:261–70.
[56] McConnell HM, Hubbell WL. Molecular motion in spin-labeled phospholipids and membrane. J Am Chem Soc 1971;93:314–26.
[57] Schimschick EJ, McConnell HM. Lateral phase separation in phospholipid membranes. Biochemistry 1973;12:2351–3236.
[58] Quinn PJ. Dennis Chapman: oiling the path to biomembrane structure. In: Semenza XG, editor. Stories of success – personal recollections. Comprehensive biochemistry, vol. 45. Netherlands: Elsevier; 2007 [chapter 4].
[59] Byrne P, Chapman D. Liquid crystalline nature of phospholipids. Nature 1964;202:987–8.
[60] Chapman D. Phase transitions and fluidity characteristics of lipids and cell membranes. Quart Rev Biophys 1975;8:185–225.
[61] Chapman D, Penkett SA. Nuclear magnetic resonance spectroscopic studies of the interaction of phospholipids with cholesterol. Nature 1966;211:1304–5.
[62] Razi Naqvi K, Gonzalez-Rodriguez J, Cherry RJ, Chapman D. Spectroscopic technique for studying protein rotation in membranes. Nat New Biol 1973;245:249–51.

CHAPTER

10

Lipid Membrane Properties

OUTLINE

An Introduction to Biological Membranes
http://dx.doi.org/10.1016/B978-0-444-63772-7.00010-5

The previous chapter (Chapter 9) investigated some of the major principles of the "Fluid Mosaic" model. While these principles form the core of understanding basic membrane structure, there are many other, more poorly understood membrane properties and concepts that must be considered. Membranes are just very complex. Several of these properties will be discussed in Chapters 10 and 11.

1. COMPLEX LIPID INTERACTIONS

1.1 Lipid Affinities

A basic hallmark of all biological membranes is their extreme heterogeneity. This topic will come up over and over again in discussions of lipid–lipid and lipid–protein interactions. Since a single biological membrane is composed of hundreds or even thousands of different lipid molecular species, it is logical to assume that they will exhibit different affinities for one another. Each lipid may prefer to be in the company of one type of lipid while avoiding another type. One well studied example of this is the lipid raft. Lipid rafts (Chapter 8) are enriched in sphingolipids and cholesterol, the components responsible for the raft's basic physical properties. In fact, cholesterol is believed to be the molecular "glue" that holds rafts together. Rafts are destroyed upon the addition of cyclodextrin (Fig. 10.1), a family of cyclic glucose oligosaccharides that are often used to extract cholesterol from membranes [1].

Through the years, cholesterol (Chapter 5) has probably received more attention than any other membrane lipid. Cholesterol was first identified as a solid in gallstones by François Poulletier de la Salle in 1769. However, the man most associated with early cholesterol studies was the French chemist, Eugene Chevreul (Fig. 10.2). Chevreul was an early pioneer of lipid components in animal fat. He was the first to isolate and name stearic and oleic acid, and also, in 1815 coined the term "cholesterine" from the Greek word *chole* for bile. Chevreul is also credited with creating margarine and, due to his longevity (102 years), was a pioneer in the field of gerontology. Therefore he was both an investigator and a subject of his own investigations!

Most of the attention given to cholesterol is due to its deleterious role in human atherosclerosis (heart disease) [2]. Just the word cholesterol has become synonymous with bad things. As if cholesterol has not been maligned enough, a recent report has even linked this unfortunate sterol to tail-chasing in dogs [3]!

But cholesterol is not all evil. It does play an essential role in some, but not all, membranes. As discussed in Chapter 5, cholesterol is by far the major polar lipid in mammalian plasma membranes, comprising up to ~60 mol% of the polar lipids, yet it is almost totally absent in other mammalian intracellular membranes (eg, the mitochondrial inner membrane). It is also hard to understand how cholesterol can play such an essential role in life processes in mammalian cells yet be essentially missing in fungi, plants, and bacteria (Chapter 5). Simplistically, cholesterol is a "membrane homogenizer." It intercalates between membrane phospholipids and sphingolipids and helps to control membrane lipid packing. Lipid packing is an essential feature in membrane "fluidity" (Chapter 9) and permeability (Chapter 19). In a tightly packed gel state, membrane insertion of cholesterol decreases lipid packing, thus increasing membrane fluidity and permeability. In these membranes cholesterol behaves as a lipid-soluble contaminant. In contrast, in a liquid crystalline or fluid state, membrane cholesterol increases lipid packing thus decreasing fluidity and permeability. In these membranes cholesterol prevents the formation of *gauche* kinks that are responsible for acyl chain melting (see Chapter 9).

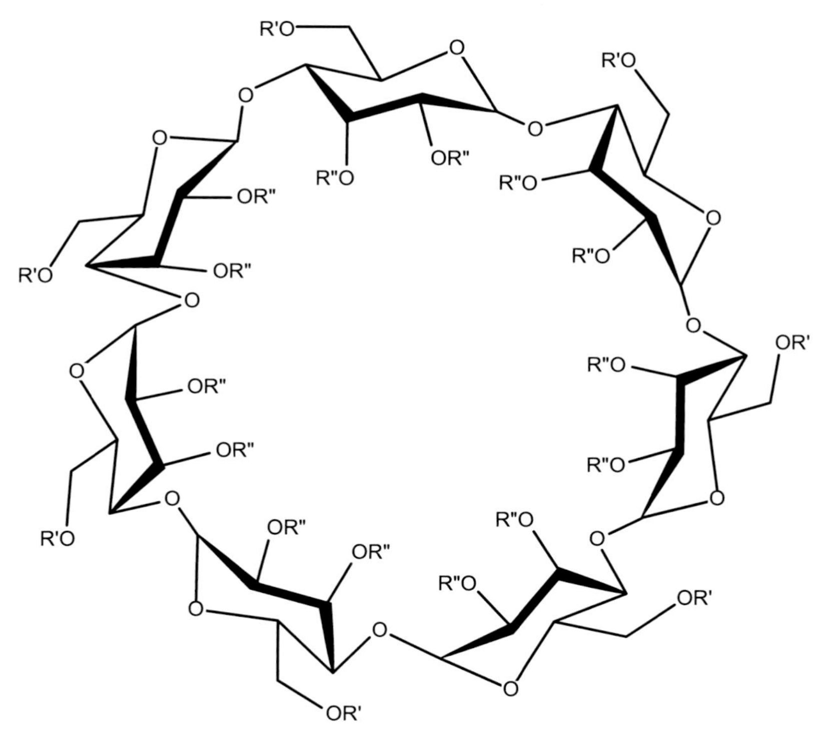

FIGURE 10.1 α-Cyclodextrin.

FIGURE 10.2 Eugene Chevreul, 1786–1889.

Mixing cholesterol and phospholipids often produces an unusual fluid membrane state called the liquid ordered (l_o) state [4–6]. The l_o state behaves as if it were half way between the gel (L_β) and liquid crystalline (L_α) states, having properties of both. In the l_o state, the acyl chains are extended (they have fewer *gauche* kinks) and so in this sense behave like a gel state. However, lateral diffusion in the l_o state is almost as great as in the liquid crystalline state. Since acyl chains are extended by cholesterol, membrane domains that are rich in cholesterol tend to be thicker than regions that are devoid of the sterol. Because lipid rafts are highly enriched in cholesterol, they are thicker than the surrounding nonraft membrane.

1.2 A Differential Scanning Calorimetry Study

In Fig. 9.23 it was shown that cholesterol slightly lowers the main melting transition (T_m) and substantially broadens the transition and diminishes the ΔH of dipalmitoyl phosphatidylcholine (DPPC) in lipid bilayers. More than a decade before the advent of lipid rafts, van Dijck et al. [7] and Demel et al. [8] realized they could use the obliteration of T_ms to monitor relative affinities of cholesterol for various phospholipids. van Dijck's 1976

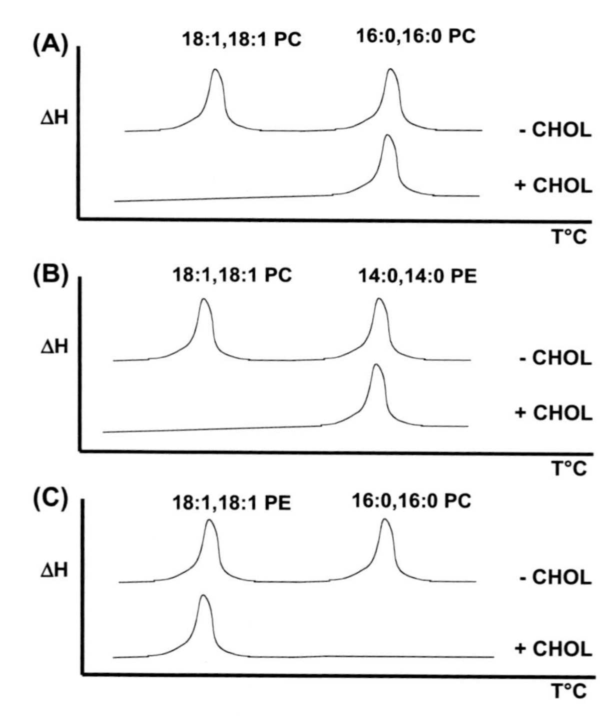

FIGURE 10.3 Relative affinity of cholesterol for PC and PE as detected by DSC. In Panel (A) cholesterol is shown to preferentially obliterate the lower melting transition (18:1,18:1 PC > 16:0,16:0 PC) when both transitions are for the same phospholipid type (PC in this example). In Panel (B) cholesterol is shown to preferentially obliterate the lower melting 18:1,18:1 PC over the higher melting 14:0,14:0 PE. In Panel (C), where the T_ms of PC and PE are reversed, cholesterol preferentially obliterates 16:0,16:0 PC even though its T_m is higher than that of 18:1,18:1 PE. This experiment was interpreted as evidence that cholesterol associates more strongly with PCs than with PEs. *Experiment is redrawn from van Dijck PWM, De Kruijff B, van Deenen LLM, de Gier J, Demel RA. The preference of cholesterol for phosphatidylcholine in mixed phosphatidylcholine-phosphatidylethanolamine. Biochim Biophys Acta 1976;455:576–87.*

experiment, outlined in Fig. 10.3, compared phosphatidylcholine (PC) to phosphatidylethanolamine (PE). The experiment was based on having two phospholipids with T_ms sufficiently far apart that their transitions are distinct and do not overlap (they are monotectic). If both the high melting and low melting phospholipids were of the same class (ie, they were both PC or both PE), low levels of cholesterol first reduced the lower melting transition. This is depicted in Fig. 10.3, panel A for 18:1,18:1 PC (low T_m) and 16:0,16:0 PC (high T_m). Higher levels of cholesterol eventually diminished the higher melting transition as well. Fig. 10.3, panel B compares a low melting PC (18:1,18:1 PC) to a high melting PE (14:0,14:0 PE). Low levels of cholesterol first obliterated the low melting PC transition. This does not indicate whether cholesterol prefers PC over PE or whether cholesterol simply prefers the lower melting transition lipid. In Fig. 10.3, panel C, the transition temperatures of the PC and PE were reversed. In this experiment, PE (18:1,18:1 PE) was the lower melting component while PC (16:0,16:0 PC) was the higher melting. Here cholesterol first obliterated the higher melting, PC component. From these differential scanning calorimetry (DSC) experiments it was concluded that cholesterol associates better with PC than PE. Later Demel et al. [8] extended these studies to compare several of the most common membrane lipids and reported that cholesterol has the following affinity:

$$\mathrm{SM} > \mathrm{PS},\ \ \mathrm{PG} > \mathrm{PC} > \mathrm{PE}$$

This DSC study was an early indication that cholesterol associates more strongly with sphingomyelin (SM) than with other membrane lipids. Two decades later, Simons proposed the existence of lipid rafts, cell signaling membrane domains that are highly enriched in SM and cholesterol (Chapter 8).

1.3 An X-ray Diffraction Study

Compatibility with cholesterol should depend not only on the type of phospholipid headgroup but also on the nature of the acyl chains. The structure of cholesterol with its four inflexible rings (Chapter 5) predicts that the sterol may not be compatible with highly flexible, polyunsaturated chains. One method that has been used to test this premise involves measuring cholesterol solubility in membranes by X-ray diffraction (XRD) [9,10]. In Chapter 9 it was noted that a different application of this technique (an electron density profile) has been used to measure membrane thickness. Cholesterol's solubility in membranes can be determined by measuring radial intensity profiles plotted against reciprocal space (I-q plots) that show distinct peaks indicating how much cholesterol has been excluded from the membrane in the form of cholesterol monohydrate crystals. In the experiments outlined in Figs. 10.4 and 10.5 and summarized in Table 10.1, Martin Caffrey et al. made lipid bilayer vesicles from different PC and PE molecular species with increasing mol fractions of cholesterol. At low levels, cholesterol was accommodated into the bilayer structure and no second order scattering peaks due to excluded cholesterol monohydrate crystals could be detected. However, at some critical concentration, cholesterol exceeds the carrying capacity of the bilayer and is excluded. The excluded monohydrate crystals are observed in certain regions of the I-q plots (Fig. 10.4). The integrated intensities of the scattering peaks 002 (0.3701 Å^{-1}), 020 (1.033 Å^{-1}), and 200 (1.044 Å^{-1}) were combined and plotted against the mol% cholesterol (Fig. 10.5). Linear extrapolation of these plots to zero gives an accurate estimate ($\pm$1.0 mol%) of cholesterol solubility limit in bilayers made from various phospholipids.

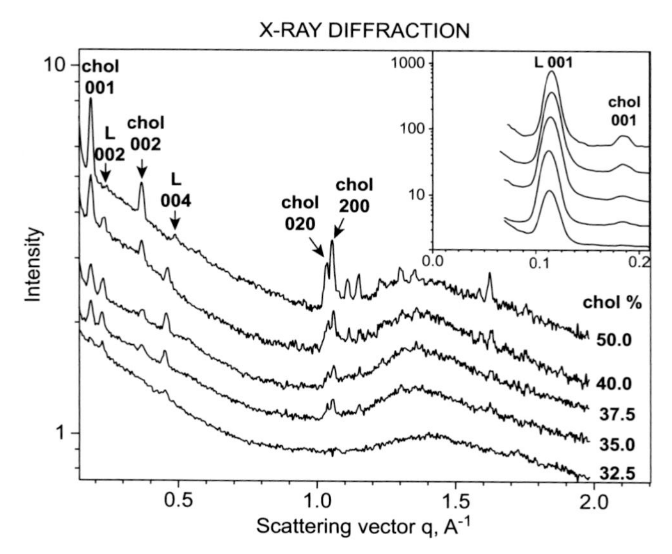

FIGURE 10.4 I-q plots for 16:0,22:6 PE membranes showing the intensities of cholesterol monohydrate peaks with added cholesterol. The integrated intensities of the scattering peaks 002 (0.3701 $Å^{-1}$), 020 (1.033 $Å^{-1}$) and 200 (1.044 $Å^{-1}$) were combined and plotted against the mol% cholesterol in Fig. 10.5 [9,10].

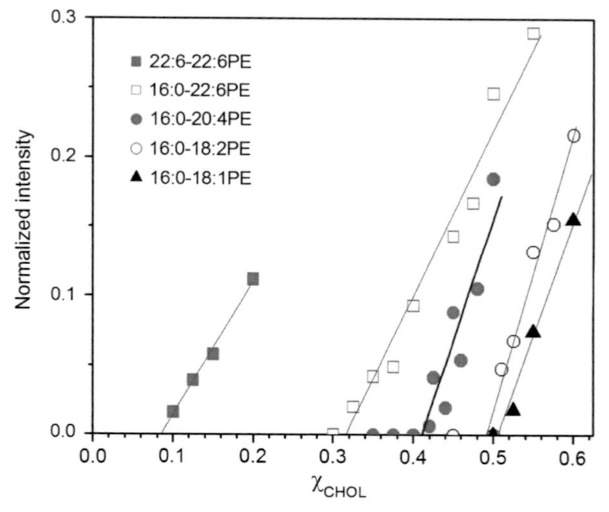

FIGURE 10.5 Plot of the normalized scattering peaks for cholesterol monohydrate crystals excluded from the membrane vs the mol fraction of cholesterol added to the membrane. The sum of the normalized cholesterol monohydrate peaks are extrapolated to 0, representing the membrane carrying capacity for cholesterol. The extrapolated values are reported in Table 10.1 [9].

TABLE 10.1 Limiting Solubility of Cholesterol in Various Phosphatidylcholine (PC) and Phosphatidylethanolamine (PE) Lipid Membranes. Values Were Obtained From X-ray Diffraction (Fig. 10.4) and I-q Plots (Fig. 10.5) [9]

Phosphatidylethanolamine	Cholesterol solubility (mol%)	Phosphatidylcholine	Cholesterol solubility (mol%)
16:0-18:1	51 ± 3	16:0-18:1	65 ± 3
16:0-18:2	49 ± 2		
16:0-20:4	41 ± 3	18:0-20:4	49 ± 1
16:0-22:6	31 ± 3	18:0-22:6	55 ± 3
22:6-22:6	8.5 ± 1	22:6-22:6	11 ± 3

The results are compiled in Table 10.1. Table 10.1 confirms that PC bilayers can accommodate more cholesterol than PE bilayers with identical acyl chain composition. Also, within a family of either PCs or PEs, cholesterol is more excluded from bilayers as a function of increasing acyl chain double bond content. Of particular interest is the extremely low solubility of cholesterol in PC or PE membranes containing two chains of docosahexaenoic acid (DHA, $22{:}6^{\Delta 4,7,10,13,16,19}$), the most unsaturated fatty acid commonly found in membranes. This experiment confirms the idea that the more flexible the acyl chain, the less accommodating it is for cholesterol.

1.4 Lipid Raft Detergent Extractions

In its initial form, the Fluid Mosaic model (Singer and Nicolson, 1972 [11]) did not recognize lipid heterogeneities. However, it soon became obvious that lipid patchiness must exist. By 1974, a series of biophysical studies started to appear supporting the basic concept of lipid microdomains in membranes [12]. Of particular importance were cholesterol- and sphingolipid-enriched domains studied in model bilayer membranes in the late 1970s by Biltonen and Thompson [13]. The lipid microdomain concept was formalized in a classic 1982 paper by Karnovsky et al. [12]. Therefore, cholesterol/sphingolipid microdomains predated the concept of lipid rafts by more than two decades.

The preraft experiments strongly implied that there was a preferential affinity of cholesterol for sphingolipids. This observation on model membranes was given a biological link with the discovery of lipid rafts in biological membranes (Chapter 8) [14,15]. Rafts can be extracted as an insoluble fraction at 4°C with nonionic detergents such as Triton X-100 or Brij-98. Isolated rafts contain about twice the amount of cholesterol and also are enriched in sphingolipids by about 50% when compared to the surrounding plasma membrane bilayer. Cholesterol is proposed to be the "glue" that holds rafts together since rafts fall apart when cholesterol is extracted by cyclodextrin. Also, cholesterol is responsible for rafts existing in the tightly packed, l_o state and being thicker than the surrounding nonraft membrane. A characteristic of most, but apparently not all raft lipids, is their having long mostly saturated acyl chains.

Fig. 10.6 shows molecular models of SM and cholesterol. In this representation, the inverted cone shape of SM fits nicely next to the cone-shaped cholesterol, enhancing affinity

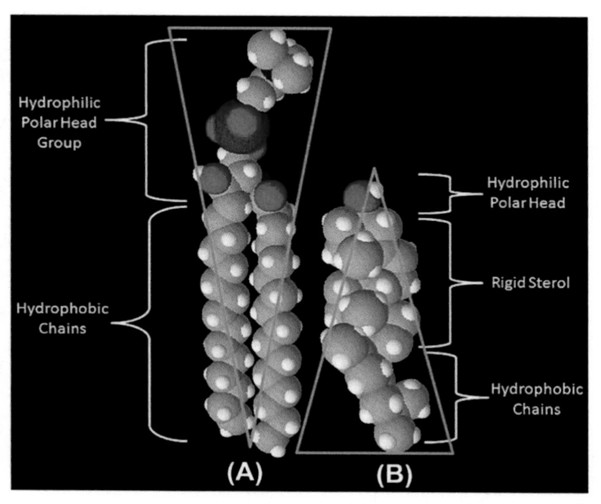

FIGURE 10.6 Space filling models of two major components of lipid rafts, SM (A) and cholesterol (B). Note, in this depiction SM is drawn as a *slight inverted cone*, while cholesterol is *cone-shaped*. The two molecules fit together, thus displaying "complementarity."

between these two lipids. Fitting lipids of different shapes together, like pieces of a jigsaw puzzle is known as "complementarity." However, it is not certain whether raft-insolubility in cold nonionic detergent results from extraction of intact rafts or is the result of similar, coincidental insolubility of sphingolipids and cholesterol in the detergent solutions. Nondetergent methods for raft extractions from membranes have now been developed [16] and are discussed in Chapter 13. In general, these newer methods support the basic conclusions derived from the older detergent methods. Sphingolipids and cholesterol do appear to have an affinity, albeit imperfectly, for one another.

An interesting medical problem related to raft structure has recently come to light [17]. Heavy, long term use of cholesterol-lowering statin drugs, primarily in the elderly, has produced a result similar to that of cyclodextrin on lipid raft stability. Statin-induced hypocholesterolemia has resulted in the functional failure of cholesterol-rich lipid rafts in processes involving exocytosis and endocytosis. For example, a five year trial showed a 30% increase in the incidence of diabetes associated with a cholesterol reduction therapy. Long-term statin use has also been linked through lipid raft failure to bone loss and fractures. These results surprisingly linking heavy statin use, cholesterol reduction by cyclodextrin, and lipid raft function in some important human maladies may represent a new paradigm for medical interventions.

2. NONLAMELLAR PHASES

When picturing a biological membrane, the first thing that comes to mind is the simple lipid bilayer, but we have seen how deceptive this can be. Even the lipid bilayer (actually the lamellar phase) can exist in a variety of very different states. In prior sections we have used the terms; bilayer, lamellar, gel, solid, S_o, L_β, $L_{\beta'}$, ripple, $P_{\beta'}$, fluid, liquid crystalline, L_α, liquid disordered, l_d, liquid ordered, and l_o to describe the "simple" lipid bilayer. But there is even more, far more, to structures of polar lipids dispersed in water. Depending on the polar lipid and conditions of the aqueous solvent (eg, temperature, water content, ionic strength, polyvalent cations, pH, pressure etc.), lipid dispersions can exist in a wide variety of additional forms.

Multiple long-range structures that amphipathic (polar) lipids can take when dispersed in water are referred to as "lipid polymorphism" [18–21]. Two of the major questions about membrane lipids concern "lipid diversity" and lipid polymorphism. Lipid diversity addresses the question why are there so many different lipids in membranes. Lipid polymorphism addresses the question why do amphipathic lipids form so many different long-range structures when dispersed in water.

While there are dozens of highly unusual long-range structures that amphipathic lipids can assume, only four are believed to be stable at biologically high water levels and only one of these, the lamellar state, predominantly, perhaps exclusively, exists in a living membrane. The four lipid phases that are stable at high water concentrations are [22]:

1. lamellar
2. inverted hexagonal (H_{II})
3. cubic Q^{224}
4. cubic Q^{227}

What phase an amphipathic lipid prefers is dependent on the lipid's molecular shape. Membrane lipids can be roughly divided into the three basic shapes shown in Fig. 10.7: (A) truncated cone; (B) cylinder; and (C) inverted cone [23]. Cylindrical lipids prefer the lamellar phase while lipids with very truncated or inverted cone shapes in pure form, prefer being in other, nonlamellar structures. All biological membranes are composed of hundreds or thousands of different lipids of all three basic shapes, yet membranes are found almost entirely in lamellar phase. This is particularly perplexing since most of the world's membrane lipids are not cylindrical. For example, the most prevalent membrane lipid found on planet Earth is the photosynthesis-related lipid monogalactosyldiacylglycerol (Fig. 5.17), which prefers H_{II} phase. This suggests that cylindrical lipids have a stronger influence on overall membrane structure than do the other noncylindrical lipids. Indeed, only 20–50% of the total lipids need be cylindrical to maintain the normal lamellar phase.

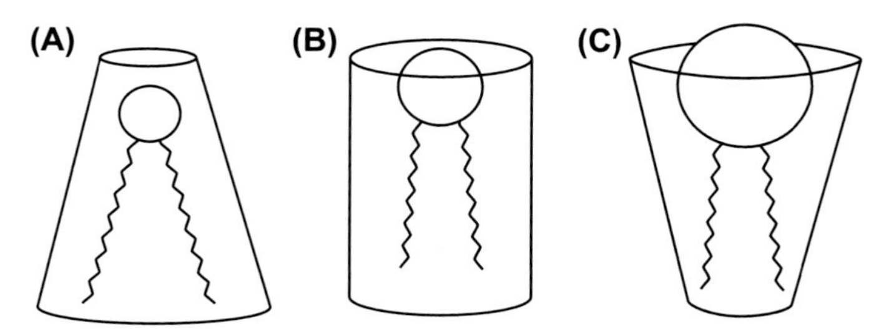

FIGURE 10.7 Basic membrane lipid shapes: (A) truncated cone; (B) cylinder; and (C) inverted cone. *Circles* represent the polar headgroup of different sizes [23].

A general description of lipid shape is not entirely satisfying as lipids actually exist at all locations on the continuum from truncated cone through cylinder to inverted cone [24]. Therefore attempts have been made to quantify lipid shape. Two of these approaches follow.

The first approach, proposed by Pieter Cullis in 1979 [19], defines lipid shape by a dimensionless parameter, S:

$$S = v/a_o l_c$$

where a_o is an optimum cross-sectional area per molecule at the lipid–water interface, v is the volume per molecule, and l_c is the length of the fully extended acyl chain. Since these parameters are not easy to obtain, a simplified version was suggested where a_o is the cross-sectional area of the head group at the aqueous interface and a_h is the cross-sectional area at the bottom of the acyl chains. Therefore if:

$a_o/a_h > 1$ then $S > 1 \rightarrow$ inverted cone
$a_o/a_h = 1$ then $S = 1 \rightarrow$ cylindrical
$a_o/a_h < 1$ then $S < 1 \rightarrow$ cone

Unfortunately, S is not a fixed parameter of the lipid's shape as it may vary with the membrane's environment (ie, bathing solution pH, ionic strength, atmospheric pressure, lateral pressure, temperature etc.).

A second approach, the equilibrium curvature (R_o) concept, was later proposed by Sol Gruner [25]. R_o is a quantitative measure of the propensity of a lipid to form nonlamellar structure. It is an estimate of the tendency of a monolayer made from a particular lipid to curl, thus resulting in hydrocarbon packing strain. R_o is normally derived experimentally from XRD of lipid films. A large negative or positive value ($>\pm 100$ Å) suggests that there is little inherent curvature in the lipid (ie, the lipid prefers a cylindrical shape). A lower absolute value ($<\pm 100$ Å) suggests that the lipid will add a curvature stress (negative or positive) and thus has a cone or inverted cone shape. Table 10.2 lists R_os of a few common membrane lipids [23]. The cylindrical lipids, 1,2-dioleoyl-*sn*-glycero-3-phospho-L-serine (DOPS) and 1,2-dioleoyl-*sn*-glycero-phosphocholine (DOPC), have $R_o\text{s} > 100$ and so form lamellar structure. Lyso PC has a small positive R_o indicating it is an inverted cone and so prefers micellar phase. 1,2-dioleoyl-*sn*-glycero-phosphethanolamine (DOPE), cholesterol, α-tocopherol, and dioleoylglycerol (DOG) all have low negative R_os indicating they are cone-shaped and so prefer a nonlamellar phase like H_{II}.

2.1 Inverted Hexagonal Phase

The most studied of the nonlamellar phases is the H_{II} phase [18,20,26]. H_{II} phase (Fig. 10.8) consists of six long, parallel cylindrical tubes surrounding a seventh cylinder of indefinite length that are distributed into a hexagonal pattern, thus giving rise to the designation H_{II} phase. Water fills the cylinders and hydrocarbon chains fill the voids between the hexagonally packed cylinders, holding the entire H_{II} structure together. The molecular shape of H_{II}-preferring lipids (wide base, narrow head) gives negative curvature strain to adjacent lipids or proteins.

A full H_{II} phase would require a substantial number of cone-shaped lipids to be recruited to one location in a biological membrane. This seems highly unlikely and many unsuccessful attempts have been made to clearly identify H_{II} structure in complex biological membranes.

TABLE 10.2 The Spontaneous Radius of Curvature (R_o) for Several Important Membrane Lipids

Lipid	Spontaneous radius of curvature (Å)
DOPS	+150
Lyso PC	+38 to +68
DOPE	−28.5
DOPC	−80 to −200
Cholesterol	−22.8
α-Tocopherol	−13.7
DOG	−11.5

DOG, dioleoylglycerol; *DOPC*, 1,2-dioleoyl-*sn*-glycero-phosphocholine; *DOPE*, 1,2-dioleoyl-*sn*-glycero-phosphethanolamine; *DOPS*, 1,2-dioleoyl-*sn*-glycero-3-phospho-L-serine; *Lyso PC*, Lyso phosphatidylcholine.
The table is from Atkinson J, Epand RF, Epand RM. Tocopherols and tocotrienols in membranes: a critical review. Free Radical Biol Med 2008;44:739–64.

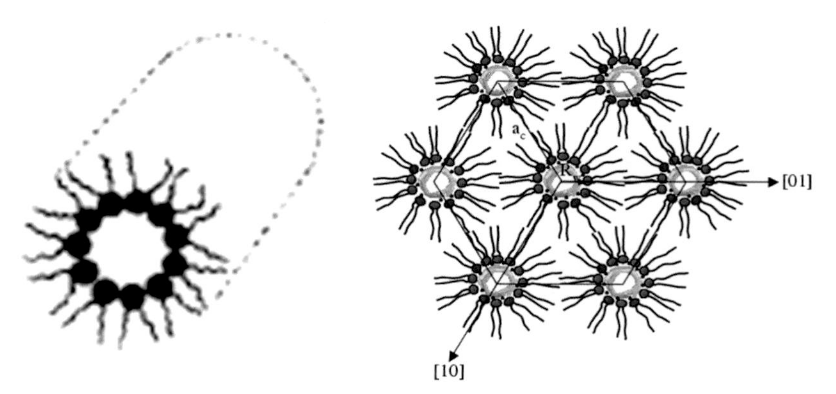

FIGURE 10.8 Left: One cylinder of H_{II} phase. Water fills the channel that is lined by the lipid polar head groups. Right: An array of six cylinders comprising the full H_{II} phase.

But this does not necessarily mean that H_{II} phase is totally absent. Perhaps, characteristic signals from the limited H_{II} phase are lost due to the "population weighted average problem." Even though it is unlikely that any significant level of H_{II} actually exists in biological membranes, this nonlamellar phase has received, and continues to receive considerable attention in the world of membrane biophysics. Primarily through model membrane studies, H_{II} phase-preferring lipids have been strongly linked to fusion and other membrane contact phenomena [27]. It was proposed that during fusion, the lipids of two stable bilayers must intermingle through some kind of unstable intermediate. This process is known as hemi-fusion and is well-documented in model membranes. In 1986, David Siegel [28] suggested that an inverted micelle structure, resembling one cylinder of H_{II} structure, represents the primary

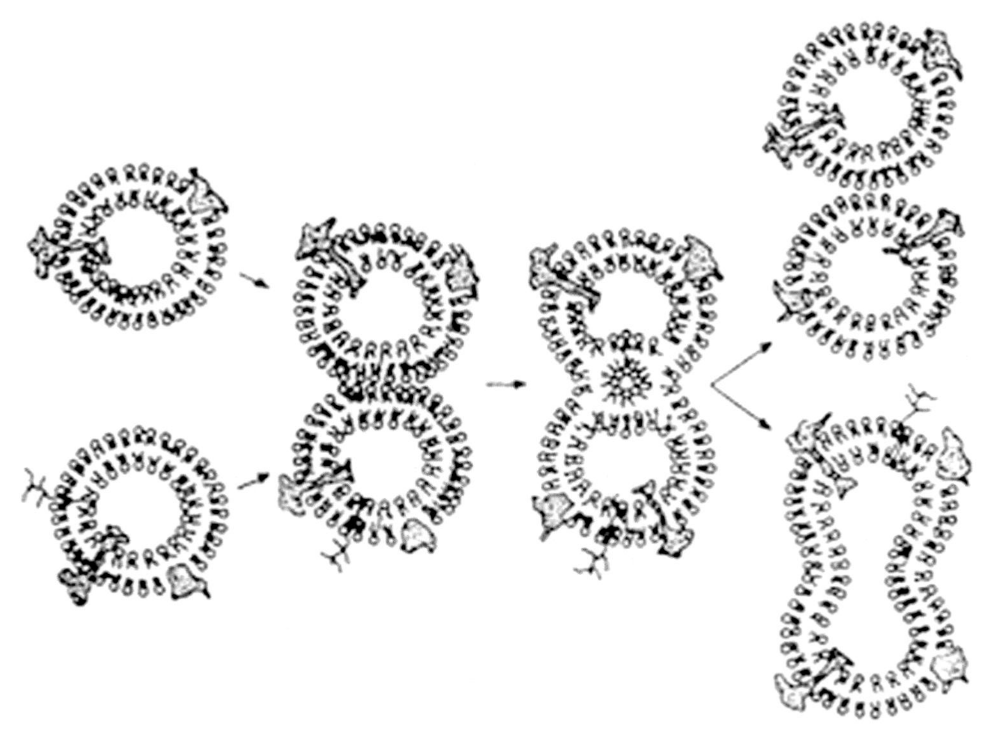

FIGURE 10.9 Diagram of the mechanism of lipid-driven membrane fusion including one central H_{II}-like cylinder [20,27].

point defect in lamellar to H_{II} transition (Fig. 10.9) [20,27]. Importantly, factors known to promote H_{II} phase also promote membrane fusion and lipid-soluble fusogens that induce cell—cell fusion in vitro, induce H_{II} formation in model membranes.

Assigning a role in membrane fusion for a nonlamellar phase was compelling since it could start to explain some aspects of the lipid polymorphism conundrum. Unfortunately, sometimes even beautiful stories disintegrate when confronted with harsh reality. Many vital processes in cells require rapid and precise control of fusion. Biological fusion occurs on the order of milliseconds and must be turned on and off in an orderly fashion. Model lipid vesicle fusion is orders of magnitude too slow to start, and once begun has no mechanism to end. In addition it is hard to envision how H_{II}-favoring lipids can overcome the large negative charge repulsion on fusing-cell surfaces. Speed and control are two problems the H_{II}-phase lipid fusion theory could not answer. In 1993, James Rothman (Fig. 10.10) and colleagues proposed a protein-based theory of membrane fusion called the SNARE hypothesis [29,30]. In the years since its proposal, this hypothesis has successfully addressed many of the complex problems associated with fusion. In a 2009 paper in *Nature* [31], Rothman reviewed the central aspects of how the proteins SNARE and Sec1/Munc18-like, in concert with Ca^{2+}, organize lipid microdomains in fusing membranes. H_{II} phase was not even mentioned!

FIGURE 10.10 James Rothman, 1950- .

2.2 Cubic Phase

Another nonlamellar phase that is stable in excess (biological) levels of water include the cubic phases, Q^{224} and Q^{227} [21,25]. Like the H_{II} phase, the cubic phase is at most a very rare component of biological membranes and remains a phase in search of a function. Q^{227} has been proposed to repair the damage of a lipolytic agent at a local spot by forming a watertight patch. Like H_{II}, cubic phase has also been proposed to be involved in fusion. However, the role of cubic phase in fusion is at best uncertain as this phase has been far less investigated than H_{II}. The cubic phase is a lipid bilayer that is convoluted into a complex bicontinuous surface. Although the role of cubic phase in biological membranes is unknown, it has found a home in a totally different arena. Q^{224} (normally composed of monoolein) has been successfully employed to nucleate and grow membrane protein crystals for X-ray crystallography [32]. The protein partitions into the cubic membrane phase while the detergent used in the protein's isolation (Chapter 13) is segregated into the aqueous phase. There is great hope that this method may provide a new paradigm for crystallizing difficult membrane proteins.

2.3 ^{31}P-Nuclear Magnetic Resonance

Perhaps the most employed and easiest to interpret technique that can distinguish membrane phase is ^{31}P-nuclear magnetic resonance (NMR). Although XRD is the classic technique used to define the parameters of model membranes, data interpretation is difficult and its application to biological membranes limited. ^{31}P-NMR is sensitive to phospholipid head group orientation and its spectra can clearly distinguish lamellar, H_{II}, and micellar phases (Fig. 10.11) [19]. Lamellar phase spectra is distinguished by a broad spectra (40–50 PPM)

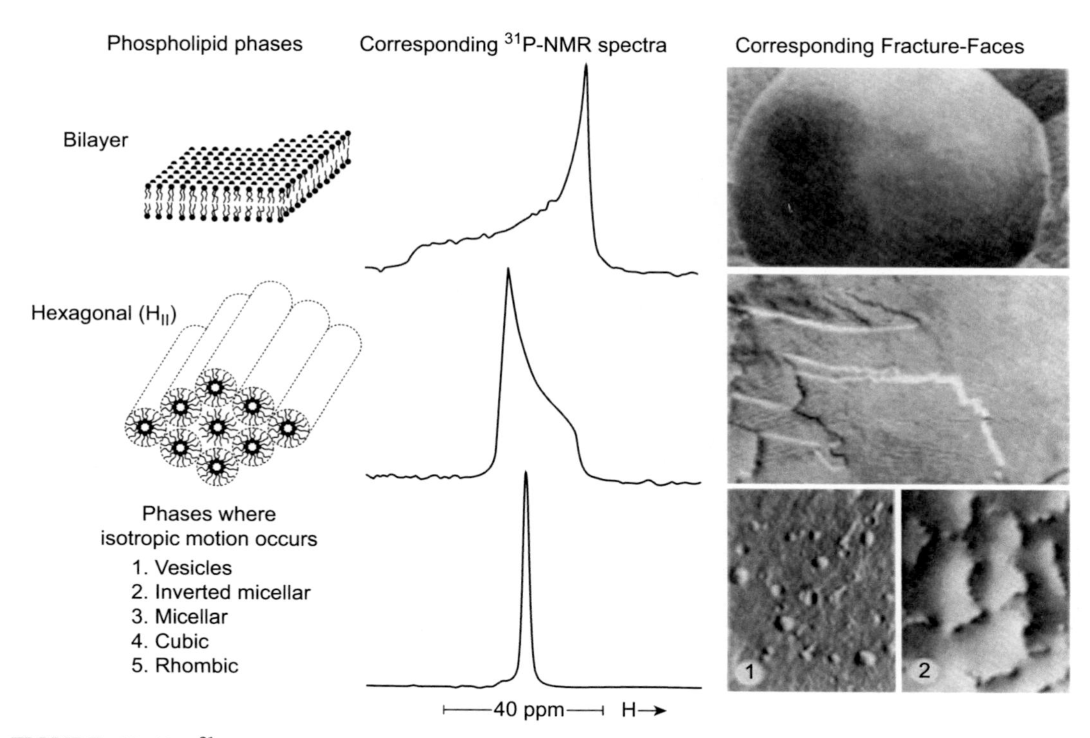

FIGURE 10.11 ^{31}P NMR-spectra of Lamellar (bilayer) phase, Inverted Hexagonal (H_{II}) phase and other (isotropic) phase. Left Column, membrane structure; Middle Column, ^{31}P-NMR spectra; Right Column, freeze fracture electron micrograph [19].

with a low-field shoulder and a high-field peak. In total contrast, the H_{II} phase spectra is narrow (~20 PPM, half the width of the lamellar phase spectra) and features a low-field peak and high field shoulder. A third type of spectra is a narrow featureless peak that may indicate the presence of micellar structure, small rapidly tumbling vesicles or even Cubic or Rhombic phase.

Unlike XRD, ^{31}P-NMR can be used to follow lamellar → H_{II} transitions. In one example, various mixtures of egg PC and soy PE were made into large unilamellar vesicles (methods discussed in Chapter 13). The vesicles were then subjected to ^{31}P-NMR analysis. The pure egg PC vesicles produced a spectra indicating lamellar phase only (ie, it had a low-field shoulder and high-field peak). The pure soy PE vesicles produced a spectra indicating H_{II} phase only (ie, it had a low-field peak and high-field shoulder). As PC was added to the PE vesicles the H_{II} phase spectra started to diminish. By 15 mol% PC, the major spectral component was a shoulderless peak, indicating a transition state. Increasing levels of PC served to amplify the lamellar spectra. By 50 mol% PC, the spectra indicated only the presence of lamellar phase. Therefore it was concluded that PC has a larger influence on lipid phase than does PE.

So we are still left with the same perplexing question. What is the biological function of lipid polymorphism? After almost four billion years of chemical and biological evolution,

TABLE 10.3 Common Membrane Lipids and Their Preferred Phase

Lipid	Comments	Phase
Phosphatidylcholine	Cylindrical under most conditions	Lamellar
	Extreme dehydration, high temperature	H_{II}
Phosphatidylethanolamine	High unsaturation	H_{II}
	Saturated chains, low temperature, gel state	Lamellar
	Saturated chains, high temperature, fluid state	H_{II}
	Unsaturated chains, pH 9	Lamellar
Sphingomyelin	Cylindrical under most conditions	Lamellar
Phosphatidylserine	pH 7, cylindrical	Lamellar
	pH < 4 (reduce negative charge repulsion)	H_{II}
Phosphatidic acid	pH 7, cylindrical	Lamellar
	pH < 4 (reduce negative charge repulsion)	H_{II}
Cholesterol	pH 7, cylindrical	Lamellar
	pH 7, $+Ca^{2+}$	H_{II}
MGDG	Cone-shaped	H_{II}
DGDG	Cylindrical	Lamellar
Cerebroside	Cylindrical	Lamellar
Ganglioside	Inverted cone	Micelle
Lyso phosphatidylcholine	Inverted cone	Micelle
MIXTURES:		
Unsaturated chains	$\rightarrow + Ca^{2+}$	
	PS-PE	Lamellar $\rightarrow$ H_{II}
	PG-PE	Lamellar $\rightarrow$ H_{II}
	PA-PE	Lamellar $\rightarrow$ H_{II}
	PI-PE	Lamellar $\rightarrow$ H_{II}
	PE-PS-cholesterol (high salt)	H_{II}

H_{II}, inverted hexagonal; *PI*, phosphatidylinositol; *PG*, phosphatidylglycerol; *MGDG*, monogalactosyldiacylglycerol; *DGDG*, digalactosyldiacylglycerol.

Nature would not produce something so profound without a very good reason. We are just at the beginning of addressing this problem. Table 10.3 lists some basic membrane lipids and their preferred phase. It appears that cylindrical lipids prefer the lamellar phase while cone-shaped lipids prefer H_{II} (or nonlamellar) phase, and inverted cone-shaped lipids prefer micellar phase. Even this is complicated by the ability of a particular lipid to have its preferred phase altered upon changes in its environment. For example, at pH 7, phosphatidylserine (PS) and phosphatidic acid (PA) have a cylindrical shape and prefer the lamellar phase. However, if the pH is reduced to less than 4, the head group negative charge is reduced, thus decreasing charge repulsion. This reduces the apparent head group diameter (a_o) without affecting the tail diameter (a_h). When reduced to pH 4, PS and PA switch from a cylindrical (lamellar)-shape to a cone (H_{II})-shape. And this is for a single pure molecular species. Biological membranes are comprised of hundreds to thousands of lipid molecular species that exhibit different and largely unknown affinities for one another. To date, only a very few lipid mixtures have been tested for lipid phase preference and this has been shown to be complicated. For example, mixtures of PS-PE, phosphatidylglycerol (PG)-PE, PA-PE, or phosphatidylinositol (PI)-PE all undergo a lamellar → H_{II} transition upon addition of Ca^{2+}. What all of this means is not known. Perhaps the preference of each molecular species for a particular phase is just a reflection of lipid molecular shape and manifests itself by imposing molecular strain on adjacent proteins. This in turn would affect the protein's conformation and thus activity. This idea is further discussed later in this chapter.

3. LIPID PHASE DIAGRAMS

Membrane lipids come in an overwhelming variety of chemical structures (lipid diversity) and each molecular species, when isolated in pure form, can aggregate into a large number of long-range structures or phases (lipid polymorphism). Environmental conditions (eg, temperature, lateral pressure, extent of hydration, aqueous solution ionic strength, presence of polyvalent metals, pH etc.) can further affect the lipid phase. The situation becomes far more complicated with multicomponent lipid mixtures, as the lipids interact with one another resulting in phase separation into discrete lipid microdomains. And all of this is before proteins are even incorporated into the membrane!

For years lipid biophysicists have attempted to pictorially describe various lipid interactions and phases through what are known as "phase diagrams" [33]. Phase diagrams can take a wide variety of forms that report lipid behavior from different perspectives. As the number of lipid components increase, the phase diagrams become ever more complex. Anyone who is not confused by three component phase diagrams, just does not understand membrane structure. There are now a countless number of phase diagrams that have been compiled [34]. Below are five examples of phase diagrams starting with simple one-component DPPC phase diagrams and progressing through to a complex, three-lipid phase diagram of a lipid raft model.

3.1 Single-Lipid Phase Diagrams

We have already encountered one single-lipid phase diagram in the form of a DSC scan of DPPC. This is depicted in Fig. 9.22. The DSC scan follows the phase transitions of DPPC from low temperature gel phase ($L_{\beta'}$) to high temperature liquid crystalline phase (L_α). With pure DPPC, there are two transitions, a lower temperature pretransition (at ~35.6°C) and a higher temperature main transition, T_m (at ~41.3°C). In the region between the pretransition and main transition, the bilayer exists in a nonflat, ripple structure ($P_{\beta'}$). Therefore the DSC-derived phase diagram shows the following transitions:

$$L_{\beta'} \rightarrow P_{\beta'} \rightarrow L_{\alpha}$$

A second phase diagram looks at the DPPC transitions from a different perspective. Fig. 10.12 shows the affect of hydration on the DPPC lipid phase [35]. The lipid solution becomes more dilute from left to right. The dotted line indicates maximum adsorption of water at the membrane surface. At any point to the right of the dotted line DPPC is saturated with water. Normal biological water levels are >99.9%. At almost all hydration levels, DPPC undergoes the same transitions reported in the DSC scan shown in Fig. 9.22 ($L_{\beta'} \rightarrow P_{\beta'} \rightarrow L_{\alpha}$). However, at very low water content and high temperature, additional nonlamellar phases including cubic and H_{II} appear. H_{II} phase is not normally reported for any PC, but can form under extreme dehydration. Importantly, dehydration may occur at small locations on a membrane surface in the presence of chaotropic agents or divalent metals including Ca^{2+}. These components are known to replace waters bound to the phospholipid headgroup.

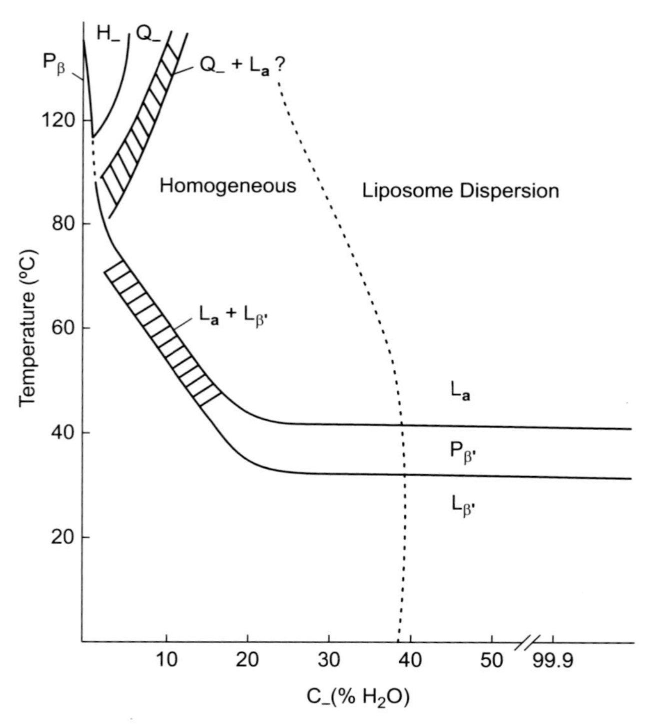

FIGURE 10.12 Phase diagram of DPPC showing the effect of water content (hydration) on lipid phase. At high (biological) levels of water, DPPC bilayers go from $L_{\beta'} \rightarrow P_{\beta'} \rightarrow L_{\alpha}$. Other unusual phases exist under conditions of extreme dehydration [35].

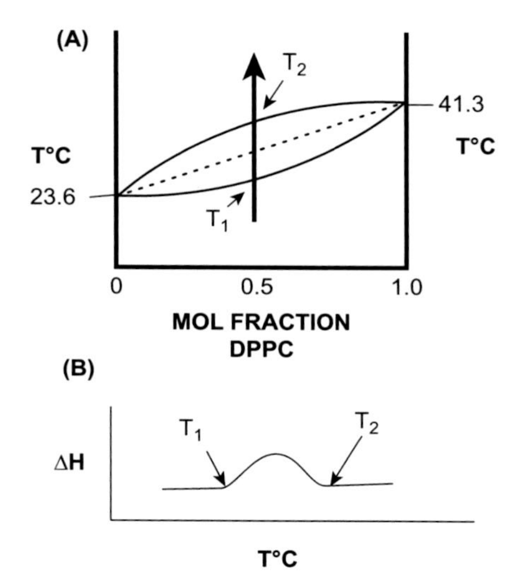

FIGURE 10.13 (A) Phase diagram for mixtures of DMPC (14:0,14:0 PC) and DPPC (16:0,16:0 PC). The T_m of DMPC is 23.6°C and DPPC 41.3°C. LUVs made from a 1:1 mixture of DMPC/DPPC is subjected to increasing temperature from low where both lipids are in the gel state to high temperature where both lipids are in the liquid crystalline state (thick arrow). (B) Corresponding DSC scan for the DMPC/DPPC mixed membrane.

3.2 Two-Component Lipid Phase Diagrams

A phase diagram for bilayers composed of mixtures of dimyristoyl phosphatidylcholine (DMPC, 14:0,14:0 PC) and dipalmitoylphosphatidylcholine (DPPC, 16:0,16:0 PC) is shown in Fig. 10.13, top panel. The T_m of DMPC is 23.6°C and DPPC 41.3°C, as depicted on the Y-axis. Since the T_ms are so close, mixtures produce a broad transition between the two single phospholipid T_m extremes. The dotted line represents the theoretical T_m for various DMPC–DPPC mixtures. The solid lines on either side of the dotted line are the experimentally derived melting values. The thick arrow follows a 1:1 (mol:mol) mixture of DMPC–DPPC as it is heated from a temperature where the lipid mixture is entirely in the gel ($L_{\beta'}$) state through the melting transition T_m until the mixture is in the liquid crystalline (L_α) state. In the region between temperatures T_1 and T_2 the membrane consists of co-existing gel and liquid crystalline states. The corresponding DSC scan is shown in the bottom panel. As the gel state mixture is heated, the membrane begins to melt at temperature T_1. The first component to melt is enriched in the lower melting lipid, DMPC. In the DSC scan, T_1 is the temperature where the membrane starts to melt and the scan leaves the baseline. As the melting continues, the mixture becomes more enriched in the higher melting DPPC. At T_m the mixture is half gel and half liquid crystal. At temperature T_2 the entire mixture melts as the membrane enters the liquid crystalline (L_a) phase where the DSC scan returns to the baseline. The last component to melt is highly enriched in DPPC, and is depleted in DMPC.

A second two-component phase diagram is shown in Fig. 10.14 for mixtures of DPPC and cholesterol [36]. Clearly this phase diagram is far more complex than that of DMPC–DPPC. At 0 mol% cholesterol the lipid melts at 41.3°C, confirming it is DPPC. At low cholesterol

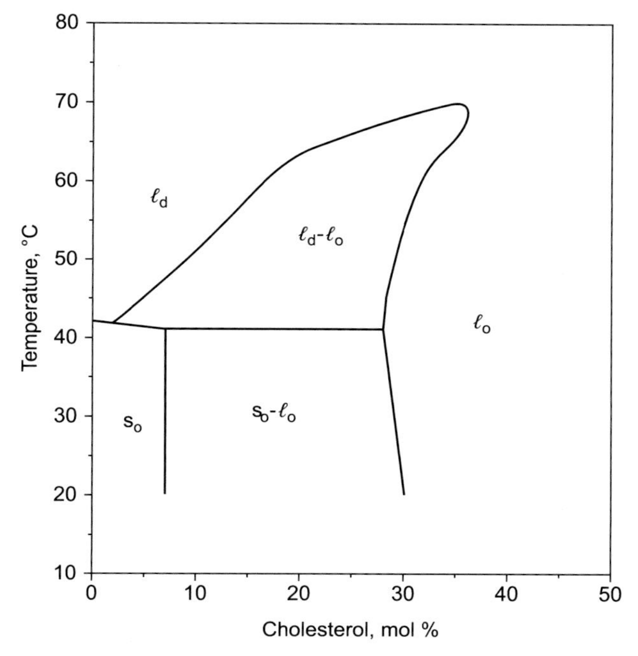

FIGURE 10.14 Phase diagram for DPPC/cholesterol mixtures. l_d is the liquid disordered (liquid crystalline) state. S_o is the solid or gel state. l_o is the liquid ordered state [36].

concentrations (<5 mol%), DPPC can simultaneously exist in either a liquid crystalline (l_d) or gel (S_o) state, depending on the temperature. At high cholesterol levels (>30 mol%), DPPC only exists in the l_o state. l_o is a state midway between liquid crystalline (l_d or L_α) and gel (S_o or $P_{\beta'}$) states, having properties of both. l_o state is characterized by having tightly packed lipids (like in gel) that nevertheless support rapid lateral diffusion (like in liquid crystalline). Between 5–30 mol% cholesterol, DPPC exists in mixed states, l_d–l_o or S_o–l_o.

3.3 A Three-Component Lipid Phase Diagram

Three-component phase diagrams are normally quite complex and not easy to interpret [37]. Fig. 10.15 shows an example of a lipid raft model phase diagram by de Almeida et al. [38]. Membranes were composed of cholesterol, 1-palmitoyl-2-oleoyl-*sn*-glycero-3-phosphocholine (POPC), and palmitoylsphingomyelin (PSM), where POPC is 16:0,18:1 PC and PSM is SM where the variable chain is palmitic acid. Assembling this complex phase diagram required a combination of various techniques from several sources. The diagram gives the composition and boundaries of the lipid rafts. Rafts are present in the dark blue, light blue and green-shaded areas. In the dark blue area l_o predominates over l_d. In the light blue area l_d predominates over l_o, while in the green l_o, l_d and S_o coexist. Raft size varies: insert A has large rafts (>75–100 nm); B has intermediate size rafts (~20 and ~75–100 nm) and C has small rafts (<20 nm).

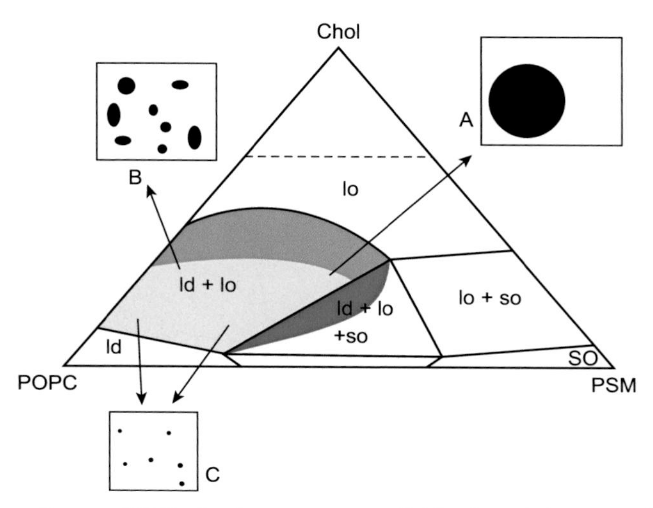

FIGURE 10.15 Phase diagram (23°C) of a three-component lipid raft model composed of cholesterol/POPC/PSM where POPC is 16:0,18:1 PC and PSM is palmitoyl-sphingomyelin. Abbreviations are the same as in Fig. 10.14. l_d is the liquid disordered (liquid crystalline) state. S_o is the solid or gel state and l_o is the liquid ordered state. Lipid rafts are present in the dark blue, light blue and green-shaded areas. In the dark blue area l_o predominates over l_d. In the light blue area l_d predominates over l_o while in the green l_o, l_d and S_o coexist. Raft size varies: insert A has large rafts (>75–100 nm); B has intermediate size rafts (~20 and ~75–100 nm) and C has small rafts (<20 nm) [38].

Since three-component phase diagrams are very hard to construct and interpret, it is hard to imagine what a biologically relevant phase diagram with hundreds of lipids would look like! Phase diagrams do have significant practical limitations.

4. LIPID–PROTEIN INTERACTIONS

4.1 Annular Lipids

One of the first, and most basic questions asked about membrane structure concerned the possible existence of a single layer of "special" lipids tightly associated with membrane proteins [39]. It is obvious that an integral protein must be solvated by lipids in the surrounding bilayer. This layer came to be known as "annular" or "boundary" lipids. But does this layer have the same lipid composition and properties as the surrounding bilayer or is it somehow different? The name annular or boundary implies a static lipid layer with some permanence surrounding the protein. This would require a significant attractive interaction between the annular lipid and the integral protein. But does such a unique layer exist and is it somehow different than the surrounding bulk membrane? An early attempt to investigate the annular lipid ring compared the rate of nearest neighbor exchange between an annular lipid and an adjacent bulk bilayer lipid, to the lateral exchange rate between two bulk bilayer lipids (Fig. 10.16).

For decades, searches have been conducted to find specific binding sites for lipids on the surface of membrane integral proteins. Of particular interest would be relatively long-lived lipid protein complexes that are stable on the order of the time required for the turnover

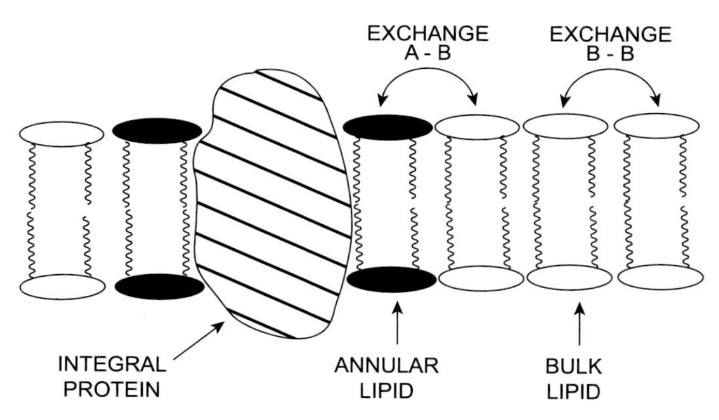

FIGURE 10.16 Nearest neighbor exchange between annular (A) and bulk (B) bilayer lipids (A − B exchange) and between two bulk bilayer lipids (B − B exchange).

of a typical enzyme (10^{-3} s). The large number of different lipid species that comprise a typical membrane (see lipid diversity discussed in Chapter 4 and Chapter 5) strongly argue for the existence of these complexes. A large number of lipid molecular species would be required to fit the many different binding sites that might exist on the surfaces of membrane proteins. Such binding sites would help to explain the perplexing problem of lipid diversity.

Some early iconic studies involving membrane structure searched for annular lipids. Initially Electron Spin Resonance (ESR) using spin-labeled phospholipids was used to distinguish annular from bulk bilayer lipids [39]. The approach is based on motional restriction of any spin probe-lipid that is adjacent to an integral protein (annular lipids). Restricted motion results in a much broader ESR spectrum than is measured for freely rotating spin-labeled phospholipids (bulk bilayer lipids) (Fig. 10.17) [40]. Therefore, addition of a spin-labeled phospholipid to a membrane containing protein(s) could produce a two component spectrum if there is a motional difference between annular and bulk bilayer lipids that exists for at least 10^{-8}–10^{-7} s, the shortest time scale detectable by ESR. A two component spectrum was indeed detected by ESR, suggesting the possible existence of annular lipids. This was corroborated by use of another rapid time technique, fluorescence quenching. Brominated phospholipids [41] were used to quench the inherent fluorescence of tryptophans in the integral protein. Fluorescence quenching is very sensitive to distances on the nm range. Brominated annular lipids, being closer to the tryptophans, quench better than do brominated lipids in the more distant bulk lipid bilayer, supporting the possibility of an annular lipid ring. However, when similar experiments were repeated using ^{2}H-NMR instead of a spin or brominated probe, no difference between annular and bulk bilayer lipids could be detected. This suggests that a distinct ring of annular lipids does not exist. Discrepancy between the experiments have been attributed to the inherently slower times that are detectable by NMR ($\sim 10^{-6}$ s) compared to ESR and fluorescence. Therefore, it was concluded that annular lipids, distinct from bulk bilayer lipids, may exist for short periods of time ($<10^{-6}$ s). If annular lipids do exist, their association with integral proteins must be transient and hence very weak. A transient existence suggests that lipid composition of annular lipids closely tracks composition of the bulk bilayer.

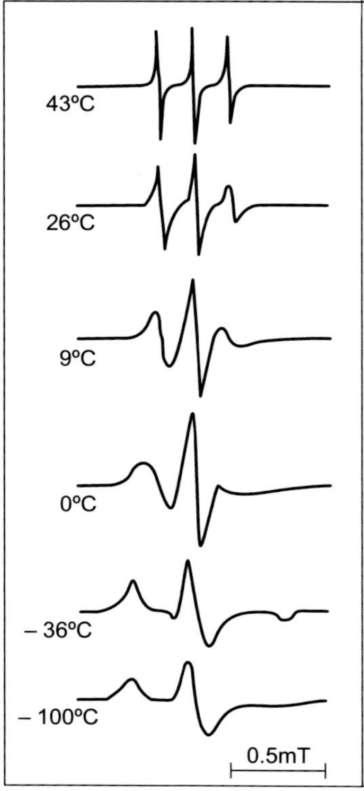

	Description of spectra	Approx, rotational lumbling times (ns)
43°C	Freely lumbling	0-1
26°C	Weakly immobilized	0-6
9°C	Moderately immobilized	2-5
0°C		5-0
– 36°C	Strongly immobilized	–300
– 100°C	Fluid glass or powder	>300
0.5mT		

FIGURE 10.17 ESR spectra of a nitroxide spin probe from freely tumbling (top) to totally restricted (glass state, bottom). Note the freely tumbling spectrum is narrow and sharp while motional restrictions broaden the spectrum.

Another related question that had to be addressed concerned the number of phospholipids required to completely solvate a specific integral protein and the lipid specificity, if any, of the solvation. In his 1988 book *Biomembranes*: *Molecular Structure and Function* [42], Robert Gennis presented a crude calculation to estimate the number of phospholipids that might surround a "typical" membrane integral protein. Some assumptions and estimates had to be employed. The "typical" protein was assigned a molecular weight of 50,000 and comprised 50% of the membrane by weight. This is well within the range of 20–75% protein for membranes (Chapter 1). Assuming an average molecular weight of ~830 for a phospholipid, the phospholipid protein ratio was 60:1 by number, again a realistic assumption. The protein was envisioned to be a cylinder that extended 10 Å above and below the bilayer surface, and the radius of the cylinder was 18 Å. Since phospholipids have a radius of ~4.4 Å, it would require ~16 phospholipids on each leaflet of the bilayer to completely solvate the protein. Therefore a total of ~32 lipids would be required to form the annulus. That would leave only 28 phospholipids left to form the next lipid layer around the protein. Since this second layer must be larger than the annular layer, there are not enough lipids to complete a second lipid shell.

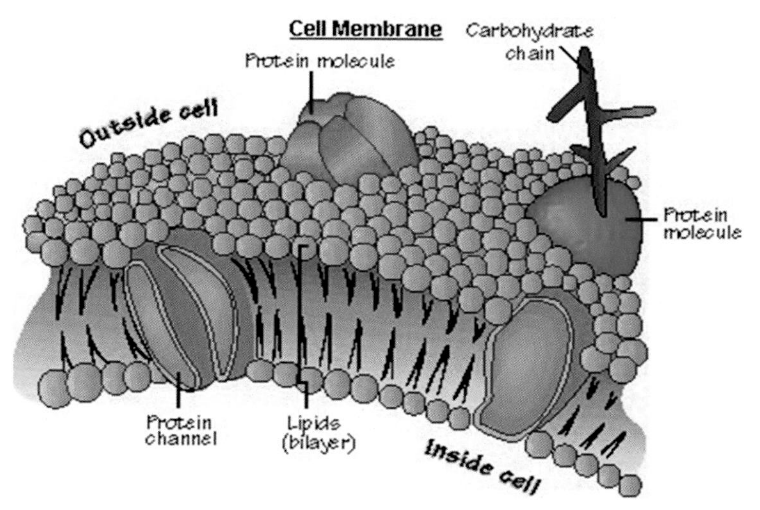

FIGURE 10.18 A cartoon drawing of the components of a "typical" plasma membrane. Similar drawings are commonly found in many textbooks. Note the unrealistically high number of phospholipid shells surrounding each protein.

Actual experiments agree with Gennis's rough estimate. For example, the integral membrane enzyme, Ca^{2+} adenosine triphosphatase (Ca^{2+}ATPase), was isolated free of phospholipids. Phospholipids were then slowly added back and the protein's activity monitored [43]. Activity did not return until the lipid protein ratio reached ~30:1 (see Section 4.2.1). The assumption was that at this ratio activity returned because the protein was completely solvated by lipid. Additional lipid added to the growing bilayer did not further increase activity. Confirming experiments were done adding a spin-labeled phospholipid to a delipidated protein [44]. The process was followed using ESR. Initially the ESP spectra were broad, indicating close association of the lipid and protein. At some lipid protein ratio all of the annular lipid sites became filled. Subsequent lipid then produced a second, narrow component to the spectrum, indicating formation of the bulk bilayer.

These and numerous other related reports call into question the classic definition of a biological membrane being "proteins floating like icebergs in a sea of phospholipids." Clearly the lipid bilayer "sea" must be limited and very, very crowded. Fig. 10.18 shows a typical example of a membrane cartoon drawing, similar to drawings found in most biology, biochemistry, physiology, and cell biology textbooks. Often in textbooks membrane proteins are depicted being solvated by >10 lipid rings! In the mitochondrial Cristae, a very busy membrane having ~75% protein by weight, a simple calculation indicates there is not enough lipid to totally surround the resident proteins even once. Many of the mitochondrial proteins must actually be touching. On average, a plasma membrane protein would have about enough lipid to form approximately two concentric layers around a protein. And one of these rings must form the annulus. An additional plasma membrane problem is related to its very high cholesterol content, ~60 mol% (or more) of the polar lipids. A report by Warren et al. [43] indicates that cholesterol is excluded from the annulus. If this is in fact true, the second lipid ring that surrounding the annulus, must be composed almost entirely of cholesterol. But cholesterol doesn't form bilayers! In the membrane with the lowest protein content, the myelin sheath, the proteins would be surrounded by several lipid rings that could keep the proteins ~11 Å apart. So how large and crowded is the bilayer "sea," and where does all that cholesterol reside?

4.2 Co-factor (Nonannular) Lipids

So far we have seen that most of the annular lipids, if distinct annular lipids even exist at all, are transient and weakly bound to integral proteins and the bilayer "sea" is very small and crowded. Despite this, the activity of many membranes is known to be tremendously affected by specific lipids [45–47]. Annular lipid selectivity is weak and may reside with either the lipid head group or hydrophobic tails. However, it does appear that at least some integral proteins, in addition to the weakly associated annular lipids, also have a few binding sites for specific lipids. These are called "nonannular" or "co-factor" lipids. The best example of strong preferential affinity for membrane proteins has been demonstrated for anionic lipids binding to the Ca^{2+} ATPase, Protein Kinase C (PKC), Na^+/K^+ ATPase, and the Potassium Channel KcsA.

4.2.1 Ca^{2+} *Adenosine Triphosphatase*

One integral membrane protein that has contributed substantially to understanding lipid–protein interactions is the Ca^{2+} ATPase, a protein that is easily isolated from the sarcoplasmic reticulum of skeletal muscle. Discovery of the enzyme and its function, to pump Ca^{2+} out of the sarcoplasmic reticulum, dates from the early 1960s. One advantage of this enzyme is that it has two measurable activities, transmembrane Ca^{2+} transport and adenosine triphosphate (ATP) hydrolysis. Importantly, the enzyme can be reconstituted using a deoxycholate dialysis procedure (Chapter 13).

In a 1975 paper in *Nature*, Warren et al. reconstituted the Ca^{2+} ATPase into liposomes made from a variety of lipid mixtures [43]. When these investigators plotted PC content versus enzyme activity they estimated that activity returned at a PC protein ratio of ~30 (Fig. 10.19). They interpreted this number as an estimate of the number of lipids required to completely surround the Ca^{2+} ATPase one time (hence the size of the annular lipid boundary). Since in mixed PC–cholesterol membranes, Ca^{2+} ATPase activity exactly followed PC content and was independent of cholesterol, they proposed that cholesterol was excluded from the annulus (Fig. 10.19). This very interesting concept was later shown via binding assays to not be true. Later work from A.G. Lee et al. [44,45,48,49] confirmed the size of the annular lipid ring, reporting a minimum number of annular phospholipid sites of 32 and 22 at 0 and 37°C, respectively. Lee also extended the lipid–Ca^{2+} ATPase interaction studies and found that anionic phospholipids PA, PS, and PI activate the protein while binding at the general annular lipid sites and are involved in binding of Mg^{2+}ATP. Also, Ca^{2+} ATPase activity is modified by bilayer acyl chain length, reaching a maximum at C-18 and falling off at C-14 and C-22. This experiment implicates specific anionic phospholipid binding sites in addition to the general annular lipid sites.

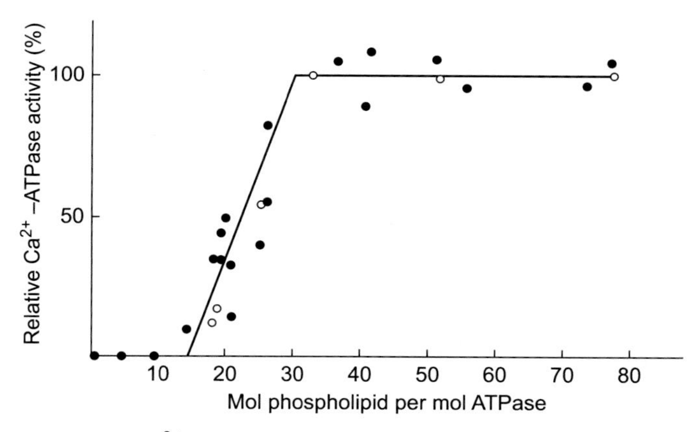

FIGURE 10.19 Activity of the Ca^{2+} ATPase isolated from the sarcoplasmic reticulum as a function of phospholipid content. It requires ~30 phospholipids to support full activity (*open circles*). This provides an estimate of the number of lipids required to completely surround the Ca^{2+} ATPase one time (hence the size of the annular lipid boundary). Since in mixed PC/cholesterol membranes Ca^{2+} ATPase activity exactly followed PC content and was independent of cholesterol (*filled circles*), it was proposed that cholesterol was excluded from the annulus [43].

4.2.2 Protein Kinase C

Perhaps the most thoroughly studied membrane protein is PKC (protein kinase C), a key enzyme in signal transduction [50]. The enzyme was only discovered in the late 1970s by Yasutomi Nishizuka [51]. PS is an essential cofactor that specifically binds to and activates PKC. In fact all of the numerous isoforms of PKC are strictly dependent on PS for activity. PKC specifically recognizes 1,2-*sn*-phosphatidyl-L-serine (the D isomer is inactive), and is independent of membrane structure [52]. This and the fact that PS is not believed to be involved in cell signaling through the formation of active PS metabolites, as is the case with PI (Chapter 5), implies that specific, nonannular, co-factor binding sites for PS must exist on PKC. Also, PKC's specificity for PS requires diacylglycerol (DAG) (Chapter 20). In the absence of DAG, PKC binds anionic membranes with no discrimination between phospholipid headgroups beyond requiring a negative charge. PKC activity is further modulated by bulk bilayer lipids related to curvature strain (discussed later). The absolute specificity for PS is remarkable for known lipid–protein interactions.

4.2.3 Na^+/K^+ ATPase

The plasma membrane-bound Na^+/K^+ ATPase is responsible for maintaining essential transmembrane Na^+ and K^+ ion gradients. Its mechanism of action is discussed in Chapter 19. The Na^+/K^+ ATPase was first described by Jens Skou (Fig. 19.13) in 1957 [53], for which he was awarded a 1997 Nobel Prize in Chemistry. The enzyme isolated from *Squalus acanthus* was shown to bind 60 phospholipids that comprise the annulus, and has a preference for negatively charged lipids ($K = 3.8$ for Cardiolipin (CL)/PC, 1.7 for PS/PC, and 1.5 for PA/PC). The affect of various lipids on the generation of the Na^+/K^+ ATPase activity has been reported by Haim et al. [54]. These authors noted that acidic phospholipids are required and PS is somewhat better than PI. Also, optimal activity is achieved with heteroacid PSs having saturated (18:0 $\geq$ 16:0) in the *sn*-1 chain and unsaturated (18:1 > 18:2) in the *sn*-2 chain. In addition they also hypothesized that PS and cholesterol interact specifically with each other near the α_1/β_1 subunit interface, thus stabilizing the protein. A general role for anionic lipids at protein–protein interfaces is becoming more evident.

4.2.4 Potassium Channel KcsA

Recently Marius et al. [55] investigated the absolute requirement of anionic lipids for potassium channel KcsA activity. The protein was isolated from *Streptomyces lividans*. High resolution X-ray crystallography showed an anionic lipid molecule bound to each of the four protein–protein interfaces comprising the channel. In addition to the annular or boundary lipids that surround the channel, there are four nonannular (co-factor) sites that bind the anionic phospholipid PG (phosphatidylglycerol). The nonannular sites are only ~60–70% occupied in bilayers composed entirely of PG. Increasing the anionic lipid content of the membrane leads to a large increase in open channel probability, from 2.5% in the presence of 25 mol% PG to 62% in 100 mol% PG. A model was proposed where three or four of the four nonannular lipid sites in the KcsA homo-tetramer have to be occupied by an anionic lipid for the channel to open.

4.3 Nonanionic Co-factor Lipids

Although the major effort to identify specific, nonannulus lipid-binding to membrane proteins has focused on the anionic-headgroup lipids, a few examples involving other types of lipid headgroups or acyl chains have appeared. The most specific of these involves rhodopsin and DHA.

4.3.1 Rhodopsin

Rhodopsin is the membrane integral protein that functions as the visual receptor of the eye rod outer segment (ROS). Its structure was discussed in Chapter 6. The protein is solvated by ~24 phospholipids and shows very little preference of PC over CL. The ROS is highly enriched in DHA, the longest and most unsaturated fatty acid commonly found in membranes (Chapter 4 and Chapter 23) [56]. In the ROS, DHA levels can approach 50% of the total acyl chains and so would be expected to have a dominant affect on this membrane's structure and function. In fact there is so much DHA in the ROS that homo-acid di-DHA phospholipid molecular species have been identified. Many years ago it was predicted that DHA may possess an unusual structure, perhaps accounting for rhodopsin activity [57,58]. Easy rotation about each methylene that separates the six double bonds results in a molecule that exists in a wide variety of highly contorted conformations. Two extreme conformers of the many possibilities, as determined by molecular dynamics (MD) simulations, are shown in Fig. 10.20 [59]. The average of all conformations makes DHA a cone-shaped molecule that, when incorporated into phospholipids, would prefer being in a nonlamellar phase. DHA phospholipids are loosely packed and can induce negative curvature strain to its immediate membrane neighborhood. These concepts will be discussed in detail later.

Combined MD simulations of rhodopsin and DHA by Scott Feller of Wabash College have resulted in a startling observation. DHA finds and fits perfectly into a groove on rhodopsin's surface (Fig. 10.21) [60]! This, of course, suggests the existence of a DHA co-factor receptor

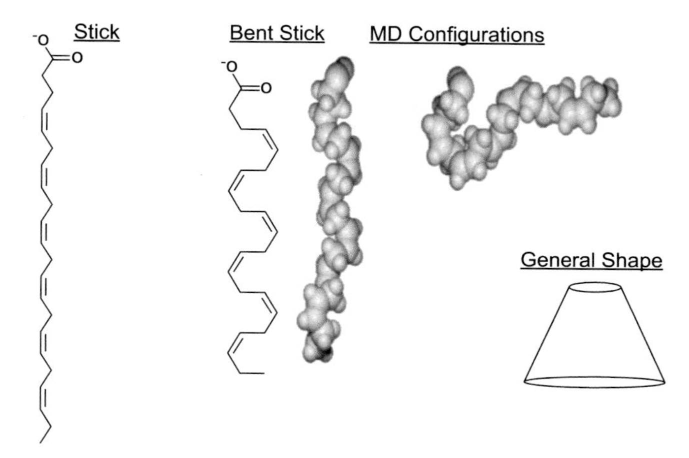

FIGURE 10.20 Several views of the structure of docosahexaenoic acid (DHA) [59].

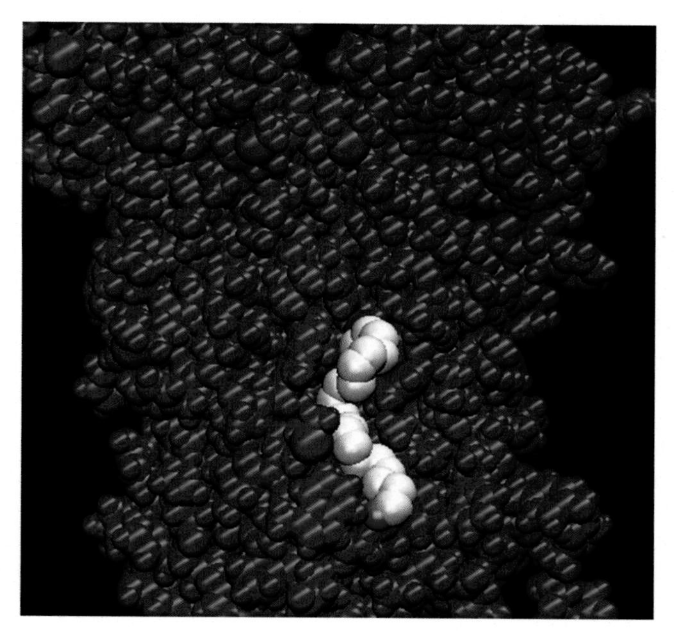

FIGURE 10.21 Molecular dynamics simulation of rhodopsin (blue) and docosahexaenoic acid (DHA, white). DHA fits into a groove on the surface of rhodopsin [60].

site on rhodopsin that may explain the need of DHA in vision. It is known that upon complete removal of DHA and its metabolic precursors from the diet, mammals will ferociously retain ROS DHA at the expense of all other lipids [56].

4.3.2 D-β-Hydroxybutyrate Dehydrogenase

D-β-hydroxybutyrate dehydrogenase is a lipid-requiring enzyme that is found on the inner surface of the inner mitochondrial membrane where it catalyzes the reaction:

$$(R) - 3 - \text{hydroxybutanoate} + NAD^+ \rightarrow \text{acetoacetate} + NADH + H^+$$

What makes this enzyme so unusual is its reputed absolute specificity for PC [61,62]. When purified free of lipid, the enzyme is inactive. Activity is regenerated by adding PC or lipid mixtures containing PC. The active form of the enzyme is the enzyme–phospholipid complex. Many membrane enzymes can be functional in bilayers made from PC, but D-β-hydroxybutyrate dehydrogenase is the only one with an absolute requirement for PC.

4.3.3 Cytochrome c Oxidase

The most logical place to look for an absolute link between a specific lipid and a mammalian function would be CL (cardiolipin) and mitochondrial electron transport and oxidative phosphorylation. CL was first isolated from beef heart in 1942 [63] and its presence was soon shown to be characteristic of bioenergetic membranes (bacterial plasma membrane and the mitochondrial inner membrane). Cytochrome c oxidase is the terminal component in electron transport where its function is to reduce O_2 to H_2O (Chapter 18). CL comprises about 20% of the total lipid composition of the mitochondrial inner membrane. The structure of CL (Chapter 5) is clearly the most unusual of all phospholipids. Essentially it is two PAs held together by a glycerol. At physiological pH, its large headgroup is a di-anion and it has four acyl chains. Due to its high negative charge density, CL can be induced by Ca^{2+} to undergo a lamellar to hexagonal ($L \rightarrow H_{II}$) phase transition. An almost limitless number of possible acyl chain combinations would imply that CL could exist in an enormous number of different molecular species. However, the vast majority of acyl chains in CL are linoleic acid ($18{:}2^{\Delta 9,12}$) with lesser amounts of oleic acid ($18{:}1^{\Delta 9}$), and linolenic acid ($18{:}3^{\Delta 9,12,15}$). This severely limits the number of possible molecular species to a manageable number.

So is CL, with all of its unusual properties, absolutely required for cytochrome c oxidase activity? Are some of the reputed ~40–55 annular lipids for cytochrome c oxidase bound to specific co-factor sites? A number of reports have linked CL levels and full electron transport activity, particularly of cytochrome c oxidase. In one report, Paradies exposed rat liver mitochondria to a free radical generating system that peroxidized the CL [64]. Concomitant with the loss of CL was a decrease in cytochrome c oxidase activity. Lipid-soluble antioxidants prevented the loss of CL and preserved the enzyme activity. External addition of CL, but not any other phospholipid, prevented the loss of cytochrome c oxidase activity. This experiment demonstrated a close correlation between oxidative damage, CL content, and reduced cytochrome c oxidase activity. In a related study these same investigators also demonstrated that lower cytochrome c oxidase activity observed in heart mitochondria from aged rats can be fully restored to the level of young control rats by exogenously-added CL [65]. Other mitochondrial diseases (eg, Tangier disease and Barth syndrome) have also been linked to CL and can be treated by controlling CL levels.

Several of the mitochondrial electron transport–oxidative phosphorylation complexes (Chapter 18) have been shown to require CL in order to maintain full enzymatic function. Examples include Complex IV (cytochrome c oxidase) that requires two CLs, Complex III (cytochrome bc1), and Complex V (F1 ATPase) that requires four CLs. While all of these various examples clearly indicate an important role for CL in mitochondrial bioenergetics, no *absolute* specificity has yet been demonstrated. Mitochondrial function can be maintained by PC without CL, albeit at reduced levels and the activity ratio for CL PC is only about 5.

4.4 Hydrophobic Match

One partial explanation for the large number of membrane lipid species (lipid diversity) is the requirement for lipids of different lengths that might be employed to properly solvate the varying hydrophobic surface areas associated with integral membrane proteins. The relationship between the length of the transmembrane protein hydrophobic surface and the hydrophobic thickness of the neighboring lipid bilayer is referred to as the "hydrophobic match" [66,67]. A basic assumption is that these two lengths should be similar and if they are not, the mismatch must be compensated for by alterations in lipid and/or protein structure.

It has been shown that the length of the hydrophobic segment on the membrane protein varies from protein to protein even within the same membrane and the length of the lipid bilayer hydrophobic interior varies with the lipid shape, acyl chain length, and number and location of any double bonds. Temperature, lateral pressure, cholesterol content, and presence of divalent metal ions can also affect thickness of the bilayer. A major membrane bilayer "thickening" agent is cholesterol. Cholesterol-rich domains, including lipid rafts, are partially characterized by being significantly thicker than the neighboring nonraft bilayer (Chapter 8). In proteins, the hydrophobic segment is related to the types and sequence of the amino acids comprising the segment. Of particular importance is the number of transmembrane α-helices present. As a result, the hydrophobic match length may vary considerably. For example, in *Escherichia. coli,* the hydrophobic length of the inner membrane leader peptidase is 15 amino acids long while the lactose permease is 24 ± 4 amino acids long. Lipids can affect protein activity, stability, orientation, state of aggregation, localization, and conformation. Proteins in turn can affect lipid chain order, phase transition, phase behavior, and microdomain formation. The effect of hydrophobic mismatch must therefore be simultaneously viewed from the perspective of both the lipid and the protein.

4.4.1 The Hydrophobic Match Length—A Differential Scanning Calorimetry Study

Determining the hydrophobic match length is not a straightforward measurement. Although X-ray crystallography can determine the precise location of every atom in a protein, it cannot identify with any certainty the location of the hydrophobic span that would accommodate the lipid bilayer. One unusual and indirect method that has been used to determine the length of the hydrophobic match involves DSC (Chapter 9). In one report, Toconne and collaborators [68] reconstituted bacteriorhodopsin isolated from *Halobacterium halobium* into lipid bilayers composed of PCs containing acyl chains of different lengths. The PCs were dilauroyl PC (DLPC, 12:0,12:0 PC), DMPC (DMPC, 14:0,14:0 PC), DPPC (DPPC,

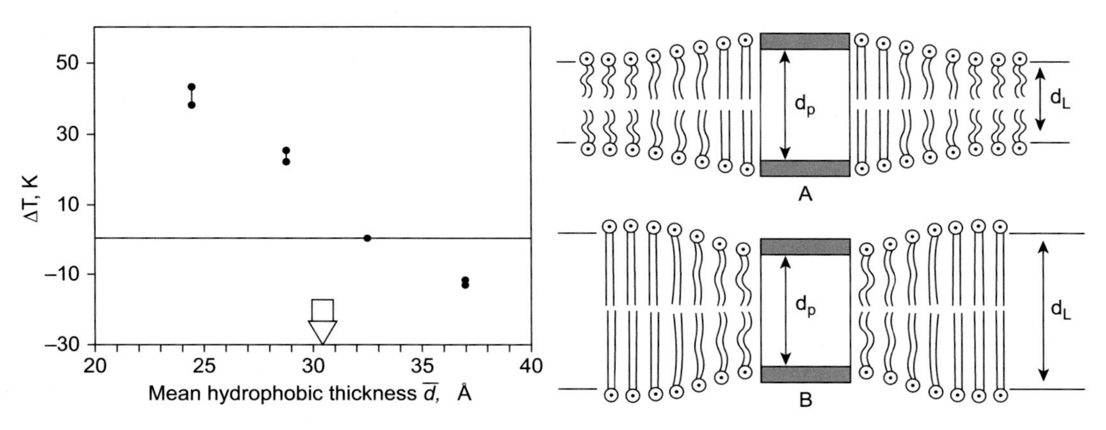

FIGURE 10.22 The Mouritsen and Bloom "Mattress" model [69] (right panel). The hydrophobic match length of an integral protein is depicted by d_P, and the length of the lipid bilayer hydrophobic interior by d_L. If $d_P > d_L$, the lipid must stretch to match the protein hydrophobic length, thus increasing its T_m (right panel, part A). If $d_P < d_L$, the lipid must shrink to match the protein hydrophobic length, thus decreasing its T_m (right panel, part B). This principle was tested by Toconne et al. [68] for bacteriorhodopsin reconstituted into bilayers made from either DLPC (12:0,12:0 PC), DMPC (14:0,14:0 PC), DPPC (16:0,16:0 PC) or DSPC (18:0,18:0 PC) (left panel). The bacteriorhodopsin-induced change in expected T_ms for each lipid was determined by DSC and plotted against the mean hydrophobic thickness. Bacteriorhodopsin reconstituted into DLPC and DMPC bilayers increased the expected T_m and so both lipids are shorter than the protein's hydrophobic match. Bacteriorhodopsin reconstituted into DSPC bilayers decreased the expected T_m and so is longer than the protein's hydrophobic match. Bacteriorhodopsin had no effect on the T_m of bilayers made from DPPC. Therefore bacteriorhodopsin's hydrophobic match length is about the same as the hydrophobic length of DPPC, ~30.5 Å [68].

16:0,16:0 PC), and distearoyl PC (DSPC, 18:0,18:0 PC). Bacteriorhodopsin-induced shifts in the expected T_m of each lipid were determined by DSC. Shifts in T_ms were consistent with Mouritsen and Bloom's "Mattress model" (Fig. 10.22, right panel) [69]. By this model, a protein with a larger hydrophobic match length than the solvating lipid can stretch the lipid to fit its hydrophobic segment. This results in an increase in lipid order that is associated with an increase in the lipid's T_m. In contrast, a protein with a shorter hydrophobic match length than the solvating lipid can shrink the lipid to fit its hydrophobic segment. This results in a decrease in lipid order that is associated with a decrease in the lipid's T_m. Tocanne found that bacteriorhodopsin reconstituted into DLPC bilayers increased the lipid's T_m +40°C (from 0 to ~40°C) (Fig. 10.22, left panel). Therefore, the length of DLPC (2.4 nm) is shorter than the hydrophobic match length of bacteriorhodopsin. The T_m of DMPC was also increased by bacteriorhodopsin, but only by +23°C (from 23.6 to ~46.6°C). Therefore, the length of DMPC (2.8 nm) is also shorter than the hydrophobic match length of bacteriorhodopsin. Upon incorporation of bacteriorhodopsin into the longer lipid DSPC (3.7 nm), T_m of the lipid *decreased* from 58 to 45°C (~13°C). This indicates that the hydrophobic match length must be shorter than 3.7 nm (DSPC), but longer than 2.8 nm (DMPC). When bacteriorhodopsin was incorporated into DPPC bilayers (3.2 nm), no shift in T_m (41.3°C) was observed. This indicates that the length of the hydrophobic match for bacteriorhodopsin is ~3.2 nm. Other methodologies agree, placing the hydrophobic match length for bacteriorhodopsin at 3.1–3.2 nm.

This experiment supports some important basic conclusions. First, proteins of the same hydrophobic match length can be accommodated into bilayers of different thickness and integral proteins can severely affect neighboring lipid structure.

Single span α-helices are particularly sensitive to hydrophobic mismatch [66,67,70]. If the length of the transmembrane α-helix exceeds the bilayer hydrophobic thickness, the helix may tilt to shorten its effective length, or it may bend or even slightly compress, or may adjust its conformation into a different type of helix altogether. An incorrect helix length may cause the protein to aggregate. If the helix is much too short, the protein may even be excluded from the membrane interior, resulting in a new surface location. Multiple span proteins are resistant to tilt and it is hard to access the size and importance of mismatch. For example, it has been reported that the multispan protein rhodopsin needs a substantial lipid mismatch of 4 Å thicker or 10 Å thinner than the protein's hydrophobic length in order to aggregate.

It is likely that all integral proteins may select lipids of appropriate match length from the myriad of lipids that comprise the surrounding bilayer. Many, perhaps most of the bilayer lipids have a cone-shape and so prefer nonlamellar structure. The presence of can-shaped lipids stabilize the lamellar structure, but the presence of so many cone-shaped lipids puts the lamellar bilayer under stress. This is referred to as a "frustrated bilayer" that is susceptible to factors that may drive the bilayer into a nonlamellar structure. Among these factors are proteins exhibiting a hydrophobic mismatch.

4.4.2 Biological Function of the Hydrophobic Match

Many studies have linked protein activity to hydrophobic mismatch [66,67]. A good example comes from the ($Ca^{2+} + Mg^{2+}$) ATPase isolated from the sarcoplasmic reticulum [71]. The enzyme was reconstituted into bilayers made from PCs with monounsaturated (*cis* Δ9) chains of length $n = 12$ to $n = 23$. Enzyme activity increased from $n = 12$ to $n = 20$, then decreased for $n = 23$. Mixtures of two of any of these lipids supported an activity that was the average of the two lipids. Also, the addition of decane to $n = 12$ increased activity. This experiment indicates that biological activity of the ($Ca^{2+} + Mg^{2+}$) ATPase was maximal at a hydrophobic match length of 18–20 carbons. These same investigators later reported similar findings for the reconstituted Na^+,K^+ ATPase [72].

Biological membrane thickness, and hence hydrophobic match length, is now believed to play a crucial role in membrane trafficking. In biological membranes, cholesterol content and membrane thickness increase from the endoplasmic reticulum to the Golgi and finally to the plasma membrane. The average plasma membrane protein's hydrophobic match length is a full five amino acids longer than those of the Golgi. It is believed that proteins with a shorter hydrophobic match remain in the Golgi while proteins with a longer hydrophobic match are transported to the plasma membrane. Therefore the plasma membrane is much thicker than the endoplasmic reticulum due to its high cholesterol and SM content, central components of lipid rafts.

A different type of hydrophobic match occurs between membrane lipids, the best documented being between cholesterol and phospholipid acyl chains. Decades ago it was noted that the rigid ring structure of cholesterol fit nicely next to a stretch of nine saturated carbons from acyl chain C-1 to C-9. Interruption of this stretch by a double bond decreases cholesterol affinity. Therefore cholesterol interacts well with long saturated fatty acyl chains (eg, palmitic and stearic acid) or with fatty acyl chains that have double bonds at position 9 (eg, oleic acid, $18{:}1^{\Delta 9}$), or farther (eg, linoleic acid, $18{:}2^{\Delta 9,12}$) down the chain. Therefore cholesterol was said to fit into a "Δ9 pocket." Cholesterol avoids fatty acyl chains with double bonds before position Δ9 (eg, arachidonic acid, $20{:}4^{\Delta 5,8,11,14}$; eicosapentaenoic acid, $20{:}5^{\Delta 5,8,11,14,17}$; and DHA, $22{:}6^{\Delta 4,7,10,13,16,19}$). Hydrophobic match between cholesterol and a saturated fatty acid is required for the formation of lamellar l_o phase, the phase of lipid rafts.

Hydrophobic match is far more complex than it would initially appear. There are many possible ways in which membranes can deal with a mismatch. Hydrophobic match is a tool to regulate local bilayer thickness and hence membrane enzyme activity and membrane trafficking.

5. LIPID INTERDIGITATION

Another unusual example of lipid polymorphism involves lipid interdigitation, where one or both acyl chains extend beyond the bilayer mid-plane [73–75]. When this occurs the penetrating chain takes up residence in both membrane leaflets, profoundly affecting basic membrane structure and hence function. Therefore interdigitation is yet another way to vary structural properties of the bilayer. Interdigitation normally occurs when one chain is significantly longer than the other (Fig. 10.23), but can also be induced in symmetrical chain phospholipids. The most common membrane lipids that normally exhibit substantial differences in their acyl chain lengths are the large family of sphingolipids. Sphingolipids have a relatively short, permanent sphingosine chain and a very long (often 24-carbons) variable chain (Chapter 5). Therefore it has been proposed that membrane domains that are enriched in sphingolipids (ie, lipid rafts) may exhibit considerable interdigitation into the opposite leaflet.

There are three basic interdigitated states termed partially interdigitated, mixed interdigitated, and fully interdigitated (Fig. 10.23, bottom). Each interdigitated state is characterized, in part, by the number of acyl chains subtended under the polar headgroup: two for the partially interdigitated, three for the mixed interdigitated, and four for the fully interdigitated state. It is clear from the cartoon in Fig. 10.23 that each state is characterized by a different bilayer thickness. The thickest membrane would be associated with the noninterdigitated state while the thinnest bilayer would characterize the fully interdigitated state. Membrane thickness is most accurately assessed by X-ray and neutron diffraction (Chapter 9), although other methodologies including DSC, Raman and Fourier transform infrared spectroscopy, NMR, Electron Microscopy (EM), ESR and Fluorescence have been employed. These other methodologies, however, must be "calibrated" with respect to XRD.

5.1 Interdigitated-Lipid Shape

Most membrane lipids that have a tendency to interdigitate do so in the gel state and have a large discrepancy in the length of the two acyl chains [75]. This discrepancy can be quantified as the Chain Inequivalence [74]:

$$\text{Chain inequivalence} = \Delta C/CL$$

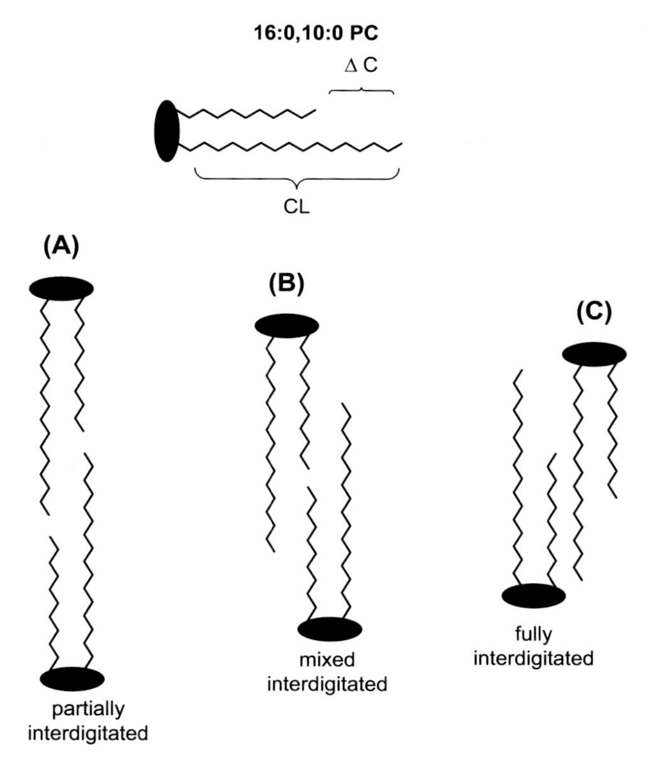

FIGURE 10.23 Lipid interdigitation. Top: An asymmetric phospholipid, 16:0,10:0 PC. Bottom: Three lipid interdigitated states: (A) partially interdigitated; (B) mixed interdigitated; and (C) fully interdigitated.

Where CL is the length of the longer chain and ΔC is the absolute difference in the chain lengths (Fig. 10.23) and is given by:

$$\Delta C = n_1 - n_2 + 1.5$$

where n_1 and n_2 are the number of carbons in the *sn*-1 and *sn*-2 acyl chains. The factor +1.5 is added to account for the *sn*-1 chain extending directly into the bilayer interior ~1.5 carbons deeper than the *sn*-2 chain (discussed below) (Fig. 10.24). The Chain Inequivalence parameter therefore represents the magnitude of the chain length inequivalence normalized by the length of the hydrocarbon region of the bilayer.

For a family of phosphatidylcholines, C18,CX PC where the *sn*-1 chain is 18-carbons and the *sn*-2 chain, X, varies from 18 to 0 carbons. C18,C18 PC is fully noninterdigitated while C18,C0 PC (a lyso PC) is fully interdigitated. Other interdigitated states are found for X between 18 and 0 carbons. The Chain Inequilicance parameter for C18,C10 PC is close to 0.5, characteristic of the mixed interdigitated state (Fig. 10.23B) Mixed interdigitation requires one chain to be about half the length of the other. Upon undergoing a thermotropic phase transition, the mixed interdigitated gel phase bilayers transform into partially interdigitated bilayers with two acyl chains subtended under the head group (Fig. 10.23A). So even in the

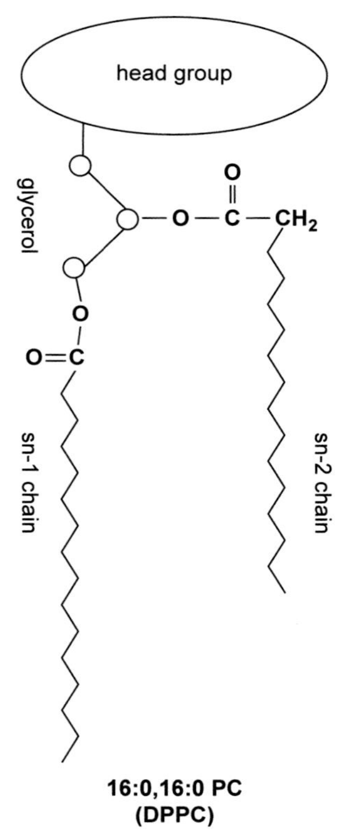

FIGURE 10.24 Symmetric acyl chain 16:0,16:0 PC (DPPC) demonstrating that the *sn*-1 chain protrudes directly into the bilayer interior while the first two carbons of the *sn*-2 chain run parallel with the membrane surface before also bending into the bilayer interior.

melted, liquid crystalline state, lipids with large asymmetries in acyl chain length may still retain some interdigitation.

The maximum possible chain asymmetry exists for lyso lipids or lipids where the *sn*-2 chain consists of either a proton or a very short acetyl ester. An example of a biologically important interdigitated lipid is C18,C2 PC, known as the platelet-activating factor. These lipids are fully interdigitated in the gel state where the short *sn*-2 chain lies parallel to the plane of the membrane, but upon undergoing a thermotropic phase transition develops membrane disrupting micelle structure.

5.2 Interdigitated Structure Inherent in Symmetrical Chain Phospholipids

Surprisingly, even phospholipids with identical *sn*-1 and *sn*-2 chains exhibit chain inequivalence. While the *sn*-1 chain drops straight down into the bilayer interior, the first two carbons of the *sn*-2 chain run parallel to the membrane plane before bending into the bilayer interior where it then runs parallel to the *sn*-1 chain (Fig. 10.24) [76]. As a result, the *sn*-2 chain behaves as if it is shorter than the *sn*-1 chain. In the gel state, the two chain terminal methyls are out of register by about 1.8 Å or 1.5 carbon bond lengths. In biological membranes this is compensated for with the *sn*-2 chain being about two carbons *longer* than the *sn*-1 chain, thus preventing interdigitation.

5.3 Methods to Detect Interdigitation

Since diffraction methods can directly measure bilayer thickness, they are the gold standard for monitoring interdigitation. As discussed in Chapter 9, the electron density profiles of all membranes look very similar by XRD. They consist of a low density bilayer interior flanked by two high density regions characteristic of the polar headgroups (Fig. 9.6). With interdigitation there is a loss of the deep depression in electron density that is characteristic of free motion associated with terminal methyls in noninterdigitated bilayers. Therefore interdigitated electron density profiles are more shallow than those associated with noninterdigitated bilayers. Also interdigitated bilayers are thinner than noninterdigitated bilayers.

Many other membrane biophysical methodologies have been used to indirectly monitor interdigitation. For example, freeze fracture EM images of noninterdigitated bilayers show a smooth surface resulting from fractionation running down the uninterrupted bilayer mid-plane. In sharp contrast are images obtained from mixed interdigitated gel phase bilayers indicating discontinuous fracture patterns where the fracture planes are interrupted by up and down steps.

Another approach measures order parameters using ESR [77]. Spin probes, particularly DOXYL attached to stearic acid at various positions on the acyl chain, were employed. This homologous series of probes has an unpaired electron (free radical) stabilized in the DOXYL structure and attached at different locations down the stearic acyl chain. Fig. 10.25 shows the structure of 5-DOXYL stearic acid.

The measured order parameters decrease from near the head group down the chain to near the highly disordered chain terminus. In one experiment order parameters at the ordered 5-position were compared to order parameters at the highly disordered 16-position. In noninterdigitated bilayers the order parameters for the two positions were distinct, with the 5-position being considerably greater than the 16-position. However, upon interdigitation, the 16- and 5-positions had similar order parameters and so reside near the same location. Therefore, the chains are interdigitated. Above T_m, both probes exhibit fluid phase isotropic motion.

5.4 Interdigitation Caused by Perturbations

There are many types of perturbations that can drive noninterdigitated bilayers into a fully interdigitated bilayers. The most studied of these perturbants are short-chain alcohols

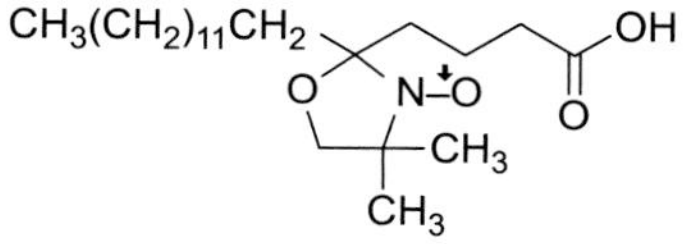

FIGURE 10.25 5-Doxyl stearic acid.

including methanol, ethanol [78], glycerol, and some anesthetics including chlorpromazine, tetracaine, and benzyl alcohol. The perturbants must be amphipathic and localize at the aqueous interface where they replace bound waters and increase the effective size of the lipid headgroups. One of the earliest interdigitation experiments involved DSC measurements of glycerol on DPPC bilayers. In the presence of glycerol (and later other alcohols), the T_m of DPPC was initially observed to decrease until a temperature was reached whereupon additional glycerol resulted in an increase in T_m. This change in T_m was referred to as the "biphasic effect" and is characteristic of a shift from a noninterdigitated to a fully interdigitated state. The biphasic effect works better for longer chain PCs but is not observed for PEs. Also observed with DSC of fully interdigitated state bilayers are hysteresis of heating and cooling scans and a shift to lower temperature for the main transition and disappearance of the pretransition. Unfortunately, the requirement of pure phospholipids and unrealistically high alcohol levels (often greater than 1M!), seriously question any biological role for alcohol-induced interdigitation.

Other perturbants include chaotropic agents (eg, thiocyanate at 1M), Tris buffer, pressure, myelin basic protein, and polymyxin with anionic lipids. Also it must be noted that ether-linked phospholipids (Chapter 5) often prefer interdigitation when compared to analogous ester-linked phospholipids. For example, in the gel state, DPPC is noninterdigitated while DHPC (hexadecyl ether linked-PC) spontaneously interdigitates.

5.5 Biological Relevance

While it is clear that a vast array of conditions can drive normal noninterdigitated state into several types of interdigitated states, it is not clear whether these transformations are merely esoteric curiosities or whether they can have some biological relevance. Suggested biological functions for interdigitation include:

1. Creation of novel membrane domains.
2. Transmembrane coupling between opposing membrane leaflets.
3. Reduction in charge density per unit area of a membrane surface.
4. Vary membrane thickness, thus altering the hydrophobic match.

6. SUMMARY

All membrane lipids exhibit different affinities and aversions for one another. An example of this is the strong affinity of cholesterol for sphingolipids that stabilizes lipid rafts. Cholesterol-rich lipid mixtures produce an l_o phase whose properties are midway between gel and liquid crystalline. Cholesterol-lipid affinity follows the sequence: SM > PS, PG > PC > PE and is stronger with saturated than polyunsaturated acyl chains. In addition to the conventional lipid bilayer (more correctly the lamellar phase), dozens of other phases whose functions are not known, exist. Various lipid mixtures prefer different phases and complex phase diagrams have been developed that pictorially describe various lipid phases and can take a wide variety of forms. For an integral protein to achieve maximal activity, the length of its transmembrane hydrophobic surface must match the hydrophobic length of the surrounding bilayer. This is known as the hydrophobic match, and is involved in membrane protein activity and trafficking. Lipids that normally exhibit substantial differences in their acyl chain lengths (eg, sphingolipids) can be involved in transmembrane coupling (interdigitation) between the inner and outer membrane leaflets.

Chapter 11 will conclude the trilogy of chapters on membrane physical properties by discussing "packing free volume," lipid area/molecule, collapse point, "lipid condensation," surface elasticity, molecular "squeeze out," long-range order, crowding, membrane domains, and homeoviscous adaptation.

References

[1] Ilangumaran S, Hoessli DC. Effects of cholesterol depletion by cyclodextrin on the sphingolipid microdomains of the plasma membrane. Biochem J 1998;335:433–40.
[2] Manolio TA, Pearson TA, Wenger NK, Barrett-Connor E, Payne GH, Harlan WR. Cholesterol and heart disease in older persons and women. Review of an NHLBI workshop. Ann Epidemiol 1992;2:161–76.
[3] Viegas J. Dog tail-chasing linked to high cholesterol. Discovery News; March 24, 2009.
[4] Quinn PJ, Wolf C. The liquid-ordered phase in membranes. Biochim Biophys Acta 2009;1788:33–46.
[5] Van der Goot FG, Harder T. Raft membrane domains: from a liquid-ordered membrane phase to a site of pathogen attack. Semin Immunol 2001;13:89–97.
[6] Brown DA, London E. Structure and origin of ordered lipid domains in biological membranes. J Memb Biol 1998;164:103–14.
[7] van Dijck PWM, De Kruijff B, van Deenen LLM, de Gier J, Demel RA. The preference of cholesterol for phosphatidylcholine in mixed phosphatidylcholine-phosphatidylethanolamine. Biochim Biophys Acta 1976;455:576–87.
[8] Demel RA, Jansen JW, van Dijck PW, van Deenen LL. The preferential interaction of cholesterol with different classes of phospholipids. Biochim Biophys Acta 1977;465:1–10.
[9] Shaikh SR, Cherezov V, Caffrey M, Soni S, Stillwell W, Wassall SR. Molecular organization of cholesterol in unsaturated phosphatidylethanolamines: X-ray diffraction and solid state ^{2}H NMR studies. J Am Chem Soc 2006;128:5375–83.
[10] Wassall SR, Shaikh SR, Brzustowicz MR, Cherezov V, Siddiqui RA, Caffrey M, et al. Interaction of polyunsaturated fatty acids with cholesterol: a role in lipid raft phase separation. In: Danino D, Harries D, Wrenn SP, editors. Talking about Colloids. Macromolecular Symposia, vol. 219. Wiley-VCH; 2005. p. 73–84.
[11] Singer SJ, Nicolson GL. The fluid mosaic model of the structure of cell membranes. Science 1972;175:720–31.
[12] Karnovsky MJ, Kleinfeld AM, Hoover RL, Klausner RD. The concept of lipid domains in membranes. J Cell Biol 1982;94:1–6.
[13] Estep TN, Mountcastle DB, Barenholz Y, Biltonen RL, Thompson TE. Thermal behavior of synthetic sphingomyelin-cholesterol dispersions. Biochemistry 1979;18:2112–7.
[14] Simons K, Ikonen E. Functional rafts in cell membranes. Nature 1997;387:569–72.
[15] Brown DA, London E. Functions of lipid rafts in biological membranes. Ann Rev Cell Dev Biol 1998;14:111–36.
[16] Smart EJ, Ying Y-S, Mineo C, Anderson RGW. A detergent-free method for purifying caveolae membrane from tissue culture cells. Proc Natl Acad Sci USA 1995;92:10104–8.
[17] Wainwright G, Mascitelli L, Goldstein MR. Cholesterol-lowering therapies and cell membranes. Arch Med Sci 2009;5.
[18] Tilcock CPS. Lipid polymorphism. Chem Phys Lipids 1986;40:109–25.
[19] Cullis PR, de Kruijff B. Lipid polymorphism and the functional roles of lipids in biological membranes. Biochim Biophys Acta 1979;559:399–420.
[20] Cullis PR, Hope MJ, Tilcock CPS. Lipid polymorphism and the roles of lipids in membranes. Chem Phys Lipids 1986;40:127–44.
[21] Gruner SM. Lipid polymorphism. The molecular basis of non-bilayer phases. Ann Rev Biophys Biophys Chem 1985;14:211–38.
[22] Luzzati V. Biological significance of lipid polymorphism: the cubic phases. Curr Opin Struct Biol 1997;7:661–8.

[23] Atkinson J, Epand RF, Epand RM. Tocopherols and tocotrienols in membranes: a critical review. Free Radical Biol Med 2008;44:739–64.
[24] Kumar VV. Lipid molecular shapes and membrane architecture. Ind J Biochem Biophys 1993;30:135–8.
[25] Gruner SM. Intrinsic curvature hypothesis for biomembrane lipid composition: a role for nonbilayer lipids. Proc Natl Acad Sci USA 1985;82:3665–9.
[26] Cullis PR, De Kruijff B. The polymorphic phase behaviour of phosphatidylethanolamines of natural and synthetic origins. A ^{31}P NMR study. Biochim Biophys Acta 1978;513:31–42.
[27] Cullis PR, de Kruijff B, Verkleij AJ, Hope MJ. Lipid polymorphism and membrane fusion. Biochem Soc Trans 1986;14:242–5.
[28] Siegel DP. Inverted micellar intermediates and the transitions between lamellar, cubic, and inverted hexagonal lipid phases. II. Implications for membrane-membrane interactions and membrane fusion. Biophys J 1986;49:1171–83.
[29] Sollner T, Whiteheart SW, Brunner M, Erdjument-Bromage H, Geromanos S, Tempst P, et al. SNAP receptors implicated in vesicle targeting and fusion. Nature 1993;362:318–24.
[30] Rothman JE, Warren G. Implications of the SNARE hypothesis for intracellular membrane topology and dynamics. Curr Biol 1994;4:220–33.
[31] Sudhof TC, Rothman JE. Membrane fusion: grappling with SNARE and SM proteins. Science 2009;323:474–7.
[32] Landau EM, Rosenbusch JP. Lipidic cubic phases: a novel concept for the crystallization of membrane proteins. Proc Natl Acad Sci USA 1996;93:14532–5.
[33] Feigenson GW. Phase diagrams and lipid domains in multicomponent lipid bilayer mixtures. Biochim Biophys Acta 2009;1788:47–52.
[34] Koynova R, Caffrey M. An index of lipid phase diagrams. Chem Phys Lipids 2002;115:107–219.
[35] Sackman E. In: Hoppe W, Lohmann W, Markl H, Ziegler H, editors. Physical foundations of the molecular organization and dynamics of membranes. New York: Springer-Verlag; 1983. p. 425–57.
[36] Sankaram HB, Thompson TE. Cholesterol-induced fluid-phase immiscibility in membranes. Proc Natl Acad Sci USA 1991;88:8686–90.
[37] Goni FM, Alonso A, Bagatolli LA, Brown RE, Marsh D. Phase diagrams of lipid mixtures relevant to the study of membrane rafts. Biochim Biophys Acta 2008;1781:665–84.
[38] de Almeida RFM, Fedorov A, Prieto M. Sphingomyelin/phosphatidylcholine/cholesterol phase diagram: boundaries and composition of raft structures. Biophys J 2003;85:2406–16.
[39] Davoust J, Schoot BM, Devaux PF. Physical modifications of rhodopsin boundary lipids in lecithin-rhodopsin complexes: a spin-label study. Proc Natl Acad Sci USA 1979;76:2755–9.
[40] Campbell LD, Dwek RA. Biological spectroscopy. Menlo Park, CA: Benjumin Cummings Publishing Co.; 1984.
[41] Jones OT, McNamee MG. Annular and nonannular binding sites for cholesterol associated with the nicotinic acetylcholine receptor. Biochemistry 1988;5:2364–74.
[42] Gennis RB. Biomembranes. Molecular structure and function. New York, NY: Springer-Verlag; 1988. 533 pp.
[43] Warren GB, Houslay MD, Metcalfe JC, Birdsall NJM. Cholesterol does not support enzyme activity and suggest that cholesterol is excluded from the phospholipids annulus surrounding an active calcium transport protein. Nature 1975;255:684–7.
[44] East JM, Melville D, Lee AG. Exchange rates and numbers of annular lipids for the calcium and magnesium ion dependent adenosine triphosphatase. Biochemistry 1985;24:2615–23.
[45] Lee AG. How lipids and proteins interact in a membrane: a molecular approach. Mol Biosyst 2005;1:203–12.
[46] Slater SJ, Kelly MB, Yeager MD, Larkin J, Ho C, Stubbs CD. Polyunsaturation in cell membranes and lipid bilayers and its effects on membrane proteins. Lipids 1996;31:S-189–92.
[47] Spector AA, Yorek MA. Membrane lipid composition and cellular function. J Lipid Res 1985;26:1015–35.
[48] Lee AG. How lipids interact with an intrinsic membrane protein: the case of the calcium pump. Biochim Biophys Acta 1998;1376:381–90.
[49] Dalton KA, East JM, Mall S, Oliver S, Starling AP, Lee AG. Interaction of phosphatidic acid and phosphatidylserine with the Ca^{2+}-ATPase of sarcoplasmic reticulum and the mechanism of inhibition. Biochem J 1998;329:637–46.
[50] Newton AC. Protein kinase c: poised to signal. Am J Physiol Endocrinol Metab 2010;298:E395–402.
[51] Takai Y, Kishimoto A, Inoue M, Nishizuka Y. Nucleotide-independent protein kinase and its proenzyme in mammalian tissues. I. Purification and characterization of an active enzyme from bovine cerebellum. J Biol Chem 1977;252:7603–9.

[52] Johnson JF, Zimmerman ML, Daleke DL, Newton AC. Lipid structure and not membrane structure is the major determinant in the regulation of protein kinase C by phosphatidylserine. Biochemistry 1998;37:12020–5.
[53] Skou J. The influence of some cations on an adenosine triphosphatase from peripheral nerves. Biochim Biophys Acta 1957;23:394–401.
[54] Haim H, Cohen E, Lifshitz Y, Tal DM, Goldshleger R, Karlish SJD. Stabilization of Na^+,K^+-ATPase purified from *Pichia pastoris* membranes by specific interactions with lipids. Biochemistry 2007;46:12855–67.
[55] Marius P, Zagnoni M, Sandison ME, East M, Morgan H, Lee AG. Binding of anionic lipids to at least three nonannular sites on the potassium channel KcsA is required for channel opening. Biophys J 2008;94:1689–98.
[56] Salem Jr N, Kim HY, Yergey JA. Docosahexaenoic acid: membrane function and metabolism. In: Simopoulos AP, Kifer RR, Martin R, editors. The health effects of polyunsaturated fatty acids in seafoods. New York: Academic Press; 1986. p. 263–321.
[57] Dratz EA, Deese AJ. The role of docosahexaenoic acid in biological membranes: examples from photoreceptors and model membrane bilayers. In: Simopoulos AP, Kifer RR, Martin R, editors. The health effects of polyunsaturated fatty acids in seafoods. New York: Academic Press; 1986. p. 319–51.
[58] Mitchell DC, Straume M, Litman BJ. Role of sn-1-saturated, sn-2-polyunsaturated phospholipids in control of membrane receptor conformational equilibrium: effects of cholesterol and acyl chain unsaturation on the metarhodopsin I – metarhodopsin II equilibrium. Biochemistry 1992;31:662–70.
[59] Feller SE, Gawrisch K, MacKerrell Jr AD. Polyunsaturated fatty acids in lipid bilayers: intrinsic and environmental contributions to their unique physical properties. J Am Chem Soc 2002;124:318–26.
[60] Feller SE, Gawrisch K, Wolfe TB. Rhodopsin exhibits a preference for solvation by polyunsaturated docosahexaenoic acid. J Am Chem Soc 2003;125:4434–5.
[61] Maurer A, McIntyre JO, Churchill S, Fleischer S. Phospholipid protection against proteolysis of D-β-hydroxybutyrate dehydrogenase, a lecithin-requiring enzyme. J Biol Chem 1985;260:1661–9.
[62] Loeb-Hennard C, McIntyre JO. (R)-3-hydroxybutyrate dehydrogenase: selective phosphatidylcholine binding by the C-terminal domain. Biochemistry 2000;39:11928–38.
[63] Pangborn M. Isolation and purification of a serologically active phospholipid from beef heart. J Biol Chem 1942;143:247–56.
[64] Paradies G, Ruggiero FM, Petrosillo G, Quagliariello E. Peroxidative damage to cardiac mitochondria: cytochrome oxidase and cardiolipin alterations. FEBS Lett 1998;424:155–8.
[65] Paradies G, Ruggiero FM, Petrosillo G, Quagliariello E. Age-dependent decline in the cytochrome c oxidase activity in rat heart mitochondria: role of cardiolipin. FEBS Lett 1997;406:136–8.
[66] Killian JA. Hydrophobic mismatch between proteins and lipids in membranes. Biochim Biophys Acta 1998;1376:401–16.
[67] Killian JA, de Planque MRR, van der Wel PCA, Salemink I, de Kruijff B, Greathonse DV, et al. Modulation of membrane structure and function by hydrophobic mismatch between proteins and lipids. Pure & Appl Chem 1998;70:75–82.
[68] Dumas F, Lebrun MC, Tocanne J-F. Is the protein/lipid hydrophobic matching principle relevant to membrane organization and function? FEBS Lett 1999;458:271–7.
[69] Mouritsen OG, Bloom M. Mattress model of lipid-protein interactions in membranes. Biophys J 1984;46:141–53.
[70] Jensen MO, Mouritsen OG. Lipids do influence protein function – the hydrophobic matching hypothesis revisited. Biochim Biophys Acta 2004;1666:205–26.
[71] Johansson A, Keightley CA, Smith GA, Richards CD, Hesketh TR, Metcalfe JC. The effect of bilayer thickness and n-alkanes on the activity of the ($Ca^{2+} + Mg^{2+}$)-dependent ATPase of sarcoplasmic reticulum. J Biol Chem 1981;256:1643–50.
[72] Johannsson A, Smith GA, Metcalfe JC. The effect of bilayer thickness on the activity of ($Na^+ + K^+$)-ATPase. Biochim Biophys Acta 1981;641:416–21.
[73] Slater JL, Huang C-H. Interdigitated bilayer membranes. Prog Lipid Res 1988;27:325–59.
[74] Slater JL, Huang C-H. Lipid bilayer interdigitation. In: Yeagle PL, editor. The structure of biological membranes. Boca Raton, FL: CRC Press; 1991. p. 175–210.
[75] Mason JT. Properties of mixed-chain length phospholipids and their relationship to bilayer structure. In: Lasic DD, Barenholz Y, editors. Handbook of nonmedical applications of liposomes: theory and basic sciences, vol. 1. Boca Raton, FL: CRC Press; 1996. p. 195–221 [chapter 8].

[76] Pascher I, Sundell S. Molecular arrangements in sphingolipids. The crystal structure of cerebroside. Chem Phys Lipids 1977;20:175–99.
[77] Boggs JM, Mason JT. Calorimetric and fatty acid spin label study of subgel and interdigitated gel phases formed by asymmetric phosphatidylcholines. Biochim Biophys Acta 1986;863:231–42.
[78] Zeng J, Smith KE, Chong PL-G. Effects of alcohol-induced lipid interdigitation on proton permeability in L-α-dipalmitoylphosphatidylcholine vesicles. Biophys J 1993;65:1404–14.

CHAPTER

11

Long-Range Membrane Properties

OUTLINE

Chapter 11 will discuss a few large-scale membrane properties including membrane lipid packing, membrane protein distribution, membrane lipid microdomains, and finally a process that keeps all of the various membrane properties in synch, homeoviscous adaptation (HVA).

1. MEMBRANE LIPID PACKING

1.1 Lipid Packing Free Volume

It has been well-documented that the large-scale state of the lipid bilayer component of membranes has a profound impact on the activity of resident membrane proteins [1,2]. One measure of this, a parameter known as the "packing free volume" (f_V), was first introduced in a series of papers in the late 1980s by Straume and Litman to quantify bulk

An Introduction to Biological Membranes
http://dx.doi.org/10.1016/B978-0-444-63772-7.00011-7

phospholipid acyl chain packing [3]. This parameter was derived from time-resolved anisotropy decay of the fluorescent lipid bilayer probe 1,6-diphenyl-1,3,5-hexatriene (DPH) (Chapter 9) which characterizes the volume available for probe reorientational motion in the anisotropic bilayer relative to that available in an unhindered, isotropic environment. The DPH-containing bilayer (where DPH displays partially hindered, anisotropic motion) is placed in the fluorescent sample cuvette while the reference cuvette contains 1,4-bis(5-phenyloxazol-2-yl)benzene (POPOP) in absolute ethanol (displaying totally unhindered, isotropic motion). Totally hindered motion has an f_V of 0.0, while the f_V of totally isotropic motion is 1.0. The measured f_V for a membrane will fall between 0 and 1. The f_V is particularly adept at measuring small changes in lipid packing that can profoundly affect integral membrane protein conformation.

Mitchell and Litman [4–6] have extensively used the f_V parameter to explore the affect of unsaturation on the visual receptor metarhodopsin I ↔ metarhodopsin II equilibrium. Metarhodopsin II is the active form of the receptor and occupies a larger volume than does metarhodopsin I. Table 11.1 presents f_Vs for several phosphatidylcholines (PCs) commonly found in the rod outer segment membrane [7,8]. At near physiological temperature (40°C), the values increase from liquid crystalline 14:0,14:0 PC with no double bonds ($f_V = 0.101$) to 22:6, 22:6 PC with 12 double bonds ($f_V = 0.201$). This demonstrates an increase in f_V with increasing phospholipid unsaturation. Also, f_V decreases for every phospholipid at lower temperature (10°C) and for increasing cholesterol content. Lower temperatures decrease the number of *gauche* kinks in the acyl chains, thus decreasing the f_V, and, as discussed in Section 1.4, cholesterol "condenses" fluid state phospholipids, also decreasing the f_V.

In addition, Mitchell and Litman measured the metarhodopsin I ↔ metarhodopsin II equilibrium, expressed as K_{eq}, in bilayer membranes made from five of the same lipids reported in Table 11.1 [7,8]. Table 11.2 shows a general relationship between the formation of the more voluminous, active metarhodopsin II (expressed as K_{eq}) and the f_V. A larger f_V

TABLE 11.1 Values of the Packing Free Volume Parameter (f_v) in Phosphatidylcholines

	f_v		
Acyl chain composition	**40°C[a]**	**10°C[a]**	**40°C[b], 30 mol% cholesterol**
di 22:6n3	0.201 ± 0.013	0.133 ± 0.01	0.122 ± 0.009
di 20:4n6	0.278 ± 0.016	0.205 ± 0.01	0.11 ± 0.01
di 18:1n9	0.147 ± 0.005	0.119 ± 0.005	0.062 ± 0.002
16:0, 22:6n3	0.154 ± 0.005	0.096 ± 0.005	0.074 ± 0.003
16:0, 20:4n6	0.155 ± 0.003	0.124 ± 0.004	0.067 ± 0.002
16:0, 18:1n9	0.130 ± 0.005	0.073 ± 0.003	0.056 ± 0.002
di 14:0	0.101 ± 0.006	–	0.011 ± 0.002

[a]*Values from Mitchell and Litman [7].*
[b]*Values from Mitchell and Litman [8].*
Values kindly provided by Drake Mitchell, Department of Physics, Portland State University.

TABLE 11.2 Packing Free Volume (f_v) and Metarhodopsin I ↔ Metarhodopsin II Equilibrium (K_{eq}) for Various Phosphatidylcholines (PCs) Commonly Found in the Rod Outer Segment

Phosphatidylcholine	Meta I ↔ Meta II (K_{eq})	Packing free volume (f_v)
22:6, 22:6 PC	6	0.201
20:4, 20:4 PC	4.7	0.278
16:0, 22:6 PC	3.3	0.154
14:0, 0 PC	1.7	0.101

drives the metarhodopsin I ↔ metarhodopsin II equilibrium to the right. These investigators also reported a reduction in K_{eq} with both lower temperature and cholesterol. From their observations, Mitchell and Litman proposed a molecular reason for the naturally high levels of docosahexaenoic acid in the rod outer segment and the requirement of this fatty acid for vision [9].

1.2 Lipid Monolayers (Langmuir Film Balance)

The first reported membrane physical properties were determined on lipid monolayers using a Langmuir Film Balance (Langmuir Trough). This device was named for Irving Langmuir, an early pioneer in the field (see Chapter 2). A schematic of a Langmuir Film Balance is depicted in Fig. 11.1. The Trough is normally made of an inert, nonwetable material such as Teflon or polytetrafluoroethylene. The monolayer is formed by adding a small volume of a volatile organic solution containing the dissolved lipid to the surface of an aqueous subphase between two teflon barriers, one or both of which can move parallel to the side walls of the trough. The barriers are in constant contact with the top of the subphase. After the organic solvent has evaporated, the barrier(s) are moved and the monolayer is compressed. Barrier movement, and hence the size of the monolayer is accurately controlled by computer. Temperature of the water subphase is also carefully controlled. During the

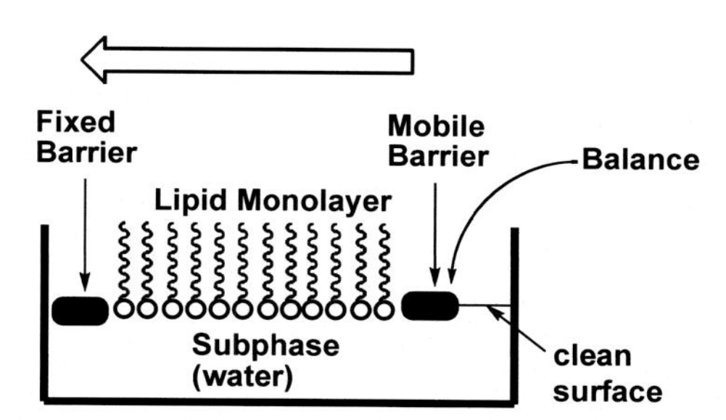

FIGURE 11.1 Diagram of a Langmuir Film Balance. A lipid monolayer is deposited on a clean aqueous surface between two barriers. Movement of the mobile barrier from right to left compresses the monolayer as the lateral pressure is continuously monitored via a sensitive electrobalance attached to a Wilhelmy plate inserted through the monolayer.

compression, surface tension of the monolayer is continuously monitored by use of a Wilhelmy plate attached to an electronic linear-displacement sensor, or electrobalance (Chapter 3). Monolayer surface pressure (π) is calculated by subtracting the surface tension of the subphase with the floating monolayer (γ) from the surface tension of the pure subphase with no monolayer (γ_o). γ_o for water, the normal subphase, is a constant at a single temperature (eg, $\approx$73 mN/m at 20°C) [10].

$$\pi = \gamma_o - \gamma$$

Surface pressure varies with the molecular area of the compressed monolayer. Surface tension measurements are exquisitely sensitive to any contamination, as contaminants will often accumulate at the water–air interface where they compete with the lipid monolayer lipids. Since even 1 ppm contaminant can radically change monolayer behavior, maximal cleanliness and purity of components must be observed. Experiments are often run in a clean room to prevent airborne contaminants and on a vibration-free table. Dilute solutions of the membrane lipid to be tested (~1 mg/mL) are made in an organic carrying solvent that must dissolve the amphipathic lipids while also being volatile. Examples of these solvents currently in use include hexane/2-propanol (3:2) or ethanol/hexane (5:95). The lipids rapidly spread over the clean aqueous interface with their polar headgroups in the water and their hydrophobic tails extended into the air. After 4–5 min to allow for the carrying solvent to dissipate, the compression is begun.

Fig. 11.2 shows pressure-area (P-A) isotherms for a simple, saturated fatty acid (stearic acid) on the left and a phospholipid (14:0, 14:0 PC) on the right. The phospholipid isotherm is far more complicated than that observed for the simple fatty acid [11]. At the lowest possible lipid densities, the lipids are not touching and the monolayer exhibits a quasi two-dimensional gas state (G). At ~78 $Å^2$/molecule, this number varies with the size of the phospholipid, the lipids come into contact with one another and enter the liquid-expanded (LE) phase. The LE phase is missing for unesterified, saturated chain fatty acids. In the LE phase, lipids are in contact but without molecular order. At ~5 mN/m a transition occurs from the LE phase to the liquid-condensed (LC) phase that exhibits order (LE to LC). Further compression will result in the solid (S) phase and eventually lead to the collapse of the monolayer (Collapse Point), where the lipids exhibit maximal possible density. The sequence of states that a phospholipid monolayer goes through upon compression is therefore:

G/G-LE/LE/LE = LC/LC/S/Collapse Point

It is estimated that the lateral pressure of a biological membrane is ~30–35 mN/m, considerably less than the **Collapse Point**, but substantially greater than the LE phase.

Fig. 11.3 shows Π-A curves for lipids of different compressibilities [12]. Curve a is the highly incompressible 22:0, 22:0 PC which remains in the solid-like phase throughout the compression; curve b is 16:0, 16:0 PC which is similar to the curve for 14:0, 14:0 phosphatidylethanolamine (PE) discussed in Fig. 11.2; and curve c is mixed chain egg PC which remains in a liquid-like phase throughout the compression. Very steep Π-A curves indicate poor compressibility (eg, long chain di-saturated phospholipids or cholesterol) while shallow curves indicate the lipid is compressible (eg, lipids containing polyunsaturated chains).

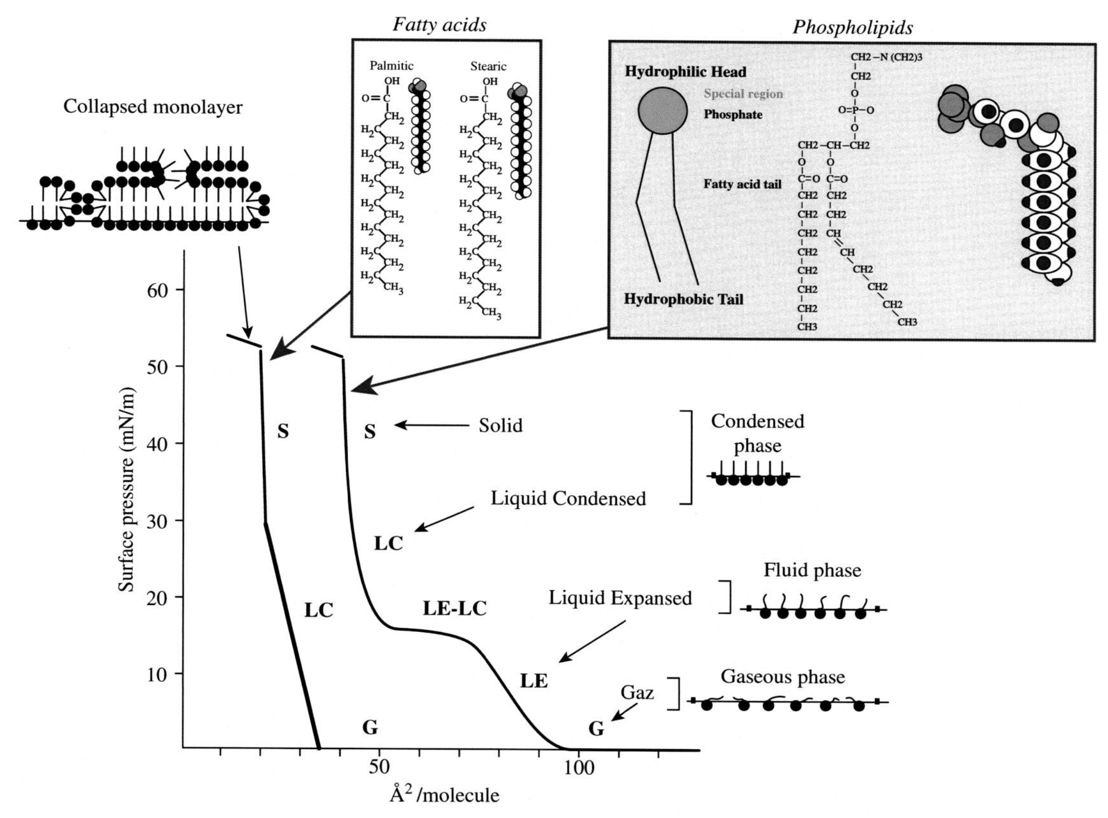

FIGURE 11.2 Pressure-area (P-A) isotherms for a simple, saturated fatty acid on the left and a heterochain phospholipid on the right. Note that the phospholipid isotherm is far more complicated than that observed for the simple fatty acid. Phases are: quasi, two-dimensional gas state (G); liquid-expanded phase (LE); liquid-condensed phase (LC); solid phase (S); and finally the Collapse Point. The sequence of states that a phospholipid monolayer goes through upon compression is: G/G-LE/LE/LE-LC/LC/S/Collapse Point.

A variety of important physical parameters can be derived from Π-A isotherms. These include the area/molecule as a function of lateral pressure, Collapse Point (maximum lipid packing density and therefore the minimal area/molecule), affect of polar head groups on the area/molecule, affect of acyl chains on the area/molecule, "condensation" between two or more lipids, surface elasticity, and molecular "squeeze out" of a component from a membrane. A few of these properties are briefly discussed in the following Sections (Sections 1.3–1.6).

1.3 Area/Molecule

The area/molecule for membrane lipids at different lateral pressures or at different temperatures can be directly read off Π-A isotherms. One interesting application of these

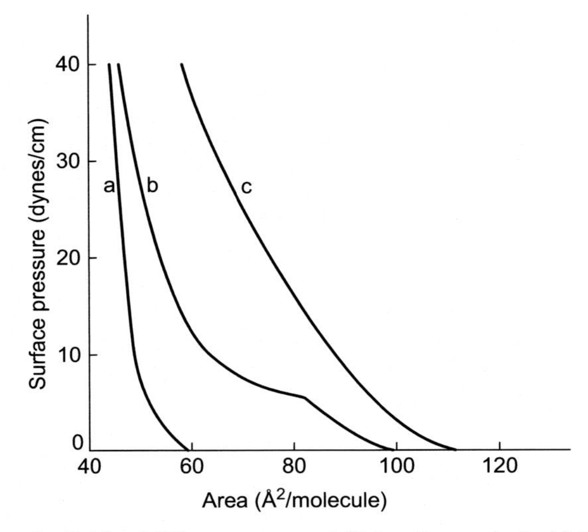

FIGURE 11.3 Π-A curves for lipids of different compressibilities. Curve a is the highly incompressible 22:0, 22:0 phosphatidylcholine (PC) which remains in the solid-like phase throughout the compression; curve b is 16:0, 16:0 PC which is similar to the curve for 14:0, 14:0 phosphatidylethanolamine (PE) shown in Fig. 11.2; and curve c is mixed chain egg PC which remains in a liquid-like phase throughout the compression [12].

measurements involves the melting (gel → liquid crystal) behavior of a di-saturated PC (DSPC). In the gel state, DSPC (18:0, 18:0 PC) has a cross-sectional area of ~48 Å^2. Upon melting to the liquid crystalline state the area increases to ~70 Å^2. Interestingly, 1,2-dioleoyl-sn-glycero-phosphatidylcholine (DOPC) (18:1, 18:1 PC) occupies almost the same cross-sectional area (~72 Å^2) as liquid crystalline DSPC. Therefore it can be concluded that the ~30° kink imparted by a *cis* double bond in an acyl chain increases the molecular area of a phospholipid approximately the same as that which arises upon increasing the *gauche* kinks upon chain melting.

1.4 Lipid "Condensation"

"Condensation" is an expression of how far two lipids behave from ideal with respect to their areas per molecule. If two molecules behave as noncompressible spheres when mixed, their combined measured areas should equal the sum of the area/molecule determined independently for each component [13].

$$A_{\text{ideal}} = \chi_1(A_1)_\pi + (1 - \chi_1)(A_2)_\pi$$

Where χ_1 is the mol fraction of component 1 and $(A_1)_\pi$ and $(A_2)_\pi$ are the mean molecular areas of components 1 and 2 at surface pressure π. Negative deviations of the experimentally measured value (A_{exp}) from A_{ideal} represent attraction ("condensation") while positive deviations represents "repulsion." If A_{ideal} = the experimentally measured value, the lipids behave ideally. The extent of nonideal behavior is expressed as % condensation:

$$\%\ \text{condensation} = [(A_{\text{ideal}} - A_{\text{exp}})/A_{\text{ideal}}] \times 100$$

where A_{ideal} is the mean molecular area calculated assuming ideal additivity and A_{exp} is that observed experimentally.

The most studied example of membrane "condensation" is with phospholipids and cholesterol [14]. For decades, cholesterol has been known to "condense" fluid state membranes, thus decreasing their permeability, fluidity, f_V, and increasing membrane thickness [15–17]. However, when cholesterol is added to gel state bilayers, membrane packing is decreased while permeability, fluidity, and f_V are increased. Bilayer thickness is also decreased. It has been proposed that cholesterol-induced "condensation," which is extremely high for sphingolipids and DSPCs may even be responsible for the formation and stability of lipid rafts (see Chapter 8). Cholesterol associates best with acyl chains that have no double bonds before position Δ9 (Δ9 pocket). In contrast, cholesterol avoids close association with chains that have double bonds before position Δ9 (eg, γ-linolenic, arachidonic, eicosapentaenoic, and docosahexaenoic acids) (Chapter 10). In the normal biological motif, where the *sn*-1 chain is saturated and the *sn*-2 chain unsaturated, cholesterol renders area "condensation" measurements relatively insensitive to structural changes in *sn*-2 chain. Therefore it has been proposed that cholesterol likely orients adjacent to the saturated, *sn*-1 side of a phospholipid.

1.5 Surface Elasticity Moduli

The surface elasticity moduli, C_s^{-1} has been proposed to provide a more accurate assessment of cholesterol–phospholipid interactions than "condensation" [18].

$$C_s^{-1} = (1 - A)(d\pi/dA)_\pi$$

Note that C_s^{-1} is a *change* in surface pressure with area and so, unlike "condensation," is dynamic.

Smaby et al. [18] used the in-plane elasticity moduli (inverse of the lateral compressibility moduli) to address the question of cholesterol interaction with mixed acyl chain PCs. The PCs had a saturated *sn*-1 chain and an *sn*-2 chain composed of either: 14:0, 18:1, 18:2, 20:4, or 22:6. At biological lateral pressure ($\geq$30 mN/m) they reported that cholesterol caused the in-plane elasticity of all of the mixed monolayers to decrease. PCs with more double bonds, however, were less affected by cholesterol and so these PC–sterol mixtures maintain relatively high in-plane elasticity. For DSPCs the decrease in interfacial elasticity is ~6–7-fold with equimolar cholesterol, while with di-polyunsaturated PCs the decrease is only ~1.7-fold. The unsaturated *sn*-2 acyl chain strongly modulates the elasticity of the mixed-chain PC films due to the poor association of the rigid α-surface of the sterol ring with the unsaturated *sn*-2 chains. These surface pressure measurements on lipid monolayers indicate that cholesterol will preferentially associate with saturated versus unsaturated acyl chains. This conclusion is in agreement with results derived from a variety of other methodologies (see Section 1 of Chapter 10).

1.6 Lipid "Squeeze Out"

In some lipid mixtures, increasing lateral pressure may force a membrane component to be excluded or "squeezed out" of the monolayer [19]. An example of this behavior is shown in

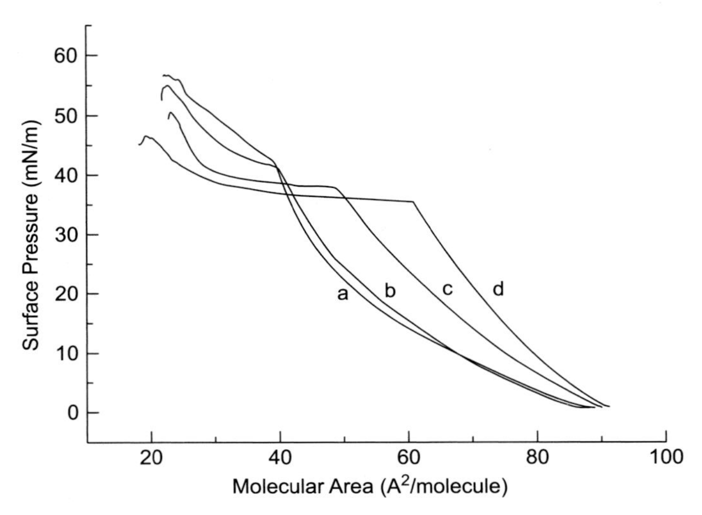

FIGURE 11.4 Π-A curves demonstrating "squeeze out" in sphingomyelin (SM)/18:0, 22:6 phosphatidylethanolamine (18:0, 22:6PE)/cholesterol monolayers. Monolayers are composed of various mixtures of SM and 18:0, 22:6 PE to which 0.05 mol fraction of cholesterol was added. The mol fraction of 18:0, 22:6 PE in SM are: curve a 0.1; curve b 0.2; curve c 0.3; and curve d 0.4. The "squeeze out" of 18:0, 22:6 PE is reflected by the plateau component that is most evident in curves b–d [13].

Fig. 11.4 [13]. Monolayers were made of mixtures of sphingomyelin (SM) and 18:0, 22:6 PE containing an additional 0.05 mol fraction of cholesterol. This lipid mixture represents a model plasma membrane–lipid raft. Upon increasing the lateral pressure, an 18:0, 22:6 PE-dependent plateau region emerged. The interpretation was that 18:0, 22:6 PE was "squeezed out" from the SM–cholesterol (lipid raft) domain. The percent "squeeze out" in the plateau portion of π/A curves is defined as:

$$L = 1 - (A_e/A_b) \times 100$$

where L is the percentage of molecules lost or "squeezed out," and A_b and A_e are the beginning and end of the surface area of the near-horizontal ("plateau") region of the curve.

2. MEMBRANE PROTEIN DISTRIBUTION

2.1 Freeze Fracture Electron Microscopy

Freeze fracture electron microscopy (EM), developed in the 1960s, is an unusual technique that has been used to investigate membrane structure from the perspective of the membrane hydrophobic interior [20]. From this technique, the distribution of membrane integral proteins in the lipid milieu can be directly estimated. Freeze fracture EM is based on rapidly freezing a membrane at a very cold temperature (<−100°C) followed by fractionation using a knife. The fraction plane follows the weak point in the frozen sample, the center of the membrane bilayer interior. This exposes the inner surface of both leaflets of the membrane (Fig. 11.5) [21]. The integral membrane proteins appear as particles of about 80–100 Å in

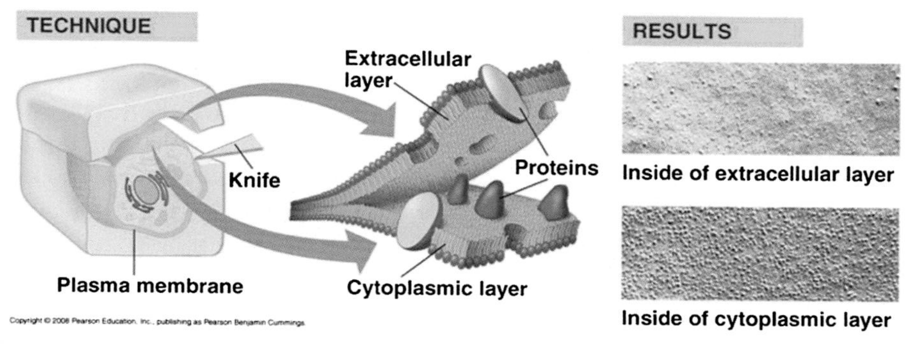

FIGURE 11.5 The technique of freeze fracture electron microscopy. The process is briefly discussed in the text and is discussed in much more detail in a 2007 review paper by Severs [20]. There are many more proteins projecting from the cytoplasmic leaflet than from the extracellular leaflet [21].

diameter. Usually they are randomly dispersed, but may be arranged in groups. Some particles are preferentially associated with one membrane leaflet face or the other (as demonstrated in Fig. 11.5). Freeze fracture EM supplies direct evidence that proteins reside inside membranes and are not just stuck on the surface. The particle distribution also indicates that there is usually no long-range order, instead observed order is only a few tenths of a micron and the membrane bilayer sea is indeed very crowded [22].

However, examples of long-range order due to cytoskeletal attachments have been observed. In an early report from 1976, Yu and Branton [23] showed that egg lecithin liposomes have very smooth membrane leaflet surfaces that are devoid of protein particles. If the erythrocyte protein band 3 is reconstituted into the liposome membranes, dispersed particles appear by freeze fracture that closely resemble what is imaged for natural erythrocyte membranes. If the pH is dropped to 5.5, the particles observed in the erythrocyte clump together due to the action of the cytoskeleton. Interestingly, if spectrin and actin, components of the cytoskeleton, are added to the reconstituted band 3-egg lecithin liposomes and the pH is decreased to 5.5, particle clumping is also observed.

Unfortunately, freeze fracture EM can be fraught with potential artifacts. Just the process of rapid freezing in liquid nitrogen produces destructive ice crystals. Therefore pretreatment with a cryoprotectant (glycerol) and gluteraldehyde is often employed. The biological membrane itself is not directly imaged. Instead a replica is made by shadowing the sample with platinum at a 45 degree angle to highlight the topography. The replica is further strengthened with a thin layer of carbon. The replica is then carefully washed off the sample and imaged by EM. Every step in a complex process is a potential source of artifacts.

3. IMAGING OF MEMBRANE DOMAINS

3.1 Macrodomains

The term "domain" has come to be used extensively in the life sciences literature. Unfortunately, "domain" means quite different things to different investigators. Funk and

Wagnalls Standard College Dictionary defines domain as "a territory over which dominion is exerted."In proteins, domains are a component of the total protein structure that exist and function essentially independent of the rest of the protein. Many proteins consist of several functional domains that can be very large, varying in length between 25 and 500 amino acids. Among the various types of protein domains are those that bind to specific portions of a nucleic acid that are also referred to as a domain. So the term domain can refer to functional patches of proteins, nucleic acids or, as we will see, membranes!

Quite different from protein and nucleic acid domains are domains that comprise membrane structure. It is universally accepted that biological membranes are not homogeneous mixtures of lipid and protein but instead consist of patches of widely differing and often rapidly changing compositions called domains. Domains exist in a bewildering array of sizes, stabilities, lipid and protein compositions and functionalities. Membrane domains can be roughly divided into large, stable macrodomains and small unstable lipid microdomains. Most macrodomains are stable for extended time periods and so are isolatable and fairly well-defined. The major part of most biological membranes, however, is likely composed of an enormous number of poorly understood and less stable lipid microdomains that are difficult to study, and so far impossible to isolate in pure form. Through the years, the reputed size and stability of lipid microdomains has continuously decreased from microns down to tens of molecules or so with associated lifetimes into the unbiological nanosecond range [24]. While large macrodomains are easily imaged, their analogous lipid microdomain counterparts in biological membranes are more elusive, far smaller and thus harder to image. In the following sections, examples of macroscopic and lipid microdomain imaging are presented.

Most macrodomains owe their discovery to the development of EM in the 1940s and 1950s. Some examples including basolateral and apical halves of epithelial cells, thylakoid grana and stroma, sperm head and tail, tight junctions, and bacteriorhodopsin 2-D patches in the purple membrane of *Halobacterium halobium* have been known for decades and will not be considered here. Images of the macroscopic domains gap junctions, clathrin-coated pits, and caveolae will be briefly discussed.

3.1.1 Gap Junctions

Gap junctions were probably first observed by the pioneering electron microscopist J.D. Robertson in 1953 [25]. Gap junctions were eventually characterized in the late 1950s and early 1960s by several investigators, most notably George Palade (Fig. 11.6) and his

FIGURE 11.6 George Palade, 1912–2008.

wife Marilyn Farquahr in 1963 [26]. George Palade is perhaps the most accomplished cell biologist of all time. As an electron microscopist, he was involved in identifying and describing many cell structures including gap junctions and Golgi apparatus. In 1974 he shared the Nobel Prize in Physiology and Medicine with Albert Claude and Christian de Duve "for discoveries concerning the functional organization of the cell that were seminal events in the development of modern cell biology."

Gap junctions are tube-like structures that connect the cytoplasms of adjacent cells [27,28]. Unlike tight junctions that press adjacent cells together, gap junctions maintain a 2–3 nm spatial gap between cells. Their function is to allow for rapid communication between cells through exchange of various small molecules (~1000 molecular weight or less) and ions. Important examples of transported solutes include the second messengers, cyclic adenosine monophosphate, inositol triphosphate and Ca^{2+}. Cells with gap junctions are therefore in direct electrical and chemical contact with each other [29]. Gap junction intercellular connections are through hollow cylinders called connexons. Each connexon is a circular arrangement of six subunits of a protein called connexin (Fig. 11.7). Since connexins come in about 15 different types, small variations in connexon structure are known. Each connexon spans the plasma membrane and protrudes into the gap between the cells where it connects to the protruded connexon of an adjacent cell, thus forming a completed channel. The channel is only about 3 nm in diameter at its narrowest, thus limiting the size of transported solutes. Gap junctions can be a single connexon or may be found in a field of hundreds or thousands of adjacent connexons called a plaque. Gap junctions are especially abundant in muscle and nerve that require rapid intracellular communication. Gap junction defects have been linked to certain congenital heart defects and deafness.

3.1.2 *Clathrin-coated Pits*

Plasma membranes are in part characterized by having structures that facilitate uptake of required solutes into the cell. An example of such a structure is the clathrin-coated pit, a structure that is involved in the process of receptor-mediated endocytosis (also referred to

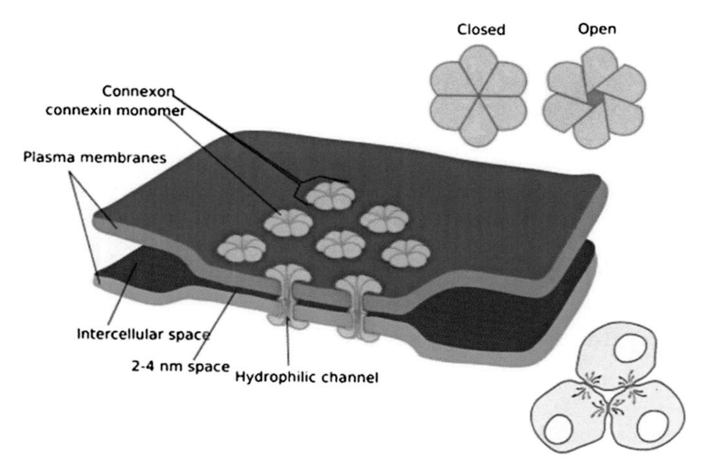

FIGURE 11.7 Schematic drawing of a field of gap junctions.

as clathrin-dependent endocytosis). Specific receptors found in the outer leaflet of the plasma membrane are involved in the internalization of macromolecules (ligands) including hormones, growth factors, enzymes, serum proteins, cholesterol-containing lipoproteins, antibodies, and ferritin–iron complexes. The best known example of this is the uptake of cholesterol-containing LDL (low-density lipoprotein) particles [29], work that won Michael Brown and Joseph Goldstein the 1986 Nobel Prize in Physiology and Medicine.

Ligand–receptor complexes diffuse in the plasma membrane until they encounter a coated pit. Proteins attached to the plasma membrane inner leaflet including clathrin and two coat proteins, COPI and COPII cause the pit to invaginate until it pinches off, forming a vesicle free in the cytoplasm (Fig. 11.8, [30]). The vesicle has three possible fates: (1) it can travel to the lysosome for degradation; (2) it can go to the Trans-Golgi Network for transport through the endomembrane system; or (3) it can return back to the plasma membrane at the opposite side of the cell where it is secreted as part of a process called transcytosis [31]. Coated pits are quite abundant in the plasma membrane. Since they occupy ~20% of the total plasma membrane surface area and the entire process of internalization only takes about 1 min, there may be 2500 coated pits invaginating into vesicles per minute in a cell. Another protein found associated with coated pits is caveolin, a major component of caveolae, a type of lipid raft. Coated pits are twice the size of caveolae, but share some properties with caveolae and lipid rafts.

3.1.3 *Caveolae*

Caveolae are complex plasma membrane structures whose properties appear to place them between coated pits and lipid rafts (Chapter 8). They are small (50–100 nm) invaginated membrane structures that superficially resemble coated pits (Fig. 11.9, [32]). In fact, in 1955, Yamada [33] proposed the descriptive name "caveolae" which is Latin for little caves. Caveolae were first described by the electron microscopist George Palade in

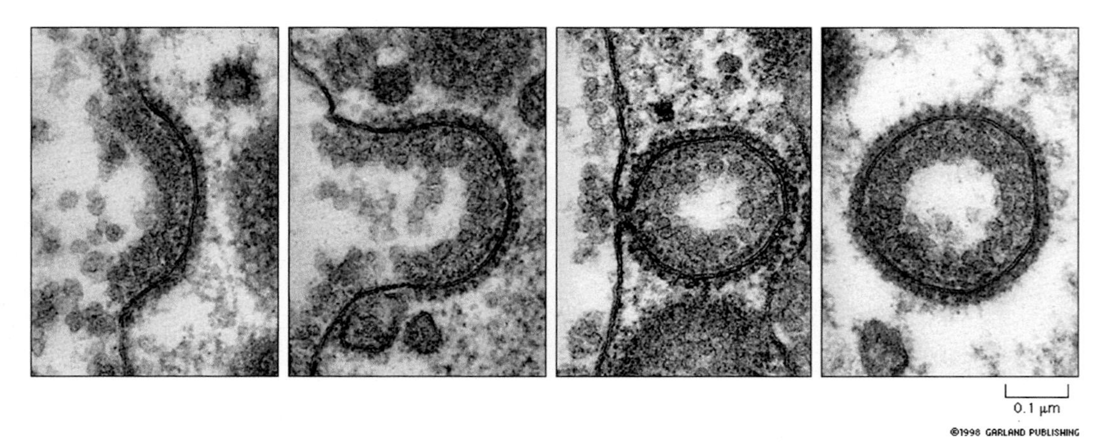

FIGURE 11.8 Sequence of formation of coated vesicles (right) from clathrin-coated pit (left) during receptor-mediated endocytosis on the surface of the plasma membrane. Pictures are electron micrographs of yolk protein on a chicken oocyte [30].

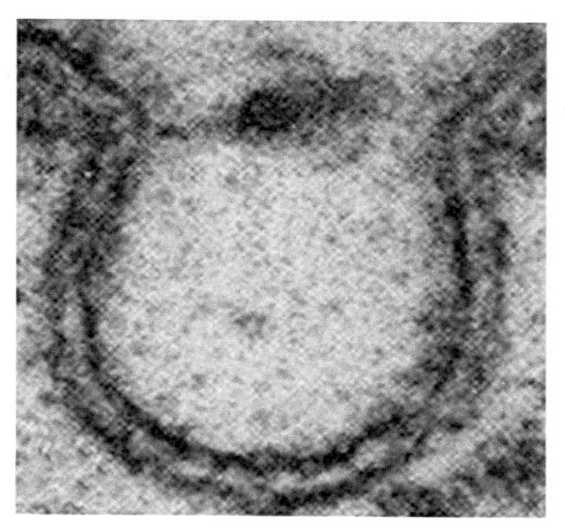

FIGURE 11.9 Transmission electron micrographs of endothelial caveolae.

1953 and are abundant in many vertebrate cell types, especially endothelial cells and adipocytes where they may account for 30–70% of the total plasma membrane surface area. Caveolae however, are not a universal feature of all cells as they are totally absent in neurons. Like lipid rafts, caveolae are partially characterized by being enriched in sphingolipids and cholesterol and also participate in signal transduction processes [34,35]. In fact caveolae are often described as being "invaginated lipid rafts" that differ primarily by the presence of a family of marker proteins called caveolins. But, similar to coated pits, caveolae may also play a role in endocytosis.

3.2 Lipid Microdomains

Although macrodomains are large, stable, and relatively easy to image, most of a biological membrane likely consists of many unstable heterogeneous patches of lipids and proteins known as lipid microdomains. These structures are well known in model membrane systems where they are driven by lipid lateral phase separations. However, if lipid microdomains exist at all in biological membranes, they are fleeting and very small. Their possible existence is based primarily on homologies with many experiments performed for decades on model lipid monolayers and bilayers. The documented phase separations have been primarily induced by changes in temperature, pressure, ionic strength, divalent cations, and cationic peripheral proteins. In contrast to macrodomains, lipid microdomains have to date been impossible to isolate and directly study as pure entities. Instead their existence is often inferred from biophysical techniques where the experimental measurements demonstrate coexisting multiple membrane populations (ie, lipid microdomains).

In early experiments with sea urchin and mouse eggs (Chapter 9), Michael Edidin employed lateral diffusion measurements using Fluorescence Recovery After Photobleaching (FRAP) to support the concept of heterogeneity in membranes [36]. Edidin's experiment followed the diffusion of two carbocyanine dyes, one with two short lipid chains (C_{10}, C_{10} DiI) and one with two long lipid chains (C_{22}, C_{22} DiI). The measured difference in diffusion rates between the two dyes was consistent with the existence of lipid microdomains.

Membrane microdomains have also been inferred from the activity of reconstituted enzymes. In one interesting model [37,38], the sarcoplasmic reticulum Ca^{2+} adenosine triphosphatase (ATPase) was reconstituted into liposomes made from either zwitterionic DOPC (18:1, 18:1 PC), anionic 1,2-dioleoyl-sn-glycero-3-phosphoric acid (DOPA) (18:1, 18:1 PA) or a DOPC−DOPA (1:1) mixture, and the enzyme activity measured. Activity was high in DOPC and low in DOPA. For the DOPC−DOPA (1:1) liposome, an intermediate activity was observed. Upon the addition of Mg^{2+} to the Ca^{2+} ATPase reconstituted in DOPC, the activity did not change and remained high. In contrast, Mg^{2+} reduced the activity of the Ca^{2+} ATPase reconstituted into DOPA to zero. The precipitous decrease in activity was attributed to Mg^{2+} binding to the anionic DOPA and inducing an isothermal phase transition resulting in the DOPA liposome being driven into the totally inactive gel state. When Mg^{2+} was added to the DOPC−DOPA mixed liposome, the Ca^{2+} ATPase activity was observed to increase as DOPA was removed from the fluid mixture as a gel. The induced isothermal phase transition resulted in the enzyme accumulating in the favorable DOPC fluid state domain. Thus the Ca^{2+} ATPase activity increased.

Direct observations (images) of lipid microdomains have routinely been reported for lipid monolayers and bilayers using fluorescent probes. It should be pointed out that lipid microdomains in model protein-free membranes are orders of magnitude larger than microdomains observed in biological membranes and so are much more amenable to imaging. A few examples follow.

Fig. 11.10 shows epifluorescence images demonstrating simple liquid crystalline−gel phase separations as a function of lateral pressure on Langmuir Trough lipid monolayers [39]. Monolayers were made from 70 mol% dipalmitoyl phosphatidylcholine (DPPC) (16:0, 16:0 PC, $T_m = 41.3°C$) and 29 mol% 1-stearoyl, 2-docosahexaenoyl phosphatidylcholine (SDPC) (18:0, 22:6 PC, $T_m = -20°C$) to which 1 mol% N-rhodamine-PE was added as the membrane imaging agent. Compressions were run and images taken at every 5 mN/m. At the experimental temperature, 23°C, DPPC is below its T_m and so would be in the gel (solid) state while SDPC is above its T_m and would be in the liquid crystalline (fluid) state. The N-rhodamine-PE fluorescent probe partitions preferentially into the fluid (SDPC) phase. The images clearly show phase separation into liquid crystalline (fluorescent, white) and gel (nonfluorescent, black) state domains that are affected by the lateral pressure.

In the late 1980s Haverstick and Glaser [40,41] investigated cation-induced phase separation of anionic phospholipids into domains. They employed a fluorescence microscope, a charge-coupled device camera, and a digital imaging processor to visualize lipid microdomains on the surface of very large (5−15 μM) lipid vesicles. The fluorophore 4-nitrobenzo-2-oxa-1,3-diazole (NBD) was attached to the *sn*-2 chain of phosphatidic acid (PA), phosphatidylserine (PS), PE, and PC. Vesicles were made from the bulk lipid DOPC (99 mol% of the total phospholipids) to which 1 mol% of the fluorescent probe (NBD−PA, NBD−PS, NBD−PE, or NBD−PC) was added. It was demonstrated that Ca^{2+}, but not Mg^{2+}, Mn^{2+}, or Zn^{2+}, induced phase separation of the anionic fluorescent phospholipids NBD−PA and NBD−PS from DOPC (Fig. 11.11, [40]). NBD−PE and NBD−PC did not phase separate. The conclusion that Ca^{2+} can induce formation of lipid microdomains composed of fluorescent anionic phospholipids was confirmed in erythrocytes, erythrocyte ghosts, and vesicles made from erythrocyte lipid extracts.

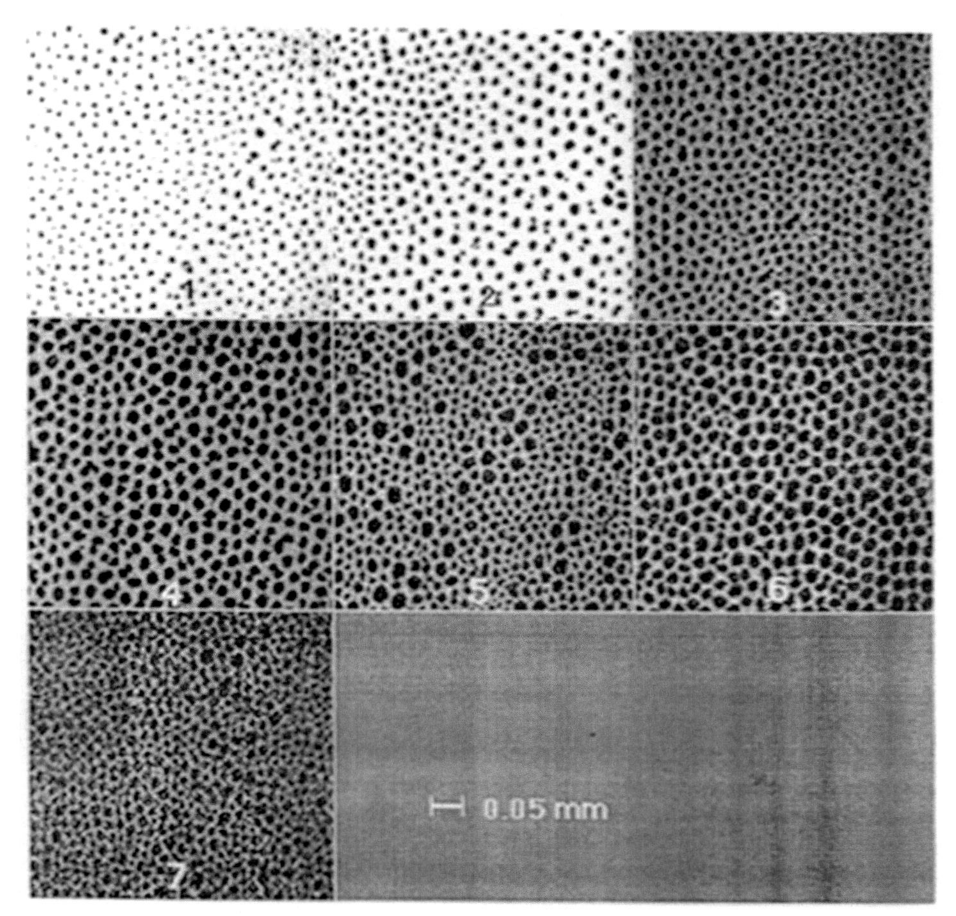

FIGURE 11.10 Epifluorescence images demonstrating simple liquid crystalline–gel phase separations as a function of lateral pressure on Langmuir Trough lipid monolayers. Monolayers were made from 70 mol% dipalmitoyl phosphatidylcholine (DPPC) (16:0, 16:0 PC) and 29 mol% SDPC (18:0, 22:6 PC) to which 1 mol% N-rhodamine-PE was added as the membrane fluorescent imaging agent. Compressions were run and images taken at every 5 mN/m (23°C). Plate 1, 5 mN/m; Plate 2, 10 mN/m; Plate 3, 15 mN/m; Plate 4, 20 mN/m; Plate 5, 25 mN/m; Plate 6, 30 mN/m; Plate 7, 35 mN/m. Gel state DPPC domains appear black while liquid crystalline state SDPC domains appear white [39].

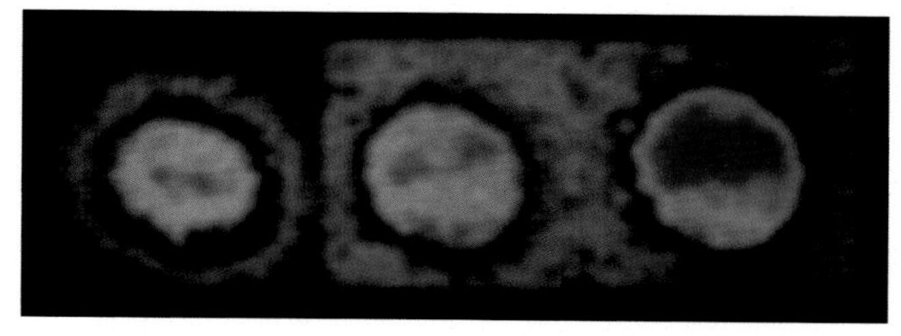

FIGURE 11.11 Fluorescence images of a Ca^{2+}-induced anionic phospholipid, phosphatidic acid (PA) domain. Vesicles were made from the bulk lipid 1,2-dioleoyl-sn-glycero-PC (DOPC) (99 mol% of the total phospholipids) to which 1 mol% of the fluorescent probe 4-nitrobenzo-2-oxa-1,3-diazole–phosphatidic acid (NBD–PA) was added. Images were taken 2, 15, and 30 min (left to right) after the addition of 2 mM $CaCl_2$. A faint fluorescent NBD-PA domain can be detected 2 min after addition of 2 mM Ca^{2+}. The domain grows and becomes more obvious after 15 and 30 min [40].

In a subsequent paper, Haverstick and Glaser extended their Ca^{2+} work to include the cationic peripheral protein cytochrome c and the transmembrane, 15-amino acid peptide gramicidin [41]. The vesicles were composed of 1 mol% NBD–PA, 4 mol% DOPA, 5 mol% danysl gramicidin with 90 mol% DOPC as the bulk lipid. The images are shown in Fig. 11.12. Panel (A) shows no discernable domains. Both NBD-PA (left) and dansyl-gramicidin (right) exhibit homogeneous dispersion of both the anionic lipid and the transmembrane peptide. Upon incubation for 30 min with 100 μM cytochrome c, both NBD–PA domains (left) and dansyl-gramicidin domains (right) were evident. This experiment demonstrated that cytochrome c can function in a similar manner to Ca^{2+} by inducing the formation of NBD–PA domains in a DOPC membrane. The experiment also showed that the cationic peripheral protein cytochrome c causes the transmembrane peptide gramicidin to be excluded from the PA-rich domain and to accumulate into the PC-rich domain. Also, measurement of the cytochrome c heme absorbance demonstrated that the peripheral protein location tracks with the PA domain. Therefore cytochrome c causes rearrangement of both lipid and protein components of a membrane.

The most relevant example of a lipid microdomain in a biological membrane is the highly controversial detergent-insoluble lipid raft. Lipid rafts were presented in Chapter 8 and their method of isolation will be discussed in Chapter 13. Rafts are an example of a lipid microdomain that may be partially isolated in crude form by extraction from the parent biological membrane. In reality the isolated raft is described by a lipid composition that is a little different than the bulk membrane, being enriched in sphingolipids and cholesterol. The raft fractions are also enriched in characteristic signaling proteins.

Since its early development in 1982 by Binnig and Rohrer [42], atomic force microscopy (AFM) has become a major imaging techniques for investigating small biological structures,

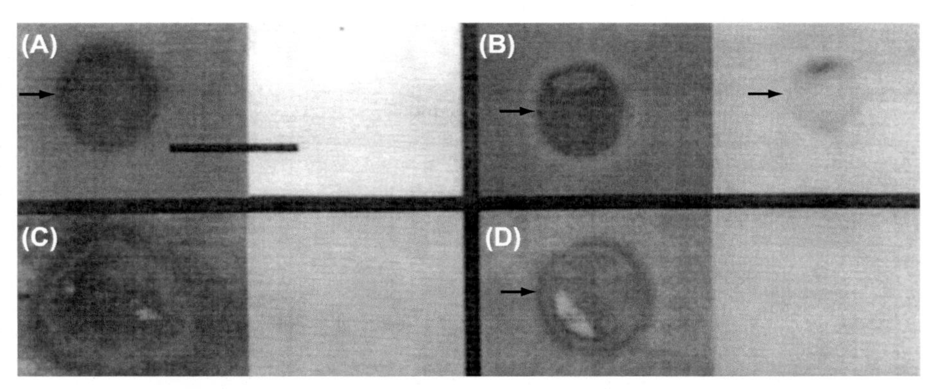

FIGURE 11.12 The cationic peripheral protein cytochrome c causes domain formation in phospholipid vesicles. Vesicles were formed containing 1 mol% fluorophore 4-nitrobenzo-2-oxa-1,3-diazole–phosphatidic acid (NBD–PA), 4 mol% 1,2-dioleoyl-sn-glycero-3-phosphoric acid (DOPA), and 95 mol% 1,2-dioleoyl-sn-glycero-phosphatidylcholine (DOPC). Vesicle domains were observed for NBD fluorescence. Bar indicates 10 μm. (A) A vesicle viewed immediately after the addition of 100 μm cytochrome c. Note no domains are visible. (B) A similar vesicle after 30 min in the presence of 100 μm cytochrome c. (C) A vesicle as in (A), viewed after 30 min in the presence of 10 μm cytochrome c. (D) A vesicle as in (B), (domain formation induced by 100 μm cytochrome c) further incubated for 30 min with 0.1 M NaCl. Fluorescent domains that are barely detectable in the presence of 10 μm cytochrome c are obvious with 100 μm cytochrome c [41].

including lipid rafts. For their discovery, Binning and Rohrer were awarded the 1986 Nobel Prize in Physics. AFM works by dragging a very sharp probe across the membrane surface while accurately measuring the vertical deflection of the probe due to changes in surface topography (Fig. 11.13, [43,44]). The probe therefore "feels" its way across the membrane surface allowing for accurate measurements with resolution down to <1 nm [45]. AFM also has the advantage of operating under water and so is applicable for use with biological membranes under near physiological conditions.

Since lipid rafts are enriched in predominately saturated, very long chain sphingolipids and cholesterol, they tend to be thicker than the surrounding nonraft membrane. Detecting small differences in surface height is exactly what AFM is best suited for. As a result, by AFM, the top of rafts appear lighter (they are higher) than the thinner and thus lower surrounding membrane domains. Fig. 11.14 shows an example of lipid raft microdomains imaged by AFM. A mica-supported lipid bilayer was made from equimolar DOPC (18:1, 18:1 PC) and brain SM to which the glycosylphosphatidylinositol (GPI)-anchored protein placental alkaline phosphatase (PLAP) was added [45]. The SM-rich raft domains appear orange and protrude above the black, DOPC-rich, nonraft domains by ~7 Å. The yellow peaks correspond to PLAP and are preferentially located in the rafts.

Before AFM, lipid rafts had historically been imaged by fluorescence microscopy. Raft gangliosides were imaged by specific labeling using fluorescently-labeled cholera toxin [46]. Model bilayer membranes consisting of various raft (eg, SM, cholesterol, ganglioside) and nonraft (eg, various unsaturated phospholipids) lipids. With fluorescent cholera toxin, raft domains light up. If the same model bilayer membranes are also imaged by AFM, overlapping domains are observed. Both techniques identify the same membrane patches as being

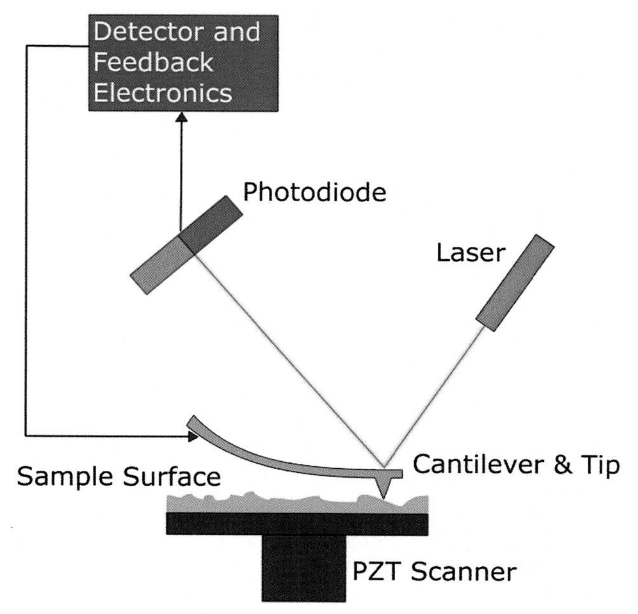

FIGURE 11.13 Schematic diagram of an atomic force microscope [43].

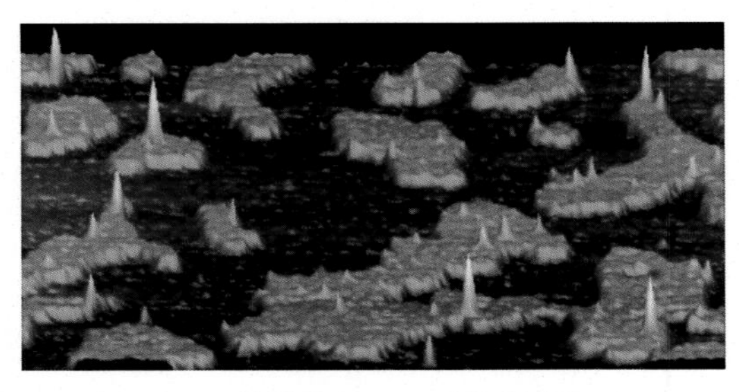

FIGURE 11.14 Atomic force microscopy (AFM) image of lipid bilayer membrane composed of equimolar 1,2-dioleoyl-sn-glycero-phosphatidylcholine (DOPC) and brain sphingomyelin (SM) to which the glycosylphosphatidylinositol (GPI)-anchored protein placental alkaline phosphatase (PLAP) is added [45]. SM-rich lipid rafts are in orange, the DOPC-rich nonrafts in black and placental alkaline phosphatase (PLAP) in yellow. The width of the scan is ~2 μm.

raft domains. Many other examples simultaneously imaging rafts by fluorescence and AFM can be found in the literature (for example see [47]).

While biophysical techniques generate many types of numbers that define membrane physical properties, these numbers must all eventually be tied in some way to membrane structure. If not, they are just free-floating, esoteric curiosities. For this reason, imaging methodologies form the foundation of membrane studies. Without an accurate depiction of a membrane, the generated numbers would be impossible to interpret. Many membrane images are discussed throughout this book. Since membranes are incredibly complex structures that are both very small and ever changing, obtaining accurate images is a difficult task.

One unusual method to generate a membrane picture is through molecular dynamics (MD) calculations that are of two types, atomistic and coarse-grained [48]. The objective of MD calculations is to use the power of computers to generate accurate depictions of each molecular component of a membrane and follow their movements and interactions over time. In essence, the MD calculations provide a motion picture of a membrane at the molecular level. Computer modeling can provide insights into the existence, structure, size, and thermodynamic stability of localized raft-like regions in membranes [48]. The atomistic MD simulations are at one end of the continuum (the most precise), while coarse-grained MD simulations are at the other end (they are the crudest). In the atomistic MD simulations, every atom in the membrane patch under investigation is included in the calculations. While this approach is ideal and is the eventual final goal of membrane studies, the required computing power is beyond current capacity. As a result, atomistic calculations are limited to small membrane patches and can only span very short times (a few nano-seconds). As computing capacity increases, membrane patch size and times will increase. The relatively small patch size causes problems since the calculations must impose unnatural boundary conditions to the patch. Also, nanoseconds are too short to observe lateral diffusion within the patch. One unsatisfactory solution is to just wait for the next generation of more powerful computers to be developed. Another solution is to "cheat" by not including each and every atom in the calculations. This would stretch the computer calculations over more molecules, generating bigger membrane patches and longer analysis times. These methods are referred to as coarse-grained MD simulations. A first approach to coarse-grained simulations is to eliminate all nonhydrogen bonding hydrogens as being nonessential. Another simplification involves treating several adjacent membrane atoms as one "pseudo-atom." These combined groups of atoms are of course artificial and are contrived only to reduce the MD calculations. By doing these and other "tricks" it is now possible to generate coarse-grained images for large membrane patches with times extending into the microsecond range, long enough for lateral diffusion to occur.

One essential feature of biological membranes is their extreme heterogeneity. Each membrane is composed of hundreds of different proteins and thousands of different lipids, all in constant flux. At any one instant in time, a membrane is composed of countless patches of different protein and lipid compositions and hence different functions. These fleeting patches are referred to as microdomains, the most renowned being lipid rafts [49]. A major focus in membrane science is understanding the composition and structure of and the forces behind microdomain formation, stability, and function. Recently, MD simulations have been employed in this endeavor [48].

One simple example of phase separation into a lipid raft-like domain was published in the *Proceedings of the National Academy of Sciences, US* by Risselada and Marrink [50]. These investigators used coarse-grained MD analysis of a three component model lipid bilayer to follow membrane phase separation of a lipid raft-like domain from a bulk fluid lipid bilayer. The results of this analysis are shown in Fig. 11.15. The ternary lipid membranes were composed of mixtures of dipalmitoyl-PC (16:0, 16:0 PC, depicted in green, Panel (A)), dilinoleyl-PC (18:2,18:2 PC, depicted in red, Panel (A)), and cholesterol (depicted in gray with a white hydroxyl group, Panel (A)). Phase separation of a raft-like liquid-ordered (L_o) phase enriched in 16:0, 16:0 PC and cholesterol was demonstrated from a nonraft phase consisting mainly of liquid-disordered (L_d) phase (18:2, 18:2 PC, depleted of cholesterol). It is well-documented from other types of imaging studies that the presence of polyunsaturated phospholipid tails (including 18:2, 18:2 PC) increases the tendency to form phase separated domains [51]. In the Risselada and Marrink report [50], phase separation was followed by the MARTINI coarse-grained model [52] and is evident in Fig. 11.15, Panels (B)−(E).

Fig. 11.15, Panel (B) follows phase separation of the completely mixed ternary lipid mixture (16:0, 16:0 PC; 18:2, 18:2 PC; cholesterol 0.42/0.28/0.30 mol%) at $t = 0$ to a de-mixed bilayer at $t = 20$ μsec. The analyzed membrane patch contains ~2000 lipid molecules and the view is from the top. $t = 0$ represents a warm temperature where complete mixing occurs. Quenching (demixing) begins upon decreasing the temperature to 295 K, inducing phase separation into raft-like (green) and nonraft, fluid (red) nanoscale domains. Fig. 11.15, Panel (C) shows the same lipid mixture as in Panel (B) except in a small, 20 nm diameter liposome. The view is also from the top and the final time represents 4 μsec. The unusual striped appearance of the domains in Panel (D), is an artifact resulting from the small size of the membrane patch that allows the domain to connect to itself across the (periodic) boundaries. Finally, the circular domains depicted in Panel (E) arises from increasing the size of the 18:2, 18:2 PC component (0.28 × 0.42 mol%) while decreasing the 16:0, 16:0 PC component (0.42 × 0.28 mol%). The raft-like (16:0, 16:0 PC−cholesterol) domain (green) can now be completely surrounded by a circular nonraft-like fluid domain composed of 18:2, 18:2 PC (red).

Importantly, structural and dynamic properties obtained from these MD simulations closely matched those from experimental data. Such matches are a target of MD simulations,

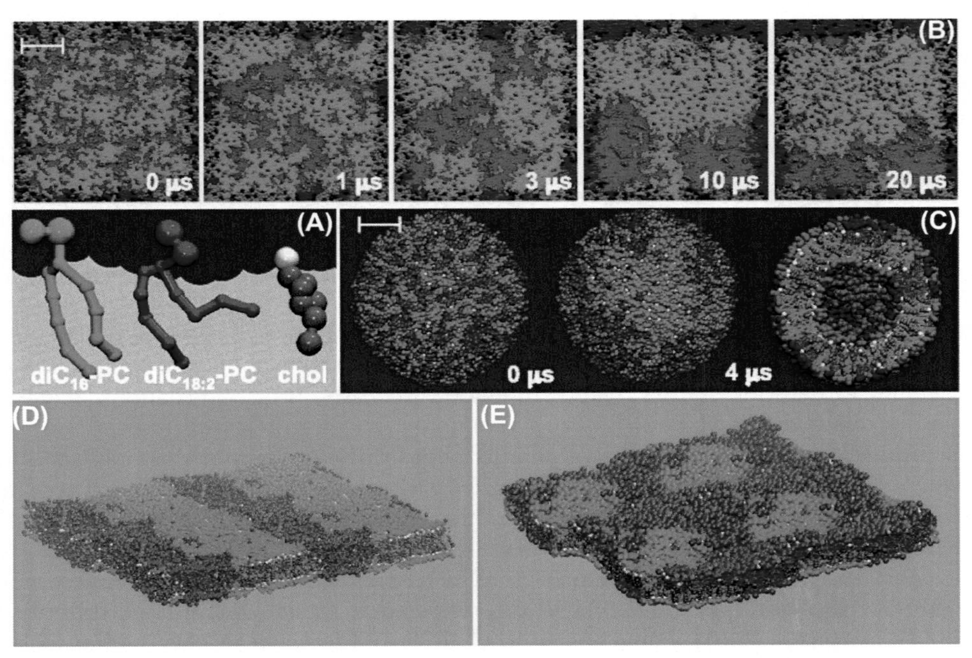

FIGURE 11.15 Formation of liquid-ordered (L_o) domains in ternary lipid mixtures followed by coarse-grained molecular dynamics (MD) calculations. Panel (A), color coding of the lipid components: 16:0, 16:0 phosphatidylcholine (PC) (*green*); 18:2, 18:2 PC (*red*); and cholesterol (*gray* with a white hydroxyl group). Panel (B), time-resolved phase segregation of a planar membrane, viewed from above, starting from a randomized mixture ($t = 0$), ending with the L_o/ liquid-disordered (L_d) coexistence ($t = 20$ μs). Panel (C), phase segregation for the same lipid mixture in a small, 20 nm diameter liposome. Initial ($t = 0$) and final ($t = 4$ μs, both top view and cut through the middle) configuration. Panel (D) and Panel (E), multiple periodic images (2 × 2) of the phase-separated 16:0, 16:0 PC /18:2, 18:2 PC–cholesterol systems show striped pattern formation in the 0.42:0.28:0.3 system, Panel (D) and circular domains in the 0.28:0.42:0.3 system, Panel (E). (Scale bar: 5 nm.) [50].

but in essence are cartoon depictions of what is already known. Occasionally however, new understandings can arise from the MD simulations that had not been observed experimentally. For example, the Risselada and Marrink report [50] predicts the existence of a small surface tension between the monolayer leaflets that drives the observed transbilayer registration of the inner- and outer-leaflet domains.

4. HOMEOVISCOUS ADAPTATION

Many of the essential membrane properties discussed in Chapters 9,10, and this chapter must be maintained within narrow constraints for life to continue. The well-studied examples of disease and aging are often associated with deleterious alterations of membrane physical properties. For warm-blooded (homeothermic) animals, membranes are naturally kept between a narrow, functional temperature range and are not exposed to harmful temperature fluctuations. However, most organisms on planet Earth have no internal temperature control mechanism (they are cold-blooded or poikilothermic) and are at the mercy of the external environment. Examples of poikilothermic organisms are reptiles, fish, plants, fungi, bacteria, protists and, even to some extent hibernating mammals. Diving mammals face similar problems, as an increase in environmental pressure affects membranes similar to a decrease in temperature. All organisms must be able to control environmental insults to assure proper membrane function. The effort to maintain proper membrane properties is known as homeoviscous adaptation (HVA), a general term where all temperature-dependent properties are lumped into "membrane viscosity." Viscosity is roughly the reverse of "fluidity" (see Chapter 9).

The concept of HVA was first proposed in 1974 by Sinensky from membrane lipid studies on *Escherichia coli* [53]. HVA caught on quickly and was supported by many studies on arctic fish, plants, fungi, and bacteria. Of particular importance was a series of papers by A.R. Cossins [54,55]. Cossins assessed membrane order (a measure of "fluidity") for a series of organisms of different body or habitat temperatures. The measured order parameters were found to directly follow the organism's temperature [Antarctic fish ($-1°C$) < perch ($15°C$) < convict cichlid ($28°C$) < rat ($37°C$) < pigeon ($42°C$)] [56]. These results indicate that "evolutionary adaptation to cold environments produces membranes of significantly lower order" [56], and this is at the heart of HVA theory.

In an excellent and thoughtful review, Jeffrey Hazel [56], agrees that lipids are indeed involved in HVA but suggests their effect is far more complex than standard HVA theory would suggest. An abbreviated list of membrane properties that are hard to incorporate into conventional HVA theory include: membrane remodeling, microdomain heterogeneity, specificity of lipid–protein interactions, and proliferation of mitochondrial and sarcoplasmic reticular membranes. Several membrane properties, closely linked to fluidity and likely involved in HVA, are briefly discussed below:

Acyl chain length: Shorter chain lengths have lower main melting transition (T_ms) than longer chain lengths (Chapter 4) and so accumulate in membranes of organisms exposed to lower temperatures.

Acyl chain double bonds: Double bonds reduce fatty acid T_ms and so also accumulate in membranes of organisms exposed to lower temperatures. For example, in cold acclimation, the proportion of di-unsaturated PC and PE in plants increases 50%. In bacteria, double bonds are often replaced by adding methyl branches or cyclopropyl rings (Chapter 4). Phytol chains also contain methyl branches and so can interfere with tight acyl chain packing associated with cold temperature.

Phospholipid head groups: Phospholipids with identical acyl chains often have widely different T_ms. Most important for HVA are the two major membrane structural phospholipids, PC and PE. The T_ms for PEs are usually $>20°C$ (or more) higher than homologous PCs (Chapters 4 and 5). Therefore another way to increase membrane fluidity in colder temperatures is to increase the membrane PC:PE ratio.

Sterols: Cholesterol is known to fluidize gel state membranes and reduce the fluidity of fluid state membranes (Chapters 5 and 9). Therefore membranes exposed to low temperature may employ cholesterol or other similar sterols to fluidize membranes.

Isoprene lipids: Although far less abundant than membrane sterols, other isoprene lipids such as vitamin E and D, could also serve to fluidize membranes at low temperatures.

Antifreeze proteins: Antifreeze proteins (AFPs) were first isolated by Arthur De Vries from Antarctic fish in 1969 [57]. These proteins have since been found in a wide variety of organisms including certain vertebrates, plants, fungi, and bacteria that survive in subzero environments. AFPs do not function as ordinary antifreeze solutes like ethylene glycol since they are effective at 300–500 times lower concentrations. They undoubtedly contribute to HVA of membranes.

Divalent metals: Divalent metals can induce anionic phospholipids to undergo isothermal phase transitions. By binding to the anionic headgroup, M^{2+}s impose order to the entire phospholipid, including the acyl chains. This can result in a substantial increase of the phospholipid's T_m, perhaps driving a liquid crystal state lipid into a gel state without any temperature change (isothermal phase transition). Table 11.3 shows the affect of Mg^{2+} and Ca^{2+} on the anionic phospholipid, dipalmitoylphosphatidylglycerol (DPPG). While 100 mM NaCl has no affect on the T_m of DPPG, 5 mM Mg^{+2} increases the T_m by 11°C and 5 mM Ca^{2+} increases the T_m a substantial 25°C. Another option for poikilothermic organisms would therefore be to decrease their membrane anionic phospholipid content or to sequester their major divalent metals.

Don Juan Pond: An extreme example of life adjusting to a very harsh environment is the Don Juan Pond in the west end of the Wright Valley in Victoria Land, Antarctica [58] (Fig. 11.16, [59]). Despite its provocative name, the pond is actually named for the two helicopter pilots, Lt. Don Roe and Lt. John Hickey, who discovered the pond in 1961. The Don Juan Pond is a small and very shallow hypersaline lake that almost never freezes despite temperatures that fall below −50°C. When it was first discovered, the pond was about 300 m long, 100 m wide and about 1 foot deep. Since then, the Pond has shrunk considerably. The Don Juan Pond is defined as being hypersaline as it is over 18 times more saline than the ocean and more than twice as saline as the Dead Sea. The major salt components of the pond are $CaCl_2$ (3.7 M) and NaCl (0.5 M). Don Juan Pond levels of NaCl have been reported to decrease the freezing point of water down to −21°C and $CaCl_2$ levels down to −50°C, accounting for the failure of the Don Juan Pond to freeze.

TABLE 11.3 Affect of Divalent Metals on the T_m of the Anionic Phospholipid Dipalmitoyl Phosphatidylglycerol (DPPG) (16:0, 16:0). The Unmodified T_m of DPPG is 42°C

Added cation	T_m (°C)
NaCl, 100 mM	42
Mg^{2+}, 1 mM	50
Mg^{2+}, 5 mM	53
Ca^{2+}, 1 mM	57
Ca^{2+}, 5 mM	67

FIGURE 11.16 The Don Juan Pond in Antarctica [54].

What is most surprising is the fact that the Don Juan Pond supports a variety of life! Microflora of yeasts, blue–green algae, fungi, and bacteria have been shown to inhabit the pond although they may lack extensive capability for continuous carbon reduction. In a 1979 report in *Nature*, Siegel et al. [60] reported an extensive, irregular pellicle or mat-like structure 2–5 mm thick inhabiting the Pond. Regardless of how abundant these Don Juan Pond organisms are, they certainly are a prime example of HVA.

5. SUMMARY

Membrane properties must support resident protein activity. One measurable property is the f_V, that quantifies the required breathing space for enzyme activity. Several important membrane properties, including area/molecule, collapse point, 'lipid condensation', surface elasticity, and molecular 'squeeze out' have been assessed with lipid monolayers using a Langmuir Trough. It was demonstrated that lipids vary in their lateral compressibility. Cholesterol and di-saturated phospholipids are poorly compressible, while polyunsaturated phospholipids are highly compressible. The lateral pressure of a biological membrane is estimated to be ~30–35 mN/m. Freeze fracture EM has supplied direct evidence that proteins indeed reside inside membranes and, without cytoskeleton involvement, there is no long-range order. Observed order is only a few tenths of a micron and the membrane bilayer sea is very crowded. Membrane domains can be roughly divided into large, stable macrodomains (eg, gap junctions, clathrin-coated pits, and caveolae) and small, unstable lipid microdomains (eg lipid rafts). In general, membrane properties are maintained by homeoviscous adaptation (HVA).

Chapter 12 will discuss the methodologies used to homogenize a cell and then isolate, purify, and analyze the various cell membrane fractions.

References

[1] Slater SJ, Kelly MB, Taddio FJ, Ho C, Rubin E, Stubbs CD. The modulation of protein kinase C activity by membrane lipid bilayer structure. J Biol Chem 1994;269:4868–71.
[2] Epand RM. Lipid polymorphism and protein-lipid interactions. Biochim Biophys Acta 1998;1376:353–68.
[3] Straume M, Litman BJ. Influence of cholesterol on equilibrium and dynamic bilayer structure of unsaturated acyl chain phosphatidylcholine vesicles as determined from higher order analysis of fluorescence anisotropy decay. Biochemistry 1987;26:5121–6.
[4] Mitchell DC, Straume M, Litman BJ. Role of sn-1-saturated, sn-2-polyunsaturated phospholipids in control of membrane receptor conformational equilibrium: effect of cholrsterol and acyl chain unsaturation on the metarhodopsin I ↔ metarhodopsin II equilibrium. Biochemistry 1992;31:662–70.
[5] Litman BJ, Mitchell DC. A role for phospholipid polyunsaturation in modulating membrane protein function. Lipids 1996;31:S193–7.
[6] Mitchell DC, Gawrisch K, Litman BJ, Salem Jr N. Why is docosahexaenoic acid essential for nervous system function? Biochem Soc Trans 1998;26:365–70.
[7] Mitchell DC, Litman BJ. Molecular order and dynamics in bilayers consisting of highly polyunsaturated phospholipids. Biophys J 1998;74:879–91.
[8] Mitchell DC, Litman BJ. Effect of cholesterol on molecular order and dynamics in highly polyunsaturated phospholipids bilayers. Biophys J 1998;75:896–908.
[9] Salem Jr N, Kim HY, Yergey JA. Docosahexaenoic acid: membrane function and metabolism. In: Simopoulos AP, Kifer RR, Martin R, editors. The health effects of polyunsaturated fatty acids in seafoods. New York: Academic Press; 1986. p. 263–317.
[10] Davies JT, Rideal EK. In: Interfacial phenomena. 2nd ed. New York: Academic Press; 1963. p. 265.
[11] Max Planck Institute for Polymer Research. Monolayers and the Langmuir-Blodgett technique – a brief review. mpip-mainz.mpg.de; 2008.
[12] Jain MK, Wagner RC. Introduction to biological membranes. New York: John Wiley and Sons; 1980. p. 61.
[13] Shaikh SR, Dumaual AC, Jenski LJ, Stillwell W. Lipid phase separation in phospholipid bilayers and monolayers modeling the plasma membrane. Biochim Biophys Acta 2001;1512:317–28.
[14] Bonn M, Roke S, Berg O, Juurlink LBF, Stamouli A, Muller M. A molecular view of cholesterol-induced condensation in a lipid monolayer. J Phys Chem B 2004;2004(108):19083–5.
[15] Yeagle P. The roles of cholesterol in biology of cells. In: Yeagle P, editor. The structure of biological membranes. Boca Raton (FL): CRC Press; 1992 [Chapter 7].
[16] Yeagle PL. The roles of cholesterol in the biology of cells. In: The structure of biological membranes. 2nd ed. Boca Raton, FL: CRC Press; 2005 [Chapter 7].
[17] Finegold LX, editor. Cholesterol in membrane models. Boca Raton (FL): CRC Press; 1993. 265 pp.
[18] Smaby JM, Momsen MM, Brockman HL, Brown RE. Phosphatidylcholine acyl chain unsaturation modulates the decrease in interfacial elasticity induced by cholesterol. Biophys J 1997;73:1492–505.
[19] Boonman A, Machiels FHJ, Snik AFM, Egberts J. Squeeze-out from mixed monolayers of dipalmitoylphosphatidylcholine and egg phosphatidylglycerol. J Colloid Interface Sci 1987;120:456–68.
[20] Severs N. Freeze-fracture electron microscopy. Nat Protoc 2007;2:567–76.
[21] Campbell N, Reece J. Biology. 7th ed. Pearson Education, Publishing as Pearson Benjamin Cummings; 2005.
[22] Ellis RJ. Macromolecular crowding: obvious but unappreciated. Trends Biochem Sci 2001;26:597–604.
[23] Yu J, Branton D. Reconstitution of membrane particles in recombinants of erythrocyte protein band 3 and lipid: effects of spectrin-actin association. Proc Natl Acad Sci USA 1976;73:3891–5.
[24] Edidin M. Shrinking patches and slippery rafts: scales of domains in the plasma membrane. Trends Cell Biol 2001;11:492–6.
[25] Robertson JD. Ultrastructure of two invertebrate synapses. Proc Soc Exp Biol Med 1953;82:219–23.
[26] Farquahr MG, Palade GE. Junction complexes in various epithelia. J Cell Biol 1963;17:375–412.
[27] Evans WH, Martin PE. Gap junctions: structure and function (Review). Mol Membr Biol 2002;19:121–36.
[28] Revel JP, Yee AG, Hudspeth AJ. Gap junctions between electrotonically coupled cells in tissue culture and in brown fat. Proc Natl Acad Sci USA 1971;68:2924–7.
[29] Goldstein JL, Brown MS. Receptor-mediated endocytosis: concepts emerging from the LDL receptor system. Annu Rev Cell Biol 1985;(1):1–39.
[30] Alberts B, et al. Essential Cell Biology. 3rd ed. Garland Science; 2010.

[31] Rappoport JZ. Focusing on clathrin-mediated endocytosis. Biochem J 2008;412:415–23.
[32] Gautier A, Bernhard W, Oberling C. Sur l'existence d'un appareil lacunaire pericapillaire du glomerule de Malpighi, revele par la microscopie electronique. C R Séances Soc Biol Ses Fil 1950;144:1605–7.
[33] Yamada E. The fine structure of the gall bladder epithelium of the mouse. J Biophys Biochem Cytol 1955:445–58.
[34] Anderson RG. The caveolae membrane system. Annu Rev Biochem 1998;67:199–225.
[35] Li X, Everson W, Smart E. Caveolae, lipid rafts, and vascular disease. Trends Cardiovasc Med 2005;15:92–6.
[36] Wolf DE, Kinsey W, Lennarz W, Edidin M. Changes in the organization of the sea urchin egg plasma membrane upon fertilization: indications from the lateral diffusion rates of lipid-soluble fluorescent dyes. Dev Biol 1981;81:133–8.
[37] Starling AP, East JM, Lee AG. Effects of gel phase phospholipid on the Ca^{2+}-ATPase. Biochemistry 1995;34:3084–91.
[38] Lee AG, Dalton KA, Duggleby RC, East JM, Starling AP. Lipid structure and (Ca^{2+})-ATPase function (review). Biosci Rep 1995;15:289–98.
[39] Dumaual AC, Jenski LJ, Stillwell W. Liquid crystalline/gel state phase separation in docosahexaenoic acid-containing bilayers and monolayers. Biochem Biophys Acta 2000;1463:395–406.
[40] Haverstick DM, Glaser M. Visualation of Ca^{2+}-induced phospholipid domains. Proc Natl Acad Sci USA 1987;84:4475–9.
[41] Haverstick DM, Glaser M. Influence of proteins on the reorganization of phospholipid bilayers into large domains. Biophys J 1989;55:677–82.
[42] Binnig G, Rohrer H, Gerber C, Weibel E. Phys Rev Lett 1982;1982(49):57.
[43] Wikipedia, the free encyclopedia. File: Atomic force microscope block diagram.svg.
[44] Morris VJ, Kirby AR, Gunning AP. In: Atomic force microscopy for biologists. 2nd ed. London: Imperial College Press; 2010. 406 pp.
[45] Henderson RM, Edwardson JM, Geisse NA, Saslowsky DE. Lipid rafts: feeling is believing. News Physiol Sci 2004;19:39–43.
[46] Saslowsky DE, Lawrence J, Ran X, Brown DA, Henderson RM, Edwardson IM. Placental alkaline phosphatase is efficiently targeted to rafts in supported lipid bilayers. J Biol Chem 2002;277:26966–70.
[47] Shaw JE, Epand RF, Epand RM, Li Z, Bitman R, Yip CM. Correlated fluorescence-atomic force microscopy of membrane domains: structure of fluorescence probes determines lipid localization. Biophys J 2006;90:2170–8.
[48] Pandit SA, Scott HL. Atomistic and coarse-grained computer simulations of raft-like lipid mixtures. Methods Mol Biol 2007;398:283–302.
[49] Jacobson K, Mouritsen OG, Anderson RGW. Lipid rafts: at a crossroad between cell biology and physics. Nat Cell Biol 2007;9:7–14.
[50] Risselada HJ, Marrink SJ. The molecular face of lipid rafts in model membranes. Proc Natl Acad Sci 2008;105(45):17367–72.
[51] Filippov A, Orädd G, Lindblom G. Domain formation in model membranes studied by pulsed-field gradient-NMR: the role of lipid polyunsaturation. Biophys J 2007;93:3182–90.
[52] Marrink SJ, Risselada HJ, Yefimov S, Tieleman DP, de Vries AH. The MARTINI force field: coarse grained model for biomolecular simulations. J Phys Chem B 2007;111:7812–24.
[53] Sinensky M. Homeoviscous adaptation—a homeostatic process that regulates the viscosity of membrane lipids in *Escherichia coli*. Proc Natl Acad Sci USA 1974;71:522–5.
[54] Cossins AR, Prosser CL. Evolutionary adaptation of membranes to temperature. Proc Natl Acad Sci USA 1978;75:2040–3.
[55] Behan-Martin MK, Jones GR, Bowler K, Cossins AR. A near perfect temperature adaptation of bilayer order in vertebrate brain membranes. Biochim Biophys Acta 1993;1151:216–22.
[56] Hazel JR. Thermal adaptation in biological membranes: is homeoviscous adaptation the explanation? Annu Rev Physiol 1995;57:19–42.
[57] DeVries AL, Wohlschlag DE. Freezing resistance in some Antarctic fishes. Science 1969;163:1073–5.
[58] Mitchinson A. Geochemistry: the mystery of Don Juan pond. Nature 2010;464:1290.
[59] Goldstein R. A cool place to work. The southern Illinoisan. Southern Illinois University, College of Science, Microbiology; 1999. http://www.micro.siu.edu/CoolPlace.html.
[60] Siegel BZ, McMurty G, Siegel SM, Chen J, LaRock P. Life in the calcium chloride environment of Don Juan pond, Antarctica. Nature 1979;280:828–9.

CHAPTER

12

Membrane Isolation Methods

OUTLINE

An Introduction to Biological Membranes
http://dx.doi.org/10.1016/B978-0-444-63772-7.00012-9

1. INTRODUCTION

Isolating biological membrane is as much of an art form as it is a science. The procedures are tedious and require patience, precision, and organization. Since every membrane has its own peculiarities, countless procedures have been published. This chapter will consider a few of the problems encountered in isolating membranes and some of the more common techniques that have been employed.

As discussed in Chapter 1, a eukaryotic cell is composed of an astonishing number of different membranes, all packed tightly into a very small volume. Membranes present a tiny and ever changing target for investigation, making their isolation particularly challenging. Every eukaryotic cell is surrounded by a relatively tough plasma membrane (PM) that surrounds countless more delicate internal membranes. This presents a dilemma. How does one break open the PM without severely impacting the tightly packed, interconnected, and delicate internal membranes? An old analogy seems appropriate. You are given a bag of different types and sizes of watches that had been smashed with a sledgehammer into thousands of intermingled, broken parts. From this you are asked to reassemble the watches and determine how they measure time!

Instantly upon breaking open a cell, the internal membranes are relieved of curvature strain by taking new morphologies, often in the form of similar-sized vesicles called microsomes [1].

1.1 Microsomes

The concept of microsomes is critical to membrane studies. Microsomes are small sealed vesicles that originate from fragmented cell membranes (often the endoplasmic reticulum [ER]). These vesicles may be rightside-out, inside-out, or even fused membrane chimeras. Microsomes may have unrelated proteins sequestered in their internal aqueous volumes or attached to their surfaces. Microsomes are therefore artifacts that arise as a result of cell homogenization and are very complex. By their very definition, microsomes per se are not present in living cells and are physically defined by an operational procedure, usually differential centrifugation (discussed below) [2]. In a centrifugal field, unbroken cells, nuclei, and mitochondria sediment out at low speed ($<$10,000g), whereas microsomes do not sediment out until much higher speeds ($\sim$100,000g). Soluble cellular components like salts, sugars, and enzymes remain in solution at speeds that pellet out microsomes.

Microsomes have been observed, if not understood, for a long time. In an early review from 1963, Siekevitz linked microsomes to remnants of the ER after cell homogenization [3]. Typically, discussions of microsome properties and isolation procedures are found buried in papers whose primary objective is to investigate a specific integral membrane protein (for example see [4]).

Once broken, all membrane fractions are instantly exposed to new osmotic stresses, divalent metal ions, unnatural pHs, and degradative enzymes including proteases, oxidases, lipases, phospholipases and nucleases [2,5–7]. Historically, sucrose has been the major osmotic component of cell homogenization buffers [8]. Sucrose is inexpensive, readily available in pure form, and is poorly permeable to most membranes. Importantly, sucrose does not destroy enzymatic activity. In many contemporary membrane isolations, Ficoll (GE Healthcare companies) has replaced most or all of the sucrose. Ficoll is an uncharged, highly branched polymer formed by the co-polymerization of sucrose and epichlorohydrin [9]. Due to its multiple (−OH) groups, Ficoll, like sucrose, is highly water-soluble. Often a little sucrose is added to the Ficoll to accurately control the density, viscosity, and osmotic strength of the buffer. Ficoll is extensively used in Density Gradient Centrifugations (discussed below). In addition to the osmoticum (eg, Ficoll, sucrose), many additional compounds are included in membrane isolation buffers. These include pH buffers, chelators to remove harmful divalent metals, antioxidants, and inhibitors of many degradative enzymes. Homogenization buffers are indeed a complex "soup".

The investigator must have many tools in his bag and work quickly to identify and separate the various membrane fractions before serious artifacts arise. At best it is impossible to obtain an absolutely pure membrane, just an enriched fraction. The general sequence of steps involved in membrane isolations are listed in Table 12.1 [5,6].

2. BREAKING THE CELL – HOMOGENIZATION

With the final objective in mind, the proper cell type is chosen and cleaned of unrelated debris such as veins. If the starting tissue is solid (eg, liver, muscle etc.), the target cells must be separated from neighboring cells. This process may involve use of proteases or other enzymes that disaggregate adjacent cells and may be aided by chelating agents that bind divalent cations. Disaggregation will alter some physical structures that reside on the cell PM outer surface (eg, gap junctions and tight junctions). Obviously if the objective is to isolate and reconstitute rhodopsin for light receptor studies, liver, brain, and muscle tissue would be inappropriate. Once sufficient quantities of the cell are obtained, the cell must be broken open by as gentle a method as possible. This process is referred to as "homogenization" [2,5,6,8]. Vigorous procedures result in damaged membranes and produce many small vesicular microsomes that are difficult to separate from one another. If the procedure is too gentle,

TABLE 12.1 Steps Involved in Membrane Isolation

BASIC STEPS
• Identify the best cell to obtain the membrane of interest. • Isolate and clean the cells. • Break open the cell (homogenization). • Separate large intact organelles (fractionation). • Separate the small microsomal cell fractions. • Identify the cell fractions by use of appropriate membrane markers (both positive and negative markers).
POSSIBLE ADDITIONAL STEPS
• Isolate and purify a specific membrane protein. • Isolate and purify membrane lipids. • Reconstitute the membrane protein into lipid vesicles. • Determine the mechanism of action of the reconstituted protein.

many cells will remain intact. Since it is estimated that even the simplest, routine membrane isolation procedures take ~15% of an investigators time [10], a variety of cell disruption methodologies have been developed for specific applications. For a brief comparison of the advantages and disadvantages of many of the methods discussed in this Chapter, see reference [11].

2.1 Homogenization Methods

2.1.1 Use of Enzymes

Cells that are encased in thick, tough cell walls (eg, plant, fungi, and some bacteria) present a particularly challenging set of problems. The cell wall must be removed while doing as little damage as possible to the more delicate underlying PM. Once the cell wall is removed, the remaining cell wall-free protoplast can be homogenized. A variety of commercially available enzymes are used to digest the cell wall [7,12,13]. These enzymes include cellulase, pectinase, lysozyme, lysostaphin, zymolase, glycanase, mannase, mutanolysin, and many more. Unfortunately, it is almost impossible to remove a cell wall without doing considerable damage to the delicate molecules residing below that are attached the outer surface of the PM. Also, the enzymatic methodologies for cell wall degradation are not applicable to large-scale preparations, and the enzymes must be inactivated or removed before the cell wall-free protoplasts are to be homogenized.

2.1.2 Shear Force (Mortar and Pestle)

The most commonly employed cell homogenization technique involves use of mortar and pestles. The basic concept involves forcing cells through a gap between two very close surfaces. The resulting shear force tears the cell apart [14]. The shear force will increase as the gap is narrowed or as the movement of the pestle relative to the fixed mortar is increased. Very accurate control of the shear force applied to the cells is essential. If the shear force is insufficient, cells will not be disrupted. If the force is too great the entire cell, including its organelles, will be broken into many different microsomes, and at extreme shear force, biochemicals may even be destroyed.

Several commercial mortar and pestle homogenizers are at the heart of most cell membrane studies. The most commonly used homogenizers are the Ten Broeck and Dounce homogenizers (Fig. 12.1), both of which are glass mortar and pestles that are purchased with preset, controlled gap sizes. Cells are forced through the gap by manual manipulation of the pestle. The process is normally performed in ice to prevent heat-induced destruction of cell components. The addition of a motor driven teflon pestle is part of the Potter-Elvehjem homogenizer (Fig. 12.1).

Although Van Rensselaer Potter (Fig. 12.2) and Conrad Elvehjem are now best known for their homogenizer, both made major unrelated contributions to the life sciences. Potter pioneered and named the field of Bioethics while Elvehjem discovered Niacin.

2.1.3 Blenders

Another very commonly employed type of homogenizer is generally referred to as a "blender". Blenders can vary from simple inexpensive, household food blenders or juice

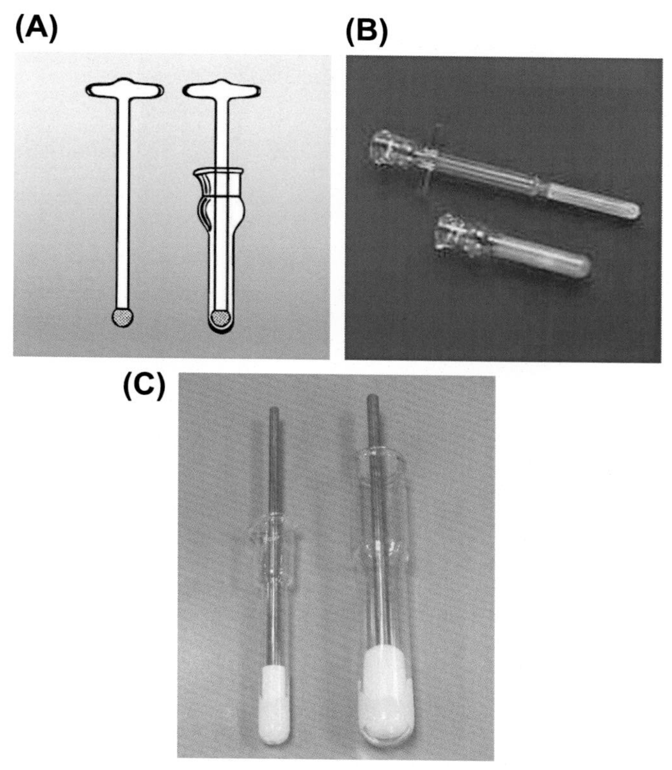

FIGURE 12.1 Commonly used mortar and pestle type cell homogenizers: (A) Dounce Homogenizer; (B) Ten Broeck Homogenizer and; (C) Potter-Elvehjem Homogenizer.

extractors to sophisticated, expensive instruments that employ very high-speed blades and specialized sample chambers.

The first simple food blender was introduced to the public in 1936 by big band leader and popular radio and television personality, Fred Waring, and was an immediate hit. However, Waring did not actually invent the blender. This was accomplished in 1922 by Stephen Poplawski and was later improved by Fred Osius in 1935. The Osius blender became the original Waring blender. In fact Fred Waring was only the shill that popularized the appliance. Originally, Waring called his blender the Miracle Mixer but later changed it to the Waring Blendor (note "or" not "er"). By 1954, 1 million Waring Blendors had been sold and even today Waring blenders are considered the standard for the field. Soon after its introduction, the Waring Blendor became a commonly used scientific instrument for cell homogenization. Jonas Salk used one in developing his polio vaccine. A Waring Blendor is shown in Fig. 12.3.

Although general food blenders are useful for many applications, they cannot efficiently disrupt microorganisms. A step-up from the simple Waring type blenders are the very high speed blade-type homogenizers, including the Brinkman Polytron and the Virtis Tissue

FIGURE 12.2 Van Rensselaer Potter, 1911–2001.

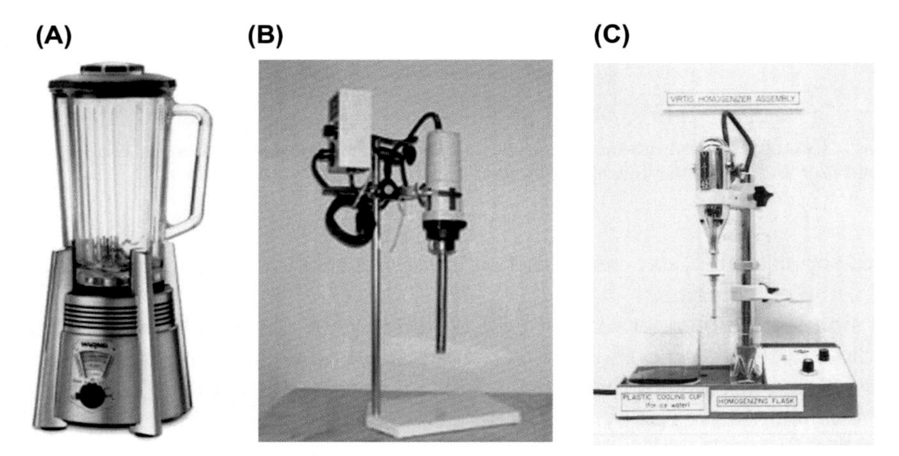

FIGURE 12.3 Types of blenders commonly used for cell homogenization: (A) Waring; (B) Brinkman Polytron Blender and; (C) Virtis Tissue Homogenizer.

Homogenizer (Fig. 12.3). These homogenizers employ a powerful electric motor to drive a shaft terminated by specially designed blades that spin at high speeds. As an example, the Virtis Cyclone IQ2 homogenizer has a 1/10HP motor mounted above a homogenizing flask/bottle. The blades rotate at between 5000 and 30,000 rpm and the sample volume can vary from 0.2 to 2000 mL.

2.1.4 Osmotic Gradients

As discussed in Chapter 2, the lipid bilayer component of a cell membrane is an excellent osmometer [15,16]. An osmometer is a device for measuring the osmotic strength of a solution. Membranes are osmometers because they are impermeable to most ions and water-soluble solutes while allowing water to readily diffuse across. The osmotic strength of one solution relative to another can be defined by three basic classifications: hypertonic, hypotonic, or isotonic. With regard to a cell, a hypertonic solution is one having a greater solute concentration outside the cell than is found in the cytosol. Water then leaves the cell and the cell shrinks (Fig. 2.1). In contrast, a hypotonic solution is one having a lesser solute concentration outside the cell than is found in the cytosol. Water then enters the cell and the cell swells. If the difference between the outside and cytosolic osmotic strength is sufficient, the cell may swell until it bursts. In the case of the erythrocyte, cell disruption releases internal hemoglobin in an easily followed process called hemolysis. As discussed in Chapter 2, hemolysis played an important role in early membrane studies (William Hewson in Chapter 2). Isotonic solutions contain equal concentrations of impermeable solutes on either side of the membrane and so the cell neither swells nor shrinks. The swelling or shrinking rate of an artificial lipid bilayer vesicle (called a "liposome", discussed in Chapter 13) is an excellent osmometer [15,16].

In most cases, intracellular organelles are more easily extracted if the cell is slightly swollen, enhancing its breakage [17]. Cells are more susceptible to breakage when placed in hypoosmotic buffers and are more stable in hyperosmotic buffers. While there are countless homogenization solutions, historically the major osmoticum has been either ~250 mM sucrose or ~120 mM KCl lightly buffered to pH 7.4.

2.1.5 Sonication

Ultrasonication employs high frequency sound waves, typically 20–50 kHz, to disrupt cells [18,19]. Ultrasonication can be used by itself for cell homogenization or more commonly in conjunction with other techniques. Light application of ultrasonication can even be used to separate cells from one another or to remove cells from a tissue culture substrate (glass or plastic).

There are three basic types of ultrasonication: tip, cup horn, and bath (Fig. 12.4). In a tip sonicator, the high frequency is generated electronically and the energy transmitted directly

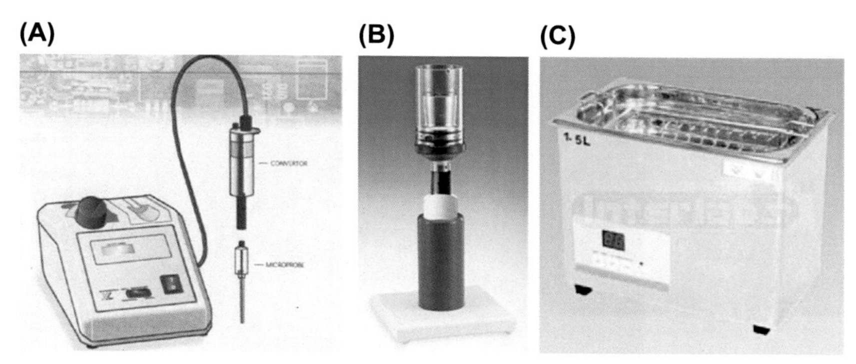

FIGURE 12.4 Common types of sonication homogenizers: (A) Tip Sonicater; (B) Cup Horn Sonicater and; (C) Bath Sonicater.

to the sample via a metal (titanium) tip that oscillates at high frequency. The rapid tip movement results in alternating very low pressure areas (cavitation) and high pressure areas (impaction) that can tear the cell apart. The technique is about 60 years old and advances continue to decrease required sample volume (now down to 200 μL or less) with more accurate control of the sonication parameters. Although tip sonication remains the most popular for cell disruption, it does have some serious problems. As the titanium tip oscillates, it slowly disintegrates. As a result the tip must be replaced regularly and the generated titanium particles must be removed from the cell sample. In addition the rapid tip oscillation generates a lot of heat that must be dissipated quickly to prevent destruction of the delicate biological material.

To avoid problems associated with inserting the titanium tip probe directly into the sample, another method has been developed where the tip is placed in a water bath adjacent to a chamber that holds the sample. A commercial device that accomplishes this is referred to as a "cup horn" (Fig. 12.4). Since the sonic energy must pass through a water bath and sample chamber wall before reaching the cell, this method is far gentler and less destructive than the direct tip method. Sonication is provided by a standard tip sonicater that is orientated upside down with the cup horn attached to the vertical sonicater shaft.

Another gentle, indirect sonication method involves use of a sonicating water bath, sometimes called a "jeweler's bath" (Fig. 12.4). No titanium tip is involved. Instead the walls of the bath supply the sonic energy and a vessel containing the cell sample is suspended in the bath. This is the least destructive of the three methods and has the advantage of providing accurate temperature control. However, its low energy, limits its membrane applications. Sonic baths are often used to suspend lipids in the preparation of liposomes.

2.1.6 Bead Beaters

Another commonly employed homogenization procedure involves "bead beaters" [20], an example of which is shown in Fig. 12.5. These devices disrupt cells by violently shaking them in the presence of small (usually) glass beads. Sometimes ceramic, zirconium, or steel beads

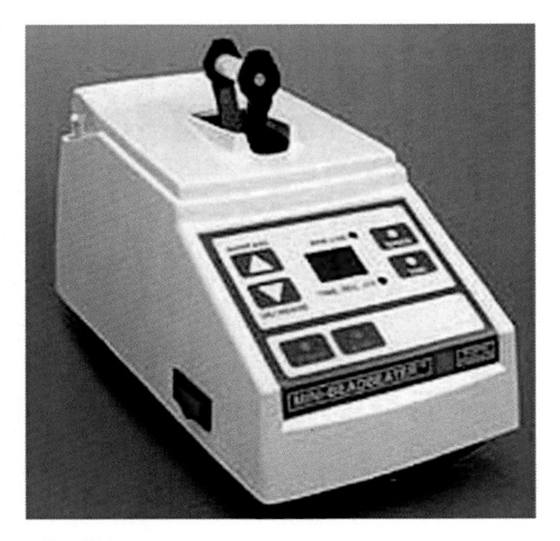

FIGURE 12.5 Bead Beater type of cell homogenizer.

are used. Bead beating is done in sealed vials typically containing 100 μL to 1 mL samples and bead sizes are typically between 0.5 and 1 mm diameter. Samples are agitated at 2000 to 5000 oscillations per minute in a specially designed clamp driven by a high energy electric motor.

2.1.7 Gas Ebullition (High Pressure "Bomb")

Very rapid cell homogenization can be achieved by use of what is known as a "High Pressure Bomb" [21]. The most commonly used pressure bomb for cell homogenization, a Parr Pressure "Bomb", is shown in Fig. 12.6. The basis of this method is similar to what divers experience as the bends (decompression sickness). The high pressures experienced by divers in deep water forces gas to dissolve into the diver's body fluids. If the diver returns to the surface too rapidly, the gasses (primarily nitrogen) leave the body fluids and return to the gaseous state as bubbles. The rapid formation of these bubbles results in a variety of painful abnormalities and even death. The first sign of pain is often in joints that are rendered stiff and cannot bend, hence the term "bends".

Cells are placed in a thick-walled chamber (Fig. 12.6) and put under high pressure (usually nitrogen or another inert gas up to about 25,000 psi) and the pressure is rapidly released. The rapid pressure drop causes the dissolved gas to be released as bubbles that lyse the cell.

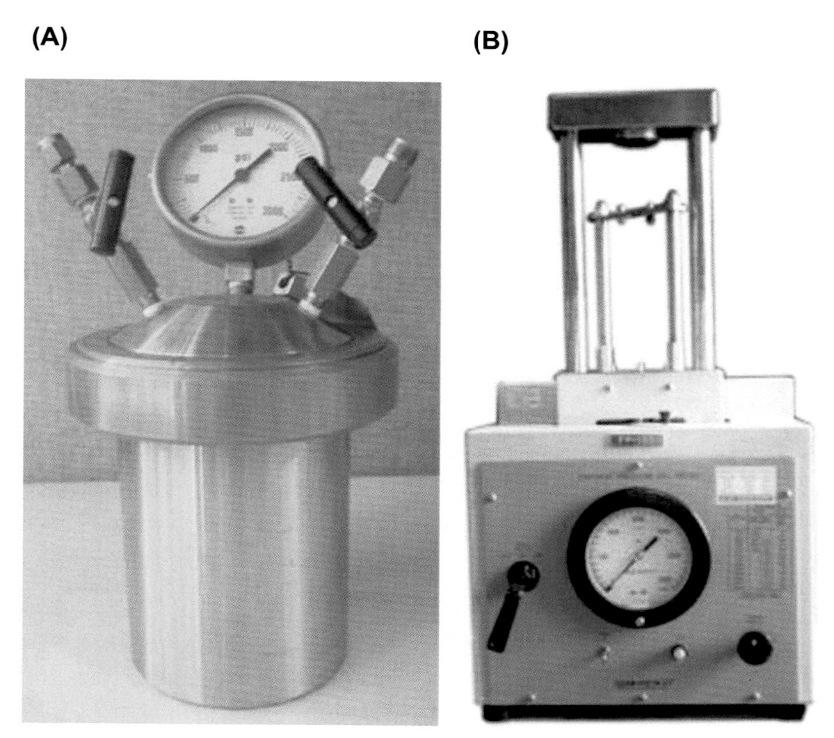

FIGURE 12.6 Parr Pressure 'Bomb' and French Press cell homogenizers: (A) Parr Pressure 'Bomb' and; (B) French Press.

2.1.8 French Press

A less frequently used type of cell homogenizer is the French Press [22] (Fig. 12.6). The device is named for its pioneer, Stacy French, a former plant physiologist at the Carnegie Institute of Washington. This apparatus functions by forcing cells through a narrow needle valve under very high pressure. The cells are exposed to large shear forces that are capable of disrupting very tough cells, even stripping off thick cell walls. It has been reported that the French Press preferentially produces inside-out microsomal vesicles that can have important biochemical applications. Disadvantages of the French press are its heavy, large size and its high cost. French presses typically are used for small samples (from 1 to 40 mL) and can reach pressures of 20,000 to 40,000 psi. By controlling the pressure and valve tension, the shear force can be regulated to optimize cell disruption.

2.1.9 Freeze/Thaw

The fact that ice occupies a larger volume than does an equal quantity of liquid water (water is denser than ice, see Chapter 3), can be used to disrupt cells [23]. Alternating rapid freezing in liquid nitrogen or in a slurry of dry ice and isopropanol, followed by thawing in warm water can homogenize cells. More often, freeze/thaw cycles are used as a preliminary step to be followed by another homogenization procedure.

3. MEMBRANE FRACTIONATION – CENTRIFUGATION

By far, the major technique that has been employed to separate the various cell fractions is centrifugation [8,24–26]. Two basic types of centrifugation exist, Differential and Density Gradient.

Differential Centrifugation is a first crude step to separate major cell fractions whose physical properties, size and density, are very different from one another. In fact at least one order of magnitude difference in the sedimentation coefficient (discussed below) is required to achieve separation, and even then separation is not complete.

After the tissue sample is homogenized the mixed cell contents are subjected to repeated centrifugations at increasing speed, a process known as Differential Centrifugation. After each centrifugation, the pellet (heavier component) is removed and the remaining cell suspension (lighter components) centrifuged at an increased speed. Separations are achieved from biggest (heaviest) to smallest (lightest) fractions (Table 12.2).

Centrifugation speeds in membrane studies vary over a wide range from ~1000g for 5 min for unbroken cells up to 1,000,000g for hours to finally pellet cytosolic macromolecules. Each step in Differential Centrifugation only enriches the target membrane. It does not completely purify the membrane.

The limitation of Differential Centrifugation results from large, heavy particles initially residing at the top of the centrifuge tube pelleting at the same time as smaller, lighter particles that are initially near the bottom of the tube. For example, consider the case of a mixture of heavy particle A and light particle B. The initial centrifugation pellets ~95% of A, but also pellets ~20% of B. If the pellet is separated, resuspended in fresh media and recentrifuged at the same speed, ~85% of the initial amount of A is found in the pellet, but only ~1%

TABLE 12.2 Differential Centrifugation of a Homogenized Cell (From Largest to Smallest Components)

Largest	Pellet 1	Unbroken cells
	Pellet 2	Nucleus
	Pellet 3	Mitochondria
	Pellet 4	Lysosome
	Pellet 5	Microsomes
	Pellet 6	Ribosomes
Smallest	Suspension	Cytoplasmic proteins, sugars, amino acids, nucleotides, salts etc.

of B. Therefore, Differential Centrifugation can only produce enriched fractions that may be further purified by repeated centrifugations. However, each centrifugation results in a significant loss of the target membrane. Final purification is often accomplished through subsequent Density Gradient Centrifugation.

Density Gradient Centrifugation is based on generating a centrifugation media that is densest at the bottom of the centrifuge tube and least dense at the top of the tube. The density gradient is used to separate particles on the basis of size and denity. How fast something moves in a centrifugal field is defined by the sedimentation coefficient (s).

$$s = 2r^2(\rho_p - \rho_m)/9\eta$$

Where, s = sedimentation coefficient, r = radius of the particle, ρ_p = density of the particle, ρ_m = density of the media, η = viscosity of the media.

From this equation it can be deduced that: bigger particles centrifuge faster than smaller particles; centrifugation rate is increased with increasing difference between density of the particle and density of the media; and centrifugation rate is slowed by an increase in media viscosity. Normally the sedimentation coefficient s has units of 10^{-13}. To avoid this cumbersome unit, the sedimentation coefficient is multiplied by 10^{13} generating the more commonly used term, the Svedberg Unit, S.

$$S = s \times 10^{13}$$

Density gradients are of two basic types, discontinuous (or step) and continuous. In most cases the cell homogenate is placed on top of the medium in the centrifuge tube and the centrifugation is started. The cell fractions will migrate down the density gradient until $\rho_p = \rho_m$ and s drops to 0. At this point the centrifugation is stopped and the now separated cell fractions individually isolated.

A discontinuous (or step) gradient is made by layering solutions of decreasing density on top of one another. The densest solution is placed at the bottom of the tube with successively lower density solutions on top (Fig. 12.7A). The solutions are carefully layered by use of a pipette or syringe. To prevent unwanted mixing of the solutions during gradient formation, the centrifuge tube is tilted ~20 degrees from perpendicular as the solutions are layered.

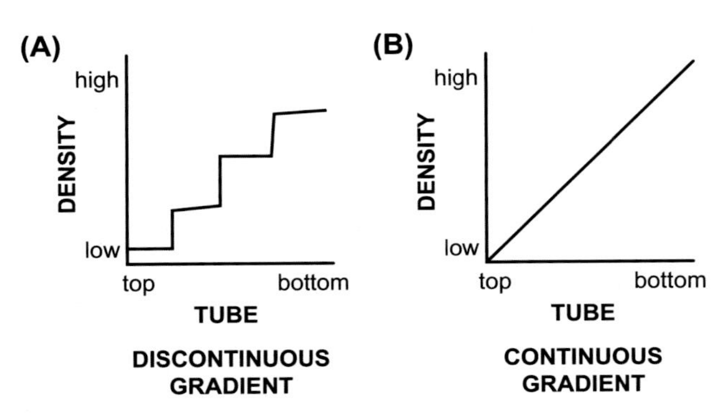

FIGURE 12.7 Density Gradient Centrifugation density profiles: (A) Discontinuous (or step) gradient and; (B) Continuous gradient.

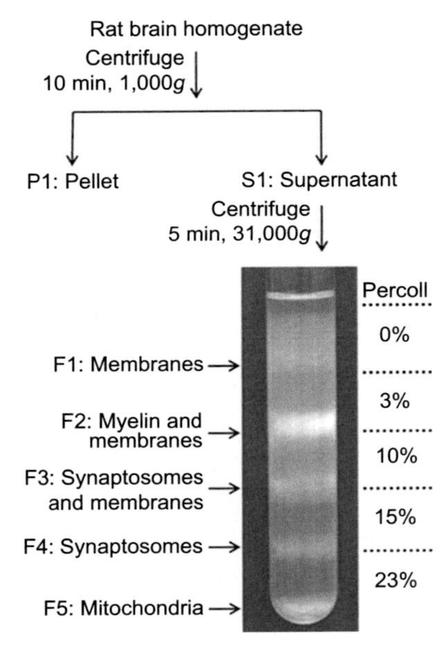

FIGURE 12.8 Discontinuous (step) gradient of various membrane fractions isolated from rat brain homogenate. Density of the five fractions was established by the amount of Percoll in each fraction (0%, 3%, 10%, 15% and 23%) [27].

This method is inexpensive and requires no special "gradient making" equipment. Since the solution volumes can vary from layer to layer and the densities need not decrease in a linear fashion, it is possible to tailor a specific gradient fraction to match the density property of the membrane of interest and maximize its yield.

Fig. 12.8 shows a discontinuous gradient of various membrane fractions isolated from rat brain homogenate [27]. Densities of the five fractions were determined by the amount of Percoll in each fraction (0%, 3%, 10%, 15%, and 23%).

A continuous gradient displays a linear decrease in solution density from the bottom to the top of the centrifuge tube (Fig. 12.7B). This gradient can be generated by either a "gradient maker" or can be self-generating during centrifugation. Fig. 12.9 demonstrates two ways of making a continuous gradient using a gradient maker, the lower density first method (A) and higher density first method (B). Both methods are based on mixing two solutions of different density before adding them to the centrifuge tube. Details of the procedures are given in the legend for Fig. 12.9.

Gradients can be shallow (the two initial solutions have similar densities) or steep (the two initial solutions have very different densities). Shallow gradients are used to separate particles of similar physical properties. While separation of large, cell homogenate components (eg, nuclei, mitochondria, chloroplasts etc.) is relatively easy and can often be done with just differential centrifugation, separation of microsomes (small cell vesicles) is difficult. All microsomes have similar physical properties including size, shape, and density but may be separated using shallow Density Gradient Centrifugation methodologies.

Some solutes have the ability to self-generate a density gradient during centrifugation. Examples include cesium chloride (CsCl), Percoll, Ficoll, and OptiPrep. The major application for CsCl is in separating very dense molecules like nucleic acids. The density of DNA is about

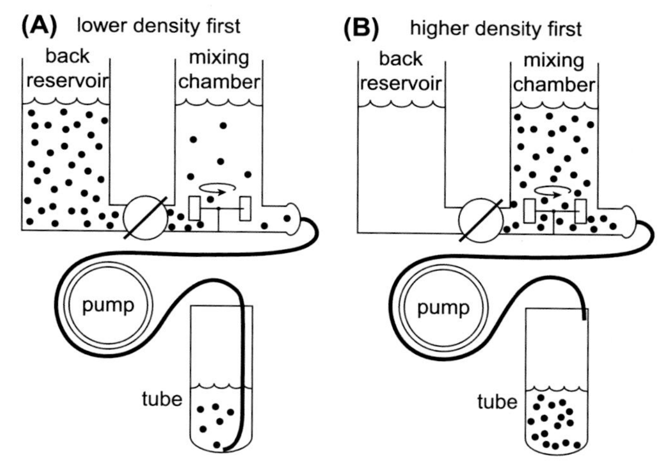

FIGURE 12.9 Formation of continuous density gradients using a density gradient maker. The device consists of two chambers, one of which contains a heavy (dense – black balls) solution and the other a light (less dense), clear solution. The densities of the two solutions will eventually define the solution density at the bottom and top of the centrifuge tube. The chamber on the right (the mixing chamber) has a mixer and an outlet where the mixed solution is pumped into the centrifuge tube. The chamber on the left holds a reservoir that is slowly added to the mixing chamber, but has no direct outlet to the centrifuge tube. In panel A (lower density first), the lower density solution, initially in the mixing chamber, is first added to the bottom of the centrifuge tube. As the process continues the denser solution, initially in the back reservoir, is slowly added to the mixing chamber, increasing its density. Therefore the initial low density solution is continuously replaced from the tube bottom by a more dense solution, creating the gradient. In the right panel (higher density first) the solutions are reversed, with the denser solution in the mixing chamber. Therefore the first solution added to the tube is the most dense. In contrast to the low density first method, the solution is added from the top of the centrifuge tube. The gradient is made by adding less dense solutions from the top.

TABLE 12.3 Densities of Some Major Biomolecules (g/mL)

Lipids	0.9–1.1
Proteins	1.25
DNA	1.7
RNA	2.0

1.7 g/mL and RNA about 2 g/mL while membrane vesicles fall between 1.1 and 1.3 g/mL, the range covered by Percoll, Ficoll, and OptiPrep. Table 12.3 lists the densities of some major biomolecules.

In Fig. 12.10, particle density (g/mL) is plotted against sedimentation coefficient in Svedberg Units (*S*) [28]. Clearly all membrane vesicle fractions (microsomes) are close together, making them very difficult to separate from one another. DNA and RNA on the other hand are much denser than the membrane fractions and can be easily separated on CsCl gradients.

Percoll (GE Healthcare [29]) consists of silica particles (15 to 30 nm diameter) coated with nondialyzable polyvinylpyrrolidone. Percoll is nontoxic, chemically inert, readily adaptable to physiological ionic strength and pH, is of low viscosity, and can be adapted to discontinuous or continuous gradients. Importantly it does not stick to membranes. It is its role in continuous gradient centrifugation that makes Percoll so attractive for membrane isolations (for example see [27]). A continuous Percoll gradient can be simply and quickly generated by moderate speed centrifugation (200 to 1000 g, for a few minutes).

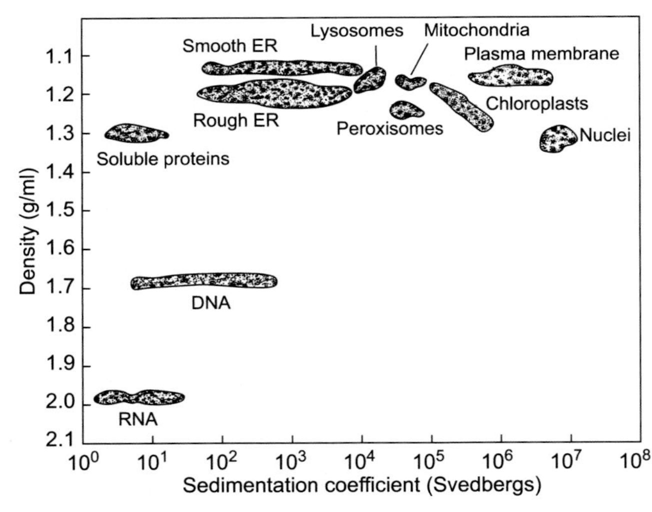

FIGURE 12.10 Particle density in g/mL is plotted against the log of the sedimentation coefficient in Svedberg Units. The values for DNA and RNA were determined from CsCl gradients [28].

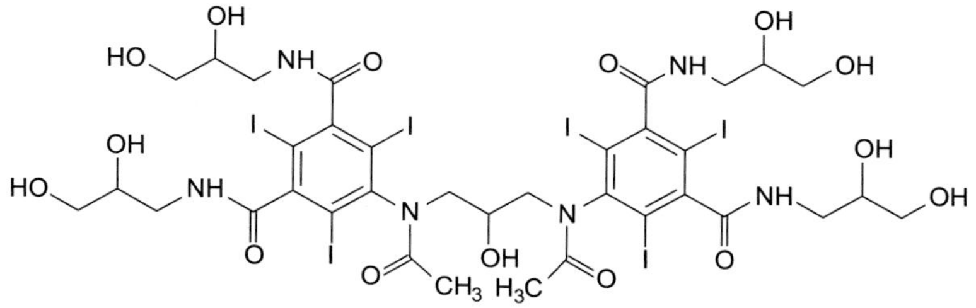

FIGURE 12.11 Iodixanol.

Ficoll (GE Healthcare) is an uncharged, highly branched polymer formed by the co-polymerization of sucrose and epichlorohydrin. Due to its many (–OH) groups, it readily dissolves in water. Like Percoll, Ficoll can be adapted to discontinuous or self-generating continuous gradients [29].

More recently a new self-generating density gradient medium, OptiPrep (Axis Shield [30]), has significantly impacted the field of membrane purification. OptiPrep is a solution of 60% iodixanol (Fig. 12.11) in water.

Iodixanol is nontoxic to cells, is metabolically inert, and has low viscosity and osmolality. OptiPrep avoids the high viscosity problems associated with sucrose and Ficoll gradients and the inconvenience of removing Percoll from membrane fractions after centrifugation. OptiPrep has been successfully applied for the isolation of PMs and their domains, including lipid rafts.

Continuous Density Gradient Centrifugations are of two basic types, isopycnic (or equilibrium) and rate zonal. In isopycnic, centrifugation is stopped when the particle no longer moves down the density gradient. At this point, density of the particle equals density of the medium and $S = 0$. Sometimes two different particles may reach the same or very similar final location in the gradient and so cannot be separated after completion of the centrifugation. However, the particles may have different physical properties (eg, size, shape) and so reach the same final location at different rates. In rate zonal, centrifugation is stopped before the particles reach final equilibrium. At this point, density of the particles being separated are greater than density of the medium and so the particles may be separated from one another.

4. MEMBRANE FRACTIONATION – NONCENTRIFUGATION METHODS

Isolation of membrane fractions remains very much an art form and so there appears to be almost as many specific procedures as there are membranes to be isolated. While most isolation procedures are centered around centrifugation, an almost limitless number of clever adaptations or "tricks" have been employed for specific applications. Often these "tricks" are combined with standard centrifugation methods to enhance final membrane purification. A brief sample of these techniques are discussed in this section.

4.1 Affinity Chromatography

Affinity Chromatography is the most used of the noncentrifugation methods [31,32]. The most commonly employed applications for the isolation of membranes use lectins or antibodies. Fortunately, lectins and antibodies are among the most commercially available biochemicals and are ideally suited for column chromatography separations. Both applications work best for PM vesicles (microsomes). As discussed in Chapters 5 and 7, carbohydrates are attached to outer leaflet PM sphingolipids (cerebrosides, globosides, and gangliosides). Most PM proteins are also heavily glycosylated (Chapter 6), with the sugars always facing outside of the cell. In fact, carbohydrates are a characteristic of the PM and so are obvious targets for membrane purification schemes.

4.1.1 Lectin-Affinity Chromatography

Lectins play a central role in membrane studies. Their history and function are thoroughly discussed in several excellent reviews by Nathan Sharon [33,34]. The term lectin is derived from the Latin word *legere*, "to select", and indeed that is what lectins do. Lectins are proteins of nonimmune origin that reversibly bind to specific sugars without modifying them. They are ubiquitous in nature, being found in animals, plants, insects, microorganisms, and even humans. Each lectin has two or more separate binding sites and so are bifunctional reagents that have the ability to clump or agglutinate cells. It is this property that led to their 1888 discovery by Herrmann Stillmark (Fig. 12.12) at the University of Dorpat in Estonia. In his PhD. thesis. Stillmark described the agglutination of erythrocytes by the lectin ricin isolated from castor bean (*Ricinus communis*).

FIGURE 12.12 Hermann Stilmark, 1860–1923.

TABLE 12.4 Carbohydrate Specificity for Some Common Lectins

Abbreviation	Name	Carbohydrate bound
Con A	Concanavalin A	Mannose, glucose
GNA	Snowdrop Lectin	Mannose
RCA	*Ricinus communis* Agglutin	Galactose, N-Acetyl glucosamine
PNA	Peanut Agglutin	Galactose, N-Acetyl Galactosamine
WGA	Wheat Germ Agglutin	Sialic Acid, N-Acetyl glucosamine
UFA	*Aleuria europaeus* Agglutin	Fucose

Countless lectins have now been identified making their classification difficult, but increasing their versatility in membrane isolations. For example, the Sigma–Aldrich catalog offers almost 500 lectins from dozens of sources! Lectins vary from those that bind to a specific simple sugar to many that bind to, di-, tri-, and even polysaccharides. Table 12.4 lists a few common lectins with their carbohydrate specificity. This Table is overly simplified as several of the lectins actually bind to a di- or tri-polysaccharide containing the listed sugars and not just the simple sugar itself. Lectins are bound to carbohydrates by many weak reversible noncovalent bonds [35,36], similar to binding of a substrate to the active site of an enzyme. Most lectins actually share little structural similarity with one another as demonstrated by their amino acid sequence, molecular size, and other molecular properties. It is hard to comprehend how these widely varying structures can all bind carbohydrates.

Many free lectins are commercially available and several are even available conjugated to inert matrices that can be packed into columns for use in lectin-affinity chromatography. Conjugation to the inert matrix, often Sepharose or Agarose, does not affect the lectin's ability to bind to specific sugars, and so is useful for membrane isolations. Sugar binding is also not affected by detergents, compounds that are essential for membrane protein isolation (Chapter 13). Lectin-affinity chromatography consists of pouring the cell homogenate over a lectin column. Membranes that do not have the requisite sugar selected by the lectin simply pass through the column and are removed. The bound membranes are washed and then released from the column by a solution containing a high concentration of the simple sugar the lectin recognizes. A single pass, over an appropriate lectin-affinity column can greatly purify a mixed membrane population in one step. However, this procedure only works for membrane fractions that have sugars on their surfaces, mostly PM microsomes.

The fact that a simple sugar can block the attachment of a lectin to its normal polysaccharide target can have important medical implications. For example, a urinary track infection can result from binding of *Escherichia coli* (*E. coli*) to a resident urinary track lectin [37]. It was noted that the *E. coli* infection can be greatly reduced by administrating mannose which dislodges the *E. coli*. This may explain why cranberry juice, rich in mannose, has historically been used to prevent bladder infections.

4.1.2 Antibody-Affinity Chromatography

Another important type of affinity chromatography employs columns with attached antibodies [32,38]. Antibody-affinity chromatography is more versatile and powerful than lectin-affinity chromatography. While use of lectins is limited to membranes possessing surface carbohydrates, appropriate antibodies can bind to any surface antigen including carbohydrates and proteins. Lipids and nucleic acids are antigenic only when bound to carbohydrates or proteins. In addition, antigen–antibody binding is much stronger than carbohydrate–lectin binding.

The term antigen was coined in 1899 by Hungarian microbiologist Ladislas Deutsch (he also had an alias, Laszio Detre). Antigen is a contraction of the original name "antisomatogen". An antigen is simply any molecule or part of a molecule that is recognized by the immune system and can bind to the antigen-binding site of an antibody. Antibodies have the ability to recognize and bind specifically to an antigen in the presence of a vast solution of molecules that are often similar in structure. The precise specificity of an antibody for an antigen makes them ideal for membrane purification by affinity chromatography. Countless custom antibodies are now commercially available for attachment to column matrix material. A crude membrane homogenate is passed over the antibody column whereupon only the membrane fractions containing the appropriate surface antigen bind. Other membrane fractions pass through the column and are discarded. After washing the column, the bound membrane is released by altering the pH of the washing solution.

4.1.3 Ligand-Receptor Affinity Chromatography

Another type of affinity chromatography employs ligand-receptors. The outer surface of the PM has a large number of proteinaceous receptors whose function is to bind specifically to external molecules referred to as ligands. Included in a long list of ligands are small signaling peptides, hormones, neurotransmitters, vitamins, and toxins. An unusual application employs zinc-affinity columns to isolate proteins that require zinc cofactors to function [39]. Ligands bind to, and dissociate from receptors according to the law of mass action. By this methodology, ligands are covalently bound to the column matrix and the cell homogenate is passed over the column. Membranes with the receptor for the ligand bind to the column and the nonbound membranes pass through the column and are discarded. Ligand-bound membrane fractions are released from the column by washing with a large excess of free ligand. While ligand affinity columns have occasionally been used to separate membrane fractions, their major use has been in isolating and purifying receptors for signal transduction studies.

4.2 Anion Exchange Chromatography

Ion exchange chromatography [40] has been an important scientific tool since its discovery in about 1850. Anion exchange columns are made of resin beads, often agarose or cellulose, to which is covalently bound a cationic functional group. The stationary phase (the resin beads) interacts with anions dissolved in a mobile, aqueous phase. Anion exchange chromatography separates compounds based on their net negative surface charge. Since biological membranes are anionic, they are amenable to this technique. Membrane phospholipids are either neutral (they are zwitterions like phosphatidylcholine) or they are anions (full anions like phosphatidic acid, phosphatidylserine, phosphatidylinositol, phosphatidylglycerol, cardiolipin, or a partial anion like phosphatidylethanolamine). In addition, gangliosides by definition contain the anionic sugar sialic acid and so provide negative membrane surface charge. As a result, all membranes have varying degrees of negative charge density and will bind to anion exchange columns. The higher the negative charge density, the tighter the membrane binding. As with affinity chromatography, a cell homogenate is passed over the anion exchange column and washed with water. Membrane fractions are released from the column with either increasing retention time in the presence of a fixed anion (normally Cl^-) concentration or by adding solutions of increasing anion concentrations. The poorest binding membrane fraction has the lowest negative charge density and is the first to be released. The most negative membrane fraction is the tightest bound and so is last to be released.

There are now many examples of commercially available anion exchange resins. The leader in the field is Dow Chemical Company's division Dionex Corporation (Dow-Ion Exchange). Membrane separation by anion exchange chromatography has serious limitations and does not have the specificity and precision of affinity chromatography.

4.3 Derivitized Beads

4.3.1 Magnetic Beads

Since the 1980s, magnetic micro- and nano-beads have been used for a wide array of biological applications [41], including membrane separations. Unmodified magnetic beads would be of little or no value to membrane isolations. However, magnetic beads have proven to be readily modified by several of the same molecules known to work so well with affinity chromatography (eg, lectins and antibodies). In addition, magnetic beads have assumed a major role in molecular biology by being modified to bind to specific nucleic acid sequences, thus allowing separation of various nucleic acids [42] and isolation of DNA-binding proteins.

Separations using magnetic beads have proven to be not only versatile but also extremely gentle. The rougher techniques of column chromatography and centrifugation can often be completely avoided. The technique is quick, easy, sensitive, inexpensive, and amenable to automation.

The leader in the field of magnetic beads is Dynabeads (Invitogen, Life Technologies [43]). Dynabeads are exceptionally uniform spheres of highly defined composition [43]. Importantly they are superparamagnetic, meaning they have magnetic properties when placed in a magnetic field but loose these properties completely when removed from the field. A uniform polymer shell surrounding the superparamagnetic core can be modified through a variety of bioreactive conjugates (eg, lectins, antibodies etc.) and separates the inert internal core from active surface components that protrude into the external aqueous environment. An appropriately modified Dynabead will attach specifically to a target membrane in a complex cell homogenate. A magnet placed on the outside of the tube containing the homogenate will attract the Dynabeads to the tube wall. The remaining, nonattached membrane fragments can then be simply discarded. After adding new media, the external magnet is removed and the desired membrane attached to the Dynabeads is left free in solution where it can be readily removed from the beads.

4.3.2 Other Types of Microbeads

There has been a recent explosion in the development and application of a wide variety of microbead technologies. One leading company, Phosphorex, Inc. (Fall River, MA), offers microspheres and nanospheres with sizes ranging from 20 nm to 1000 μm. The beads are made from poly-(styrene), poly-(methyl methacrylate), poly-(acrylic acid), nylon, poly-(lactic acid), (poly-(lactic-co -glycolic acid)), chitosan, and many other synthetic and natural polymers. They can be modified through attachment of several biomolecules including lectins and antibodies. These microspheres and nanospheres have the additional property of being able to sequester molecules into their aqueous interiors, making them suitable as drug delivery vehicles (Chapter 23). Membrane separations with modified microspheres and nanospheres parallel methods discussed previously for affinity chromatography (Section 4.1).

4.4 Other Methodologies

4.4.1 Two Phase Partitioning

This unusual methodology is based on membranes partitioning between two co-existing, but immiscible liquid phases. The phases are somewhat analogous to what is seen in some household novelty items that are commonly found in department stores. One is a slowly rocking device that generates what resembles highly colored ocean waves. The second is the more popular "lava lamp" [44]. The lava lamp is a recent addition to pop culture, invented by Edward Craven-Walker (Fig. 12.13) in 1963. The immiscible lava lamp solutions are often water−glycerol and wax−carbontetrachloride. A heat source in the form of an incandescent light bulb decreases the wax phase density, causing it to rise. The lighter wax then rises as a blob and slowly cools, sinking back to the bottom in a lava-like display. For membrane separations, partitioning is between two high molecular weight immiscible

FIGURE 12.13 Edward Craven-Walker, 1918−2000.

liquids, often polyethylene glycol and dextran [45]. In one example Yoshida and Kawata [46] reported that "mung bean tonoplast has a high partition coefficient for the polyethylene glycol-enriched upper phase and the smooth endoplasmic reticulum (SER) has a high partition coefficient for the Dextran-enriched lower phase". Two phase separation of membrane fractions is at best poorly understood and is presently only a curiosity that will likely never receive wide application.

4.4.2 Silica Particles

Silica particles have been used to isolate PM fractions [47]. Affinity is provided by charge attraction as silica microbeads are positively charged while PM fractions are negatively charged. Silica microbeads are also dense and can be easily centrifuged away from other, nonbound and hence lighter, membranes. The procedure begins by binding whole cells to the dense silica particles. Upon cell lysis, large open sheets of PM remain bound to the silica and are isolated away from the unbound internal membranes by centrifugation. Using this procedure, Chaney and Jacobson reported that PM from *Dictyostelium discoideum* was obtained in high yield (70 to 80%) and purified 10- to 17-fold [47].

4.4.3 Separation by Size

It is relatively easy to separate organelles from intact cells and even from one another on the basis of size and this is normally accomplished by differential centrifugation. However separation on the basis of size can also be achieved by common filtration techniques including Millipore Filters, Sephadex (size exclusion chromatography) and PAGE electrophoresis (polyacrylamide gel electrophoresis). These techniques have much more important applications in other aspects of bioscience, but have occasionally been used for membrane separations. For example, Millipore Filters are extensively used to remove contaminants from laboratory water and to size liposomes (Chapter 13). Sephadex is used to separate nonsequestered solutes from solutes sequestered inside liposomes (Chapter 13) and PAGE electrophoresis for protein purification.

4.4.4 Membrane-Specific "Tricks"

Membrane purification is as much an art as it is a science. A large number of what could be best described as "tricks" have been employed for very specific applications. These applications have little or no general use in membrane purification procedures. Three examples are:

1. It is often difficult to separate microsomes derived from the rough endoplasmic reticulum (RER) from those of the smooth endoplasmic reticulum (SER). However, in the presence of 15 mM CsCl, RER microsomes aggregate, greatly facilitating the separation.
2. Rat liver lysosomes can be hard to separate from mitochondria and peroxisomes due to their similar sedimentation coefficients. Nobel Prize recipient Christian de Duve (in Physiology or Medicine, 1974) reported that if the detergent Triton WR 1339 was injected into the rat prior to organelle extraction, lysosomes accumulated the detergent making them lighter [48]. With detergent, lysosome density decreased from 1.21 to 1.12 facilitating their separation from the unmodified mitochondrial and peroxisomal fractions.
3. Finally, an ER separation has been reported based on the presence of an ER enzyme, glucose 6-phosphatase. Glucose 6-phosphate, the substrate for the enzyme, enters the

ER microsomes via a transmembrane transporter protein. Once inside the vesicle, glucose 6-phosphate is cleaved yielding glucose and free phosphate. It was noted that if the membrane fractions were bathed in Pb^{2+}, insoluble lead phosphate ($Pb_3(PO_4)_2$) accumulated in the ER microsome interior, significantly increasing their density. The ER microsomes could then be easily centrifuged away from the other microsomes.

5. MEMBRANE MARKERS

For each of the many steps in isolating and purifying a membrane, tests must be run not only to demonstrate that the target membrane is indeed present and being purified but also that possible contaminating membranes are being eliminated. For the initial run through a multistep procedure this may require a tremendous amount of very tedious assays. Positive identification of a membrane is achieved by following two basic types of membrane markers: structural markers and marker enzymes [49]. Both types of markers must uniquely reside in the membrane being tested. To follow the outcome of each membrane purification step, relatively rapid and quantitative procedures, mostly enzyme assays, are preferred over slower and more qualitative procedures like electron microscopy. After the investigator has achieved what he hopes is high purity, a variety of confirming enzyme assays, structural measurements, and compositional analysis are performed. Once the membrane purification scheme is shown to be accurate and reproducible, subsequent purifications need not be so detailed. Some of these markers are listed in Table 12.5 (membrane structural markers) and Table 12.6 (membrane marker enzymes) these Tables are based on Table 1.2 in [49].

TABLE 12.5 Partial List of Membrane Structural Markers

Membrane	Structural marker
Nucleus	Chromosomes
	DNA (Feulgen stain)
Plasma membrane	Gap junctions
	Brush border glycocalyx
	Microvilli
	Bind lectins
Mitochondrial inner	F1-ATPase "knob", cardiolipin
RER	Ribosomes
Golgi	Stains well with osmium
Thylakoid	Chlorophyll
Any membrane	Specific antibodies

TABLE 12.6 Partial List of Membrane Marker Enzymes

Membrane	Marker enzyme
Plasma membrane	5′-Nucleosidase
	Alkaline phosphodiesterase
	Na^+/K^+ ATPase
	Adenylate cyclase
Mitochondria inner	Cytochrome c oxidase
	Succinate-cytochrome c oxidoreductase
Mitochondria outer	Monoamine oxidase
Lysosome	Acid phosphatase
	β-Galactosidase
Peroxisome	Catalase
	Urate oxidase
	D-Amino acid oxidase
Golgi	Galactosyltransferase
	α-Mannosidase
Endoplasmic reticulum	Glucose-6-phosphatase
	Choline phosphotransferase
	NADPH-cytochrome c oxidoreductase
Cytosol	Lactate dehydrogenase

6. SUMMARY

Purifying a membrane is as much an art as it is a science. The first step in purifying a membrane is breaking open (homogenizing) the cell. Once broken, the many cell membrane components must be fractionated. Methods discussed include: homogenization methods (enzymes to break thick cell wall, shear force (mortar and pestle), food blenders, high speed blade-type homogenizers, osmotic gradients, sonicators, gas ebullition, French Press, and freeze/thaw); centrifugation fractionation methods (differential centrifugation and Density Gradient Centrifugation); Noncentrifugation fractionation methods (lectin-affinity chromatography, antibody-affinity chromatography, ligand-receptor affinity chromatography, anion exchange chromatography, magnetic beads, microbeads, two-phase partitioning, silica particles, size filtration, and size exclusion chromatography). After each step in membrane purification, tests must be run to ensure that the step resulted in purification of the target membrane and that other membranes had been excluded. To achieve this, a large number of structural markers and marker enzymes characteristic of a particular membrane are employed.

Membrane reconstitution is often a necessary step in determining a membrane protein's mechanism of action. Chapter 13 will review the methodologies to purify membrane proteins and lipids and to reconstitute these components into a functional liposome. Lipid raft isolation will also be discussed.

References

[1] Voet D, Voet JG. Biochemistry. 3rd ed. Wiley; 2004. p. 1309.

[2] Evans WH. Isolation and characterization of membranes and cell organelles. In: Rickwood D, editor. Preparative centrifugation – a practical approach. Oxford University Press; 1992. p. 233–70.

[3] Siekevitz P. Protoplasm: endoplasmic reticulum and microsomes and their properties. Ann Rev Physiol 1963;25:15–40.

[4] Heinemann FS, Ozols J. Isolation and structural analysis of microsomal membrane proteins. Rev Front Biosci 1998;3:d483–93.

[5] Graham JM. Homogenization of tissues and cells. In: Graham JM, Rickwood D, editors. Subcellular fractionation – a practical approach. Oxford University Press; 1997. p. 1–29.

[6] Garcia FAP. Cell disruption and lysis. Encyclopedia of Bioprocess Technology John Wiley & Sons, Inc.; 2002.

[7] Garcia FAP. Cell wall disruption and lysis. Encyclopedia of industrial bioprocess: bioprocesses, bioseparation and cell technology. John Wiley & Sons, Inc.; 2009. p. 1–12.

[8] Graham JM, Rickwood D. Subcellular fractionation: a practical approach. 339 pp. Oxford (UK): Oxford University Press; 1997.

[9] Amersham Bioscience. Ficoll PM 70, Ficoll PM 400. Cell separation. Data file 18-1158-27 AA. 2011. p. 1–6.

[10] Camara C. Sample preparation for speciation. Anal Bioanal Chem 2005;381:277–8.

[11] Kido H, Micic M, Smith D, Zoval J, Norton J, Madou M. A novel, compact disk-like centrifugal microfluidics system for cell lysis and sample homogenization. Colloids Surf B 2007;58:44–51.

[12] Hopkinson J, Mouston C, Charlwood KA, Newbery JE, Charlwood BV. Investigation of the enzymatic digestion of plant cell walls using reflectance Fourier Transform Infrared Spectroscopy. Plant Cell Rep 1985;4:121–34.

[13] Kalia A, Rattan A, Chopra P. A method for extraction of high-quality and high-quantity genomic DNA generally applicable to pathogenic bacteria. Anal Biochem 1999;275:1–5.

[14] Urban K, Wagner G, Schaffner D, Roglin D, Ulrich J. Rotor-stator and disc systems for emulsification processes. Chem Eng Technol 2006;29:24–31.

[15] de Gier J. Osmotic behaviour and permeability properties of liposomes. Chem Phys Lipids 1993;64:187–96.

[16] de Gier J. Liposomes recognized as osmometers. J Liposome Res 1995;5:365–9.

[17] Srinivastan PR, Miller-Faures A, Frrera M. Isolation techniques for HeLa-cell nuclei. Biochim Biophys Acta 1982;65:501–5.

[18] Diagonistics LLC. Sonication 101. 2003.

[19] Chandler DP, Brown J, Bruckner-Lea CJ, Olson L, Posakony GJ, Stults JR, et al. Continuous spore disruption using radially focused, high-frequency ultrasound. Anal Chem 2001;73:3784–9.

[20] Burgmann H, Pesaro M, Widmer F, Zeyer J. A strategy for optimizing quality and quantity of DNA extracted from soil. J Microbiol Methods 2001;45:7–20.

[21] Goldberg S. Mechanical/physical methods of cell disruption and tissue homogenization. In: Methods in: molecular biology, vol. 424, I. New York: Springer Publishing; 2008. p. 3–22.

[22] Rinker AG, Evans DR. Isolation of a chromosomal DNA from a methanogenic archaebacteria using a French pressure cell. Biotechniques 1991;11:612–3.

[23] Harju S, Fedosyuk H, Peterson KR. Rapid isolation of yeast genomic DNA: bust n' grab. BMC Biotechnol 2004;4:8.

[24] Mikkelsen SR, Corton E. Chapter 13. Centrifugation methods. In: Bioanalytical chemistry. Hoboken (NJ): John Wiley & Sons, Inc.; 2004. p. 247–67.

[25] Rickwood D. Preparative centrifugation: a practical approach. 399 pp. Oxford (UK): IRL Press at Oxford University Press; 1992.

[26] Graham JM. Biological centrifugation (the basics). 210 pp. Oxford (UK): Bios Scientific Publishing; 2001.

[27] Dunkley PR, Jarvie PE, Robinson PJ. A rapid Percoll gradient procedure for preparation of synaptosomes. Nat Protoc 2008;3:1718–28.
[28] Price CA. Centrifugation in density gradients. New York: Academic Press; 1982.
[29] GE-Healthcare. Percoll™, www.gelifesciences.com.
[30] Shield A. Biological separations. News bulletin for axis-shield density gradient media. Issue 2009;2. OptiPrep™ Preparation of gradient solutions.
[31] Hage, DS, Cazes, J, editors. Handbook of affinity chromatography. 2nd ed. Chromatographic Science Series, Marcel Dekker; 2005.
[32] Bailon P, editor. Affinity chromatography: methods and protocols methods in molecular biology. Totowa (NJ): Humana Press; 2000. 230 pp.
[33] Sharon N. History of lectins: from hemagglutinins to biological recognition molecules. Glycobiology 2004;14:53R–62R.
[34] Sharon N, Lis H. Lectins. 2nd ed. Dordrecht (The Netherlands): Springer; 2007. 454 pp.
[35] Kennedy JF, Palva PMG, Corella MTS, Cavalcanti MSM, Coelho LCBB. Lectins, versatile proteins of recognition: a review. Carbohydr Polym 1995;26:219–30.
[36] Van Damme EJM, Peumans WJ, Pusztai A, Bardocz S. Handbook of plant lectins: properties and biomedical applications. 452 pp. New York: John Wily & Sons; 1998.
[37] Liu Y, Gallardo-Moreno AM, Pinzon-Arango PA, Reynolds Y, Rodriguez G, Camesano TA. Cranberry changes the physicochemical properties of *E. coli* and adhesion with uroepithelial cells. Colloids Surf B 2008;65:35–42.
[38] Nisnevitch M, Firer MA. The solid phase in affinity chromatography: strategies for antibody attachment. J Biochem Biophys Methods 2001;49:467–80.
[39] Ogut O, Jin J-P. Expression, zinc-affinity purification, and characterization of a novel metal-binding cluster in troponin T: metal-stabilized α-helical structure and effects of the NH_2-terminal variable region on the conformation of intact troponin T and its association with tropomyosin. Biochemistry 1996;35:16581–90.
[40] Yamamoto S, Nakanishi K, Matsuno R. Ion exchange chromatography of proteins (chromatographic science series). 401 pp. New York: Marcel Dekker, Inc.; 1988.
[41] Haukanes B-I, Kvam C. Application of magnetic beads in bioassays. Nat Biotechnol 1993;11:60–3.
[42] Levison PR, Badger SE, Dennis J, Hathi P, Davies MJ, Ian J, et al. Recent developments of magnetic beads for use in nucleic acid purification. J Chromatogr A 1998;816:107–11.
[43] Invitogen, L. Technologies. Dynabeads® magnetic separation technology. Carlsbad (CA).
[44] Bellis, M. The history of lava lamps: Craven Walker – How Lava Lamps Work, About.com Inventors.
[45] Persson A, Jergil B. Purification of PMs by aqueous two-phase affinity partitioning. Anal Biochem 1992;204:131–6.
[46] Yoshida S, Kawata T. Isolation of smooth endoplasmic reticulum and tonoplast from mung bean hypocotyls (*Vigna radiata* [L.] Wilczek) using a Ficoll gradient and two-polymer phase partition. Plant Cell Physiol 1988;29:1391–8.
[47] Chaney LK, Jacobson BS. Coating cells with colloidal silica for high yield isolation of PM sheets and identification of transmembrane proteins. J Biol Chem 1983;258:10062–72.
[48] Leighton F, Poole B, Beaufay H, Baudhuin P, Coffey JW, Fowler S, et al. The large-scale separation of peroxisomes, mitochondria, and lysosomes from the livers of rats injected with triton WT-1339. J Cell Biol 1968;37:482–513.
[49] Gennis RB. Biomembranes. Molecular structure and function. (Table 1.2, p.20). New York, NY: Springer-Verlag; 1989. 533 pp.

CHAPTER

13

Membrane Reconstitution

OUTLINE

Membranes are unimaginably complex. They consist of hundreds of different proteins floating in a bewildering and crowded sea of a 1000 or more different lipids and this mixture is in constant flux. A membrane is not homogeneous but exists in fleeting patches that exhibit both lateral and transmembrane asymmetry. Also, countless reactions occur simultaneously in and around the membrane. Can the membrane problem be simplified? In this chapter we will examine how one membrane protein can be isolated away from the others and reconstituted into a model lipid bilayer membrane in order to determine how the protein functions. Finally, we will investigate the isolation and properties of an important membrane domain involved in cell signaling, the lipid raft.

An Introduction to Biological Membranes
http://dx.doi.org/10.1016/B978-0-444-63772-7.00013-0

1. MEMBRANE DETERGENTS

It is hard to overstate the importance of detergents in life science studies, particularly those involving biological membranes. At present, detergents form the cornerstone of membrane studies at the molecular level. As a result, a large detergent industry has flourished, producing well over a 1000 different products. Detergents are amphipathic molecules, simultaneously containing polar and nonpolar ends (Chapter 3) [1]. The first detergent was also the first soap (Chapter 5), namely the potassium salt of free fatty acids resulting from the saponification of animal fat.

Detergents are a class of molecules that have the ability to disrupt hydrophobic–hydrophilic interactions like those that drive the stability of membranes (see the hydrophobic effect, Chapter 3) [1]. Detergents are used to lyse cells and to extract membrane proteins and lipids for reconstitution experiments. Many additional functions have been assigned to detergents including protein crystallization, immunoassays, and electrophoresis. While detergents' amphipathic structure may grossly resemble that of membrane phospholipids, a closer examination shows that detergents are many orders of magnitude more water-soluble (Table 13.1). For example, solubility of the phospholipid dipalmitoyl phosphatidylcholine (DPPC) is only $\sim 4.7 \times 10^{-7}$ mM, much lower than for membrane detergents. This trait of resembling a phospholipid while being much more water-soluble is the reason detergents can be successfully employed to extract and purify proteins from membranes.

Detergents that are used in membrane studies can be divided into five basic categories: anionic, cationic, nonionic, zwitterionic, and bile salts. The structures of some of the more commonly used detergents in membrane studies are depicted in Fig. 13.1 The first four of these classes are synthetic detergents and are classified by the net charge residing on their polar moiety. The fifth classification (bile salts) is based on it being a biologically produced

TABLE 13.1 Detergents Commonly Used in Membrane Studies, With Their Hydrophile–Lipophile Balance (HLB) and Critical Micelle Concentration (CMC) Values

Type	Detergent	HLB	CMC (w/v%)
Anionic	SDS		0.17
Cationic	Cetylpyridinium bromide	16.7	0.042
Nonionic	Brij-35	16.9	0.09
	Octyl-glucoside	13.1	0.7
	Triton X-100	13.6	0.0155
	Tween 20	16.7	0.0987
Zwitterionic	CHAPS	13.1	0.5
Bile salt	Cholate	18	0.388
	Deoxycholate	16	0.083
Phospholipid	DPPC	Solubility	$\sim 3.5 \times 10^{-7}$ g/L

CHAPS, 3-[(3-cholamidopropyl) dimethylammonio]-1-propanesulfonate; *DPPC*, dipalmitoyl phosphatidylcholine; *SDS*, sodium dodecylsulfate.

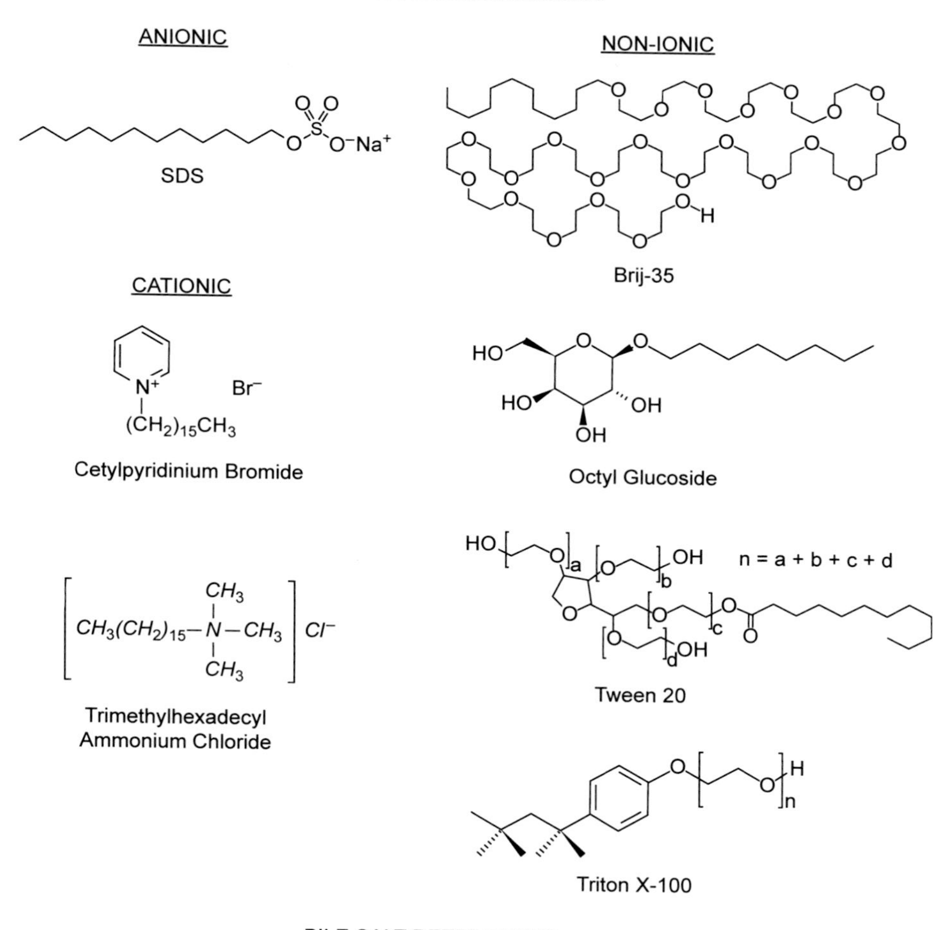

FIGURE 13.1 Examples of basic types of detergents most commonly used in studies of biological membranes.

type of detergent. Besides being separated into these five classes, each detergent is given two numbers that may be predictive of its use in membrane studies, the critical micelle concentration (CMC) and the hydrophile–lipophile balance (HLB). CMC is a measured parameter, while the HLB is an empirically derived number.

The CMC is the detergent (or any surfactant) concentration above which micelles are spontaneously formed [2]. Below the CMC, each detergent molecule will either reside at the aqueous interface or, if water-soluble enough, be solvated. Above the CMC, additional surfactant will only increase the number of micelles in solution. At the CMC, micelles form, changing several basic parameters of the solution. The abrupt solution changes that occur at the CMC are commonly followed by an increase in solution turbidity (light absorbance), a change in electrical conductivity, or a change in surface tension. Determination of CMC using surface tension measurements is shown in Fig. 13.2 [3]. At low concentrations, the detergent will preferentially accumulate at the air–water interface greatly reducing the surface tension. As detergent accumulates at the interface, the increasingly crowded condition forces some detergent into the aqueous subphase resulting in the generation of micelles. The detergent concentration where this occurs is the CMC. CMCs for the common membrane detergents (structures shown in Fig. 13.1) are listed in Table 13.1.

A second number often used to characterize a detergent is the HLB [4,5]. The HLB is a number that is an empirical measure of how hydrophilic or hydrophobic the detergent is. There are several ways to calculate the HLB. The most commonly used method to calculate HLBs for membrane studies was developed by Griffin in papers published in 1949 [4] and 1954 [5]. The Griffin method calculates the net HLB from combining the number of atoms comprising the hydrophilic and hydrophobic portions of the detergent:

$$\mathrm{HLB} = 20 \times M_{\mathrm{h}}/M$$

where M_{h} is the molecular mass of the hydrophilic portion of the molecule and M is the molecular mass of the entire molecule. By this method, HLBs can range from 0 to 20 with

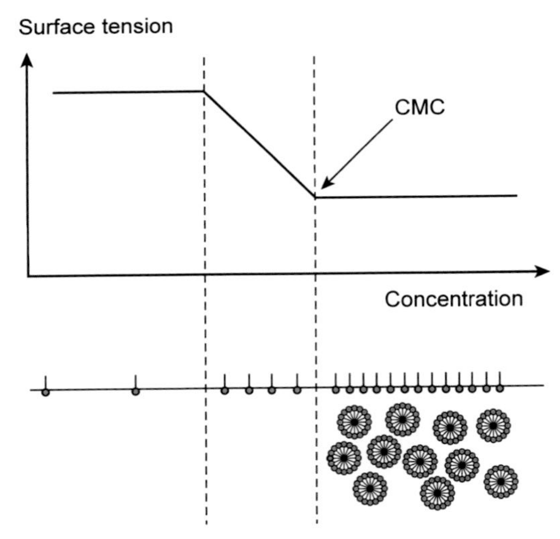

FIGURE 13.2 Determination of a detergent's CMC by surface tension measurements. At low concentrations, the detergent will preferentially accumulate at the air/water interface greatly reducing the surface tension. As detergent accumulates at the interface, the ever increasing crowded condition forces some detergent into the bathing solution generating micelles at the CMC. At this point there is no further decrease in surface tension with additional detergent [3].

0 being a molecule that is completely hydrophobic and 20 a molecule that is completely hydrophilic. Since HLB is entirely empirical, other formulations can span over different ranges. A high HLB indicates that the detergent will partition into the aqueous phase, while a low HLB indicates the detergent will prefer remaining in the membrane, and extraction from the membrane will be difficult. Detergents used for membrane studies generally have HLBs between 13 and 15. Structures of several detergents commonly used in membrane studies are depicted in Fig. 13.1 and their CMCs and HLBs listed in Table 13.1. It should be mentioned that CMCs and HLBs can vary significantly with pH and solvent ionic strength.

Membrane Detergents: The most suitable detergents for extraction of functional membrane proteins are classified as nonionic, zwitterionic, or bile salts (detergent structures are shown in Fig. 13.1). A neutral detergent that has received a great deal of use in isolating integral membrane proteins is n-octyl-β-D-glucoside (octyl glucoside). Variations of this basic detergent are commercially available with mannose replacing glucose and with decyl (10-carbon) or lauryl (12-carbon) chains replacing octyl (8-carbon). **Octyl glucoside** is hydrophilic enough to be readily removed from extracted membrane proteins. 3-[(3-cholamidopropyl) dimethylammonio]-1-propanesulfonate (**CHAPS**) is a popular zwitterionic detergent used to solubilize integral membrane proteins without causing denaturation. **Brij-35** (a polyoxyethyleneglycol lauryl ether) is another nonionic detergent involved in membrane integral protein extractions. Brij-35 is also used to prevent nonspecific protein binding to gel filtration and affinity chromatography supports. A major drawback to Brij-35 is its difficulty in being removed by dialysis. However, it has been reported that Brij-35 can be removed by Extracti-Gel D Detergent Removing Gel. **Triton X-100** is also a nondenaturing, nonionic detergent. However, its application in membrane studies is generally different than octyl glucoside or Brij-35. The major use for Triton X-100 is to lyse or permealize cells. It is not as useful in solubilizing functional membrane proteins and has difficulty breaking protein–protein interactions. Triton X-100 is also hard to remove from cell membranes. Anyone who has ever tried to pipette Triton X-100 will remember the unpleasant experience. Triton X-100 is very viscous! Often Triton X-100 is used in conjunction with zwitterinic detergents (such as CHAPS) that are better suited for membrane protein extractions.

Cholate is the major bile salt comprising ~80% of all bile salts produced in the human liver. This detergent is a natural product synthesized from cholesterol. Another closely related bile salt that is popular as a membrane protein extraction detergent is **deoxycholate** whose structure differs from cholate only by lacking a single (–OH) (Fig. 13.1). Deoxycholate is considered to be a secondary bile acid and is a metabolic byproduct of intestinal bacteria. Both cholate and deoxycholate have been proposed to function in a similar fashion (Fig. 13.3). Each has a smooth hydrophobic surface that interacts with acyl chains of the membrane phospholipids. The other surface is polar, being dominated by (–OH) groups (three for cholate and two for deoxycholate). A cylinder is formed with the membrane integral protein forming the core, surrounded by a ring of phospholipids and finally a second ring of cholate or deoxycholate. The outer ring is stabilized by polar (–OH) interactions with water. The charged carboxylate on the detergents further stabilize the cylinder.

Sodium dodecyl sulfate (**SDS**) is by far the best known of a family of related anionic detergents that bind to and denature proteins. Similar detergents can have different

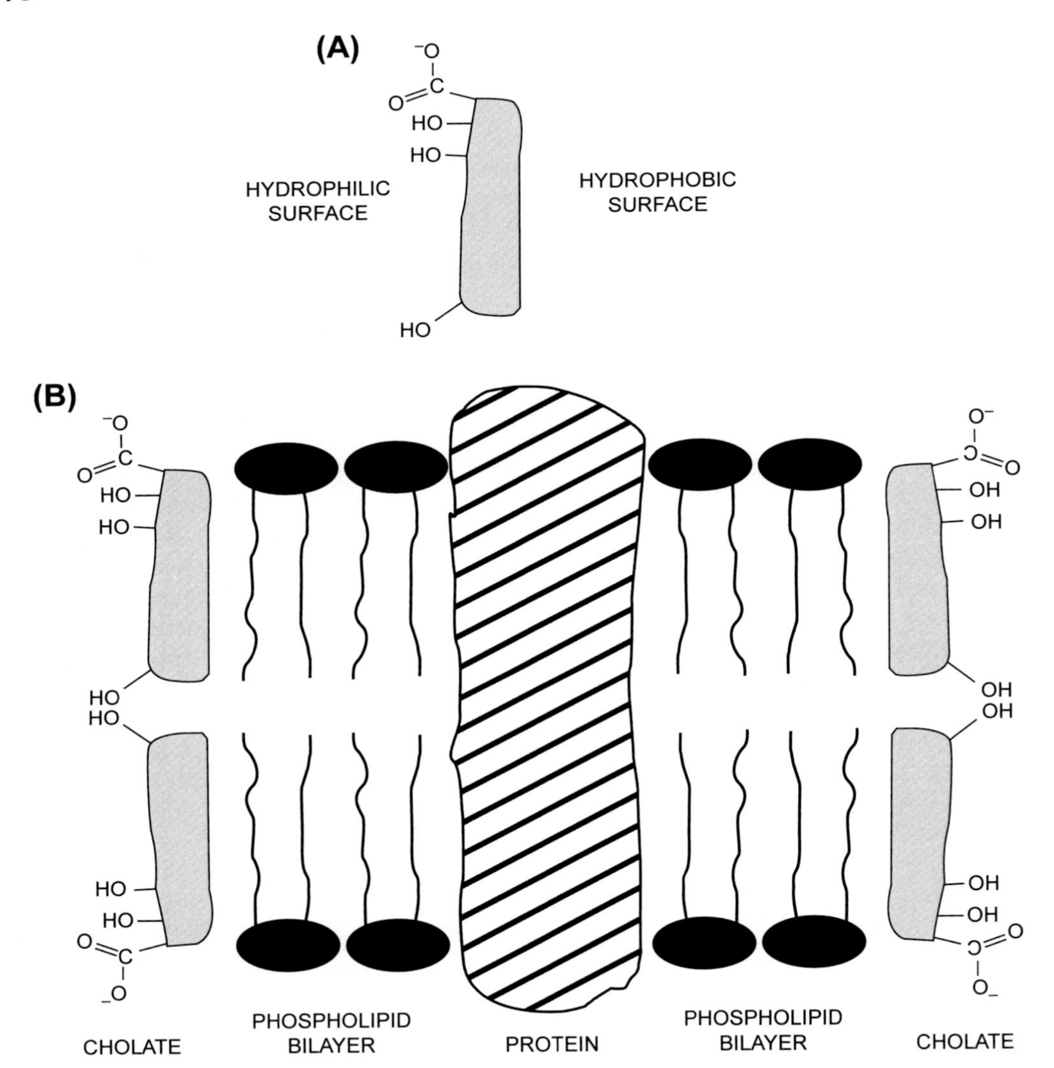

FIGURE 13.3 Mechanism by which cholate solubilizes an integral membrane protein. The structure of cholate is shown in Fig. 13.1. Fig. 13.3, Panel A is a cartoon drawing of cholate demonstrating its hydrophilic surface and its hydrophobic surface. Fig. 13.3, Panel B shows how cholate may extract a plug from the membrane that contains an integral protein and phospholipids.

alkyl chain lengths and different anionic head groups (eg, sulfonate, carboxylate, phosphate etc.). Below the CMC, SDS monomers bind to both membrane and cytosolic proteins in an equilibrium-driven process. Binding is cooperative meaning the binding of a first SDS facilitates binding of a second and so forth. Upon SDS binding, the protein structure is altered into rigid rods whose length is proportional to its molecular weight. SDS binds to the protein chain at a ratio of one SDS for every two amino acid residues. Since the SDS head is negatively charged, the net charge on the denatured protein is proportional to the protein's molecular weight. SDS-ladened proteins can then be separated in an electric field generated on a stable support. The process is referred to as SDS-polyacrylamide gel electrophoresis. The function of SDS in membrane studies is therefore to determine the number and size of membrane proteins. Since proteins are denatured by SDS, this detergent is of little value in reconstituting functional proteins into membranes. SDS has many more applications outside the world of science. It is a powerful surfactant that in high concentrations is the major ingredient in many industrial cleaners and at low concentrations is found in toothpaste, shampoo, shaving cream, bubble bath formulations, and even laxatives.

2. MEMBRANE PROTEIN ISOLATION

Methodologies for the isolation and purification of integral membrane proteins have always lagged far behind those for water-soluble cytosolic proteins. A water-soluble protein can often be precipitated by ammonium sulfate (salting-out) in relatively pure form. In sharp contrast, integral membrane proteins must first be extracted from the membrane and come attached to membrane lipids making them water-insoluble goos that defy purification.

An early attempt to isolate membrane proteins involved what is known as "acetone powders." Tissues are ground up and the aqueous fraction removed by filtration or centrifugation. The nonsoluble fraction is then mixed with a large excess of cold acetone that dissolves the membrane lipids. The solution is then filtered to remove the insoluble proteins. The filtered protein powder is washed several times with cold acetone and dried producing the acetone powder. Unfortunately, most proteins comprising the residual protein powder are denatured and therefore useless for almost all functional studies. The method of producing acetone powders is just too harsh for delicate biological membrane material, however, it has been successfully employed in cytoskeletal protein isolations [6].

A gentler method to isolate membrane proteins is by use of "chaotropic agents" [7]. This method has also been shown to have unwanted complications, thus limiting its application. Chaotropic agents act by disrupting intermolecular interactions including hydrogen bonding, van der Waals forces, and hydrophobic effects that are responsible at the molecular level for most biological structure. Chaotropic agents include urea, guanidinium chloride, lithium perchlorate, sodium thiocyanate, NaBr, and NaI. When dissolved in water at very high levels (~4–8 M) chaotropic agents cause membranes to fall apart by eliminating the stabilizing hydrophobic effect (Chapter 3). Unfortunately, chaotropic agents can also denature proteins, and their enormous concentrations makes their complete removal difficult, limiting their usefulness.

The methods discussed in Chapter 12 for isolation and purification of membrane fractions are also applicable for the isolation and purification of membrane integral proteins. Many of the protein isolation methods are based on centrifugation and affinity chromatography. The major difference is that extraction of an integral protein from its membrane in nondenatured form generally involves the use of mild detergents. Obtaining a

TABLE 13.2 Affect of Detergent Levels on Membranes

Detergent:lipid Ratio	Affect on membrane
0.1:1 to 1:1	Bilayer remains intact. Some proteins extracted
1:1 to 2:1	Bilayer solubilized into mixed micelles
10:1	All lipid–protein interactions are replaced by detergent–protein interactions

noncontaminated membrane fraction in high yield greatly enhances the chance of subsequently obtaining a particular resident protein in pure form. For example, if the objective is to extract and purify cytochrome c oxidase, a first step would involve obtaining an intact mitochondria by simple differential centrifugation.

Moderate concentrations of mild detergents are used to compromise membrane integrity and facilitate the extraction of integral proteins in water-soluble, nondenatured form. The extracted proteins are found in mixed micelles composed of the protein, membrane phospholipids, and the detergent. The affect of detergent levels on membranes is concentration-dependent (Table 13.2).

The membrane solubilization buffer should contain enough detergent to provide greater than one micelle per membrane protein in order to assure that individual proteins are isolated in separate micelles. The required amount of detergent will depend on the detergent's physical properties (CMC, HLB, and aggregation number) as well as the aqueous buffer pH, salt concentration, and temperature.

The complex solution containing many membrane proteins (mostly located in mixed micelles), can then be passed over a lectin or antibody affinity column (Chapter 12). Only the selected protein binds to the column and the remaining, unbound proteins are discarded. The bound protein can then be removed by addition of excess free sugar for a lectin affinity column or by altering the pH or salt concentration for an antibody affinity column.

If the desired protein is hard to isolate or is present in low amounts, recombinant DNA technology may be employed, where the protein is expressed in large quantities in another cell type. Also the newly expressed protein can be tagged by something, often a polyhistidine tag (His-tag), that can greatly facilitate the protein's isolation.

The His-tag technique usually involves engineering a sequence of 6–8 histidines to either the C- or N-terminal of the protein [8,9]. The His-tag binds strongly to divalent metal ions like nickel or cobalt. Therefore, if a complex mixture of proteins is passed through a column containing immobilized nickel, only the His-tag protein binds and all other proteins can be discarded. The His-tag protein can be eluted from the column by an imidazole-containing solution. Imidazole, the side chain of histidine, competes for the poly-histidine binding site. The His-tag protein can also be released from the nickel column by lowering the pH to ~4.5. It should be mentioned that any protein with a high affinity for divalent metals can be separated by an immobilized-nickel column without the need of a special His-tag.

Another approach to purification of a protein through use of an engineered tag involves addition of an antigen peptide tag. Separation is achieved using an appropriate

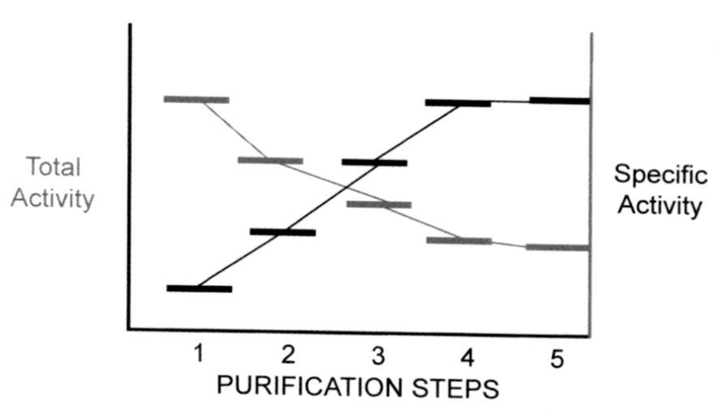

FIGURE 13.4 With each step in the purification process, the total enzyme activity decreases while specific activity increases until the enzyme is pure.

antibody-attached resin column. This methodology is highly specific to a single protein. After the desired protein is isolated, the engineered tags can be easily removed by a protease.

After the desired protein is obtained in pure form, it is usually so dilute it must be substantially concentrated for subsequent reconstitution studies. The most commonly used technique is lyophilization or freeze drying. The sample is frozen and the water content reduced by sublimation under vacuum. Another procedure used to increase the protein concentration is ultrafiltration. Selectively permeable membranes allow water and small molecules to cross the membrane while retaining the much larger protein. The solution is forced through the filter by mechanical or gas pressure or by centrifugal force in a specially designed centrifuge tube.

At each step in a multistep protein purification, both the total amount of all proteins present and activity of the desired protein must be determined. Although the total protein and total activity of the desired protein will decrease after each step, the specific activity will increase if any purification is achieved (Fig. 13.4). Enzyme activity is equal to the moles of substrate converted to product per unit of time. The standard unit of specific activity is the katal where:

$$1 \text{ katal} = 1 \text{ mol substrate converted per second}$$

However, a katal is such a large unit, specific activity is more usually expressed as:

$$1 \text{ enzyme unit (U)} = 10^{-6} \text{ mol substrate converted/min}$$

Therefore, 1 U corresponds to 16.67×10^{-9} katals.

The specific activity (SA) is the activity of an enzyme per mg of total protein.

$$\text{SA} = 10^{-6} \text{ mol converted/mg min}$$

The SA of a pure enzyme is constant and is a characteristic of the enzyme. Therefore, the purity of an enzyme at any step in the purification, expressed as % purity, can be determined by:

$$\%\text{Purity} = \text{measured SA/SA of pure enzyme} \times 100$$

FIGURE 13.5 Oliver Lowery (right), 1910–1996.

Until purity is achieved, the measured SA will always be less than that of the pure enzyme because other proteins besides the desired enzyme are still present in the solution.

In addition to enzyme activity, total protein must be determined at each purification step. Many methods to accurately quantify total protein are currently in use. The simplest is to measure solution absorbance at 280 nm. This method is not sensitive and suffers from many nonprotein compounds in the purification buffers that can also absorb at 280 nm. A revolutionary colorimetric procedure to quantify proteins was published by Oliver Lowery (Fig. 13.5) in 1951 [10]. For decades this paper was the most cited in all scientific literature. A more sensitive method to quantify proteins was described by Bradford in 1976 [11] and further improved by Zor and Selinger in 1996 [12]. A Bradford Protein Assay kit, commercially available from Bio-Rad, is perhaps the most popular protein assay in use today. Another very popular protein assay is the Pierce BCA (bicinchoninic) Protein Reagent Assay that is reputed to be the best detergent-compatible protein assay available.

3. MEMBRANE LIPID ISOLATION

During the early years of biochemistry, knowledge of lipids far exceeded that of proteins. Isolation of the first protein was not until 1925 when Sumner reported his work on urease from jack bean [13]. In comparison, the pioneering lipid extractions by Chevreul were done in the early 19th century [14]. Modern membrane polar lipid extraction procedures had their beginnings with the classic 1957 paper by Folch (Fig. 13.6), Lees and Stanley [15]. With slight modifications, this method is still in use today. Lipid extraction requires that the tissue must be fresh and devoid of lysophospholipids, oxidized products, large quantities of mono- or di-acyl glycerol and free fatty acids. The tissue is homogenized with chloroform–methanol (2:1) at a ratio of 1 g tissue to 20 ml solvent. The tissue naturally contains a small amount of water that adds to the mixture and sometimes must be augmented to improve phase separation. After filtration or centrifugation to remove solid cellular debris, the liquid mixture is removed and washed with 0.2 vol of 0.9% NaCl in water. Low speed centrifugation separates the mixture into two phases. The upper phase, primarily methanol and water, is enriched in very polar lipids like gangliosides while the

FIGURE 13.6 J. Folch-Pi, 1911–1979.

chloroform-rich lower phase contains the phospholipids and nonpolar lipids. This phase is isolated and evaporated under vacuum and further extracted with cold acetone. The acetone extraction removes the nonpolar lipids leaving the membrane phospholipids in the acetone-insoluble pellet. Recently, dichloromethane has occasionally been used in place of chloroform to avoid adverse health effects. Another very similar procedure involving extraction with chloroform–methanol–water was later proposed by Bligh and Dyer in 1959 [16].

After extraction by the classic Folch method, the lipids are distributed between three fractions: the methanol–water upper phase containing gangliosides and other very polar lipids; the chloroform phase containing the membrane phospholipids; and the acetone phase containing nonpolar lipids like triacylglycerols and cholesterol. Each of these fractions can be dried and subjected to additional fractionations. Although there are a variety of methodologies used to separate membrane lipid fractions, the two most common are thin layer chromatography (TLC) and high-performance liquid chromatography (HPLC) [17].

3.1 Thin Layer Chromatography

TLC was discovered by Ukranian scientists, N.A. Izmailov and M.S. Shraiber in 1938 [18]. At first, TLC was difficult to use since the TLC plates all had to be homemade. Now, a wide variety of premade TLC plates are commercially available. TLC is so fast, simple, and inexpensive it is routinely employed and is considered to be "a poor man's HPLC" (Section 3.2). TLC plates consist of a supporting sheet of glass, aluminum, or plastic to which is adhered a thin coating of a solid adsorbent, usually silica or alumina. Aluminum or plastic plates have the advantage of being easily cut to a desired shape by a pair of scissors. A small amount of a volatile solvent containing the dissolved membrane lipid mixture is spotted near the bottom of the plate [19,20]. The spot is dried and the plate then placed inside a sealed developing chamber containing a shallow pool of a developing solvent known as the eluent. The eluent constitutes the mobile phase and percolates up through the stationary phase by capillary action. For lipid separations, the stationary phase is usually silica. The spotted lipids are bound to a different extent to the stationary silica depending on their physical properties of polarity and hydrophobicity. As the eluent passes through the bound lipids, an equilibrium is established for each lipid in the mixture between the stationary silica and the mobile liquid phase. The lipids that are most soluble in the mobile phase will spend more time in that phase and move farther up the TLC plate. When the mobile solvent front approaches the end of the plate, the plate is removed from the developing chamber and dried, thus ending the separation. The separated lipids are then visualized by a number of reagents that can be sprayed on the plate [21]. Table 13.3 lists some of the most commonly used sprays to visualize lipids on TLC plates. The relative distance the lipid travels from the original spot (the origin) is quantified by its R_f value (retention factor). R_f is defined as the distance the lipid travels up the plate divided by the total distance the eluent travels (the front). Determination of R_f values is depicted in Fig. 13.7 [20].

Lipids can be separated by 1-dimensional or 2-dimensional TLC as shown in Fig. 13.8 [22]. Sequential 1-dimensional TLC is shown in the left panel. In the chromatogram, the polar lipids were first separated by two successive runs with chloroform–methanol–water (60:30:5 by volume). In the third run, the nonpolar lipids were separated using hexane-diethyl ether-acetic acid (80:20:1.5 by volume). In the first two runs, the nonpolar lipids ran at the front and were not separated. In the third run, the polar lipids did not migrate,

TABLE 13.3 Visualization Methods for Identifying Lipids on Thin Layer Chromatography (TLC) Plates [21]

Reagent	Reacts with	Lipid identified
Ultraviolet light	Nonspecific	All lipids
H_2SO_4 heat	Carbon charring	All lipids
Iodine	Double bonds	Lipids with double bonds
Primuline	Nonspecific	All lipids
Rhodamine 6G	Nonspecific	All lipids
Molybdate	Phosphate	All phospholipids
Malachite green	Phosphate	All phospholipids
Ninhydrin	Primary amines	PE, PS
Dragendorf reagent	Choline	PC
2,4-Dinitrophenylhydrazine		Plasmalogens
Resorcinol	Sugars	Gangliosides
Ferric chloride		Cholesterol

PC, phosphatidylcholine; *PE*, phosphatidylethanolamine; *PS*, phosphatidylserine.

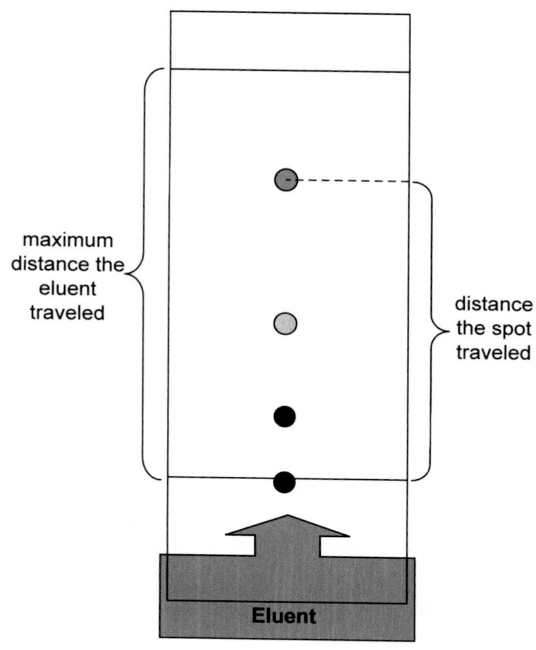

FIGURE 13.7 Determination of R_f values [20]. R_f is the distance the spot travels divided by the maximum distance the eluent travels (the front).

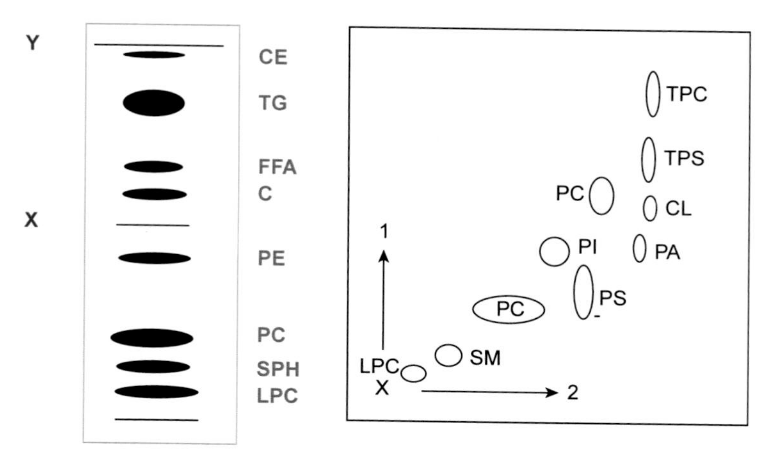

FIGURE 13.8 Lipid separation by one dimensional (left panel) and two dimensional (right panel) TLC [22]. Left Panel: One dimensional TLC with three successive solvent systems. The first two solvents (up to X, chloroform-methanol-water (60:30:5 by volume)) are used to separate the polar lipids, while the third system (up to Y, hexane-diethyl ether-acetic acid (80:20:1.5 by volume)) separates the non-polar lipids. *CE*, cholesterol esters; *TG*, triacylglycerols; *FFA*, free fatty acids; *C*, cholesterol; *PE*, phosphatidylethanolamine; *PC*, phosphatidylcholine; *SPH*, sphingomyelin; *LPC*, lysophosphatidylcholine. Right Panel: Two dimensional TLC. First dimension (1, chloroform/methanol/conc ammonia (65/35/4, by volume). Second dimension (2, butanol/acetic acid/water (60/20/20, by volume). *PI*, phosphatidylinositol; *PS*, phosphatidylserine; *SM*, sphingomyelin; *PC*, phosphatidylcholine; *PE*, phosphatidylethanolamine.

but the nonpolar lipids did, resulting in their separation. The right panel of Fig. 13.8 demonstrates a 2-dimensional TLC lipid separation. In the first direction, the eluent was chloroform–methanol–concentrated ammonia (65:35:4, by volume). This was followed by a second eluent, 2, butanol–acetic acidwater (60:20:20, by volume) run at right angles to the first. Better separation is normally achieved by 2-dimensional TLC. Lipid-derived spots on the TLC plates can be directly quantified by scanning densitometry. Several manufacturers (eg, Camag, Shimadzu, X-Rite, and Tobias) offer computer-controlled densitometers. Digital camera pictures can also be used to quantify the spots on TLC plates. In addition, the lipid components can be removed from the TLC sheet using an eluent solvent that results in a high R_f for the lipid (ie, the lipid is very soluble in the eluent). The isolated lipid can then be subjected to further analysis like determining the fatty acid content of a phospholipid (see Section 3.3). Since the number of possible eluent solutions is almost infinite, it is theoretically possible to separate almost any lipids by TLC.

3.2 High-performance Liquid Chromatography

In recent years, HPLC (high performance liquid chromatography) has supplanted TLC as the method of choice for membrane lipid separations [23–25]. The basic principles behind HPLC are essentially the same as TLC. While TLC does not involve any elaborate equipment, HPLC uses expensive instrumentation. Therefore HPLC is the "rich man's" TLC. HPLC uses a variety of stationary phases and mobile phases that are selected to identify and separate the desired analyte (a substance undergoing analysis). The mobile phase is pumped at extremely high pressure (up to 400 atm) through a column tightly packed with the stationary phase. Also required is a sensitive detector. The analytes dissolved in the mobile phase are injected into the HPLC column at time zero. The time required for an analyte to produce its maximum peak as indicated by the detector is referred to as the retention time and is characteristic of the analyte under conditions of the experiment. Separation conditions include the employed stationary and mobile phases and the column length, pressure, and temperature. Initially the HPLC stationary phase was silica gel and the separation process resembled that described for TLC (Section 3.1). The more polar analytes bind better to the silica gel while the more nonpolar analytes partition more readily into the nonpolar mobile phase and so reach the detector quicker and have a shorter retention times.

While silica gel HPLC is termed "normal phase" HPLC, it is not the most commonly used procedure. Reversed phase HPLC (RP-HPLC or RPC) is by far the most popular type of HPLC used in biological membrane studies [25–27]. RP-HPLC employs a nonpolar stationary phase and a polar mobile phase that contains water. The RP-HPLC stationary phase is often silica gel that has been modified by RMe_2SiCl where R is either a C-8 or C-18 saturated hydrocarbon chain. Therefore the RP-HPLC stationary phase is polar silica with an oily, hydrophobic coat. Opposite to normal phase HPLC, with RP-HPLC retention time is longer for nonpolar analytes and can be further increased by adding more water to the mobile phase and is decreased by adding more organic solute (eg, methanol, acetonitrile). Analytes with a large number of nonpolar C–H or C–C bonds have longer retention times, while those with polar –OH, $-NH_2$ or, COO^- groups have short retention times. Similarly, retention times decrease with double bond content (single bond > double bond > triple bond). Table 13.4 presents relative retention times in RP-

TABLE 13.4 Relative Retention Times in Reversed Phase-High-Performance Liquid Chromatography (RP-HPLC) of Various Phospholipids in the Most Commonly Used Mobile Phases, Acetonitrile—Water and Propan-2-ol—water. Retention Times Increase From Top to Bottom

Acetonitrile—water	Propan-2-ol—water
PA	CL
CL	PE
PI	PI
PS	PS
PE	PA
PC	PC
SM	SM
Lyso PC	Lyso PC

CL, cardiolipin; *PA*, phosphatidic acid; *PC*, phosphatidylcholine; *PE*, phosphatidylethanolamine; *PI*, phosphatidylinositol; *PS*, phosphatidylserine; *SM*, sphingomyelin.

HPLC of various phospholipids in the most commonly used mobile phases, acetonitrile—water and propan-2-ol—water.

3.3 Fatty Acid Analysis

As discussed in Sections 3.1 and 3.2, the various membrane lipid and phospholipid classes can be isolated, primarily by TLC and RP-HPLC. Most membrane lipids are classified as "complex lipids" (Chapter 4) that contain esterified fatty acids. Usually the nature of these fatty acids constitutes essential information as they comprise the bulk of a membrane's hydrophobic interior. The thin oily center of a lipid bilayer is responsible for stability and basic membrane properties. The hydrophobic interior constitutes the diffusion barrier and provides the proper environment for membrane protein activity (Chapter 6). Determining membrane fatty acid composition is therefore at the heart of membrane studies and their analysis is usually achieved by gas chromatography (GC) or more properly gas—liquid chromatography (GLC).

GC is used to separate and identify compounds that can be vaporized without decomposing [28]. GC was first described by the Russian scientist Mikhail Semenovich Tswett in 1903 and was fully developed as GLC in 1950 by A.J.P. Martin (Fig. 13.9), the 1952 Nobel Laureate. Martin was the pioneer of both liquid—liquid and paper chromatography. The basic principle for GC is analogous to that of TLC and HPLC with a few very important differences. In GC, the mobile phase is a carrier gas and the analytes partition between an immobile liquid stationary phase and a mobile gas phase. This accounts for the more correct term GLC. The gas is usually an inert gas like helium or an unreactive gas like nitrogen. The stationary phase is a thin coating of various materials lining the inside of a long thin tube referred to as a column. The most common stationary phase used in fatty acid analysis is Carbowax (polyethylene glycol, $H-(OCH_2CH_2)_n-OH$). The analytes

Lavoslav Ružička
Juli 1970

FIGURE 13.9 A.J.P. Martin, 1910–2002.

partition to a different extent between the mobile gas phase and the stationary phase. Analytes that favor the gas phase leave the column faster and so have shorter retention times. The column is positioned in an oven to assure accurate, programmable temperature control. It is an elevated temperature that volatizes the fatty acid methyl esters and makes GC feasible. The final components of a functional GC, depicted in Fig. 13.10 [29], are an injection port and a detector. There are many types of detectors used in GC, the most common being the flame ionization detector and the thermal conductivity detector.

Free fatty acids as well as fatty acids esterified to other biochemicals (eg, acyl glycerols, cholesterol esters, phospholipids) are not volatile and if heated to a high enough temperature would disintegrate. However, methyl esters of fatty acids are volatile and can be used in GC analysis [30]. Preparation of fatty acyl methyl esters is perhaps the most commonly used reaction in lipid analysis. Fatty acid methyl esters can be made from free fatty acids that have been hydrolyzed (saponified) from their biological ester source (Chapter 5) [31]. However, it is not necessary to actually obtain free fatty acids before methylation. Fatty acid methyl esters can be directly obtained from biological esters by a process known as transesterification [32]. A method that can transesterify can also directly methylate free fatty acids. Methylation can be achieved in either acidic (by methanolic hydrogen chloride or boron trifluoride in methanol) or basic (sodium methoxide in methanol) conditions.

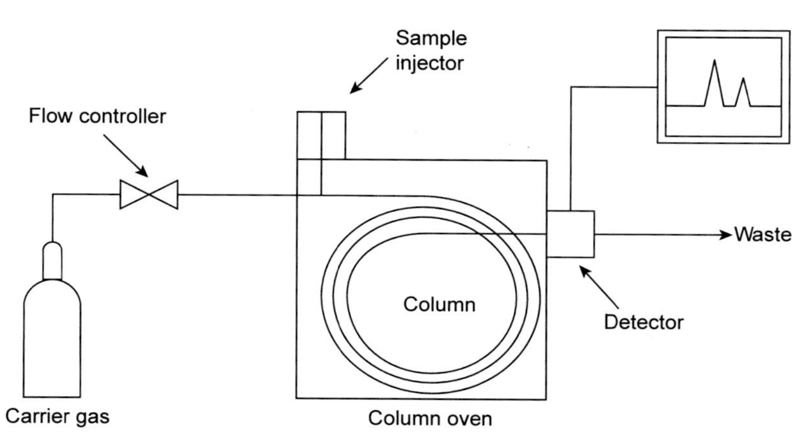

FIGURE 13.10 Diagram of a gas chromatograph [29].

An emphasis must be placed on anhydrous conditions as even traces of water can be detrimental to esterification. The methoxide method is extremely fast (complete in a few minutes) and is accomplished at room temperature [33].

Fig. 13.11 shows a fatty acid profile of liver phospholipids determined by GC [34]. Each fatty acid is identified by three numbers: number of carbons, number of double bonds, and the series (omega-family). Elution time is on the abscissa and the detector intensity on

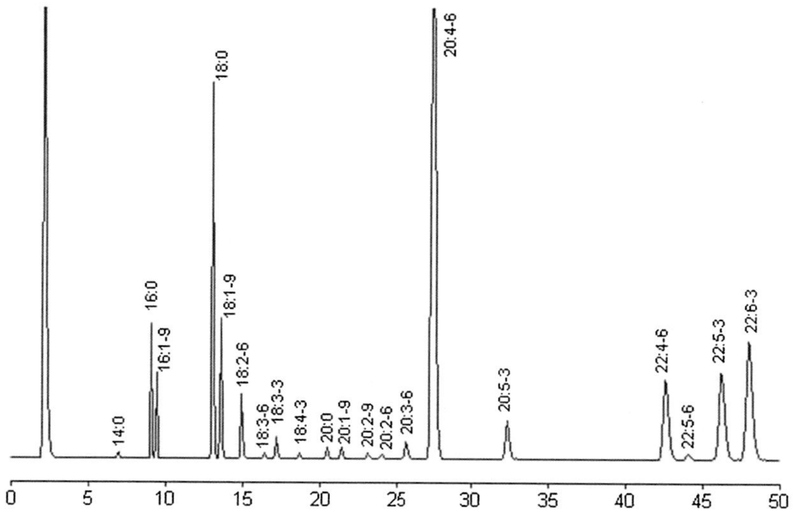

FIGURE 13.11 Fatty acid profile of liver phospholipids determined by GC on a Carbowax stationary phase column (Alltech) [34]. The mobile gas phase was helium and the temperature profile was 100°C for 2 min, 195°C for 33 min and 220°C up to the end. Temperature ramp was 40°C/min. Each fatty acid is identified by three numbers: number of carbons, number of double bonds, and the series (omega-family). Elution time is on the abscissa and the detector intensity on the ordinate. The area under each curve represents the relative amount of each fatty acid.

the ordinate. The area under each curve represents the relative amount of each fatty acid in the sample. A fatty acid methyl ester data base has been established [35] that contains retention times and mass spectral information on ~100 common fatty acids.

4. LIPID MODELS USED IN MEMBRANE STUDIES

Biological membranes are so complex that they often defy comprehensive analysis. For this reason a variety of relatively simple model membrane systems have been developed (Table 13.5).

4.1 Nonbilayer Membrane Models

Biological membranes are so complex that they often defy comprehensive analysis. For this reason a variety of relatively simple model membrane systems have been developed. The simplest model consists of membrane lipids dissolved in organic solution. Obviously this model is extremely limited as it has no 3-dimensional structure that in anyway resembles a membrane. Its primary application is in following very short distance molecular interactions, primarily lipid–lipid associations. The oldest model membrane is the lipid monolayer, discussed in Chapter 2. Lipid monolayers have their origins in a very distant, murky past where it was used by ancient mariners. Although monolayers are only one half of a bilayer, they formed the foundation for modern membrane biophysical studies. The first stable lipid bilayer was reported by Mueller et al. in 1962 [36]. This model membrane was made by spreading an organic solution of membrane polar lipids across a small (~1 mm) hole in a teflon partition separating two aqueous chambers (Fig. 13.12) [37]). Initially a thick glob of lipid in organic solution occupied the hole. This rapidly thinned as the organic solvent was squeezed out leaving a thin membrane that occupied most of the hole (Chapter 9). Electrical and 90 degree reflected light interference measurements proved this suspended membrane was indeed a lipid bilayer. It was therefore called a bimolecular lipid membrane (or black

TABLE 13.5 Lipid Models Used in Membrane Studies

1. Organic solutions
2. Micelles
3. Lipid monolayers
4. Planar bimolecular lipid membranes
Regular bimolecular lipid membranes
Supported lipid bilayers
Tethered bilayer membranes
5. Lipid vesicles
Multilamellar vesicles
Small unilamellar vesicles
Large unilamellar vesicles
Giant unilamellar vesicles
6. Bicelles

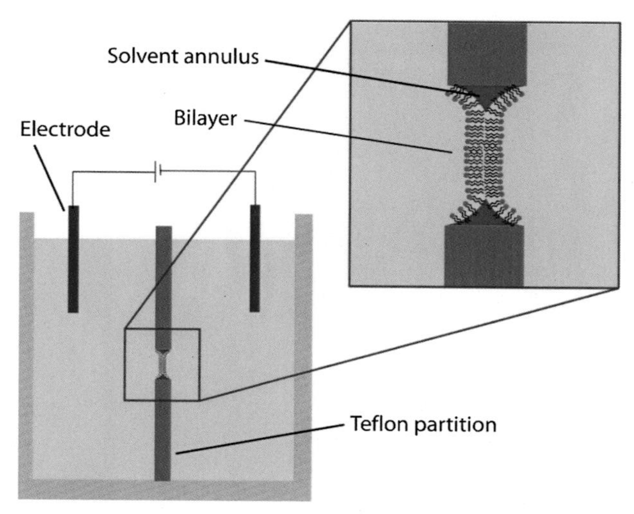

FIGURE 13.12 The planar bimolecular lipid membrane is also referred to as a black lipid membrane or bimolecular lipid membranes (BLM). It is formed by spreading an organic solution of membrane polar lipids across a small hole in a Teflon partition separating two aqueous chambers [37].

lipid membrane due to total destructive interference when viewed at 90 degree from incident light) or simply BLM. Although a BLM is well suited for transmembrane electrical measurements, it is very susceptible to vibrations and even slight organic contaminates cause its demise. Also, proteins often denature at BLM—water interfaces and most transmembrane solute flux measurements are not possible. A more realistic and rigorous model membrane that could be used to reconstitute integral proteins was required. This came in the form of spherical lipid bilayer vesicles, more commonly referred to as lipid vesicles or liposomes.

4.2 Model Lipid Bilayer Membranes—Liposomes

The discovery of liposomes is attributed to Alec D. Bangham (Fig. 13.13) in 1961 (published in 1964 [38]). For years, Bangham had been working on what he referred to as "multilamellar smectic mesophases." The word "smectic" is derived from the Greek word for soap. Although smectic mesophases had been observed since the 1930's, it was Bangham who recognized them for what they were, concentric lipid bilayers and so initially they were called Bangosomes. A few years later, Gerald Weissmann changed their name to liposomes, and it stuck. Alec Bangham is considered to be the "Father of Liposomes." It is amazing how easy it is to prepare multilamellar liposomes. Phospholipids are dissolved in a volatile organic solvent and transferred to a round-bottom flask. The solvent is evaporated using nitrogen or argon gas to prevent lipid oxidation. As the flask is rotated during the evaporation process, a thin layer of phospholipid coats the flask's inside wall. The flask is then placed under a vacuum overnight to eliminate traces of the solvent. An aqueous buffer at a temperature above the phospholipid's melting point (T_m) (Chapter 5) is added to the flask with a few glass beads to aid in dispersion. As the solution is vortexed, it immediately becomes milky, indicating the formation of multilamellar liposomes.

FIGURE 13.13 Alec D. Bangham, 1921–2010.

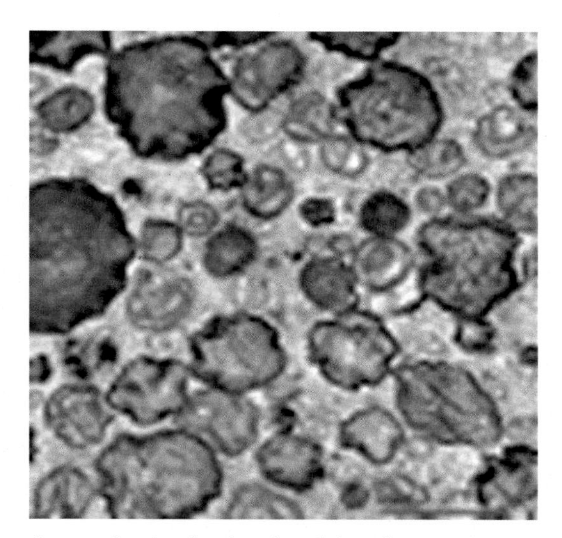

FIGURE 13.14 Electron micrograph of a family of multilamellar vesicles (MLVs) made by the method of Bangham [38].

An electron micrograph of a family of multilamellar vesicles (MLVs) is shown in Fig. 13.14. Depending on the type of phospholipid, the process is complete in ~1–2 min. The process is much slower with poorly hydrated lipids such as phosphatidylethanolamine (PE) and with phospholipids containing long saturated chains with high T_ms. Liposome versatility makes them important for many types of scientific investigations. Liposomes can sequester water-soluble solutes in their aqueous interiors and house membrane integral proteins in their bilayers. Since liposomes can also be made from nonimmunogenic membrane lipids, they have obvious potential as drug delivery vehicles, and indeed thousands of examples of such vehicles can be found in the medical literature (reviewed in Chapter 23, [39,40]). Besides the very simple method for making multilamellar liposomes described previously, there are a wide variety of alternate methodologies used in the production of other types of liposomes. Before we proceed to some of these methodologies, a brief description of size exclusion chromatography (also known for life science applications as gel filtration chromatography) is appropriate.

One important use for liposomes is in measuring transmembrane solute diffusion. Since lipid bilayers are poorly permeable to most solutes, solute gradients can be readily established across the liposomal membranes. One important way to create solute gradients is by use of size exclusion (gel filtration) chromatography, a process that allows rapid separation of big from small particles (Fig. 13.15). The technique was first described in the mid-1950s by Lathe and Ruthven [41,42] and requires at least a 10% difference in size between the two particles. This criterion is easily met when very large liposomes are separated from very small solutes [43]. The most commonly used stationary phase for gel filtration is the resin Sephadex. Sephadex is a trademark for cross-linked dextran gels made by Pharmacia since 1959. Sephadex comes in several discrete bead sizes (Sephadex G-10, G-15, G-25, G-50, G-75, and G-100) with properties altered by the extent of cross-linking. The stationary phase is riddled with tiny pores that are large enough to be penetrated by the solute, but small enough to exclude the liposome. As a result, as the solute–liposome mixture passes down the gel filtration column, the solute enters the pores and so travels a longer, more convoluted path to reach the end of the column than does the large, excluded liposomes. Liposomes leave the column before the nonliposome-sequestered solute. Of

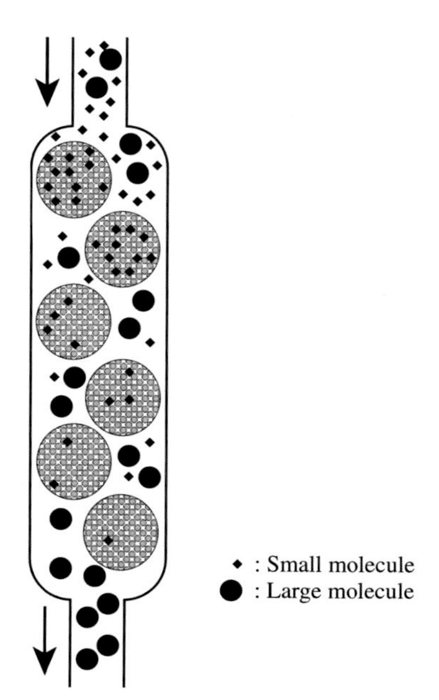

FIGURE 13.15 Separation of small from large particles using gel filtration (size exclusion) chromatography.

course, solutes initially sequestered inside the liposomes pass through the gel and exit the column with the liposome fraction. The solute "sees" ~80% of the total column volume, while the liposome only "sees" ~35% of the volume. In one example, liposomes could be made in a 100 mM KCl buffer and passed through the gel filtration column using a large volume of a 100 mM NaCl buffer. The liposomes exiting the column would have an interior concentration of 100 mM KCl and 0 mM NaCl, while the extra-liposomal solution would be 0 mM KCl and 100 mM NaCl. The established gradients of K^+ and Na^+ would parallel those of living cells, high K^+ inside and high Na^+ outside.

Another method that can accomplish many of the same objectives as gel filtration chromatography is filtration using dialysis tubing. Like gel filtration, dialysis tubing separates large from small particles. Dialysis tubing is a thin semipermeable sheet made of cellulose or cellophane (a form of cellulose). The sheet has microscopic holes that allow small solutes to pass through while retaining large particles. Dialysis tubing can be obtained of different overall size and with different cut-off pore diameter. Liposomes are made in a buffer containing a solute and sealed in a dialysis tube. The tube is floated in a large excess of buffer devoid of the sequestered solute but containing a different solute of the same osmotic strength. During dialysis the nonliposome sequestered solute goes down its concentration gradient through the dialysis tube pores into the bathing solution where it can be readily washed away. Liposomes containing the sequestered solute cannot squeeze through the pores and are then removed from the dialysis tube for further experimentation. Similar to the example described for gel filtration, liposomes made in 100 mM KCl can be sealed in a dialysis tube and floated in a large excess of the same buffer but containing 100 mM NaCl in place of 100 mM KCl. Upon completion of dialysis, the tube contains liposomes with 100 mM KCl sequestered inside and 100 mM NaCl outside. Although gel filtration and dialysis theoretically result in the same final product, dialysis is many, many times slower. While complete dialysis may take days, gel filtration is complete in minutes. During extended dialyses, lipids can oxidize and hydrolyze, producing products that are deleterious to lipid bilayer structure and, in addition, the entire sample is ripe for bacterial growth.

Measurement of leakage rates from liposomes has a wide variety of applications and as a result many methods are available. One of the most common and sensitive methods uses ^{14}C-glucose (or any other small, water-soluble, radiolabeled molecule) as the solute. After the initial dialysis that produces the ^{14}C-glucose sequestered liposomes, the final product is then sealed into a fresh dialysis tube and placed in a new bathing solution devoid of ^{14}C-glucose. As ^{14}C-glucose leaks out of the liposomes, it can be detected at low levels outside the dialysis bag by a scintillation counter. Similarly, gel filtration can be used to measure ^{14}C-glucose leakage. Aliquots of the final ^{14}C-glucose-sequestered liposomes can be taken at different times and run through a small gel filtration column. The ^{14}C-glucose still sequestered in the liposomes comes off the column first while the leaked (nonsequestered) ^{14}C-glucose elutes in a later fraction. The ratio of nonsequestered to sequestered ^{14}C-glucose as a function of time represents the leakage rate.

Gel filtration and dialysis account for the production of most solute-sequestered liposomes. However, in some specific examples the second dialysis or gel filtration steps used to monitor solute leakage can be avoided. For example, carboxyfluorescein (CF^-) leakage can be directly monitored by fluorescence in cuvettes [44]. This technique works because at very high concentration (~160 mM), CF^- is self-quenching and appears a dark brown, but at low concentrations fluoresces a brilliant yellow. CF^- is sequestered inside the liposomes at 160 mM where it appears brown and starts to fluoresce yellow as CF^- leaks out its external concentration drops drastically and as a result fluoresces yellow. Therefore liposome leakage is directly proportional to fluorescence intensity.

In another example, glucose leakage is followed by a linked series of enzymatic reactions in a cuvette. The reactions are followed by light absorbance at 340 nm [45]. Liposomes are made with sequestered glucose. The bathing solution contains adenosine triphosphate (ATP), Mg^{2+}, nicotinamide adenine dinucleotide phosphate ($NADP^+$), and the enzymes hexokinase and glucose-6-phosphate dehydrogenase. As glucose leaks out of the liposomes, the following reactions occur in the bathing solution:

$$\text{Glucose} + \text{ATP} + \text{Mg}^{2+} \xrightarrow{\text{(Hexokinase)}} \text{glucose-6-phosphate} + \text{ADP} + \text{Mg}^{2+}$$

$$\text{Glucose-6-phosphate} + \text{NADP}^+ \xrightarrow{\text{(Glucose-6-phosphate dehydrogenase)}} \text{6-Phosphoglucanate} + \text{NADPH}$$

NADPH absorbs at 340 nm while $NADP^+$ does not. Therefore glucose leakage from the liposomes is proportional to an increase in absorbance of NADPH at 340 nm.

Unilamellar Vesicles: The original liposome preparation method of Bangham produces MLVs of varying size [38]. MLVs have several important shortcomings that limit their use. They are not of uniform size and a large portion of their total lipid is sequestered inside the onion-like vesicles as internal bilayers and is not exposed to the external environment. This severely limits the internal aqueous space available to solutes and makes it impossible to accurately assess the amount of lipid comprising the external membrane surface. Also evaluating the net leakage rate is very complex as solutes must traverse several bilayers before making their way through to the external bathing solution. Therefore a different type of liposome, one having only a single surrounding bilayer and of uniform size was required. This type of liposome is generally referred to as a unilamellar vesicle (ULV), in contrast to an MLV (multilamellar vesicle). ULVs are further subdivided by size. Table 13.6 lists the most common types of liposomes.

SUV by Sonication: There are three basic ways to make SUVs: by sonication, by extrusion, and by use of detergents. The sonication methods discussed in Chapter 12 for cell homogenization can also be used to convert MLVs into SUVs. High energy tip type sonicators are preferred. MLVs are a mixed population of large (0.1–5 μm), partially collapsed onion-like

TABLE 13.6 The Most Common Types of Liposomes, Listed by Size

Abbreviation	Name
MLV	Multilamellar vesicle
ULV	Unilamellar vesicle
SUV	Small unilamellar vesicle
LUV	Large unilamellar vesicle
GUV	Giant unilamellar vesicle

structures that in solution have an opaque, milky appearance. Upon sonication a striking change in the solution appearance occurs. The MLVs are broken down into SUVs that appear translucent with a slight bluish sheen. The transition indicates that the sonication is complete. Sonicated SUVs are ~200 Å in diameter. After sonication, the vesicles are single walled with two-thirds of the lipids in the more loosely packed outer bilayer leaflet and one-third comprising the tightly packed inner leaflet. These vesicles cannot get any smaller and continued sonication will only lead to lipid oxidation and hydrolysis. Sonication is a vigorous procedure, producing local heating that must be dissipated by sonicating on ice with an on–off duty cycle. Sonication will also generate unwanted titanium particles and will denature any proteins that may have been reconstituted into the starting MLVs. The very tight radius of curvature also makes small sonicated vesicles highly susceptible to fusion, thus increasing their size over time. Extensively fused, sonicated vesicles can then be used as a starting population to make single large lamellar vesicles (LUVs).

Vesicles by Extrusion: By an imprecise definition, SUVs range between 200–500 Å in diameter. ULVs larger than this are either defined as LUVs (large unilamellar vesicles) or if they are very large, GUVs (giant unilamellar vesicles). SUVs and perhaps some smaller LUVs can be made by extruding MLVs several times through a series of Nuclepore filters. Nuclepore filters are a plastic (polycarbonate) filter that is perforated with perfectly round, uniform-sized holes. The holes are made by exposing the plastic filter disc to radiation by a process called Nuclear Particle Track Etching. The round holes can be made of many sizes. Nuclepore filters can be obtained with holes of 10, 8, 5, 3, 2, 1, 0.8, 0.6, 0.4, 0.2, 0.1, 0.05, and 0.015 μm. MLVs are forced through these filters under pressure, normally 10–20 times for each chosen pore size. Pressure is provided through devices known as extruders (Fig. 13.16).

Two types of extruders are available (Fig. 13.16). With large-pore filters it is easy to force MLVs through the filter and a manual dual syringe-type extruder will suffice (Fig. 13.16, Top, Avestin LipsoFast-Basic Extruder). In order to force the MLVs through small-pore filters, very high pressures are required; far more than can be obtained by a manual syringe extruder. High pressure extruders (Fig. 13.16, Bottom, Lipex Extruder) force the MLVs through the Nuclepore filters under extremely high nitrogen pressure where a special gas regulator is even required. Usually the MLVs are first passed through a large-pore filter, then a medium-pore filter, and finally a small-pore filter. Approximately 20 passes are run for each filter. The final product should be a uniform-sized population of LUVs. Unfortunately, unless the vesicles are very small, they may be of uniform size and much smaller then the parent MLVs, but still contain some sequestered lipid bilayers (they are still MLVs). A 2007 paper by Lapinski et al. [46] compares the basic physical properties of LUVs made by sonication versus extrusion.

Vesicles by Detergent Dilution: Detergents are essential for membrane studies. They are amphipathic molecules that in some ways resemble membrane polar lipids but are orders of magnitude more water-soluble (see Section 1). If added to membranes at high enough concentration, they can replace the resident membrane lipids while maintaining normal protein conformation and activity. Detergents make functional protein isolation and reconstitution into proteoliposomes possible, but also provide another path to LUV construction. If membrane polar lipids are hydrated in a buffer containing a large excess of detergent, they become solubilized into detergent micelles. The water-soluble detergents can then be easily removed by dialysis or by gel filtration. When the detergent–membrane lipid mixture is sealed in a

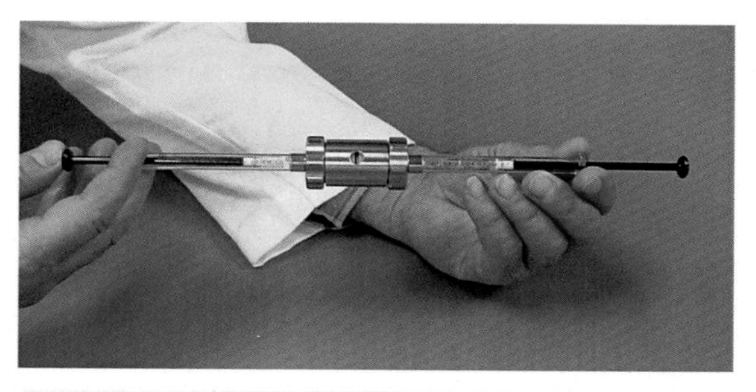

FIGURE 13.16 Two types of extruders. Top: manual dual syringe-type extruder (Avestin LiposoFast-Basic Extruder). Bottom: high pressure extruder (Lipex Extruder). http://www.avestin.com/lf.html.

dialysis tube, the detergent diffuses down its concentration gradient eventually crossing the dialysis tubing and entering the external bathing solution. As the detergent concentration decreases inside the dialysis tube, the membrane polar lipids fall out of solution as LUVs that are too large to fit through the dialysis tubing pores. These vesicles do not have sequestered lipids and so are unilamellar. Detergents can be rapidly eliminated by passing the detergent–membrane lipid mixture through a gel filtration column. As the mixture passes down the column, the detergent is extracted into the small spaces in the gel matrix, while the polar membrane lipids come out of solution as LUVs that are excluded from the gel. The newly formed LUVs leave the column before the free detergent and so can be isolated largely devoid of detergent. Vesicle size is dependent on the detergent used and can be partially altered by changing the dialysis or gel filtration rate. Affect of detergents on vesicle size can be dramatic [47]. The relationship between detergent and vesicle size is shown in the following table:

Detergent	Vesicle size
Sodium cholate	SUV (small)
n-Octyl (or hexyl or heptyl) β-D-glucopyranoside	LUV (medium)
POE4 (lauryl dimethylamine oxide)	GUV (very large)

SUV, small unilamellar vesicle; *LUV*, large unilamellar vesicles; *GUV*, giant unilamellar vesicle.

Unfortunately it is very difficult to remove all of the detergent from new liposomes prepared by dialysis or gel filtration. This is particularly true for detergents with a low CMC (eg, Triton X-100). Another detergent-removal process, extraction using hydrophobic adsorption onto nonpolar polystyrene beads [48], has proven to be very effective and has become commonly used in LUV preparations. The best known polystyrene product is Bio Beads from Bio-Rad Laboratories. Polystyrene beads can be used to clean up LUVs made by any detergent process or can be used *in lieu* of dialysis or gel filtration. Polystyrene beads are responsible for major advances in membrane integral protein reconstitutions into proteoliposomes discussed in Section 5.

LUVs: Dilution Methods. LUVs can be made by variations of methods that are all based on a similar theme; rapid dilution of a concentrated membrane lipid solution. By these methods, membrane polar lipids are dissolved in a small volume of an organic solvent. Solvents include ethanol, ether, Freon ($CHFCl_2$), or a detergent–in-water solution. The concentrated membrane lipid solution is injected into a large excess of a rapidly mixing aqueous bathing solution. The organic solvent or detergent is instantly dispersed into a large excess of the bathing solution as the water-insoluble membrane polar lipids separate into LUVs. Ethanol works well as the lipid solvent since it is infinitely miscible in water. Another method, known as "ether vaporization," is demonstrated in Fig. 13.17 [49]. While ether is totally insoluble in water, it does have a high vapor pressure and can readily be removed from water by vacuum. The membrane lipids are dissolved at high concentration in ether and slowly injected into an aqueous bath that is both warmed and under vacuum. The ether is instantly vaporized and pumped away while the membrane lipids are left behind as LUVs. The Freon method is similar to the "ether vaporization" method. Membrane lipids are dissolved at high concentration in Freon and slowly injected into an aqueous bath maintained at 37°C. Since Freon boils at ~9°C, it is instantly vaporized leaving behind LUVs. Finally, membrane lipids, dissolved in an aqueous detergent solution can be injected into a large aqueous bath where the water-soluble detergent is rapidly dispersed leaving LUVs. The detergent dilution method has a very important advantage over the ethanol, ether vaporization and Freon methods. Functional integral proteins, isolated in detergent/membrane lipid micelles, can be incorporated into LUVs as they form, a process known as "membrane reconstitution."

4.3 Bicelles

Most membrane protein reconstitution studies have historically employed detergents and some form of lipid bilayers (see Section 5). While these studies have led to spectacular advances in our understanding of membrane structure and function, they have presented some serious problems. Major artifacts have arisen because of the detergents required to solvate the proteins for isolation from biological membranes. Detergents just do not closely mimic natural bilayer lipids, thus resulting in isolated proteins with altered, unnatural conformations. As a result, other methodologies employing much lower levels (or even no) detergent have been sought. One unusual, but highly successful, system employs what is referred to as "bicelles" [50–52]. Like ULVs, bicelles contain a central lipid bilayer (Fig. 13.18). However, ULVs and bicelles are quite different with respect to both their

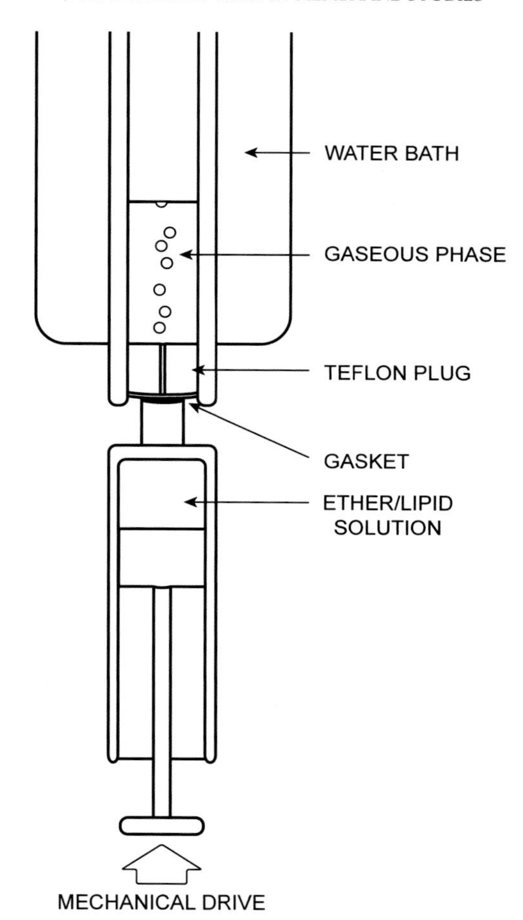

FIGURE 13.17 Ether vaporization method for production of large unilamellar vesicles (LUVs) [49].

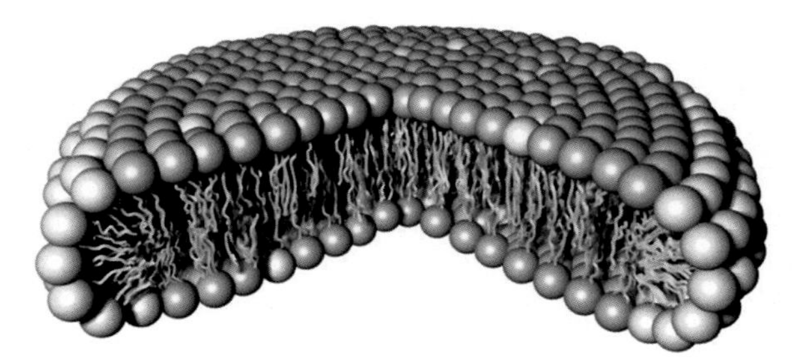

FIGURE 13.18 Drawing of a bicelle. http://www.cbmn.u-bordeaux.fr/client/gfx/photos/contenu/bicelle_580.JPEG.

composition and structure. In addition to having a bilayer component, bicelles also have a micelle (Chapter 5) component.

Bicelles are composed of two different types of lipids, one a long-chain, lipid bilayer-preferring polar lipid [usually a phosphatidylcholine (PC)] and the other a micelle-preferring amphipathic lipid (often a detergent but sometimes a short chain polar lipid) [50–52]. The bicelle center is composed of a lipid bilayer patch whose edges are stabilized by a micelle ring that keeps water from penetrating into the bilayer interior (Fig. 13.18). Bicelles can vary in size, but are always much smaller than their lipid vesicle cousins. Bicelles can be thought of as intermediate structures bridging the gap between micelles and lipid bilayers, and combining some of the attractive properties of both model membrane systems [50,52].

Bicelles were initially developed in the early 1990s [53,54]. Importantly, it was found that these systems can align in even low magnetic fields [55] while simultaneously exhibiting liquid crystalline-like (L_α phase) bilayer properties [50]. In the 1990s, Sanders and Prestegard introduced the concept of aligned bicelles that, when coupled with their inherent small size and low detergent levels, made these model membranes ideally suited for use in solid-state nuclear magnetic resonance (NMR) studies of membrane orientation, structure, and dynamics [51,55]. Of particular importance are the many studies that have been performed on membrane peptides and proteins [55,56]. The original bicelles were mixtures of the phospholipid, dimyristoylphosphatidylcholine (DMPC, 14:0,14:0 PC) with either a micelle-generating short chain organic amphiphile, dihexanoylphosphatidylcholine (DHPC) [54]) or the bile-salt derivative, 3-(cholamidopropyl)dimethylammonio-2-hydroxy-1-propanesulfonate (CHAPSO) [53]. Since their original discovery, many types of bicelles have been described. Some are DMPC bilayers that have been doped with additional phospholipids of different chain lengths or charged head groups. Bicelles can vary over a wide range of amphiphile (detergent) compositions, amphiphile:phospholipid ratios (roughly 1:2–1:5), water content (roughly 60–97%), buffer composition, pH, and temperature (~30–50°C).

With an increase in detergent concentration, bicelles are broken down into smaller and smaller discs, and these smaller discs exhibit reduced order. The important features of bicelles for membrane studies include: their ease of construction; their compositional versatility; their ability to study both function and structure simultaneously in the same system (by solid-state NMR); their ability to maintain enzyme activity for much longer periods of time than lipid vesicles; their excellence as substrates for lipolytic enzymes (eg, PLA_2); and their use in isolating functional membrane proteins for reconstitution experiments.

It has always been difficult to obtain membrane protein crystals for X-ray determination of 3-dimensional structure [57]. Water-soluble protein crystals are relatively easy to come by, but not membrane protein crystals. The problem stems from the detergents used in membrane studies. Unfortunately, many membrane proteins are not stable enough in detergent to produce high quality crystals [57]. Since DMPC-bicelles closely mimic many of the physical properties of natural membranes, and are generally low in detergent, they have proven to be an excellent tool to extract membrane proteins in active (native) form [58]. As a result, a number of membrane proteins have now been successfully crystallized using the bicelle method. Included in this growing list is the multispan transmembrane protein bacteriorhodopsin [59] whose structure was discussed in Chapter 6.

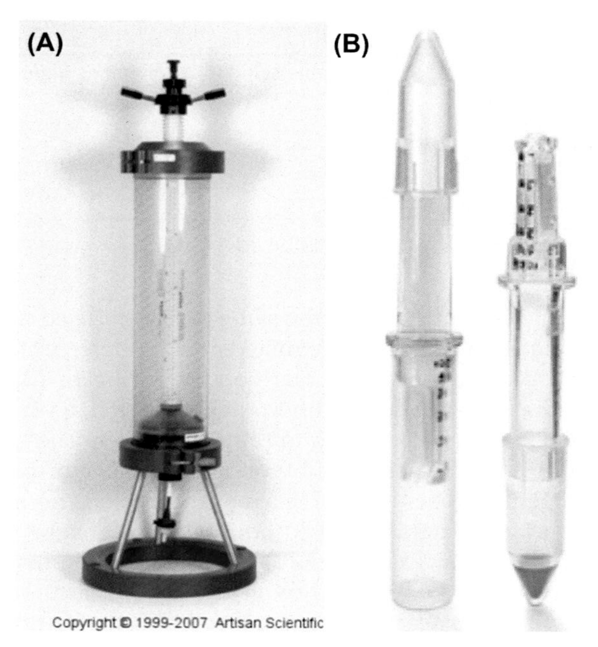

FIGURE 13.19 Two types of Amicon concentration filter devices. (A) High pressure is applied by nitrogen gas from the top. (B) Centrifugation drives the buffer through the filter pores.

Concentration Methods: The dilution methods for making LUVs have one major disadvantage, the LUVs are formed in very dilute solution and must be concentrated. This is routinely done using two types of filters, the best known being Amicon concentrators (Fig. 13.19). For both systems pressure must be applied to the diluted LUV sample to force the buffer through small holes crossing the filter. LUVs are too large to traverse the pores and so remain behind at ever increasing concentration. In one type of concentrator, high pressure is applied by nitrogen gas from the top (Fig. 13.19A) while the second type of concentrator relies on centrifugation to drive the buffer through the filter pores (Fig. 13.19B).

Using dilution methodologies, liposome size, and hence sequestered volume, can be altered by changing the injection rates. This becomes very important if the LUVs are to be used for drug delivery. The sequestered volume can vary tremendously from one type of liposome to another. Although MLVs are large, their available interior volume is surprisingly small due to the sequestered multilipid bilayers. Table 13.7 lists the sequestered volumes for several types of liposomes.

5. MEMBRANE RECONSTITUTION

In a previous section of this Chapter, it was discussed how integral membrane proteins could be isolated into detergent micelles while maintaining their correct (active) conformation (see Section 4.3). It was also shown that membrane polar lipids could be isolated and

TABLE 13.7 Sequestered Volumes for Several Types of Liposomes

Liposome type	Available sequestered volumes (μg/nmol lipid)
MLV	1.8
SUV (sonication)	0.8
LUV (ether vaporization)	14.0

LUV, large unilamellar vesicles; *MLV*, multilamellar vesicles; *SUV*, small unilamellar vesicles.

dissolved into detergent micelles, often using the same detergent used for protein isolation. If these two populations of detergent micelles are mixed and the detergent removed by dialysis, gel filtration, or preferably by binding to nonpolar polystyrene beads (Bio-Beads), proteoliposomes can be directly produced [60,61]. In addition to this method, a variety of other methods rely on adding the protein–detergent micelles to premade LUVs. These methods depend on first, partially destabilizing the LUV bilayers by either detergent or brief sonication before adding the protein–detergent micelles. Partial bilayer disruption is necessary to prevent protein denaturation at the surface of tightly packed lipid bilayers. Several detergents have been used as destabilizing agents, primarily Triton X-100. Tip sonication must be very brief to prevent protein denaturation or may be replaced by gentle bath sonication.

Ideally the proteoliposomes are nearly detergent-free LUVs that contain the active integral protein spanning the membrane lipid bilayer. However, one potential problem concerns determining the protein's orientation in the membrane. Four orientations of the newly reconstituted protein are possible. Suppose in the biological membrane, the transmembrane protein exists with an A-end exposed to the aqueous cell interior and a B-end exposed to the aqueous cell exterior. The possible orientations for the reconstituted proteoliposome are:

1. A-end inside, B-end outside (correct orientation)
2. B-end inside, A-end outside (reverse orientation)
3. A-end inside, B-end inside (bent orientation)
4. A-end outside, B-end outside (bent orientation)

During the process of reconstitution some orientations are more favorable than others. For example, if the protein is rigid like those with multiple span helices, the protein will not be flexible enough to exhibit the two bent conformations (3 and 4). Many plasma membrane proteins have an extensive sugar component that in a living cell will always face the external aqueous environment (the B-end in the example). If the sugar component is sufficiently large, steric hindrance in the small LUV interior will favor proteoliposomes with the sugars on the external surface (1 and 4). A sugar attachment containing sialic acid presents an additional problem, high negative charge density. Multiple negative charges require more spacing between the sugar components, favoring an external orientation. It is also possible to isolate and reconstitute a protein with a very large antibody attached at one end. This end will exclusively favor external orientations. Of course, once reconstituted, the antibody can be removed, thus restoring function.

There are now countless numbers of integral membrane proteins that have been successively reconstituted into proteoliposomes. Single protein proteoliposomes allow for close inspection of any lipid specificity and molecular mode of action that may be difficult to determine in complex, multicomponent biological membranes. A few examples follow.

5.1 Ca^{2+} Adenosine Triphosphatase

Ca^{2+} adenosine triphosphatase (ATPase) is one of the most studied of all integral membrane proteins [62]. This transport protein (pump) was first discovered in 1966 and was successfully reconstituted into proteoliposomes by the early 1970s [63]. The pump comes in two basic types, the plasma membrane Ca^{2+} ATPase (PMCA) and the sarcoplasmic reticulum Ca^{2+} ATPase (SERCA). Both pumps belong to the large family of enzymes known as P-Type ATPases because they form a phosphorylated intermediate during transport. Their job is to pump Ca^{2+} from inside to outside the cell using ATP as the energy source. The pumps maintain constant, low intercellular Ca^{2+} levels that are required for proper cell signaling. In 1972, Efraim Racker [63] reported a successful reconstitution of the sarcoplasmic reticulum Ca^{2+} ATPase into soybean phospholipid proteoliposomes. Importantly, the reconstituted vesicles pumped Ca^{2+} at the expense of ATP and were sensitive to chlorpromazine, a known inhibitor of calcium ATPases. Two years later, Warren et al. [64] reconstituted SERCA into proteoliposomes where >99% of the native membrane lipids had been replaced by synthetic dioleoyl phosphatidylcholine (DOPC). However, full expression of the capacity to accumulate Ca^{2+} into the proteoliposomes required the presence of internal oxalate (Fig. 13.20), a Ca^{2+} chelator. The explanation of these observations is that a truly functional calcium ATPase had been reconstituted into a proteoliposome devoid of native phospholipids but comprised of a synthetic phosphatidylcholine (PC). However, both the right side out and inside out protein conformations were found in the proteoliposomes. ATP caused Ca^{2+} to be pumped into the vesicle interior by one conformation, whereupon the opposite conformation ATPase would pump the Ca^{2+} back out, thus limiting net Ca^{2+} uptake. In the presence of internal oxalate, Ca^{2+} that was pumped into the vesicle was chelated as water-insoluble calcium oxalate (the major component of kidney stones) and so remained sequestered, resulting in a large increase in net Ca^{2+} accumulation.

5.2 Bacterial Proline Transporter

Reconstitution studies with the bacterial proline transporter into proteoliposomes demonstrate how a variety of biochemical "membrane tricks" can be used to deduce the transporter's molecular mechanism [65,66].

Proteoliposome Production: The bacterial proline transporter was isolated using a 17 amino acid polyhistidine tag attached to the C-terminal and a nickel-affinity column

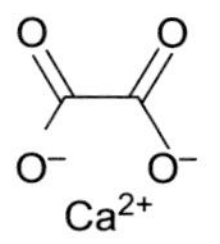

O O
O⁻ O⁻
Ca^{2+}

FIGURE 13.20 Calcium oxalate.

TABLE 13.8 Essential Features of the Bacterial Proline Transporter Mechanism

- Need a nonleaky membrane
- Need a negative interior potential
- Need a Na^+ gradient, outside to inside
- The transporter is an example of active symport

[65]. The purified transporter was reconstituted into preformed detergent (Triton X-100) destabilized LUVs. Triton X-100 was removed by polystyrene beads. Proline transport was supported by PE and phosphatidylglycerol (PG), but not PC and cardiolipin (CL). Energetics: Experiments with reconstituted proteoliposomes proved the mechanism of action for the bacterial proline transporter [66]. Proline uptake was shown to be an example of active symport (a form of co-transport, discussed in Chapter 19). The bacterial proline transporter was reconstituted into proteoliposomes in a KCl buffer and diluted into a NaCl buffer. The proteoliposomes were therefore "K^+-loaded" with a K^+ gradient that was high inside and low outside and a Na^+ gradient that was high outside and low inside. Upon addition of the K^+-ionophore valinomycin, K^+ moved down its concentration gradient generating a net negative potential in the proteoliposome interior. The negative interior drove Na^+ down its gradient from outside to inside carrying with it proline. This mechanism is active symport. Formation of the transmembrane electrical gradient was followed by fluorescence of the lipophilic cationic fluorescent dye carbocyanine. As carbocyanine entered the proteoliposome it became concentrated, resulting in fluorescence self-quenching. The decrease in net fluorescence was a direct indicator of the generation of a net negative potential in the proteoliposome interior. Leakage of K^+ was very slow indicating the proteoliposomes were surrounded by an intact, nonleaky membrane. Upon addition of valinomycin, K^+ leakage was dramatically enhanced. When ^{14}C-proline was added to the NaCl bathing solution, the addition of valinomycin corresponded with an uptake of ^{14}C-proline. If Na^+ was not included in the external solution or the Na^+ gradient was collapsed by a sodium ionophore (eg, gramicidin D), ^{14}C-proline was not taken up into the proteoliposomes. A pH gradient was also ineffective at driving proline uptake. The stoichiometry of the symport system was approximately one Na^+ for one proline. Table 13.8 summarizes the essential features of the bacterial proline uptake mechanism.

6. LIPID RAFTS

Although the basic concept of a lipid raft is compelling, it does not necessarily mean it is correct. It has been known for decades that biological membranes are not homogenous dispersions of lipids and proteins, but instead exist in a bewildering array of patches known as micro-domains [67]. Lipid rafts are an attempt to identify one of these lipid–protein patches and to assign it an essential biochemical function, namely cell signaling [68]. But do lipid rafts actually exist in biological membranes or are they just a figment of membranologists' fertile imaginations?

Lipid rafts can be defined by two basic operational definitions [69]. First, they are fractions of a plasma membrane that are enriched in cholesterol, and (predominately) saturated phospholipids and sphingolipids. These lipids, when reconstituted into lipid bilayer vesicles, prefer being in a tightly packed, liquid ordered (l_o) state. Raft fractions also contain a characteristic set of proteins that are important in cell signaling [70]. Second, lipid rafts can be dispersed by extracting cholesterol, usually with β-cyclodextran (Chapter 10). Intuitively it makes sense to compartmentalize a related set of functions into distinct membrane domains (eg, cell signaling and lipid rafts) that are separate from the remaining membrane that surrounds them. The implication has been that lipid–lipid affinities create the lipid raft that then attracts the appropriate membrane proteins. However, there are many unsolved problems associated with the raft model [71,72]. Michael Edidin [69] has made a strong argument that it is more likely raft proteins select the appropriate lipids from the surrounding lipid sea than *visa versa*. Perhaps the actual answer is that the lipid–protein associations responsible for raft formation and stability are the result of *simultaneous* lipid–lipid, lipid–protein, protein–lipid, and protein–protein interactions. A complex set of such interaction would be very hard to observe in biological or even model membranes.

One of the major problems with the concept of lipid rafts is getting a handle on their size and stability [73]. In model membranes, lipid rafts can be very large (microns in diameter). However, the size of lipid rafts in biological membranes has yet to be fully determined, but it is estimated to be very small (1–1000 nm in diameter). The estimated size of biological lipid rafts continues to decrease as imaging methodologies improve [73]. Many of the more recent studies have estimated lipid raft size at the lower end of the range, ~5 nm in diameter. The vanishingly small size of lipid rafts in biological membranes raises the possibility that even if rafts do exist, they may not be stable enough (they are just too small) to have any biological function. Rafts may not exist for a long enough stretch of time to attract the appropriate signaling proteins.

Lipid rafts are of two basic types, planar rafts and invaginated, flask-shaped rafts called caveolae [71,74]. Planar rafts and caveolae can be easily distinguished. Although both have essentially the same typical raft lipid composition, planar rafts have flotillin proteins while caveolae have a characteristic flask-shape and caveolin proteins. Flotillins and caveolins both function in recruiting signaling molecules into lipid rafts and thus provide the raft with a skeletal frame. In a parallel fashion it has been proposed that cholesterol is the "molecular glue" that holds the raft lipids together. It has been known for decades that cholesterol has a strong preference for sphingolipids over other phospholipids (Chapter 10) [75]. Cholesterol also has a stronger affinity for saturated acyl chains and avoids polyunsaturated chains [76,77]. Finally, a hallmark of lipid rafts is their dissociation upon removal of cholesterol by β-cyclodextrin [78].

The earliest support for lipid rafts came from cold temperature detergent extractions, but it is not certain what these experiments actually mean. At 4°C, the nonionic detergent Triton X-100 (later success was also reported with Brij-98 and CHAPS) solubilized much of the membrane, while the insoluble components, cholesterol, (predominately) saturated phospholipids, sphingolipids, and characteristic signaling proteins, were left behind as lipid rafts [70]. The insoluble fraction is often referred to as detergent resistant membranes (DRMs) or sometimes detergent-insoluble glycolipid-enriched complexes. Later advances eliminated the need for detergents, reducing possible artifacts. There are now countless published methods to isolate lipid rafts, most of which are slight variations on a basic theme. There follows an abbreviated outline of one of the most commonly used procedures, known as the "Smart Prep" for its inventor Eric Smart of the University of Kentucky [79]. The procedure is for the isolation of caveolae, a type of lipid raft.

Basic Steps in the Smart Prep [79].

- All steps done at 4°C
- Wash cells in 250 mM sucrose (isotonic) buffer
- Low speed centrifugation
- Homogenize cells with Dounce homogenizer (20 times) and pass through a 21 gauge needle (20 times) in sucrose
- Low speed centrifugation to remove debris and obtain a post nuclear supernatant
- High speed centrifuge on 30% Percoll self-generating gradient
- Band at middle is the plasma membrane
- Isolate and sonicate the plasma membrane fraction
- High speed centrifugation on Optiprep to isolate caveolae

It is now clear that if noncaveolae lipid rafts exist, they must be extremely small and fleeting. Problematically, the process of isolating rafts takes many orders of magnitude longer than the likely raft lifetime. Perhaps the membrane components that are isolated together as lipid rafts only share similar physical properties but never actually exist together in the intact membrane.

6.1 Experimental Support of Rafts

There is considerable experimental support for the existence of lipid rafts, primarily phase separations in model membranes, cholesterol depletion experiments, and multiple component experiments linking DRM analysis, cholesterol depletion, and microscopy. Micron sized domains have been visualized on the surface of GUVs using fluorescent membrane lipid tags that preferentially partition into l_o (raft) or l_d (nonraft) phases [80]. In one report, Dietrich et al. [81] demonstrated phase separation into l_o and l_d domains using fluorescence microscopy. Importantly, Triton X-100 was shown to selectively solubilize the l_d nonraft domain while not perturbing the l_o raft domain. Similar model membrane studies demonstrated that phase separation could be achieved at both cold temperature (4°C) and physiological temperature (37°C), supporting the possibility of rafts existing in living cells.

Hammond et al. [82] reported a study using a complex model membrane that supported the feasibility of lipid rafts. GUVs were made from DOPC, dioleoylphosphatidylglycerol (DOPG), sphingomyelin (SM), cholesterol, the ganglioside GM1, and the rhodamine-labeled LAT transmembrane peptide. This lipid mixture formed a single phase that was poised close to a separation boundary (Fig. 13.21, panel A). Upon addition of the cholera toxin B subunit, a dramatic phase separation occurred (Fig. 13.21, panel B). The fluorescent rhodamine-peptide preferentially partitioned into the l_d phase. Phase transition was induced by addition of the pentavalent cholera toxin B subunit that binds to and cross-links the GM1s. The top panel (panel A) is before the addition of cholera toxin while the bottom panel (panel B) is after cholera toxin addition. The dark regions in panel B are domains in the l_o (raft) phase.

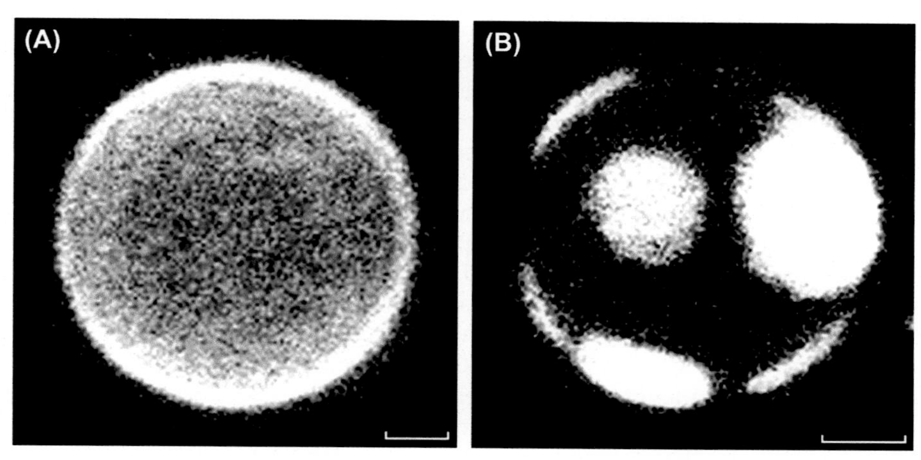

FIGURE 13.21 Fluorescence imaging of micron-sized domains on the surface of giant unilamellar vesicles (GUVs) [73]. The model bilayer membranes were composed of dioleoyl phosphatidylcholine (DOPC), dioleoylphosphatidylglycerol DOPG, sphingomyelin, cholesterol, the ganglioside GM1 and a rhodamine-labeled LAT trans-membrane peptide. This lipid mixture formed a single phase (Top Panel, A) that was poised close to a separation boundary. Upon addition of the cholera toxin B subunit, a dramatic phase separation occurred (Bottom Panel, B). The fluorescent rhodamine-peptide preferentially partitioned into the l_d phase. The dark regions in panel B were domains in the l_o (raft) phase.

One of the hallmarks of lipid rafts in biological membranes is a loss of physiological function, primarily related to signal transduction or membrane trafficking, upon cholesterol depletion [83]. By far the major agent used to extract cholesterol from membranes is β-cyclodextrin although the polyene antibiotics filipin and nystatin (Chapter 19) can effectively remove cholesterol from rafts through complexation in the membrane. One of the strongest pieces of evidence in support of lipid rafts is a strong correlation between DRM-associated proteins, cholesterol depletion, loss of physiological function, and microscopy. Although it now appears that DRMs are not the same as preexisting rafts [70], a surprisingly large number of examples of these correlations exist in the lipid raft literature.

6.2 Questions Concerning Rafts

Many unanswered questions concerning the nature and even existence of lipid rafts abound [71]. The major trouble concerns the extremely small apparent size of lipid rafts in biological membranes (perhaps as small as 5 nm and still shrinking) while lipid rafts in model membranes can be in the micron range. How small is too small for lipid rafts to have any biological significance? Glycosylphosphatidylinositol (GPI)-anchored proteins are routinely found in DRMs, indicating that they are common raft components, yet by fluorescence resonance energy transfer, GPIs are shown to be evenly distributed on the cell surface or are present in at most very small (nanoscale) clusters containing only a few proteins. Also, lateral diffusion measurements using fluorescence recovery after photobleaching (FRAP) show that lipid raft components (GPI-anchored proteins, other raft proteins, and raft lipids) do not diffuse as a single large unit in the plasma membrane. If raft components diffuse at different rates, how can they be found together in a raft? It remains to be shown if lateral membrane diffusion rates are sufficient to allow proteins time to diffuse into lipid rafts. The preponderance of lipid raft isolations into DRM fractions use cold temperatures (4°C) and nonionic detergents, both of which can produce misleading artifacts. Using model membranes it is hard to access the potential importance of lipid asymmetry and cytoskeletal involvement in rafts. Also, it is not certain if intracellular membrane rafts even exist.

Reputed raft proteins present another group of problems. Since raft lipids exist in the l_o state where they are extended, membrane lipid raft domains should be thicker than the surrounding nonraft l_d state regions. Indeed this has been confirmed by atomic force microscopy in model membranes [84]. Therefore one would expect that raft proteins should exhibit a longer hydrophobic match surface (Chapter 10) than nonraft proteins. However, to date a protein's hydrophobic span length has not been shown to be related to its preferential location in rafts. The rough surface associated with the membrane-spanning portion of an integral protein should exclude transmembrane proteins from l_o state raft lipids. Indeed, most transmembrane proteins do associate poorly with DRMs. However, rafts are known to accommodate a few characteristic transmembrane proteins. The question remains, what makes raft integral proteins different from nonraft integral proteins and why are they concentrated in rafts?

There are many reports where polyunsaturated fatty acyl chains including arachidonic acid ($20{:}4^{\Delta 5,8,11,14}$) and docosahexaenoic acid ($22{:}6^{\Delta 4.7.10.13.16.19}$) have been found at significant levels among raft lipids [85]. This seems contrary to the basic precept that rafts exist in the tightly packed l_o state. How does a raft accommodate a highly contorted polyunsaturated fatty acid? To justify this contradiction it has been proposed for $PI(4,5)P_2$, a common component of rafts, that the *sn*-2 arachidonic chain fits into a groove on a protein's surface while the saturated stearic *sn*-1 chain inserts into the bilayer [86]. The affect of cholesterol depletion on transmembrane signaling functions has become a cornerstone in defining rafts. Yet even this has its problems. Cholesterol levels can affect membrane structure and function by ways that are unrelated to raft structure. For example, Pike and Miller [87] have made a strong case that cholesterol depletion affects both raft structure and signaling through $PI(4,5)P_2$.

It is now clear that DRMs are not exactly the same as lipid rafts that are resident in the plasma membrane [70]. The current lipid raft model is in flux and will continue to be adjusted to match new experimental findings.

6.3 Lipid Raft Summary

The fundamental importance of the lipid raft model was to propose a type of domain that could serve as a simple paradigm for general membrane structure–function. Unfortunately, the original concept of a lipid raft has continued to evolve, becoming ever more complex to where it is now hard to define exactly what a lipid raft actually is and is not. Raft dynamics is poorly understood and fixed rules remain elusive. Table 13.9 lists some recently suggested characteristics of lipid rafts.

TABLE 13.9 Some Characterstics of Lipid Rafts

- Rafts are highly regulated and may occur only in response to stimulation.
- In unstimulated cells, rafts are very small and unstable, if they exist at all.
- Lipid rafts have a high affinity for ordered lipids.
- Plasma membrane lipid composition is maintained close to that required for phase separation so that small changes in protein or lipid can have a large influence on lipid rafts.
- Caveolins and flotillins probably organize rafts into specialized membrane domains.
- Some raft proteins are linked to the actin cytoskeleton network.
- The myristate–palmitate (lipoprotein) motif appears to be an especially efficient raft-targeting signal.
- Particularly important is palmitoylation for raft signaling.
- Polyunsaturated fatty acyl chains are hidden in clefts on a raft protein surface.
- It will be essential to determine the actual structure of lipid rafts as they exist in a biological membrane.

7. SUMMARY

Since it is so difficult to study a single protein in a complex biological membrane, methods have been devised to purify and reconstitute it into a liposome of known composition. These methods rely on detergents. Detergents are amphipathic molecules that resemble membrane polar lipids, but are orders of magnitude more water-soluble. Hundreds of detergents are used to lyse cells, extract membrane proteins and lipids, and reconstitute them into membranes. After isolation, the purified membrane is dissolved in an appropriate detergent solution and the target integral protein purified, usually by some form of affinity chromatography. After purification, the protein is concentrated and mixed with a desired lipid that is also solubilized by the same detergent. Upon rapid removal of the detergent (by dialysis, size-exclusion chromatography, rapid dilution, or polystyrene beads), polar lipids and integral proteins fall out of solution as proteoliposomes of known composition. An unusual model membrane known as a "bicelle" has recently been developed to isolate, investigate, and crystalize membrane proteins. At present, the hot topic in membrane structure/function is lipid rafts, either planar rafts or caveolae. It is now clear that if planar lipid rafts exist, they must be extremely small (<5 nm) and fleeting.

This chapter concludes the first part of this book (**Part I. Membrane Composition and Structure**). Chapter 14 will begin the second part of the book (**Part II. Membrane Biological Functions**) by discussing the biogenesis of membrane fatty acids.

References

[1] Helenius A, Simons K. Solubilization of membranes by detergents. Biochim Biophys Acta 1975;415:29–79.
[2] Rosen MJ. Surfactants and interfacial phenomena. New York: John Wiley & Sons; 1989. 431 pp.
[3] Kunjappu JT. Advancing the chemical sciences. Chemistry World. RSC; March 2003.
[4] Griffin WC. Classification of surface-active agents by "HLB". J Soc Cosmet Chem 1949;1:311.
[5] Griffin WC. Calculation of HLB values of non-ionic surfactants. J Soc Cosmet Chem 1954;5:259.
[6] Wilson L. Methods in cell biology: the cytoskeleton. Cytoskeletal proteins, isolation and characterization. vol. 24, Part 1, p. 275. New York: Academic Press; 1982. 445 pp.

[7] Hatefi Y, Hanstein G. Solubilization of particulate proteins and nonelectrolytes by chaotropic agents. Proc Natl Acad Sci USA 1969;62:1129–36.
[8] Hochuli E, Bannwarth W, Döbeli H, Gentz R, Stüber D. Genetic approach to facilitate purification of recombinant proteins with a novel metal chelate adsorbent. Bio/Technology 1988;6:1321–5.
[9] Hengen P. Purification of His-Tag fusion proteins from *Escherichia coli*. Trends Biochem Sci 1995;20:285–6.
[10] Lowery OH, Rosebrough NJ, Fan AL, Randall RJ. Protein measurement with the Folin phenol reagent. J Biol Chem 1951;193:265–75.
[11] Bradford MM. Rapid and sensitive method for the quantitation of microgram quantities of protein utilizing the principle of protein-dye binding. Anal Biochem 1976;72:248–54.
[12] Zor T, Selinger Z. Linearization of the Bradford protein assay increases its sensitivity: theoretical and experimental studies. Anal Biochem 1996;236:302–8.
[13] Sumner JB. The isolation and crystallization of the enzyme urease. J Biol Chem 1926;69:435–41.
[14] The AOCS Lipid Library. Giants of the past. Michel Eugène Chevreul (1786–1889). 2010.
[15] Folch J, Lees M, Stanley GHS. A simple method for the isolation and purification of total lipids from animal tissues. J Biol Chem 1957;226:497–509.
[16] Bligh EG, Dyer WJ. A rapid method of total lipid extraction and purification. Can J Biochem Physiol 1959;37:911–7.
[17] St John LC, Bell FP. Extraction and fractionation of lipids from biological tissues, cells, organelles and fluids. BioTechniques 1989;7:476–81.
[18] Berezkin VG. History of analytical chemistry: contributions from N.A. Izmailov and M.S. Schraiber to the development of thin-layer chromatography (on the 70th anniversary of the publication of the first paper on thin-layer chromatography). J Anal Chem 2008;63:400–4.
[19] Sherma J, Fried B. Practical thin-layer chromatography: a multidisciplinary approach. Boca Raton (FL): CRC Press, Inc.; 1996. 275 pp.
[20] Preparing your own thin layer chromatography plates (and then using them). 2011. www.instructables.com.
[21] Vaskovsky VE, Svetashev VI. Phospholipid spray reagents. J Chromatogr 1972;65:451–3.
[22] Christie WW, The AOCS Lipid Library. Thin-Layer chromatography of lipids. The American Oil Chemists Society; 2015.
[23] Hax WMA, Geurts van Kessel WSM. High-performance liquid chromatographic separation and photometric detection of phospholipids. J Chromatogr 1977;142:735–41.
[24] Bell MV. Separations of molecular species of phospholipids by high-performance liquid chromatography. In: Christie WW, editor. Advances in lipid methodology – four. Dundee, Scotland: Oily Press; 1977. p. 45–82.
[25] Dong MW. Modern HPLC for practicing scientists. Hoboken (NJ): John Wiley & Sons, Inc; 2006. 286 pp.
[26] Amersham Biosciences. Reversed phase chromatography. Principles and methods. Piscataway, NJ. 1999. 84 pp.
[27] Corran PH. Reversed-phase chromatography of proteins. In: Oliver RWA, editor. HPLC of macromolecules, a practical approach. IRL Press: Oxford Press; 1989. p. 127–56.
[28] Grob RL, Barry EF. Modern practice of gas chromatography. New York: John Wiley & Sons; 1995. 888 pp.
[29] Chromatographer. Resolution matters. Gas chromatography.
[30] Christie WW. Gas chromatography and lipids: a paractical Guide. Dundee, Scotland: The Oily Press; 1989.
[31] Christie WW. Preparation of ester derivatives of fatty acids for chromatographic analysis. In: Christie WW, editor. Advances in lipid methodology – two. Dundee, Scotland: Oily Press; 1993. p. 69–111.
[32] Christie WW. A simple procedure for rapid transmethylation of glycerolipids and cholesteryl esters. J Lipid Res 1982;23:1072–5.
[33] Bannon CD, Breen GJ, Craske JD, Hai NT, Harper NL, O'Rourke KL. Analysis of fatty acid methyl esters with high accuracy and reliability: III. Literature review of and investigations into the development of rapid procedures for the methoxide-catalysed methanolysis of fats and oils. J Chromatogr A 1982;247:71–89.
[34] Leray C. Separation of fatty acids by GLC. Cyberlipid Center. ResourceSite For Lipid Studies; 2011. www.cyberlipid.org/fattyt/fatt0003.htm.
[35] Bicalho B, David F, Rumplel K, Kindt E, Sandra P. Creating a fatty acid methyl ester database for lipid profiling in a single drop of human blood using high resolution gas chromatography and mass spectrometry. J Chromatogr A 2008;1211:120–8.

[36] Mueller P, Rudin DO, Tien HT, Westcott WC. Reconstitution of excitable cell membrane structure in vitro. Circulation 1962;26:167–1171.

[37] Tien HT. Bilayer lipid membranes (BLM): theory and practice. New York: Marcel Dekker; 1974. 655 pp.

[38] Bangham AD, Horne RW. Negative staining of phospholipids and their structural modification by surface active agents as observed in the electron microscope. J Mol Biol 1964;8:660–8.

[39] Lasic DD, Papahadjopoulos, editors. Medical applications of liposomes. New York: Elsevier Publishing; 1998. 779 pp.

[40] Florence AT. Liposomes in drug delivery (drug targeting and delivery). Harvard Academic Publishers; 1993. 256 pp.

[41] Lathe GH, Ruthven CR. The separation of substances on the basis of their molecular weights, using columns of starch and water. Biochem J 1955;60(4):xxxiv.

[42] Lathe GH, Ruthven CR. The separation of substances and estimation of their relative molecular sizes by the use of columns of starch in water. Biochem J 1956;62:665–74.

[43] Mori S, Barth HG. Size exclusion chromatography. New York: Springer Science; 1999. 234 pp.

[44] Weinstein JN, Blumenthal R, Klausner RD. Carboxyfluorescein leakage assay for lipoprotein-liposome interabtion. Methods Enzymol 1986;128:857–68.

[45] Gould RM, London A. Specific interaction of central nervous system myelin basic protein with lipids. Effects of basic protein on glucose leakage from liposomes. Biochim Biophys Acta 1972;290:200–18.

[46] Lapinski MM, Castro-Forero A, Greiner AJ, Ofoli RY, Blanchard GJ. Comparison of liposomes formed by sonication and extrusion: rotational and translational diffusion of an embedded chromophore. Langmuir 2007;23:11677–83.

[47] Rhoden V, Goldin SM. Formation of unilamellar lipid vesicles of controllable dimensions by detergent dialysis. Biochemistry 1979;18:4173–6.

[48] Rigaud JL, Levy D, Mosser G, Lambert O. Detergent removal by non-polar polystyrene beads. Eur Biophys J 1998;27:305–19.

[49] Deamer D, Bangham AD. Large volume liposomes by an ether vaporization method. Biochim Biophys Acta 1976;443:629–34.

[50] Durr UHN, Gildenberg M, Ramamoorthy A. The magic of bicelles lights up membrane protein structure. Chem Rev 2012;112(11):6054–74.

[51] Sanders CR, Prosser RS. Bicelles: a model membrane system for all seasons? Curr Biol Struct 1998;6:1227–34.

[52] Cavagnero S, Dyson HJ, Wright PE. Improved low pH bicelle system for orienting macromolecules over a wide temperature range. J Biomol NMR 1999;13(4):387–91.

[53] Sanders CR, Prestegard JH. Magnetically orientable phospholipid bilayers containing small amounts of a bile salt analog, CHAPSO. Biophys J 1990;58:447–60.

[54] Sanders CR, Schwonek JP. Characterization of magnetically orientable bilayers in mixtures of DHPC and DMPC by solid state NMR. Biochemistry 1992;31:8898–905.

[55] Marcotte I, Auger M. Bicelles as model membranes for solid- and solution-state NMR studies of membrane peptides and proteins. In: Semelka R, editor. Current clinical imaging series. Concepts in magnetic resonance Part A, vol. 24A(1). Wiley Press; 2005. p. 17–37.

[56] Whiles JA, Deems R, Vold RR, Dennis EA. Bicelles in structure–function studies of membrane-associated proteins. Bioorg Chem 2002;30:431–42.

[57] Michel H. Crystallization of membrane proteins. Trends Biochem Sci 1983;8:56–9.

[58] Uiwal R, Bowie JU. Crystallizing membrane proteins using lipidic bicelles. Methods 2011;55(4):337–41.

[59] Faham S, Bowie JU. Bicelle crystallization: a new method for crystallizing membrane proteins yields a monomeric bacteriorhodopsin structure. J Mol Biol 2002;316:1–6.

[60] Silvius JR. Solubilization and functional reconstitution of biomembrane components. Ann Rev Biophys Biomol Struct 1992;21:323–48.

[61] Rigaud J-L. Membrane proteins: functional and structural studies using reconstituted proteoliposomes and 2-D crystals (review). Braz J Med Biol Res 2002;35:753–66.

[62] Carafoli E. Calcium pump of the plasma membrane (review). Physiol Rev 1991;71:129–53.

[63] Racker E. Reconstitution of a calcium pump with phospholipids and a purified Ca^{++}adenosine triphosphatase from sarcoplasmic reticulum. J Biol Chem 1972;247:8198–200.

[64] Warren FB, Toon PA, Birdsall NJM, Lee AG, Metcalf JC. Reconstitution of a calcium pump using defined membrane components. Proc Nat Acad Sci USA 1974;71:622–6.
[65] Jung H, Tebbe S, Schmid R, Jung K. Unidirectional reconstitution and characterization of purified Na^+/proline transporter of *Escherichia coli*. Biochemistry 1998;37:11083–8.
[66] Chen CC, Wilson TH. Solubilization and functional reconstitution of the proline transport system of *Escherichia coli*. J Biol Chem 1986;261:2599–604.
[67] Karnovsky MJ, Kleinfeld AM, Hoover RL, Klausner RD. The concept of lipid domains in membranes. J Cell Biol 1982;94:1–6.
[68] Simons K, Ikonen E. Functional rafts in cell membranes. Nature 1997;387:569–752.
[69] Edidin M. The state of lipid rafts: from model membranes to cells. Ann Rev Biophys Biomol Struct 2003;32:257–83.
[70] Brown DA. Lipid rafts, detergent-resistant membranes, and raft targeting signals. Physiology 2006;21:430–9.
[71] Pike LJ. The challenge of lipid rafts. J Lipid Res 2009;50(Suppl.):S323–8.
[72] Shaw AS. Lipid rafts: now you see them, now you don't. Nat Immunol 2006;7:1139–42.
[73] Edidin M. Shrinking patches and slippery rafts: scales of domains in the plasma membrane. Trends Cell Biol 2001;11:492–6.
[74] van Meer G. The different hues of lipid rafts. Science 2002;296:855–7.
[75] Estep TN, Mountcastle DB, Barenholz Y, Biltonen RL, Thompson TE. Thermal behavior of synthetic sphingomyelin-cholesterol dispersions. Biochemistry 1979;18:2112–7.
[76] Shaikh SR, Cherezov V, Caffrey M, Soni S, Stillwell W, Wassall SR. Molecular organization of cholesterol in unsaturated phosphatidylethanolamines: X-ray diffraction and solid state ^2H NMR studies. J Am Chem Soc 2006;128:5375–83.
[77] Wassall SR, Shaikh SR, Brzustowicz MR, Cherezov V, Siddiqui RA, Caffrey M, Stillwell W. Interaction of polyunsaturated fatty acids with cholesterol: a role in lipid raft phase separation. In: Talking about colloids; 2005.
[78] Ilangumaran S, Hoessli DC. Effects of cholesterol depletion by cyclodextrin on the sphingolipid microdomains of the plasma membrane. Biochem J 1998;335:433–40. In: Danino D, Harries D, Wrenn, SP, editors. Macromolecular Symposia, vol. 219. Wiley-VCH. pp. 73–84.
[79] Smart EJ, Ying Y-S, Mineo C, Anderson RGW. A detergent-free method for purifying caveolae membrane from tissue culture cells. Proc Natl Acad Sci USA 1995;92:10104–8.
[80] Schroeder R, London E, Brown D. Interactions between saturated acyl chains confer detergent resistance on lipids and glycosylphosphatidylinositol (GPI)-anchored proteins: GPI-anchored proteins in liposomes and cells show similar behavior. Proc Natl Acad Sci USA 1994;91:12130–4.
[81] Dietrich C, Bagatolli LA, Volovyk ZN, Thompson NL, Levi M, Jacobson K, et al. Lipid rafts reconstituted in model membranes. Biophys J 2001;80:1417–28.
[82] Hammond AT, Heberle FA, Baumgart T, Holowka D, Baird B, Feigenson GW. Crosslinking a lipid raft component triggers liquid ordered-liquid disordered phase separation in model plasma membranes. Proc Natl Acad Sci USA 2005;102:6320–5.
[83] Kabouridis PS, Janzen J, Magee AL, Ley SC. Cholesterol depletion disrupts lipid rafts and modulates the activity of multiple signaling pathways in T lymphocytes. Eur J Immunol 2000;30:954–63.
[84] Henderson RM, Edwardson JM, Geisse NA, Saslowsky DE. Lipid rafts: feeling is believing. News Physiol Sci 2004;19:39–43.
[85] Li Q, Wang M, Tan L, Wang C, Ma J, Li N, et al. Docosahexaenoic acid changes lipid composition and interleukin-2 receptor signaling in membrane rafts. J Lipid Res 2005;46:1904–13.
[86] Hope HR, Pike LJ. Phosphoinositides and phosphoinositide-utilizing enzymes in detergent-insoluble lipid domains. Mol Biol Cell 1996;7:843–51.
[87] Pike LJ, Miller JM. Cholesterol depletion delocalizes phosphatidylinositol bisphosphate and inhibits hormone-stimulated phosphatidylinositol turnover. J Biol Chem 1998;273:22298–304.

PART II

MEMBRANE BIOLOGICAL FUNCTIONS

CHAPTER

14

Membrane Biogenesis: Fatty Acids

OUTLINE

1. INTRODUCTION

The first half of this book focused on the composition and structure of biological membranes (**Part I. Membrane Composition and Structure**). The last half of the book will concentrate on membrane function (**Part II. Membrane Biological Functions**). Since biological membranes are such extremely complex entities, composition, structure and, function are so completely integrated that it is impossible to discuss one aspect in the absence of the other

An Introduction to Biological Membranes
http://dx.doi.org/10.1016/B978-0-444-63772-7.00014-2

two. This last part of the book will first investigate membrane biogenesis with discussion of the following topics:

Present chapter. Membrane Biogenesis: Fatty Acids
Chapter 15. Membrane Biogenesis: Phospholipids, Sphingolipids, Plasmalogens, and Cholesterol
Chapter 16. Membrane Biogenesis: Proteins

Membrane biogenesis is a "chicken and egg" question—which came first? For example, a membrane requires a lipid bilayer to exist yet the bilayer lipids require membrane proteins (enzymes) to be synthesized! Lipid synthesis therefore requires a lipid bilayer to make more lipids and this can only occur by membrane-bound proteins. It is obvious that the synthesis of a new membrane cannot be initiated with a lipid bilayer alone to which proteins are later added nor can it begin with a protein that then synthesizes a bilayer around it. Instead a new membrane must be put together piece by piece from a preexisting, functional membrane template.

This discussion of membrane biogenesis will begin with the most basic component of the lipid bilayer, the fatty acid. Importantly, fatty acids are the major component providing the "hydrophobic effect" (Chapter 3) that dictates the size, stability and physical properties of the membrane lipid bilayer (Chapters 9, 10, and 11). And it is within the lipid bilayer that membrane proteins flourish.

2. FATTY ACID BIOSYNTHESIS

2.1 Saturated Fatty Acids—Palmitic Acid

Biological membrane fatty acids (Chapter 4) are comprised predominantly of even numbered carbon chain lengths. This is the result of their mode of synthesis through linking two carbon acetyl coenzyme A (acetyl CoA) units end to end. Since acetyl CoA is the source of carbons that comprise fatty acid chains, a case can be made that acetyl CoA is the paramount biochemical in fatty acid biosynthesis. An acetyl (2-carbon, acetyl CoA) or acyl (unspecified number of carbons, acyl CoA) is attached to coenzyme A through a thio-ester linkage. The structure of acetyl CoA and the thio-ester link are depicted in Fig. 14.1 (Panel A) and (Panel B) respectively. Although here we are only interested in acetyl CoA's role in fatty acid biosynthesis, this molecule plays a number of other essential roles in biochemistry [1].

Emphasizing the biological significance of CoA are the two Nobel Prizes in Physiology or Medicine awarded to Fritz Lipmann (Fig. 14.2) in 1953 and Konrad Bloch (Fig. 14.3) and Feodor Lynen (Fig. 14.4) in 1964. Lipmann's Nobel Prize was for his 1946 discovery of CoA [2]. Lipmann's other major contribution was his 1941 discovery of the bioenergetic function of adenosine triphosphate (ATP), introducing the concept of an "energy-rich phosphate bond" and the designation squiggle P (~P) to denote such bonds [3]. Lipmann shared the 1953 Nobel Prize in Physiology or Medicine with Hans Krebs for his discovery of the citric acid (Krebs) cycle. Konrad Bloch and Feodor Lynen received the 1964 Nobel Prize for their discoveries linking acetyl CoA and fatty acid metabolism [4].

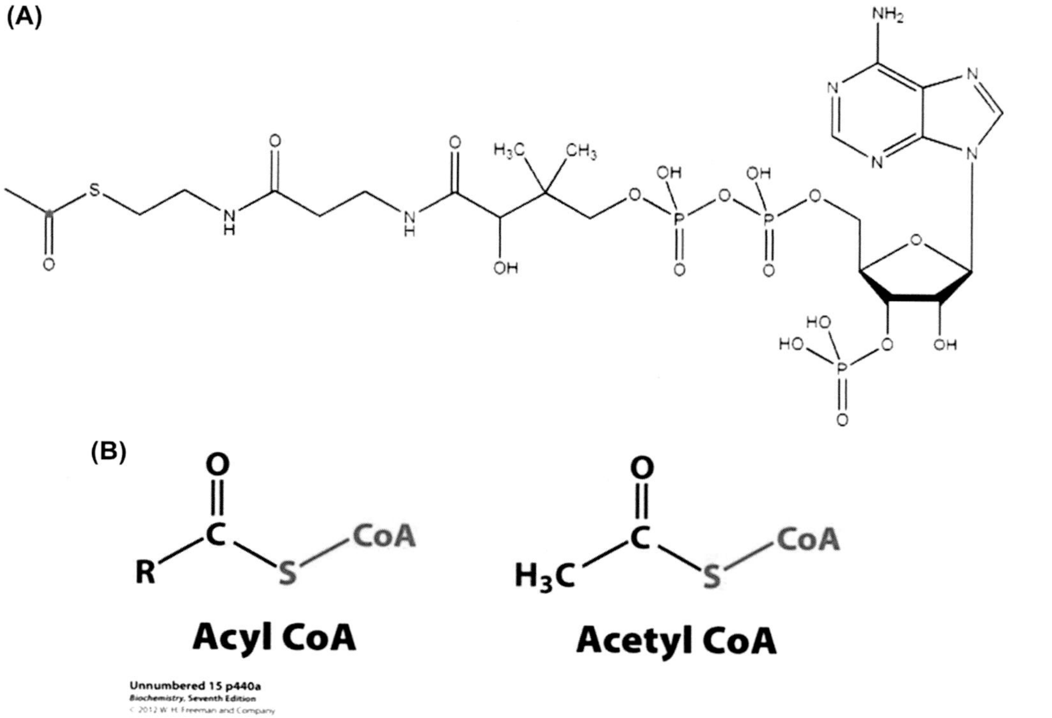
(A)
(B)
Acyl CoA
Acetyl CoA
Unnumbered 15 p440a
Biochemistry, Seventh Edition
© 2012 W. H. Freeman and Company

FIGURE 14.1 Structure of acetyl CoA (Panel A) and the thio-ester linkage that connects coenzyme A to acetyl (2-carbon, acetyl CoA) or acyl (unspecified number of carbons, acyl CoA), (Panel B). **Panel A**. http://www.moravek.com/Structures/MC269.gif. **Panel B**. http://www.studyblue.com/notes/note/n/chapter-15/deck/3163157.

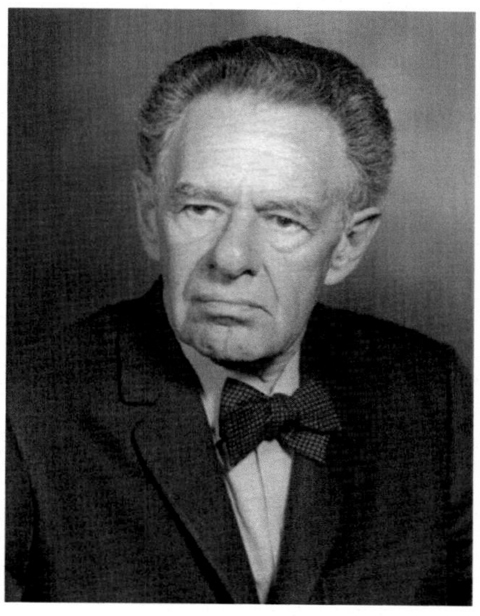

FIGURE 14.2 Fritz Lipmann, 1899–1986. http://en.wikipedia.org/wiki/Fritz_Albert_Lipmann.

FIGURE 14.3 Konrad Bloch, 1912–2000. http://exploreuk.uky.edu/catalog/xt7sf7664q86_4097_1.

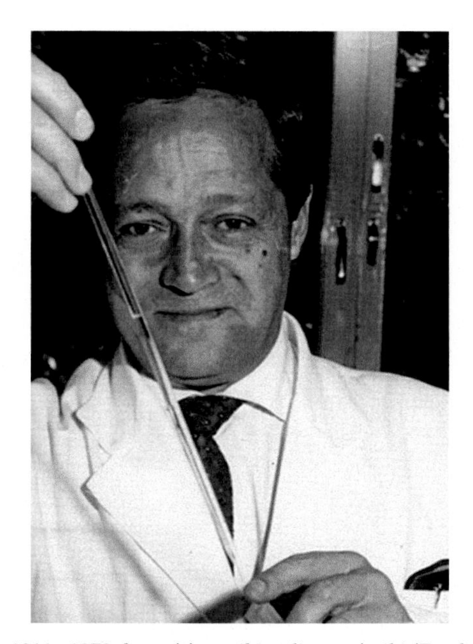

FIGURE 14.4 Feodor Lynen, 1911–1970. http://en.wikipedia.org/wiki/Feodor_Felix_Konrad_Lynen.

Fatty acid biosynthesis begins by coupling two acetyl units to produce a four-carbon chain [5,6]. Additional energy is required to create the new C—C bond. As with most biosynthetic mechanisms, the energy originates with ATP. The coupling is not direct, but instead involves the enzyme acetyl CoA carboxylase (Fig. 14.5). This enzyme uses the hydrolytic energy stored in ATP to carboxylate acetyl CoA storing some of the energy as the new bond connecting carboxylate to acetyl CoA producing malonyl CoA. The facile decarboxylation of malonyl CoA is the energy used to couple the two acetyl units.

A long chain fatty acid is built up 2-carbons at a time from malonyl CoA units. Six sequential reactions complete one cycle of the synthesis coupling the new C-2 unit to the growing chain. All six of the individual enzymes are sequestered together in a large complex referred to in animals as fatty acid synthase I (FASI). At the heart of FAS function lies a protein known as an acyl carrier protein (ACP). Both CoA and ACP have a phosphopantheine (vitamin B5) moiety with a distal sulfhydryl terminal that can form a thioester ester with a fatty acid (Fig. 14.6). Of the six steps in the sequence, the first involves converting malonyl CoA to malonyl ACP in the FAS complex.

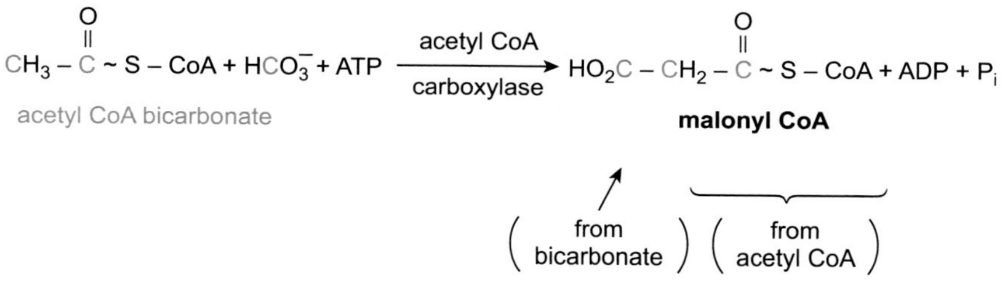

$$CH_3-\overset{O}{\overset{\|}{C}}\sim S-CoA + HCO_3^- + ATP \xrightarrow[\text{carboxylase}]{\text{acetyl CoA}} HO_2C-CH_2-\overset{O}{\overset{\|}{C}}\sim S-CoA + ADP + P_i$$

acetyl CoA bicarbonate

malonyl CoA

(from bicarbonate) (from acetyl CoA)

FIGURE 14.5 Formation of malonyl CoA from acetyl CoA catalyzed by the enzyme acetyl CoA carboxylase. http://library.med.utah.edu/NetBiochem/FattyAcids/5_2a.html.

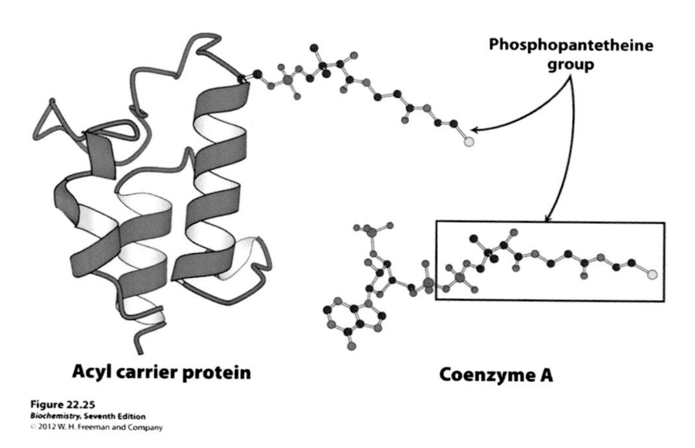

FIGURE 14.6 Structures of acyl carrier protein (ACP) and coenzyme A (CoA). Note both molecules have a phosphopantheine (vitamin B5) moiety with a distal sulfhydral terminal that can form a thioester with a fatty acid. http://oregonstate.edu/instruct/bb451/spring14/lectures/fametabolismoutline.html.

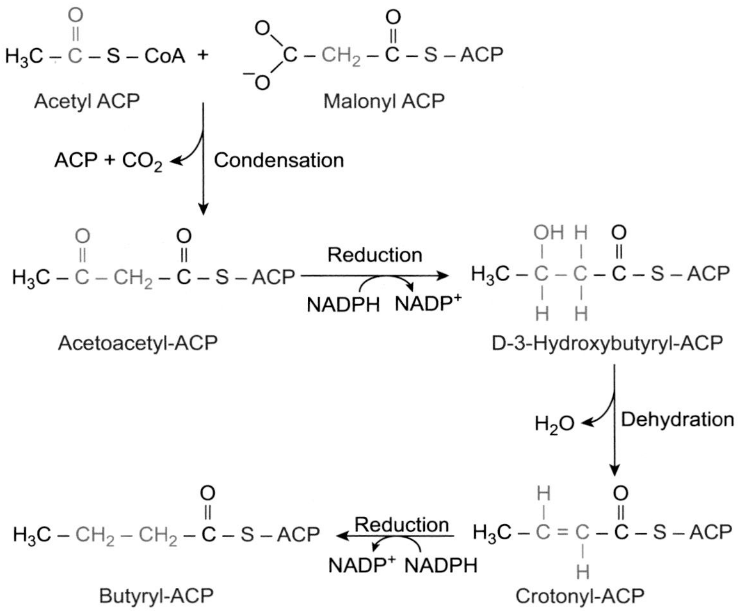

FIGURE 14.7 Individual reactions involved in fatty acid biosynthesis on the fatty acid synthase complex. (FAS). http://www.natuurlijkerwijs.com/english/Fatty_acid_metabolism.htm. *ACP*, acyl carrier protein; *NADP*, nicotinamide adenine dinucleotide phosphate.

The ACP serves as an extended, flexible arm that carries a covalently-linked (thio-ester) acetoacyl group from enzyme to enzyme in the FAS complex (Fig. 14.7). The phosphopantethheine arm fulfills two basic functions in the FAS complex. It covalently binds the fatty acid biosynthesis intermediates to a hydrolysable, energy-rich linkage (a thio-ester) that brings the intermediates from enzyme site to enzyme site in a highly efficient process. Also, flexibility of the ~2 nm long phosphopantethine arm enhances accessibility of the covalently-linked intermediates to the distinct enzyme sites in an assembly line fashion.

The sequence of reactions catalyzed by the FAS complex are depicted in Fig. 14.7. The sequential steps involve: condensation to produce the keto acetoacetyl ACP; reduction of the ketone to an alcohol; dehydration to form a (−C=C−) and reduction of the double bond to a saturated acyl chain. The steps are repeated over and over until a saturated C-16 (palmitoyl) chain is attained still attached to the ACP on the FAS complex. At this point, the palmitoyl chain is released as palmitoyl-CoA.

While palmitic acid is synthesized by this mechanism, a biological membrane is composed of hundreds of different acyl chains (Chapter 4). How are these nonpalmitoyl chains synthesized?

2.1.1 Medium Chain Saturated Fatty Acids

Medium chain fatty acids (eg, myristic acid (14:0), lauric acid (12:0)) can be made by animals in the FASI complex by early chain termination.

2.1.2 *Longer Chain Saturated Fatty Acids*

Saturated acyl chains longer than C-16 (eg, stearic (18:0), arachidic (20:0), and behenic (22:0)) can be made from palmitoyl CoA by elongation on FASIII. The mechanism of synthesis is identical to that for palmitic acid on FASI shown in Fig. 14.7.

$$\text{palmitoyl CoA} + \text{malonyl CoA} \rightarrow \text{stearoyl CoA} + CO_2$$

2.1.3 *Odd and Branched Chain Saturated Fatty Acids*

The vast majority of saturated fatty acids in animals are straight-chain compounds with 14, 16, and 18 carbon atoms, but all possible odd- and even-numbered chains from 2 to 36 carbon atoms have been found in nature in esterified form. The shortest odd chain fatty acid is propanoic acid (3:0) that plays several important biochemical roles, but, with the exception of platelet-activating factor, is rarely found esterified to lipids. Longer, odd chain fatty acids, particularly C13:0 to C19:0, can be found at trace levels in most animal tissues where they are obtained through the diet. Their major dietary source is ruminant animals where odd chain fatty acids comprise >5% of the total fatty acid population. In addition, odd chain fatty acids can be synthesized using propionyl CoA as the primer molecule in place of acetyl CoA and as products of even chain length fatty acids by alpha-oxidation.

Propionyl CoA can also be carboxylated to D-methylmalonyl CoA by the enzyme propionyl CoA carboxylase (Fig. 14.8). Upon substituting D-methylmalonyl CoA for normal malonyl CoA, a methyl branch is inserted into the growing fatty acyl chain [7].

2.2 Unsaturated Fatty Acids

Unsaturated fatty acids are as prevalent in the bilayer interior of membranes as are saturated fatty acids. Their primary function is to maintain the proper membrane "fluidity", and thus the hydrophobic environment for the resident proteins (Chapter 9). Unsaturated fatty acids are both synthesized and taken up through the diet. Fatty acids that cannot be

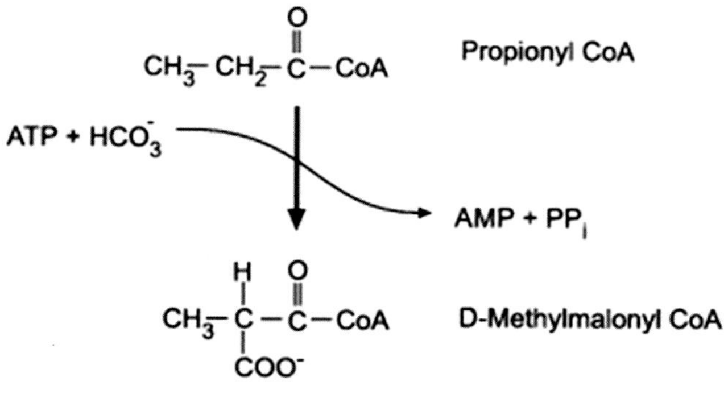

$CH_3-CH_2-C(=O)-CoA$ Propionyl CoA

$ATP + HCO_3^-$ → $AMP + PP_i$

$CH_3-CH(COO^-)-C(=O)-CoA$ D-Methylmalonyl CoA

FIGURE 14.8 Carboxylation of propionyl CoA to D-methylmalonyl CoA by the enzyme propionyl CoA carboxylase. http://learning.covcollege.ac.uk/content/Jorum/MET_Lipid-breakdown-and-ketone-bodies_LM-1.2-3FEB08/page51.htm. *ATP*, adenosine triphosphate.

synthesized by humans are referred to as "essential fatty acids" and must be included in the diet [8]. Only two fatty acids are absolutely essential in the human diet, linoleic acid ($18{:}2^{\Delta 9,12}$) and alpha-linolenic acid ($18{:}3^{\Delta 9,12,15}$). A few other unsaturated fatty acids are classified as being "conditionally essential", meaning they can become essential under some situations. Two examples of this type of fatty acid are known to play important anticancer roles; docosahexaenoic acid (DHA, $22{:}\ 6^{\Delta 4,7,10,13,16,19}$) and gamma-linolenic acid ($18{:}3^{\Delta 6,9,12}$) (Chapter 23). When these two fatty acids were discovered in 1923, they were designated "vitamin F", but were later reclassified as fatty acids [9].

Double bonds are added to full length, saturated fatty acids by a process known as aerobic desaturation [10]. The process oxidizes

$$-CH_2-CH_2- \rightarrow -CH{=}CH-$$

and is catalyzed by enzymes known as desaturases. Desaturation is a multicomponent process that involves the saturated fatty acid, O_2, nicotinamide adenine dinucleotide phosphate hydrogen, and cytochrome b_5 (Fig. 14.9).

Desaturases place a double bond at a precise location in the fatty-acyl chain. The most common desaturase reactions insert a double bond between carbons 9 and 10. It is therefore a Δ9 desaturase, eg:

$$\text{PalmiticAcid}(16:0) \rightarrow \text{PalmitoleicAcid}(16:1^{\Delta 9})$$

$$\text{StearicAcid}(18:0) \rightarrow \text{OleicAcid}(18:1^{\Delta 9})$$

Other types of human desaturases can add double bonds at positions Δ4, Δ5, and Δ6. In humans, desaturation does not occur beyond position Δ9. To achieve double bonds beyond

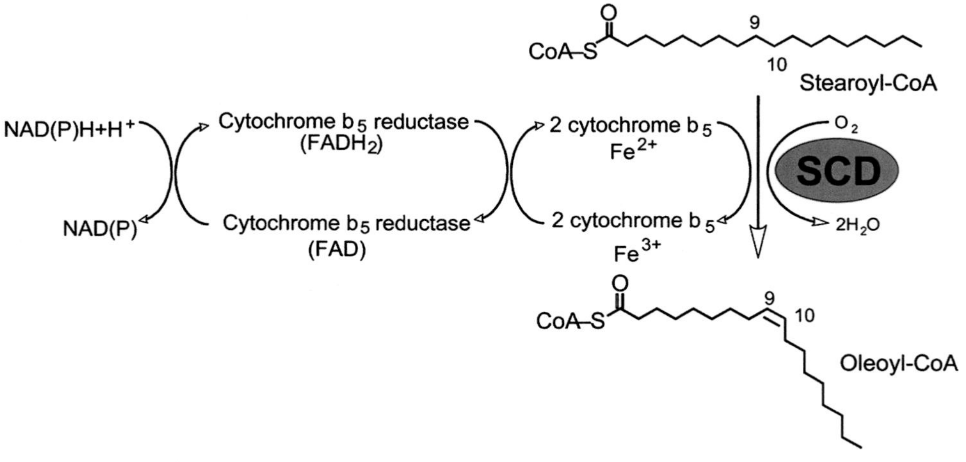

FIGURE 14.9 Aerobic desaturation of stearoyl-CoA to Oleoyl CoA. *NADP*, Nicotinamide adenine dinucleotide phosphate; *NADPH*, nicotinamide adenine dinucleotide phosphate hydrogen; *SCD*, stearoyl-CoA desaturase. *Paton CM, Ntambi JM. Am J Physiol Endocrinol Metabol, Published July 1, 2009;297:E28–37. doi:10.1152/ajpendo.90897.2008 http://ajpendo.physiology.org/content/297/1/E28.*

this position, a double bond must first be added at position Δ9 or lower and then the chain is elongated. Of course, fatty acids with double bonds beyond Δ9 can also be obtained through the diet.

2.2.1 *Biosynthesis of Polyunsaturated Fatty Acids*

In human membranes, the three most important long chain polyunsaturated fatty acids (PUFAs) are: arachidonic acid (AA or ARA, $20{:}4^{\Delta 5,8,11,14}$), eicosapentaenoic acid (EPA, $20{:}5^{\Delta 5,8,11,14,17}$), and docosahexaenoic acid (DHA, $22{:}6^{\Delta 4,7,10,13,16,19}$) (Chapter 4). PUFAs are made by a series of desaturases and elongases [11]. Fig. 14.10 (left column) depicts synthesis of the n-6 fatty acid, AA from its precursor linoleic acid. Fig. 14.10 (right column) depicts synthesis of the n-3 fatty acids EPA and DHA from their precursor alpha-linolenic acid. Other fatty acids shown in the pathway include gamma-linolenic acid (GLA), dihomo-gamma-linolenic acid (DGLA), and docosapentaenoic acid (DPA).

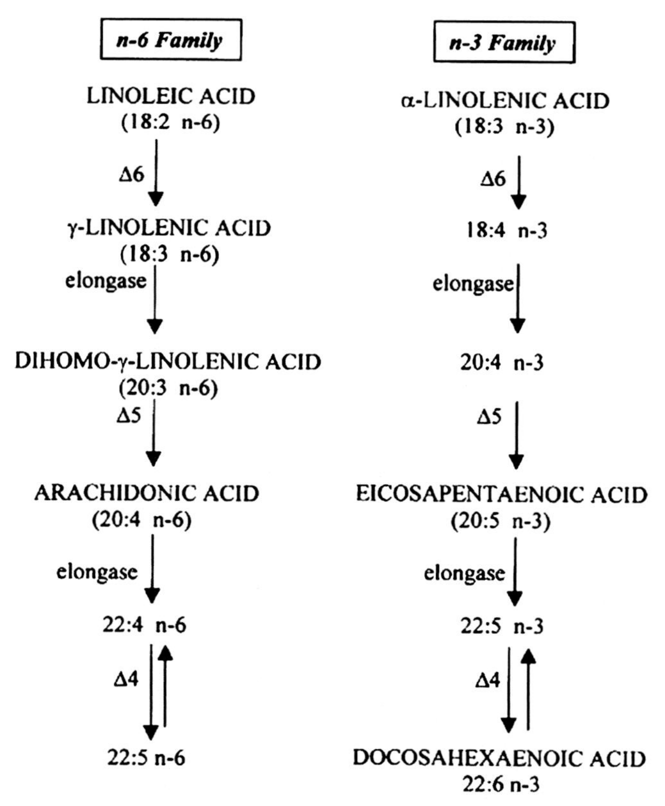

FIGURE 14.10 Synthesis of the n-6 fatty acid arachidonic acid (AA) from its precursor linoleic acid (left column), and synthesis of the n-3 fatty acids eicosapentaenoic acid (EPA) and docosahexaenoic acid (DHA) from their precursor alpha-linolenic acid (right column). http://www.clinsci.org/cs/105/0519/cs1050519.htm.

3. FATTY ACID STORAGE, RELEASE, AND TRANSPORT

Since free fatty acids can be harmful to membrane structure and function (see "detergent effect" in Chapter 13), they must be handled carefully. Normally fatty acids are rendered safe by esterfying them to alcohol moieties on a wide variety of biomolecules including lipids, sterols, proteins, and glycerol. In esterified form, fatty acids can be safely stored in the human body at high levels for future use, primarily as an energy source or as the hydrophobic component of membranes. Fatty acids' long, hydrophobic chains are largely responsible for them being a highly concentrated source of biochemical energy as well as providing the primary role in stabilizing the membrane lipid bilayer (the "hydrophobic effect", Chapter 3). The energy yield from a gram of fatty acids is approximately 9 kcal compared to 4 kcal/g for carbohydrates. Partially accounting for this difference, carbohydrates are highly polar and so are tightly bound to water while fats are anhydrous. The high water content of carbohydrates greatly reduces their energy/g in aqueous solution. As a result, if a person entirely relied on carbohydrates to store energy, he would need to carry 67.5 lb of hydrated glycogen to have the energy equivalent to 10 lb of fat.

Fatty acids are primarily ingested from the diet and stored as triacylglycerols. Triacylglycerols are composed of three fatty acids esterified to each of the three alcohols of glycerol (Chapter 5). Most triacylglycerols have seemingly random combinations of fatty acids that were available during their biosynthesis. The higher the content of unsaturated fatty acids found in a triacylglycerol, the more fluid is the molecule. If a triacylglycerol is liquid at room temperature it is referred to as an oil. If it is solid at room temperature, it is a fat. Plant triacylglycerols tend to be oils (eg, peanut oil) while animal triacylglycerols are usually fats (eg, bacon grease).

Fatty acids are usually ingested as preformed triacylglycerols that cannot be directly absorbed by the intestine. Instead, they must first be broken down by a pancreatic lipase—colipase (1:1) complex into free fatty acids and monoacylglycerol. The enzyme complex also requires an emulsion, a property that is provided by bile salts. Most of the breakdown components of triacylglycerols are absorbed as free fatty acids and 2-acyl-monoacyglycerol, although some free glycerol and diacylglycerols are also absorbed. Once across the intestinal membrane, the components are reconstituted back into triacylglycerols and packaged into chylomicrons or lipoproteins. Transport methods include: as unbound fatty acids, bound to serum albumin, or bound to lipoproteins (chylomicrons, very low-density lipoproteins (VLDL), intermediate-density lipoproteins (IDL), low-density lipoproteins (LDL), and high-density lipoproteins (HDL)). By all methods, fatty acids can arrive at cellular membranes where they can be incorporated into membrane lipids and proteins.

3.1 Unbound (Free) Fatty Acids

Free fatty acids (also monoacylglycerols and diacylglycerols) can be found at low levels in the aqueous spaces surrounding the cell (the external plasma and the internal cytosol) and are also natural components of cellular membranes. In the unbound state, free fatty acids can diffuse rapidly from place to place whereupon they can be incorporated into membrane lipids and proteins. However, the solubility of free fatty acids in water is very low ($\sim 1\ \mu M$), making simple aqueous diffusion impractical to support metabolic processes.

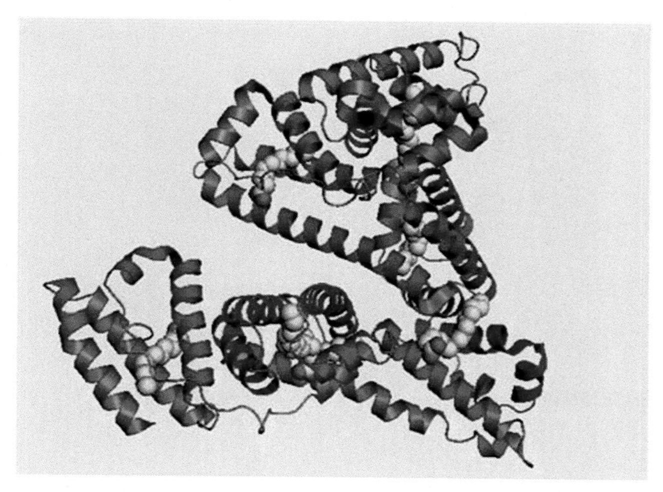

FIGURE 14.11 Structure of serum albumin. http://en.wikipedia.org/wiki/Serum_albumin.

Also, high levels of free fatty acids in membranes can emulsify the membrane into micelles (the "detergent effect").

3.2 Serum Albumin

Serum albumin (Fig. 14.11) is the most abundant protein in blood and is also the major carrier of free fatty acids in blood. Upon binding to serum albumin, the concentration of total fatty acids in blood is ~1 mM while its concentration in the surrounding aqueous solution is a 1000-fold less. In addition to fatty acids, serum albumin can nonspecifically bind steroids, thyroid hormones, hemin, and other odd molecules [12]. Serum albumin is a water-soluble, anionic globular protein of molecular weight ~65,000. The protein's structure is dominated by several long α-helices that make the protein rigid (Fig. 14.11). Serum albumin houses 11 distinct hydrophobic binding domains and so is capable of simultaneously carrying multiple fatty acids.

3.3 Lipoproteins

There are five basic types of lipoproteins found in blood: chylomicrons, VLDL (very low density lipoprotein), IDL (intermediate density lipoprotein), LDL (low density lipoprotein), and HDL (high density lipoprotein) [13,14]. Each type of lipoprotein is composed of different proportions of the same basic compounds (triacylglycerols, phospholipids, cholesterol, cholesteryl esters, and proteins) (Fig. 14.12). For example, LDLs are very high in cholesterol but are low in protein, while HDLs are high in protein but low in cholesterol. Therefore HDL is denser and smaller than LDL. Lipoprotein particle size is not a fixed number, but in general decreases in size from: chylomicron > VLDL > IDL > LDL > HDL. Even each type of lipoprotein particle varies over a range of sizes, but are approximately 20–40 nm in diameter. Particle size also changes after consuming a large meal!

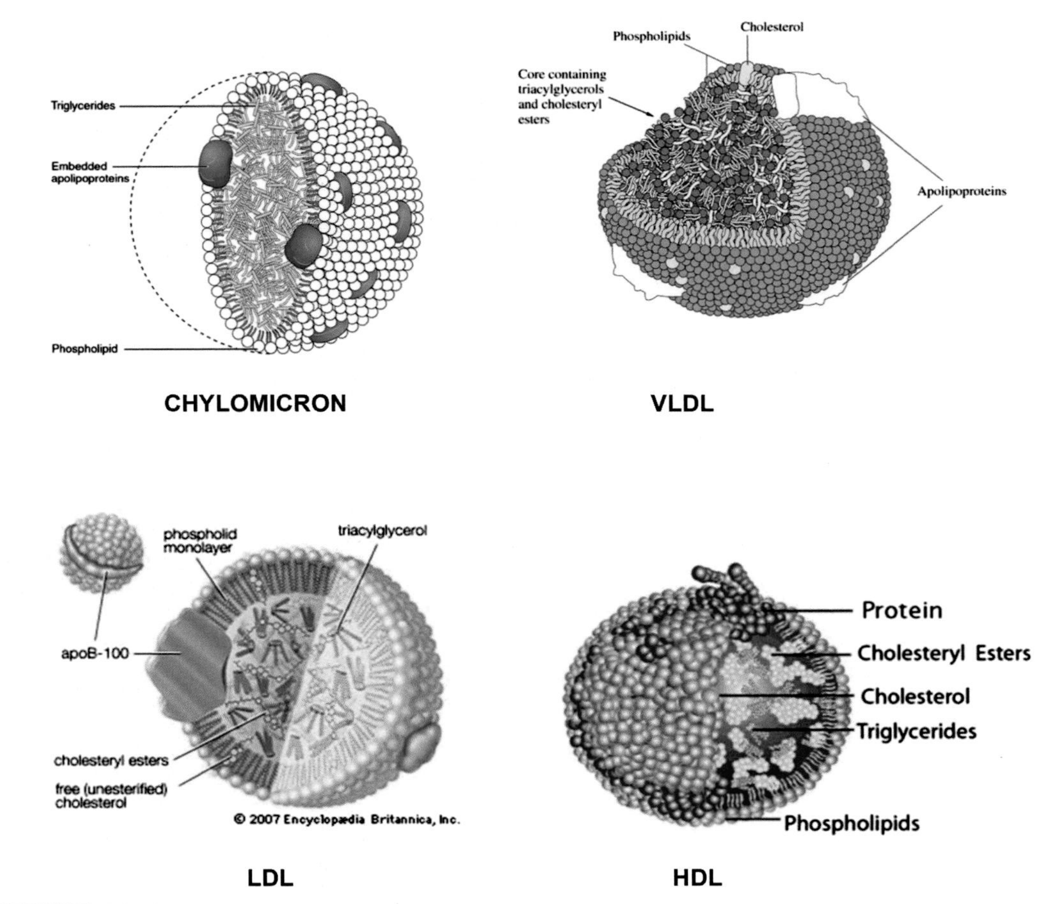

FIGURE 14.12 Structure of the lipoprotein particles: chylomicrons; Very Low-Density Lipoproteins (VLDL); Low-Density Lipoproteins (LDL); High-Density Lipoproteins (HDL). *This figure has 4 different sources:* ***Chylomicron*** *http://en.wikipedia.org/wiki/Chylomicron.* ***VLDL*** *http://apbrwww5.apsu.edu/thompsonj/Anatomy%20&%20Physiology/2020/2020%20Exam%20Reviews/Exam%201/CH18%20Lipoproteins.htm.* ***LDL*** *http://kids.britannica.com/comptons/art-149526/Cutaway-view-of-a-low-density-lipoprotein-complex-The-LDL.* ***HDL*** *http://www.scientificpsychic.com/health/lipoproteins-LDL-HDL.html.*

Chylomicrons: The function of chylomicrons (Fig. 14.12) is to carry diet-derived (exogenous) lipids from the intestines to target cells around the body [15]. They are composed of triglycerides (85–92%), phospholipids (6–12%), cholesterol (1–3%), and proteins (1–2%).

Very Low Density Lipoproteins: Whereas chylomicrons carry diet-derived (exogenous) lipids, VLDLs (Fig. 14.12) transport lipids newly synthesized by the liver (endogenous lipids) to various target cells throughout the body [16]. VLDLs are assembled in the liver from triacylglycerols containing newly synthesized fatty acids, cholesterol, cholestryl esters, and several apo-lipoproteins. After being released into the blood, additional apo-lipoproteins are acquired, generating a mature VLDL. In the blood, enzymes housed in plasma membranes of the target tissue, lipoprotein lipase and cholesteryl ester transfer protein, digest part of the VLDL triglycerides, releasing free fatty acids. The depleted VLDL then accumulates additional cholesterol which converts VLDL into IDL and eventually into LDL. The free fatty acids can then be incorporated into target tissue membrane lipids and proteins.

Intermediate-Density Lipoproteins: As with other lipoproteins, IDL particles are primarily composed of triacylglycerols and cholesterol esters and are responsible for carrying fatty acids before their eventual release at the target membrane [17]. IDLs are either cleared from the plasma by the liver through receptor-mediated endocytosis (RME) (Chapter 17) or are further degraded to LDLs (see the following discussion). About half of the IDLs are removed by REM in liver cells. When the IDL cholesterol content exceeds that of triacylglycerol, the particle is designated an LDL.

Low-Density Lipoproteins: The major function of LDLs (Fig. 14.12) is to carry cholesterol around the body. Cholesterol is so very poorly soluble in water ($\sim 2.6 \times 10^{-8}$ g/ml, ~67 nM), that a lipoprotein transport system is required. LDL particles are composed of ~50% cholesterol (free and esterified), 25% protein, 20% phospholipid, and only 5% triacylglycerol [18]. LDLs are therefore very high in cholesterol and low in triacylglycerol and so are often referred to as "bad cholesterol". The same basic "hydrophobic effect" principles that stabilize the membrane lipid bilayer in aqueous solution (Chapter 3), have been invoked to describe LDL structure. LDL particles (Fig. 14.12) are basically spherical with the very hydrophobic triacylglycerols and cholesteryl esters sequestered away from water in the LDL interior [19]. A polar phospholipid monolayer coats the LDL surface interacting positively with the aqueous plasma. Free cholesterol intercalates between the phospholipid acyl chains, providing structural rigidity to the LDL particle. LDLs are absorbed into cells via LDL receptors [18]. Once inside the cell, LDLs are degraded producing free fatty acids and cholesterol for new membrane construction. The liver controls cholesterol concentration in the blood by removing LDL. High LDL levels result in cholesterol build up in arterial walls, forming plaques that result in reduced blood flow and heart disease.

High-Density Lipoproteins: The function of HDLs (Fig. 14.12) is to collect cholesterol from around the body and take it primarily to the liver (but also, to a lesser extent, the adrenals, ovaries, and testes) for breakdown and excretion [20]. Since HDLs' function is to reduce cholesterol levels, they are referred to as "good cholesterol" and support healthy heart functions. HDL is the smallest and densest of the lipoprotein particles. HDLs begin as essentially cholesterol-free flattened spherical phospholipid—lipoprotein particles. The cholesterol is converted to very apolar cholesterol esters by the plasma enzyme lecithin-cholesterol acyltransferase. HDLs grow into a larger, spherical shape by accumulating the cholesterol esters into their hydrophobic interior. HDLs are removed from circulation by HDL receptors on the plasma membrane of target cells.

3.4 Lipolysis

Lipolysis is the enzymatic breakdown of lipids through hydrolysis [13,21]. With respect to lipoproteins, lipolysis primarily involves conversion of triacylglycerols to glycerol and fatty acids and the conversion of cholesteryl esters to cholesterol and fatty acids. Several hormones are known to trigger lipolysis. These include epinephrine, norepinephrine, ghrelin, growth hormones, testosterone, and cortisol (Chapter 20) that trigger lipolysis through G-protein coupled receptors belonging to the structural class of integral membrane proteins that have seven transmembrane α-helices (Chapters 6 and 18). These hormones activate adenylate cyclase, initiating intracellular c-AMP (cyclic-adenosine monophosphate) production (Chapter 18). c-AMP activates protein kinase A which in turn activates lipases involved in lipolysis.

The neutral lipids, triacylglycerols and cholesterol esters, are stored primarily in adipocytes, but to a lesser extent in steroid synthesizing cells of the adrenal cortex, ovary, and testes, as lipid droplets. When hormones such as epinephrine are secreted or when insulin levels drop in response to low blood glucose levels, the intracellular lipases are activated through the c-AMP-initiated phosphorylation cascade. Triacylglycerols are then hydrolyzed to glycerol and free fatty acids. For many years it was believed that a single hormone-sensitive lipase (HSL) was responsible for the entire hydrolysis (lipolysis). It is now believed that three enzymes are involved:

1. Adipose triacylglycerol lipase converts triacylglycerol to diacylglycerol and one free fatty acid.
2. HSL converts diacylglycerol to monoacylglycerol and one free fatty acid.
3. Monoacylglycerol lipase converts monoacylglycerol to glycerol and one free fatty acid.

The free fatty acids are then moved to the blood stream where they are bound to serum albumin and carried to the appropriate cell.

4. SUMMARY

A new membrane must be put together piece by piece from a preexisting, functional membrane template. Fatty acids are the most basic component of the lipid bilayer and form the membrane hydrophobic interior. Fatty acids can be synthesized in the liver or absorbed through the diet. Fatty acids are built 2-carbon units at a time from acetyl CoA. Their biosynthesis occurs on the FAS, a complex of six enzyme activities that produce palmitoyl (16:0) CoA. In subsequent steps palmitic acid can be elongated and desaturated, forming the rich variety of fatty acids found in membranes. Dietary fatty acids are ingested and stored primarily in adipocytes as triacylglycerols (fats and oils) until required for energy production or membrane biosynthesis. Free fatty acids are carried around the body bound to serum albumin. As triacylglycerols, they are packaged into lipoproteins (chylomicrons, VLDLs, IDLs, LDLs, and HDLs). After transport to the target tissue, triacylglycerols are broken down by lipolysis.

Chapter 15 will discuss the biosynthesis of membrane polar lipids, phospholipids, sphingolipids, plasmalogens, and cholesterol.

References

[1] Heinrichs A. Connected clues? Nat Rev Mol Cell Biol 2006;7:624.
[2] Lipmann F, Kaplan NO. A common factor in the enzymatic acetylation of sulfanilamide and of choline. J Biol Chem 1946;162:743.
[3] Lipmann F. Metabolic generation and utilization of phosphate bond energy. Adv Enzymol 1941;1:99–162.

[4] Bergstrom S. Award Ceremony Speech. The Nobel prize in Physiology or Medicine. 1964 [Konrad Bloch, Feodor Lynen].
[5] Dijkstra AJ, Hamilton RJ, Wolf Hamm W. Fatty acid biosynthesis. Trans fatty acids. Oxford: Blackwell Pub.; 2008.
[6] Lipid Library – Lipid Chemistry, Biology, Technology and Analysis. Fatty acids: straight-chain saturated, structure, occurrence and biosynthesis. 2011. Web, http://lipidlibrary.aocs.org/lipids/fa_sat/index.htm.
[7] Lipid Library – Lipid Chemistry, Biology, Technology and Analysis. Branched-chain fatty acids, phytanic acid, tuberculostearic acid Iso/anteiso- fatty acids. 2011. Web, http://lipidlibrary.aocs.org/lipids/fa_branc/index.htm.
[8] Whitney E, Rolfes SR. Understanding nutrition. 11th ed. California: Thomson Wadsworth; 2008. p. 154.
[9] Burr GO, Burr MM, Miller E. On the nature and role of the fatty acids essential in nutrition. J Biol Chem 1930;86:587.
[10] Aguilar PS, Mendoza D. Control of fatty acid desaturation: a mechanism conserved from bacteria to humans. Mol Microbiol 2006;62(6):1507–14.
[11] Uttaro AD. Biosynthesis of polyunsaturated fatty acids in lower Eukaryotes. Critical review. IUBMB Life 2006;58(10):563–71.
[12] Fasano M, Curry S, Terreno E, Galliano M, Fanali G, Narciso P, et al. The extraordinary ligand binding properties of human serum albumin (critical review). IUBMB Life 2005;57(12):787–96.
[13] Vance DE, Vance JE. Biochemistry of lipids, lipoproteins and membranes. 5th ed. New York: Elsevier Scientific Press; 2008. 631 pp.
[14] Miles B. Review of lipoproteins. Texas A&M University; 2003. https://www.tamu.edu/faculty/bmiles/lectures/Lipid%20Transport.pdf.
[15] Hussain MM. Review article: a proposed model for the assembly of chylomicrons. Arterosclerosis 2000; 148:1–15.
[16] Gibbons GF, Wiggins D, Brown AM, Hebbachi AM. Synthesis and function of hepatic very-low-density lipoprotein. Biochem Soc Trans 2004;32:59–64.
[17] Tatami R, Mebuchi H, Ueda K, Udea R, Haba T, Kametani T, et al. Intermediate-density lipoprotein and cholesterol-rich very low density lipoprotein in angiographically determined coronary artery disease. Circulation 1981;64(6):1174–84.
[18] Jeon H, Blacklow SC. Structure and physiologic function of the low-density lipoprotein receptor. Annu Rev Biochem. Annu Rev Biochem 2005;74:535–62.
[19] Kumar V, Butcher SJ, Katrina O, Engelhardt P, Heikkonen J, Kaski K, et al. Three-dimensional cryoEM reconstruction of native LDL particles to 16Å resolution at physiological body temperature. PLoS One May 2011;6(5):e18841.
[20] Toth P. The "good cholesterol" high-density lipoprotein. Circulation 2005;111(5):e89–91.
[21] Baldwin K, Sutherland D, Brooks GH, Fahey TD. Exercise physiology: human bioenergetics and its applications. New York: McGraw-Hill; 2005.

CHAPTER

15

Membrane Biogenesis: Phospholipids, Sphingolipids, Plasmalogens, and Cholesterol

OUTLINE

1. PHOSPHOLIPIDS

By number, the major component of mammalian membranes is the multitude of phospholipids that comprise the lipid bilayer (Chapter 5). Phospholipids are composed of two fatty acids, one glycerol, one phosphate, and one of several alcohols that are covalently linked by dehydrations (Chapter 5). The previous chapter (Chapter 14) discussed the source and availability of the membrane fatty acids that comprise the hydrophobic interior of membranes (Chapter 4). This chapter will first turn attention to the polar groups that form the aqueous–bilayer interface by discussing how the various phospholipids are synthesized.

An Introduction to Biological Membranes
http://dx.doi.org/10.1016/B978-0-444-63772-7.00015-4

1.1 Incorporation of Fatty Acids Into Phospholipids

There are two basic ways that fatty acid (acyl) chains can be incorporated into phospholipids [1]. The first involves acylating glycerol-3-phosphate during the biosynthesis of phospholipids (Fig. 15.1). The second involves remodeling existing phospholipids by acyltransferases and transacylases.

A typical mammalian membrane is composed of well over 100 different phospholipid molecular species. Controlling the nature and distribution of diverse molecular species is essential for maximal membrane function. Phospholipid distribution is accomplished by a process known as remodeling [2,3]. Remodeling of phospholipids is controlled by enzymes including: Acyl-CoA transferases, lysophospholipid acyltransferases, CoA-dependent transacylase, CoA-independent transacylase, and lysophospholipase/transacylase. One well known example of remodeling involves the preference for saturated fatty acids in the *sn*-1 position and unsaturated fatty acids in the *sn*-2 position of mammalian phospholipids [4]. The enzyme acyl-CoA:1-acyl-2-lysophospholipid acyltransferase prefers polyunsaturated fatty acyl-CoAs as acyl chain donors while acyl-CoA:2-acyl-1-lysophospholipid acyltransferase prefers saturated fatty acyl-CoAs. These related enzymes are therefore partially responsible for the *sn*-1 chain being saturated and the *sn*-2 chain unsaturated. Many other examples of acyl chain specificity in transacylation exist. For example, the CoA-dependent transacylation system transfers esterified fatty acids from diacyl phospholipids to lysophospholipids. The system is specific for only 20:4, 18:2, and 18:0 fatty acids. In sharp contrast, the CoA-independent transacylase catalyzes the transfer of C:20 and C:22 polyunsaturated fatty acids (PUFAs) from diacyl phospholipids to various lysophospholipids. There is a large family of acyltransferases and transacylases that are continuously remodeling the acyl chains of membrane phospholipids [2].

1.2 Phosphatidylcholine, Phosphatidylethanolamine, and the Kennedy Pathway

The major phospholipids found in mammalian membranes are phosphatidylcholine (PC) and phosphatidylethanolamine (PE) (Chapter 5). These two lipids comprise over 50% of the total membrane phospholipids and are thus the major structural component of the membrane lipid bilayer. Both PC and PE are synthesized by an amino alcohol phosphotransferase reaction (also known as the cytidine diphosphate (CDP)-Choline Pathway), which uses

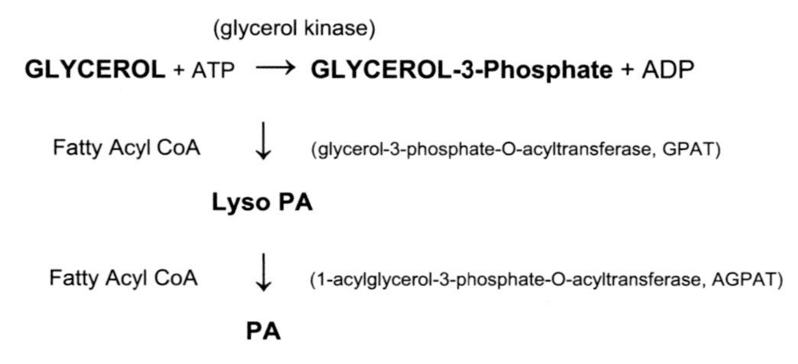

FIGURE 15.1 Incorporation of fatty acids into phospholipids through glycerol-3-phosphate.

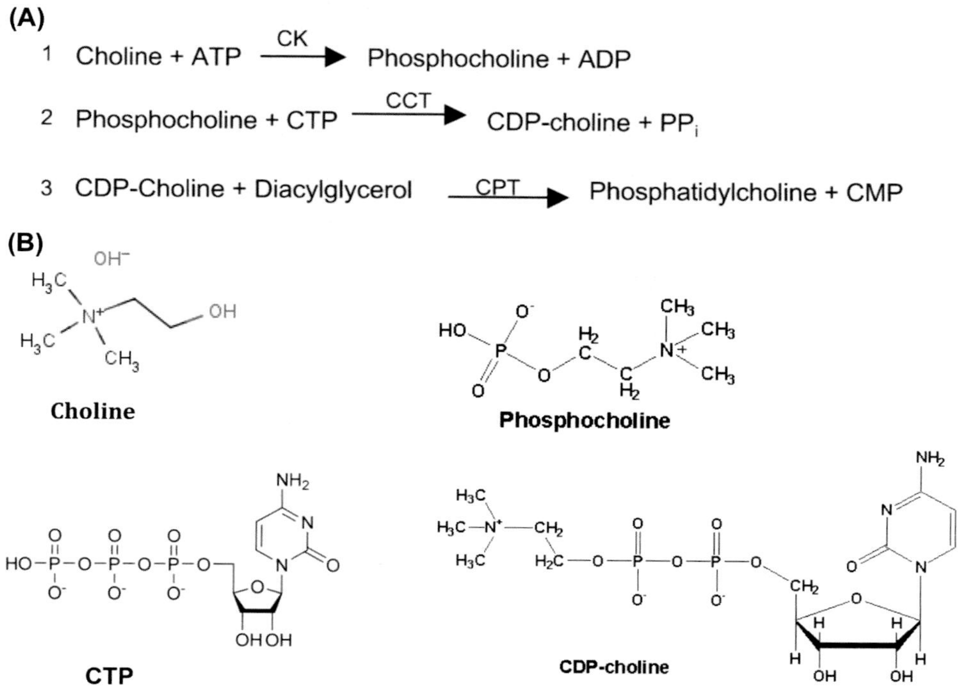

(A)

1 Choline + ATP $\xrightarrow{CK}$ Phosphocholine + ADP

2 Phosphocholine + CTP $\xrightarrow{CCT}$ CDP-choline + PP_i

3 CDP-Choline + Diacylglycerol $\xrightarrow{CPT}$ Phosphatidylcholine + CMP

(B)

Choline

Phosphocholine

CTP

CDP-choline

FIGURE 15.2 **Panel A**. The amino alcohol phosphotransferase reactions (also known as the CDP-Choline pathway) for the production of PC (and PE). Enzymes include CK (choline kinase), CCT (CTP: phosphocholine cytidyltransferase) and CPT (choline-phosphate cytidyltransferase). **Panel B**. The structures of choline, phosphocholine, CTP (cytidine triphosphate) and CDP-choline. **Panel A**. http://jem.rupress.org/content/200/1/99/F1.expansion.html. The structures in **Panel B** came from the following sources: Choline http://www.cholineinfo.org/healthcare_professionals/overview.asp. Phosphocholine http://www.pearsonhighered.com/mathews/ch19/phoschol.htm. CTP http://en.wikipedia.org/wiki/Cytidine_triphosphate. CDP-Choline http://www.pearsonhighered.com/mathews/ch19/cdpcholn.htm.

sn-1,2-diacylglycerol and either CDP-choline or CDP-ethanolamine. The steps for PC biosynthesis by this pathway are shown in Fig. 15.2. The same steps employing ethanolamine instead of choline produce PE.

The CDP–choline pathway for the synthesis of PC (and PE) was discovered by Eugene Kennedy (Fig. 15.3) beginning in the 1950s and is part of the phospholipid biosynthetic pathway that bears his name [5].

There is another parallel pathway for synthesizing PC and PE that interconnects with the CDP-Choline pathway. It is called the CDP-DAG Pathway because it uses CDP attached to diacylglycerol (DAG) forming CDP-DAG (Fig. 15.4) as its coupling intermediate [5]. The CDP-Choline pathway is depicted in Fig. 15.5, left side and the CDP-DAG Pathway depicted in Fig. 15.5, right side. These pathways are dependent on the close structural similarity between PE, PC, and phosphatidylserine (PS) that was discussed in Chapter 5 (Fig. 5.10). PC is PE that has been methylated three times (ie, it is trimethyl PE) and PE is decarboxylated PS (Fig. 15.6).

FIGURE 15.3 Eugene Kennedy, 1919–2011. http://www.asbmb.org/uploadedfiles/aboutus/asbmb_history/past_presidents/1970s/1970Kennedy.html.

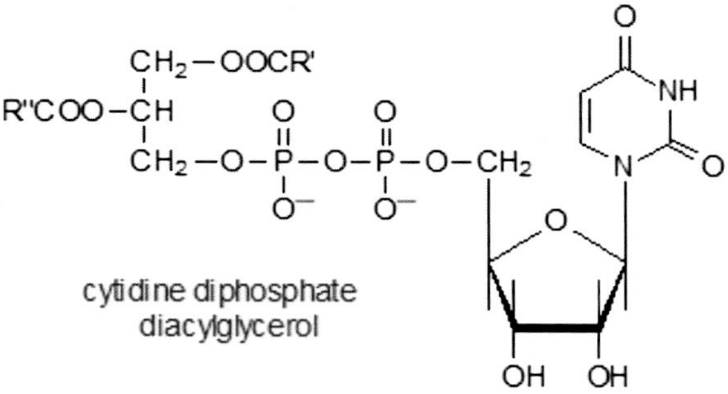

cytidine diphosphate diacylglycerol

FIGURE 15.4 Structure of CDP-DAG (cytidine diphosphate-diacylglycerol). http://lipidlibrary.aocs.org/Lipids/cdp-dg/index.htm.

In addition to interconversions of PC and PE, these pathways also produce phosphatidic acid (PA) (by the action of phospholipase D on PC), PS (by the action of PS Synthase I on PC and PS Synthase II on PE) and also produce phosphatidylinositol (PI). PS is also decarboxylated by PS decarboxylase to PE (Fig. 15.7).

1.3 Phosphatidic Acid and Beyond

The simplest phospholipid is phosphatidic acid, the only phospholipid without an alcohol moiety attached to the phosphate (Fig. 15.8) (Chapters 5 and 20). It is from this parent

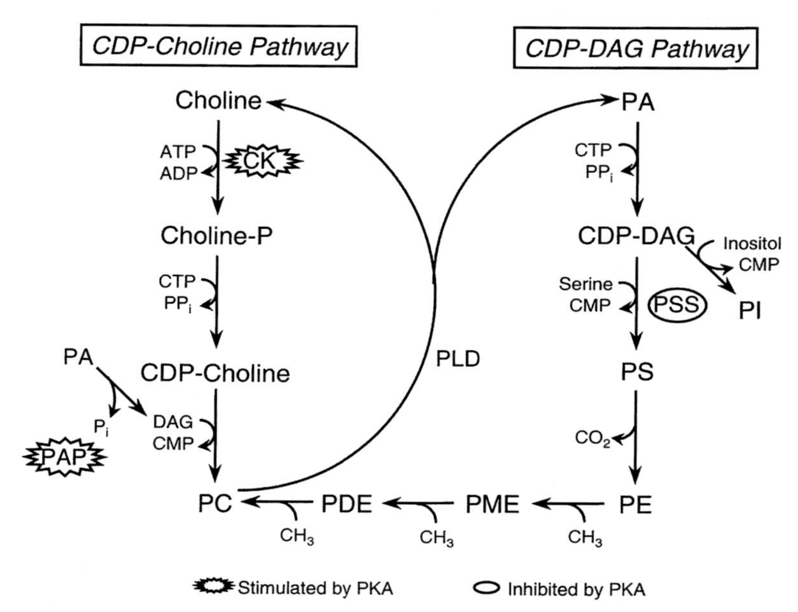

FIGURE 15.5 The cytidine diphosphate (CDP)-Choline (left) and cytidine diphosphate-diacylglycerol (CDP-DAG) (right) Pathways involving phosphatidylcholine (PC), phosphatidylethanolamine (PE), phosphatidylserine (PS), phosphatidic acid (PA), and phosphatidylinositol (PI) biosynthesis. PSS is the phosphatidylserine synthase I. PLD is phospholipase D. The three methylations involve the co-factor S-adenosylmethionine (SAM). http://www.biomed-search.com/search?q=phosphatidylserine&s=810&r=9.

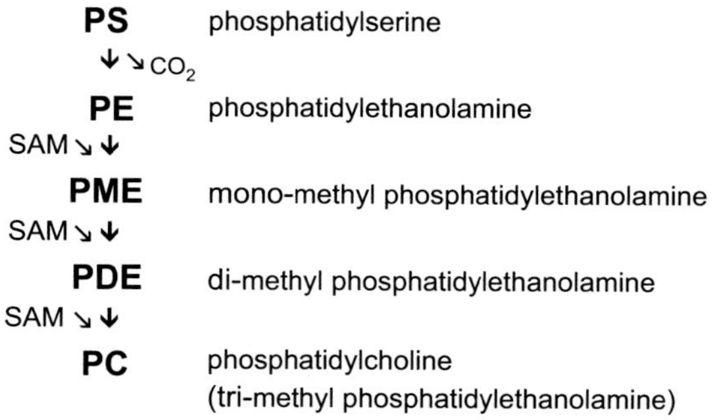

PS phosphatidylserine
↓ ↘ CO_2
PE phosphatidylethanolamine
SAM ↘ ↓
PME mono-methyl phosphatidylethanolamine
SAM ↘ ↓
PDE di-methyl phosphatidylethanolamine
SAM ↘ ↓
PC phosphatidylcholine
(tri-methyl phosphatidylethanolamine)

FIGURE 15.6 Conversion of phosphatidylserine (PS) to phosphatidylcholine (PC) through a phosphatidylethanolamine (PE) intermediate, by co-factor S-adenosylmethionine (SAM). *PS*, Phosphatidylserine; *PE*, phosphatidylethanolamine; *PME*, mono-methyl phosphatidylethanolamine; *PDE*, di-methyl phosphatidylethanolamine; *PC*, phosphatidylcholine (tri-methyl phosphatidylethanolamine).

structure that all of the other, more complex phospholipids can originate [6]. Fig. 15.5 shows how PA can be converted to PC, PE, PS and PI. In Fig. 15.9 the phospholipid biosynthetic pathway is extended to include phosphatidyl glycerol (PG) and cardiolipin (CL, also called diphosphatidyl glycerol).

PC → (phospholipase D) PA + choline

PC + L-serine → (PS Synthase I) PS + choline

PE + serine → (PS Synthase II) PS + ethanolamine

PS → (PS Decarboxylase) PE + CO_2

FIGURE 15.7 Alternate paths for the production of phosphatidylserine PS (from phosphatidylcholine (PC) and phosphatidylethanolamine (PE)), PE (from PS), and PA (from PC).

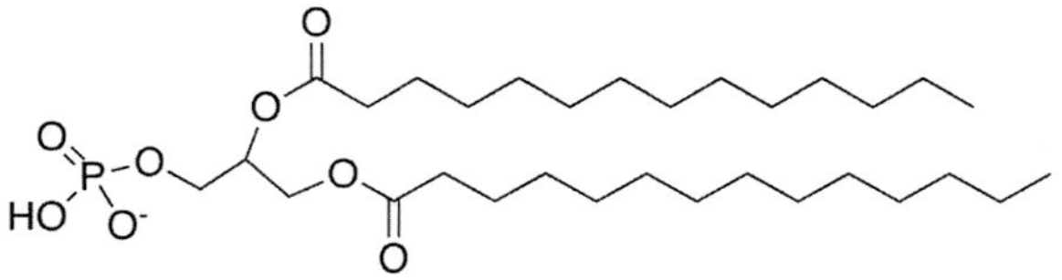

Phosphatidic Acid

FIGURE 15.8 Structure of 14:0,14:0 phosphatidic acid (dimyristoyl phosphatidic acid (PA)). http://oasys2.confex.com/acs/238nm/techprogram/P1286222.HTM.

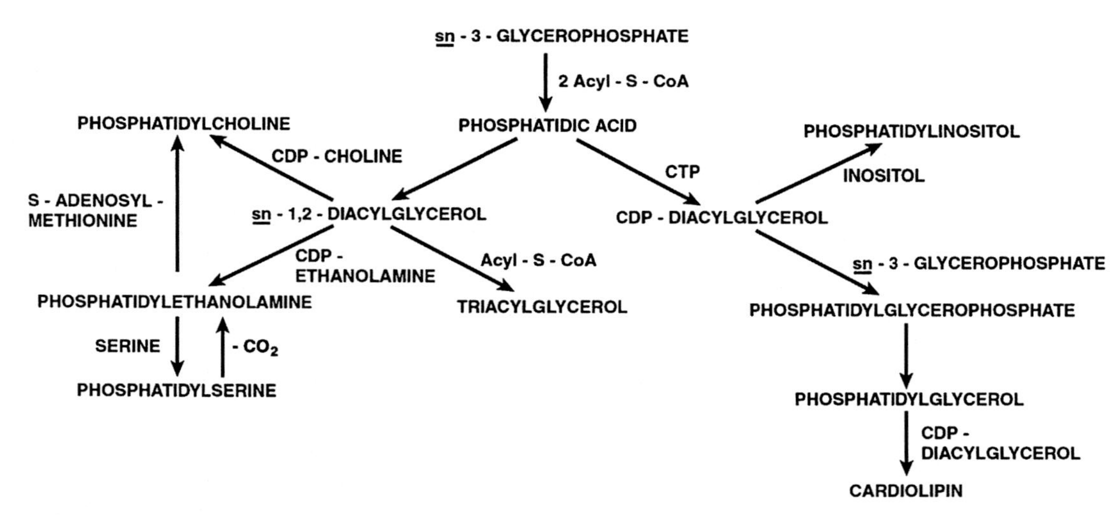

FIGURE 15.9 Skeletal outline for the complete biosynthetic pathways of the seven major mammalian membrane phospholipids: phosphatidic acid (PA), phosphatidylethanolamine (PE), phosphatidylcholine (PC), phosphatidylserine (PS), phosphatidylinositol (PI), phosphatidyl glycerol (PG), and cardiolipin (CL) [6]. http://ajplung.physiology.org/content/284/1/L1. *From reference R 15-6. Nanjundan M, Possmayer F. Pulmonary phosphatidic acid phosphatase and lipid phosphate phosphohydrolase. Am. J. Physiol Lung Cell Mol Physiol 2003;284:L1–23.*

Except for CL, phospholipid biosynthesis occurs in the cytosol adjacent to the endoplasmic reticulum (ER) membrane. It should be noted that, in mammals, CL is found almost exclusively in the mitochondrial inner membrane where it enhances the activity of proteins involved in electron transport and oxidative phosphorylation (Chapters 5 and 22) [7–9]. For this reason, CL synthase, the enzyme that couples CDP-DAG and PG to produce CL is localized in the mitochondria.

The ER membrane houses the enzymes for phospholipid biosynthesis and for their transmembrane distribution (flippases and floppases). This large collection of related enzymes eventually produce a phospholipid bilayer membrane vesicle that buds off from the ER and migrates to the appropriate cytoplasmic cellular membrane on its exterior leaflet and the plasma membrane on its inner leaflet.

2. SPHINGOLIPIDS

Sphingolipids are a large class of membrane polar lipids that are characterized by containing the unusual C-18 amino alcohol sphingosine (Fig. 5.13) or, to a lesser extent, dihydrosphingosine in their structure [10]. The sphingolipids were introduced in Chapter 5, and some of their medical implications discussed there and in Chapter 20. Several sphingolipids are classified as "bioactive lipids," having signaling and hormone-like activities (Chapter 20). Others, known as glycosphingolipids, are significant membrane surface markers that identify the cell to surrounding cells and define human blood group types. The most abundant sphingolipid in mammalian membranes is the structural polar lipid sphingomyelin (SM) (Fig. 5.14). SM is known to interact strongly with cholesterol and plays an important structural role in stabilizing lipid rafts (Chapters 8 and 10). Sphingolipids are indeed a very diverse group of important membrane lipid components.

As discussed in Section 1.3 (Fig. 15.9), PA plays a central role in phospholipid biosynthesis. The seven major classes of phospholipids can all be made from this simple phospholipid. A similar case can be made for ceramide (Fig. 15.10) in sphingolipid biosynthesis. Sphingosine, sphingosine-1-phosphate, ceramide-1-phosphate, SM, and the family of glycosphingolipids can all be derived from ceramide [11,12].

Synthesis of the large class of sphingolipids begins in the ER from nonsphingolipid precursors. The first step is catalyzed by the enzyme serine palmitoyltransferase that couples palmitoyl-CoA and serine, producing 3-ketodihydrosphingosine (Fig. 15.11). Next, 3-ketodihydrosphingosine is

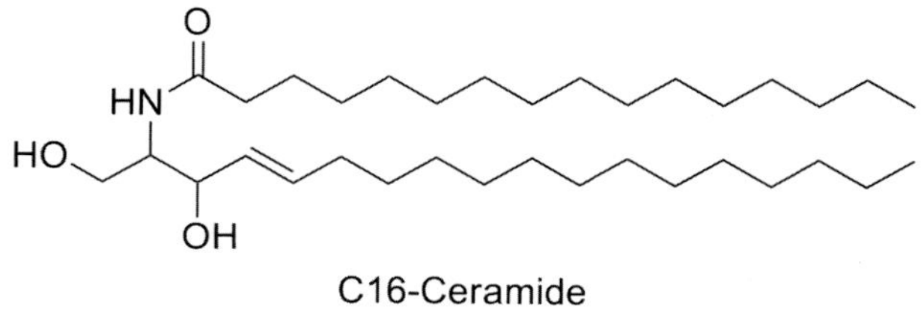

O
HN
HO
OH
C16-Ceramide

FIGURE 15.10 Structure of a "typical" ceramide (C16-Ceramide). http://www.labmuffin.com/2012/10/ceramides-what-eff-are-they-and-why-are.html.

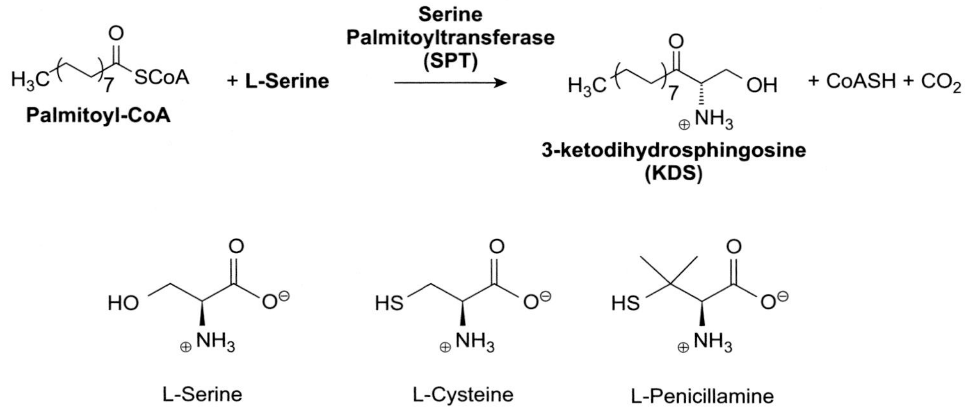

Serine Palmitoyltransferase (SPT)

H_3C–$(CH_2)_7$–C(=O)SCoA + L-Serine → 3-ketodihydrosphingosine (KDS) + CoASH + CO_2

Palmitoyl-CoA

L-Serine L-Cysteine L-Penicillamine

FIGURE 15.11 The initial step in sphingolipid biosynthesis is catalyzed by the enzyme serine palmitoyltransferase that couples palmitoyl-CoA and serine, producing 3-ketodihydrosphingosine. The preferred substrates for the reaction are palmitoyl-CoA and L-serine although L-cysteine and L-penicillamine may replace L-serine. http://pubs.rsc.org/en/content/articlelanding/2012/md/c2md20020a#!divAbstract.

reduced to form dihydrosphingosine which is then acylated by ceramide synthase producing dihydroceramide (Fig. 15.12). Dihydroceramide is then de-saturated to form ceramide.

Fig. 15.12 demonstrates that sphingolipid metabolism (both synthesis and degradation) is an array of interconnected pathways accounting for sphingolipid diversity. Further complicating the basic types of sphingolipids is the variety of fatty acids that can be esterified to the C-2 nitrogen of ceramides by ceramide synthases.

2.1 Sphingolipid Biosynthesis Passes Through Ceramide

Ceramide is the central intermediate in sphingolipid metabolism and has several different possible fates (Fig. 15.12). Ceramide may be phosphorylated by ceramide kinase to form the bioactive lipid ceramide-1-phosphate. It may also be glycosylated by glucosylceramide synthase or galactosylceramide synthase to form a single sugar-containing cerebroside with either glucose or galactose, respectively, attached to the 1-hydroxyl group of ceramide. Cerebrosides with galactose are found in the plasma membrane of neuronal cells while those with glucose are in the plasma membrane of nonneuronal cells. Cerebrosides can be further glycosylated to produce the complex, polysaccharide-containing family of gangliosides. There are more than 60 different human gangliosides that vary by the number, kinds, and order of the carbohydrates comprising the sugar chains (Chapter 5). Consistent with the rule of membrane carbohydrate topography (sugars always face out, Chapter 8), gangliosides always face the cell exterior. Ceramide can also be converted to SM by the addition of phosphorylcholine (or less commonly phosphorylethanolamine) using SM synthase. SM is the most abundant of the sphingolipids and plays an important role in maintaining proper membrane structure, particularly lipid rafts (Chapters 8 and 10). Ceramide may also be degraded by ceramidase to sphingosine that can be phosphorylated to the bioactive lipid sphingosine-1-phosphate.

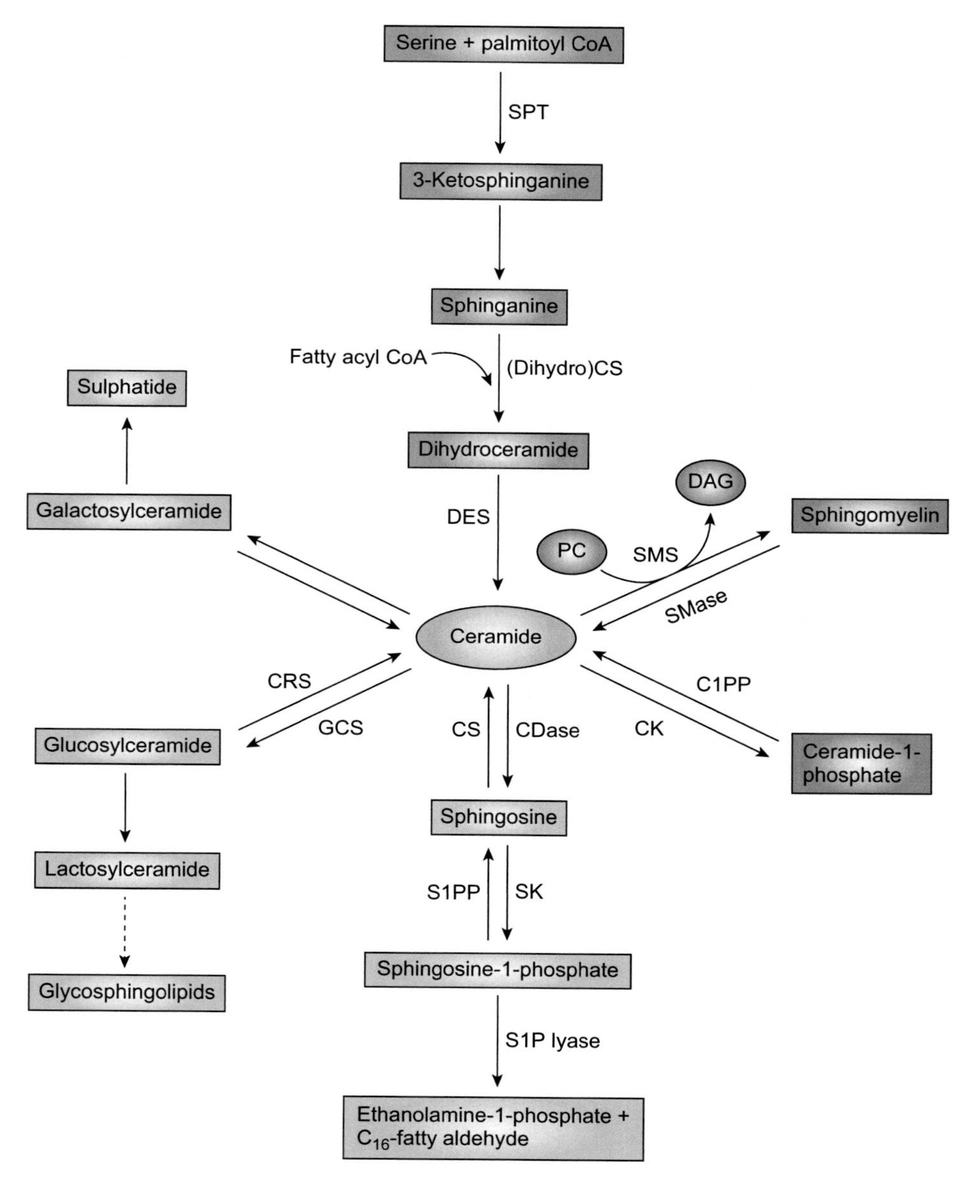

FIGURE 15.12 The central role of ceramide in the biosynthesis of sphingolipids. Sphingosine, sphingosine-1-phosphate, ceramide-1-phosphate, sphingomyelin and the family of glycosphingolipids can all be derived from ceramide [11]. http://www.nature.com/nrc/journal/v4/n8/fig_tab/nrc1411_F1.html. *This figure is from R 15-11. Ogretmen B, Hannun YA. Biologically active sphingolipids in cancer pathogenesis and treatment. Nat Rev Cancer 2004;4:604–16.*

3. PLASMALOGENS

Plasmalogens are one of the most unusual types of phospholipids commonly found in membranes [13,14]. While their two possible head groups, phosphorylcholine (designated plasmenylcholines) and phosphorylethanolamine (designated plasmenylethalomines), are the same as the two predominant mammalian membrane phospholipids (PC and PE) and they also have two long, hydrophobic chains attached to a glycerol, they introduce an unusual quirk into their structure. Their *sn*-1 chains are nonhydrolyzable ether-linked chains, while their *sn*-2 chains are connected by a conventional ester link (Fig. 5.4). In mammals, the *sn*-1 (ether) position is typically derived from C16:0, C18:0, or C18:1 fatty alcohols while the *sn*-2 (ester) position is most commonly occupied by PUFAs. This odd feature of their structure means plasmalogens must be synthesized differently than phospholipids, and indeed this is correct [15].

Although plasmalogens contain an unusual mixture of ether and ester linkages, and their function remains a mystery, in certain membranes, primarily those of nervous, immune, and cardiovascular cells, they are highly abundant. For example, in human heart tissue, nearly 30–40% of choline glycerophospholipids are plasmalogens. Plasmalogens were discovered by R. Feulgen and K. Voit in 1924 [16]. Today Robert Feulgen (Fig. 15.13) is far better known for the chromosomal and DNA stain that bears his name.

The pathway for plasmalogen biosynthesis has an unusual location for its initiation, the peroxisome. The first step in the pathway is catalyzed by the enzyme glycerol phosphate acyl transferase (GNPAT) that acylates the glycolysis intermediate dihydroxyacetone phosphate (DHAP) at its *sn*-1 position producing 1-acyl-DHAP on the luminal side of the peroxisomal membrane (Fig. 15.14). This is followed by exchange of the acyl group for an alkyl

FIGURE 15.13 Picture of Robert Feulgen, 1884–1955. http://schaechter.asmblog.org/schaechter/2010/08/plasmalogens-have-evolved-twice.html.

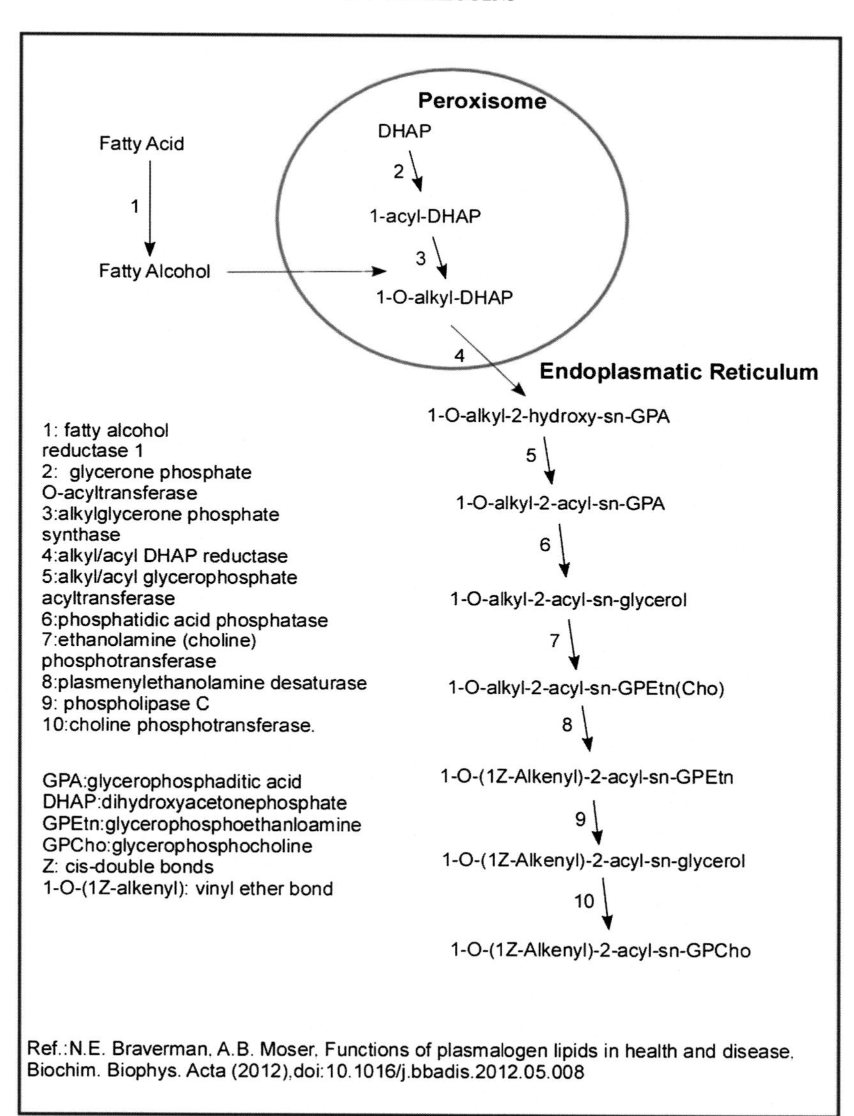

FIGURE 15.14 The complete pathway for biosynthesis of plasmalogens from dihydroxyacetone phosphate (DHAP) and a fatty acid [14]. http://en.wikipedia.org/wiki/Plasmalogen. *This figure originally appeared in: R 15-14. Braverman NE, Moser AB. Functions of plasmalogen lipids in health and disease. Biochim Biophys Acta 2012;1822(9):1442–52.*

group by the enzyme alkylglycerone phosphate synthase (AGPS). At this early point the *sn*-1, ether-linked chain is already in place.

The next step represents the pathway transition from the peroxisome to the ER. The 1-O-alkyl-DHAP is then reduced to 1-O-alkyl-2-hydroxy-*sn*-glycerophosphate (GPA) by an acyl/alkyl-DHAP reductase located in both the peroxisomal and ER membranes. All subsequent steps occur in the ER. These steps include: placing an acyl chain at the *sn*-2 position by an alkyl/acyl GPA acyltransferase; removing the phosphate by phosphatidic acid phosphatase forming 1-O-alkyl-2-acyl-*sn*-glycerol; and using CDP-ethanolamine and a phosphotransferase to add the phosphorylethanolamine to form the PE-plasmalogen (Fig. 15.14). Finally, the *sn*-1 ether chain is modified by a plasmenylethanolamine desaturase that creates the vinyl ether bond between positions 1 and 2. Since there is no plasmenylcholine desaturase, choline plasmalogens are made using a different pathway, involving PE, choline phosphotransferase, and CDP-choline.

Although the mechanisms of action for plasmalogens remain clouded in mystery, they are starting to receive medical interest as they are now being linked to Alzheimer's Disease, Down Syndrome, molecular signaling abnormalities, and cancer [14].

4. CHOLESTEROL

Blood cholesterol level remains one of the most significant, measurable indicators of human health, particularly cardiovascular health. Although far less appreciated by the general public, cholesterol is also a precursor for the biosynthesis of steroid hormones (Chapter 20), bile salts (Chapter 13), vitamin D (Chapter 21) and its biosynthetic pathway has branches that leads to production of dolichol (Chapter 16), ubiquinone (Chapter 18), and prenylated proteins (Chapter 6) [17]. The sterol is also a major structural component of the plasma membrane where it helps maintain proper membrane heterogeneity (lateral organization), stability, permeability, and "fluidity" (Chapters 8–10) [18]. Cholesterol and its biosynthetic pathway intermediates are as integrated into life processes as any biochemical can be. Yet, cholesterol is not homogeneously distributed throughout the body. A full one-quarter of cholesterol is located in the brain while most cholesterol biosynthes ($\sim$70%) occurs in the liver. Cell plasma membrane lipids are more than 50% cholesterol, while the nearby mitochondrial inner membrane has only trace amounts of the sterol.

Accurately controlling net cholesterol level and distribution is an essential task for all mammals [19]. Low cholesterol levels can be enhanced by activation of key enzymes through DNA transcriptional regulation of sterol regulatory element-binding proteins-1 and -2 [20]. Two major, unrelated proteins are upregulated by this process: 3-hydroxy-3-methylglutaryl-CoA (HMG-CoA) reductase (Fig. 15.15) and increasing lipoprotein uptake through the low-density lipoprotein (LDL)-Receptor (Chapter 17) [21].

HMG-CoA reductase is the key regulatory enzyme found in the Mevalonate Pathway for cholesterol biosynthase (Fig. 15.16). This pathway is also referred to as the HMG-CoA reductase Pathway. Downregulating the enzyme HMG-CoA reductase inhibits cholesterol biosynthesis, thus lowering blood cholesterol and preventing cardiovascular disease (CVD). Cholesterol reduction has captured attention of the pharmaceutical industry for decades. Research in this area has led to the development of statin drugs that reduce cholesterol biosynthesis by inhibiting HMG-CoA reductase.

A large number of statins are now commercially available, representing many of the most heavily advertised and financially successful drugs of all time (Table 15.1). By 2003 Lipitor became the best-selling pharmaceutical in history and in the year 2008 alone, had made Pfizer $12.4 billion.

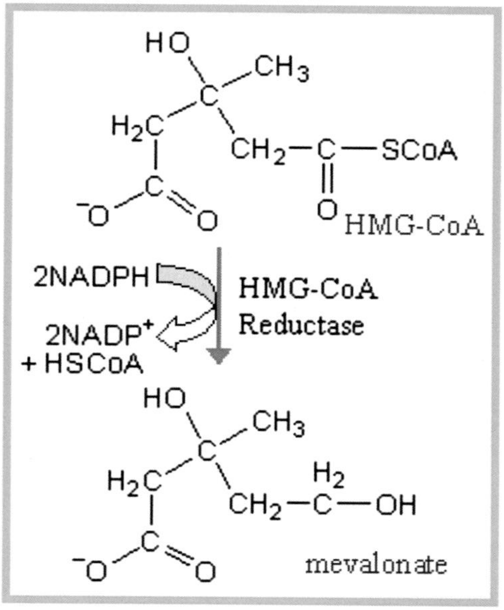

FIGURE 15.15 3-hydroxy-3-methyl-glutaryl-CoA (HMG-CoA) reductase is the key regulatory enzyme in the cholesterol biosynthetic pathway (the Mevalonate Pathway) (Fig. 15.14). HMG-CoA reductase converts HMG-CoA to mevalonate. Inhibition of this enzyme is the method by which statin drugs reduce cholesterol biosynthesis, preventing cardiovascular disease (CVD). http://www.rpi.edu/dept/bcbp/molbiochem/MBWeb/mb2/part1/cholesterol.htm.

In addition to decreasing cholesterol levels by downregulating or inhibiting HMG-CoA reductase, blood cholesterol levels can be reduced by increasing the ability of LDL receptors to take up cholesterol-rich LDLs from the blood into cells. LDLs were discussed in the previous chapter (Chapter 14). The major function of LDLs (Fig. 14.12) is to carry cholesterol around the body. Since the cholesterol level in LDLs is so high (about half of an LDL particle by weight is cholesterol), LDLs represent a potentially harmful storage source of cholesterol and so are often referred to as "bad cholesterol." Keeping LDL–cholesterol low is therefore beneficial for preventing CVD.

The LDL receptor is a large, mosaic protein composed of 860 amino acids including a 21-amino acid signal peptide [22]. It is a cell surface receptor that recognizes the B100 protein that is embedded in the outer phospholipid monolayer of LDL particles. For their discovery of the LDL receptor and its participation in cholesterol metabolism via receptor-mediated endocytosis (RME) (Chapter 17) [23], Michael S. Brown and Joseph L. Goldstein were awarded the 1985 Nobel Prize in Physiology or Medicine.

LDL receptors are present in invaginations on the plasma membrane surface called clathrin-coated pits (Chapter 17). The LDL receptor has an amino acid signal sequence on its C-terminal end that locates the receptor to the clathrin-coated pits, initiating the LDL receptor pathway [24]. After binding the cholesterol-loaded LDL particles to their receptors, the coated pits are pinched off (internally) forming clathrin-coated vesicles inside the cell. This process known as RME, occurs in all eukaryotic cells but is particularly prevalent in the liver where ~70% of circulating LDLs are internalized. Once inside the cell, the coated

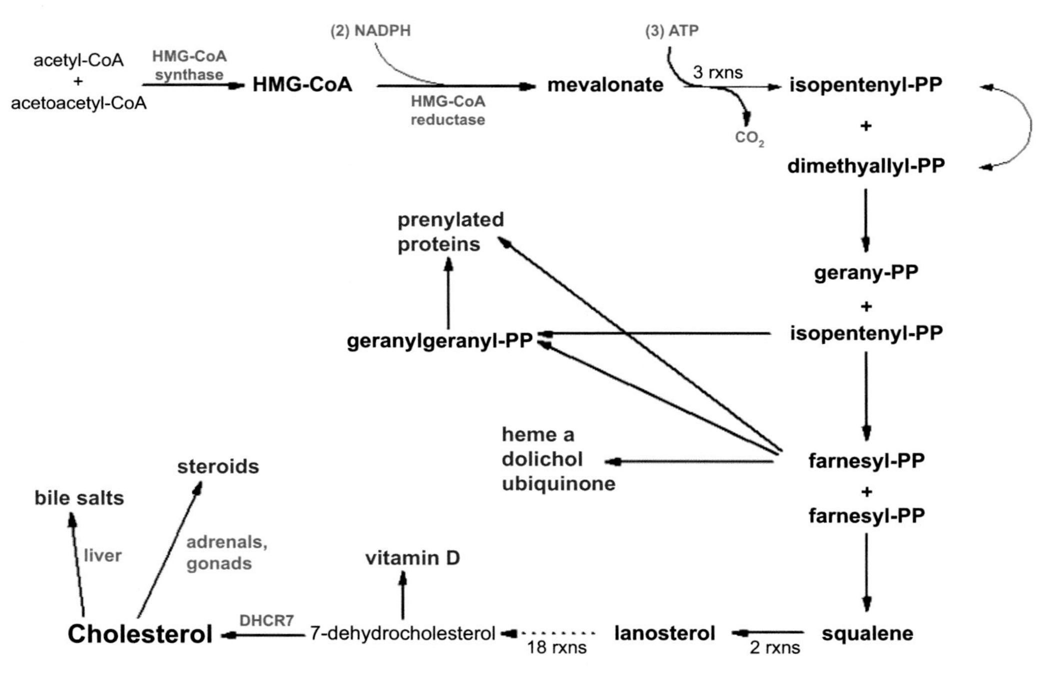

FIGURE 15.16 The Mevalonate Pathway (also referred to as the HMG-CoA Reductase Pathway) for the biosynthesis of cholesterol. Note that many important biochemical products are generated from this pathway including cholesterol, vitamin D, bile salts. Steroid hormones, dolichol, ubiquinone and prenylated proteins. http://themedicalbiochemistrypage.org/cholesterol.php.

vesicle sheds its clathrin coat and fuses with an acidic late endosome. When exposed to the new acidic conditions, the LDL receptor undergoes a conformational change, releasing the bound LDL. The LDL receptors are then either destroyed or are recycled via the endocytic cycle. The level of free cholesterol in the cell controls the synthesis of LDL receptors that are either increased or decreased by transcription of the receptor gene.

TABLE 15.1 Partial List of Commercially Successful Statins

Commercial name	Scientific name
Lipitor	(Atorvastatin)
Lescol	(Fluvastatin)
Mevacor	(Altocor)
Livalo	(Pitavastatin)
Pravachol	(Pravastatin)
Crestor	(Rosuvastatin)
Zocor	(Simvastatin)

Plant Sterols: While cholesterol stands alone at the center of animal sterols, the plant world houses a plethora of unusual sterols called phytosterols. To date more than 250 phytosterols have been identified. Included in this list of very similar looking phytosterols is cholesterol. An obvious question is why phytosterols, a normal human dietary component, do not appear to replace at least some of the animal cholesterol with plant sterols [25,26]. In animals, cholesterol and phytosterols must compete with one another at some level, with cholesterol winning out. Cholesterol in some (but not all) animal membranes (primarily the plasma membrane) plays critical structural roles including maintaining proper membrane heterogeneity, lateral organization, stability, permeability, and "fluidity" (Chapters 8–10) [18]. These same functions must be accomplished in plants by phytosterols. So why then can plant sterols not replace cholesterol in animals? Perhaps this could provide a partial solution to the human heart disease problem. Just replace cholesterol with a phytsterol.

For many decades medical studies have attempted to link serum levels of plant sterols to prevention of CVD [27]. Results of these studies, however, are not clear. Some studies with β-sitosterol (the major plant sterol) alone and in combination with other phytosterols have successfully demonstrated a related decrease in blood levels of cholesterol. Based on these studies, β-sitosterol has been used in treating hypercholesterolemia. It has been demonstrated that phytosterols are systematically excluded from the human body, primarily at the level of the intestinal epithelium. This may in part account for the dearth of phytosterols in animal membranes. Interestingly, it has been well-documented that plant sterols, particularly β-sitosterol, do compete with cholesterol uptake, thus providing a potentially beneficial role of reducing human cholesterol levels. When β-sitosterol is absorbed in the intestine, it is transported by LDL particles and can enter cell membranes in small amounts. Therefore at least some studies have indicated that phytosterols can reduce the level of LDL and total serum cholesterol. However, a thorough systematic review and metaanalysis of relevant studies from 1950 to 2010 did not support this conclusion [27]. Bernd Genser et al. (2012) concluded that: "Our systematic review and meta-analysis did not reveal any evidence of an association between serum concentrations of plant sterols and risk of CVD." Clearly further research will be required.

5. SUMMARY

There is a vast diversity of polar lipids comprising the hydrophobic component of membranes. This chapter (Chapter 15) discusses the biosynthesis of phospholipids, sphingolipids, plasmalogens, and cholesterol.

For *phospholipids*, fatty acid (acyl) chains can be incorporated into phospholipids by acylating glycerol-3-phosphate during phospholipid biosynthesis and by remodeling existing phospholipids. Remodeling includes placing a saturated fatty acid in the *sn*-1 position and an unsaturated fatty acid in the *sn*-2 position of phospholipids. Phospholipid biosynthesis involves several overlapping pathways. The two major pathways are the Kennedy Pathway and the CDP-DAG Pathway that produce PC and PE. The simplest phospholipid, PA, can be a starting point for the production of all membrane phospholipids, PE, PC, PS, PI, PG, and CL.

Sphingolipids are a very diverse group of membrane polar lipids that are characterized by containing sphingosine at their core. Sphingolipids are structural membrane lipids that stabilize lipid rafts. Several sphingolipids are "bioactive lipids," having signaling and hormone-like activities. Sphingolipids are synthesized through an array of interconnected pathways that account for sphingolipid diversity. At the heart of these pathways lies ceramide, whose metabolism has several different possible fates.

Plasmalogens are a highly unusual type of membrane structural polar lipid. These lipids are highly abundant in nervous, immune, and cardiovascular cells where their function remains a mystery. Plasmalogens are characterized by having a nonhydrolyzable, ether-linked *sn*-1 chain and a conventional, ester-linked *sn*-2 chain. The biosynthetic pathway for plasmalogens is much different than that of phospholipids or sphingolipids. The pathway begins in the peroxisome with insertion of the *sn*-1 chain ether linkage.

Cholesterol is the last major membrane structural lipid, whose essential membrane functions include: maintaining proper membrane heterogeneity, lateral organization into domains (including lipid rafts), stability, permeability, and fluidity. The proper level of cholesterol must be rigidly maintained by a balance between dietary intake and biosynthesis. Cholesterol biosynthesis occurs through the Mevalonate Pathway and its key regulatory enzyme HMG-CoA reductase, the target of cholesterol-lowering drugs. Uptake of cholesterol from the diet is through LDL ("bad cholesterol") uptake by RME.

Chapter 16, will discuss the biosynthesis and translocation of membrane proteins.

References

[1] Kent C. Eukaryotic phospholipid biosynthesis. Ann Rev Biochem 1995;64:315–43.
[2] Yamashita A, Sugiura T, Waku K. Acyltransferases and transacylases involved in fatty acid remodeling of phospholipids and metabolism of bioactive lipids in mammalian cells. J Biochem 1997;122(1):1–16.
[3] Butler PL, Mallampalli RK. Cross-talk between remodeling and de Novo Pathways maintains phospholipid balance through ubiquitination. J Biol Chem 2010;285:6246–58.
[4] Shindou H, Shimizu T. Acyl-coa: lysophospholipid acyltransferases. J Biol Chem 2009;284:1–5.
[5] Gibellini F, Smith TK. The Kennedy pathway—De novo synthesis of phosphatidylethanolamine and phosphatidylcholine. IUBMB Life 2010;62(6):414–28.
[6] Nanjundan M, Possmayer F. Pulmonary phosphatidic acid phosphatase and lipid phosphate phosphohydrolase. Am J Physiol Lung Cell Mol Physiology 2003;284:L1–23.
[7] Leray C. CYBERLIPID CENTER. Fats and Oils. Diphosphatidylglycerol (Cardiolipin). Creative Commons. http://www.cyberlipid.org/index.htm; 2016.
[8] Schlame M, Rua D, Greenberg ML. The biosynthesis and functional role of cardiolipin. Progr Lipid Res 2000;39(3):257–88.
[9] Houtkooper RH, Vaz FM. Cardiolipin, the heart of mitochondrial metabolism. Cell Mol Life Sci 2008;65:2493–506.
[10] Christie W. Introduction to Sphingolipids and Rafts. AOCS. The Lipid Library; 2012.
[11] Ogretmen B, Hannun YA. Biologically active sphingolipids in cancer pathogenesis and treatment. Nat Rev Cancer 2004;4:604–16.
[12] Gault CR, Obeid LM, Hannun YA. An overview of sphingolipid metabolism: from synthesis to breakdown. Adv Exp Med Biol 2010;688:1–23.
[13] Kanno S, Nakagawa K, Eitsuka T, Miyazawa T. Plasmalogen: a short review and newly-discovered functions. In: Yanagita T, Knapp H, editors. Dietary in fats and risk of chronic diseases. AOCS Press; 2006 [chapter 14].
[14] Braverman NE, Moser AB. Functions of plasmalogen lipids in health and disease. Biochim Biophys Acta 2012;1822(9):1442–52.
[15] Nagan N, Zoeller RA. Plasmalogens: biosynthesis and functions (Review). Progr Lipid Res 2001;40:199–229.
[16] Feulgen R, Voit K. Uber einen weiterbreiteten festen Aldehyd. Pflüger's Archiv für die gesamte Physiologie des Menschen und der Tiere 1924;206:389–410.
[17] Parish EJ, Nes WD, editors. Biochemistry and function of sterols. CRC Press; 1997. 278 pp.

[18] Yeagle PL. The roles of cholesterol in the biology of cells. In: The structure of Biological membranes. 2nd ed. Boca Raton, FL: CRC Press; 2005 [chapter 7].
[19] Brown MS, Goldstein JL. A proteolytic pathway that controls the cholesterol content of membranes, cells and blood. Proc Natl Acad Sci U S A 1999;96(20):11041–8.
[20] Wang X, Sato R, Brown MS, Hua X, Goldstein JL. SREBP-1, a membrane-bound transcription factor released by sterol-regulated proteolysis. Cell 1994;77(1):53–62.
[21] Brown MS, Goldstein JL. Receptor-mediated endocytosis: insights from the lipoprotein receptor system. Proc Natl Acad Sci U S A 1979;76(7):3330–7.
[22] Südhof TC, Goldstein JL, Brown MS, Russell DW. The LDL receptor gene: a mosaic of exons shared with different proteins. Science 1985;228(4701):815–22.
[23] Goldstein JL, Brown MS. History of discovery. The LDL receptor. Arterioscler Thromb Vasc Biol 2009;29:431–8.
[24] Lagor WR, Millar FS. Overview of the LDL receptor: relevance to cholesterol metabolism and future approaches for the treatment of coronary heart disease. Journal of Receptor. Ligand Channel Res 2010;3:1–14.
[25] Dutta PC, editor. Phytosterols as functional food components and nutraceuticals nutraceutical science and technology. New York: Marcel Dekker; 2004. 406 pp.
[26] Ling WH, Jones PJ. Dietary phytosterols: a review of metabolism, benefits and side effects. Life Sci 1995;57(3):195–206.
[27] Genser B, Silbernagel G, De Backer G, Bruckert E, Carmena R, Chapman MJ, et al. Plant sterols and cardiovascular disease: a systematic review and meta-analysis. Eur Heart J 2012;33(4):444–651.

CHAPTER

16

Membrane Biogenesis: Proteins

OUTLINE

1. INTRODUCTION

Membranes are a complex mixture of primarily lipids and proteins with surface-attached carbohydrates that are neither found inside the membrane core nor attached to the membrane inner leaflet (Chapter 7). Lipids (Chapters 4 and 5) provide the basic membrane structure and

An Introduction to Biological Membranes
http://dx.doi.org/10.1016/B978-0-444-63772-7.00016-6

associated biophysical properties, while the proteins (Chapter 6) are responsible for all biochemical activity. Each membrane type has its own characteristic combination of lipids and proteins. Chapters 14 and 15 discussed biosynthesis of the membrane lipids while this chapter (Chapter 16) will consider biosynthesis of the proteins and their assembly into a functional membrane.

Chapter 6 introduced the large variety of protein structures found in membranes. Indeed, protein structural diversity implies the necessity of a large variety of modes of biosynthesis. Fortunately there is a common theme for the biosynthesis of many of the membrane proteins. This common theme is known as the "signal hypothesis" where newly synthesized (nascent) proteins have intrinsic signals that govern their synthesis, orientation, transport, and ultimate localization in the cell. The signal hypothesis was proposed in 1971 by Gunter Blobel (Fig. 16.1) of Rockefeller University for which he was awarded the 1999 Nobel Prize in Physiology or Medicine [1].

2. SYNTHESIS OF SECRETORY PROTEINS

Synthesis of all eukaryotic, nonmitochondrial proteins begins the same. mRNA is transcribed in the nucleus and transported to the cytosol for transcription (into a protein) on a free cytoplasmic ribosome. Although this chapter concerns membrane proteins, we will first investigate synthesis of secretory proteins that, although not incorporated into membranes, begin their life similarly to membrane proteins and must cross membranes on their way to the extracellular aqueous phase (Chapter 19) [2]. The nascent protein grows from its N-terminal. The N-terminal begins with a short hydrophobic peptide of 5–30 amino acids in length called the signal peptide (or sometimes the signal sequence, leader sequence, or

FIGURE 16.1 Gunter Blobel, 1936–. http://www.prix-nobel.org/EN/Medicine/blobel.htm.

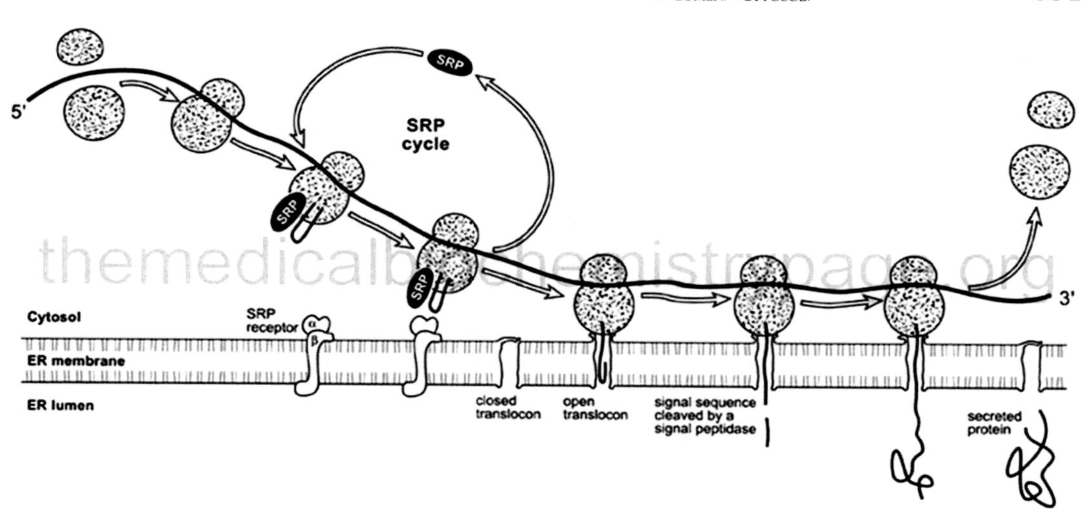

FIGURE 16.2 Steps involved in the synthesis of a secretory protein on the endoplasmic reticulum (ER). http://themedicalbiochemistrypage.org/protein-modifications.php.

leader peptide) that binds to a cytoplasmic protein called the Signal Recognition Particle (SRP), stopping the translation process [1,3]. The steps involved in secretory protein biosynthesis are depicted in Fig. 16.2.

The SRP aids in transport of the mRNA-ribosome complex to an SRP receptor found in the endoplasmic reticulum (ER) membrane. Upon binding to the SRP receptor the signal sequence is transferred to a channel protein called the translocon (also known as the Sec61 channel) that carries the nascent protein to the ER lumen. The lumen is the sequestered aqueous phase of the ER and is separated from and distinct from the cytoplasm. The SRP dissociates from the mRNA-ribosome complex allowing translation of the secretory protein to continue. The signal sequence is cleaved and translation continues. The newly synthesized chain moves through the translocon into the ER lumen in a process known as co-translational translocation. The new secretory protein then traffics through the endomembrane system to eventually be released outside the cell plasma membrane by a process known as secretion (Chapter 19).

3. SYNTHESIS OF SINGLE-SPAN PROTEINS ($N_{LUMEN}/C_{CYTOSOL}$)

The simplest transmembrane protein crosses the membrane only once (single-span or bitopic membrane protein), as an α-helix (Chapter 6). A membrane-spanning α-helix is composed of about 20–25 amino acids that are stabilized by the lipid bilayer. The lipid bilayer also anchors the protein into the membrane. The single-span family of proteins originates via a synthetic pathway almost identical to that discussed previously for secretory proteins [4,5]. Once made, the single-span proteins maintain their same orientation with respect to the bilayer plane throughout transport to their target membrane location. Protein flip-flop does not occur (Chapter 8). Thus, orientation of these membrane proteins in their final sites is established during their biosynthesis on the ER membrane. Since all proteins are translated in an N-terminal to C-terminal direction, many (but not all) (Section 4) single-span transmembrane proteins have their N-terminal hydrophilic segment on the exoplasmic side (the ER lumen) and their C-terminal hydrophilic segment on the cytoplasmic side. This orientation is referred to as $N_{lumen}/C_{cytosol}$.

Single-span proteins, therefore, possess two signals, an N-terminal signal sequence that is responsible for targeting the nascent protein to the ER surface and an internal stop-transfer membrane-anchor sequence (a topogenic sequence) that defines and locks in the membrane spanning α-helix. Topogenic sequences are up to 25 amino acids long and assures that the protein acquires the correct orientation in the membrane.

Fig. 16.3 depicts events involved in the synthesis of the single-span transmembrane insulin receptor protein ($N_{lumen}/C_{cytosol}$). Note the striking similarities between synthesis of this single-span membrane protein (Fig. 16.3) and a typical secretory protein (Fig. 16.2). In both cases, the nascent chain and attached ribosome associates with the translocon in the ER membrane. The N-terminal signal sequence is cleaved and the chain continues to elongate. In the case of the single-span transmembrane protein, once the internal stop-transfer membrane-anchor (topogenic) sequence of about 22 amino acids is synthesized, it remains in the translocon and prevents the nascent chain from extending farther into the ER lumen.

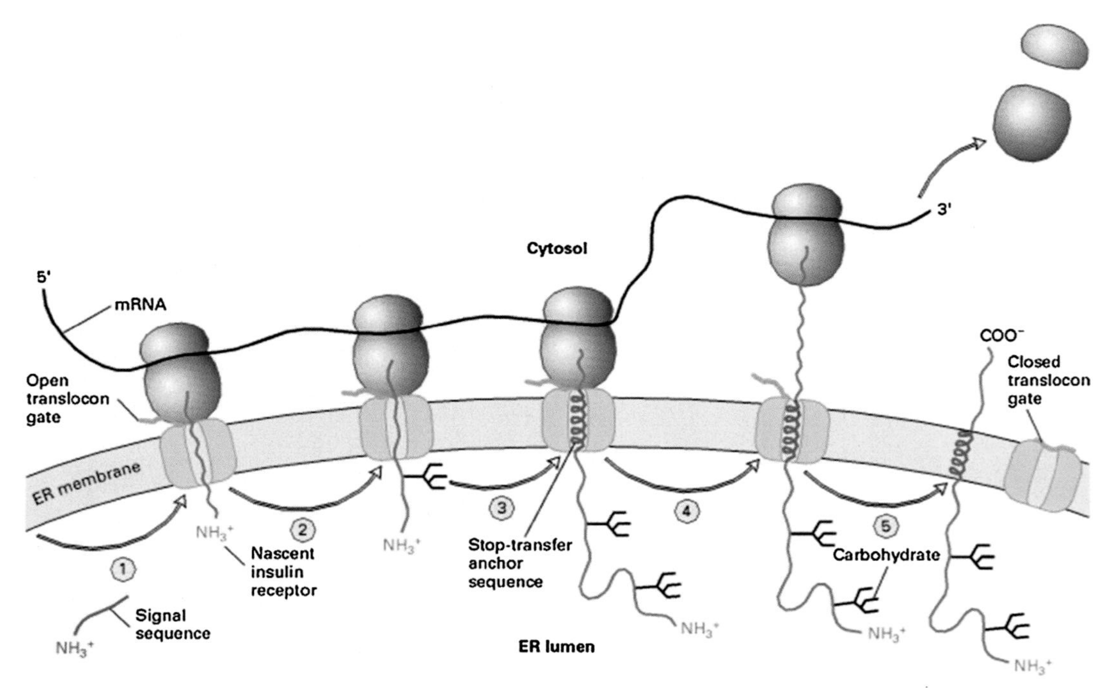

FIGURE 16.3 Synthesis of the single-span trans-membrane insulin receptor protein ($N_{lumen}/C_{cytosol}$). *http://www.1cro.com/mcb/bv.fcgi@call=bv.view..showsection&rid=mcb.figgrp.d1e80444.htm; Fig. 17-22 in Lodish H, Berk A, Zipursky SL, Matsudaira P, Baltimore D, Darnell J. Section 17.3. Overview of the Secretory Pathway. Molecular Cell Biology. 4th ed. New York: W.H. Freeman; 2000.*

The protein is therefore locked in place. The ribosome then dissociates from the translocon and protein synthesis is completed in the cytoplasm. After completion of protein synthesis, the translocon gate closes, awaiting arrival of the next nascent chain with attached ribosome.

4. SYNTHESIS OF A SINGLE-SPAN PROTEIN ($N_{CYTOSOL}/C_{LUMEN}$)

The insertion of some single-span transmembrane proteins is opposite to that described previously for the insulin receptor. These proteins have their N-terminal in the cytosol and their C-terminal in the lumen of the ER ($N_{cytosol}/C_{lumen}$). An example of this type of protein is the asialoglycoprotein receptor that derives its inverse orientation from a single internal topogenic (signal) sequence [6,7].

$N_{cytosol}/C_{lumen}$ proteins have a single internal signal-anchor sequence that functions as both an ER signal sequence and as a membrane-anchor sequence. The internal signal-anchor is unusual in inserting the nascent chain into the ER membrane with the N-terminal signal sequence facing the cytosol (Fig. 16.4). The internal signal-anchor sequence

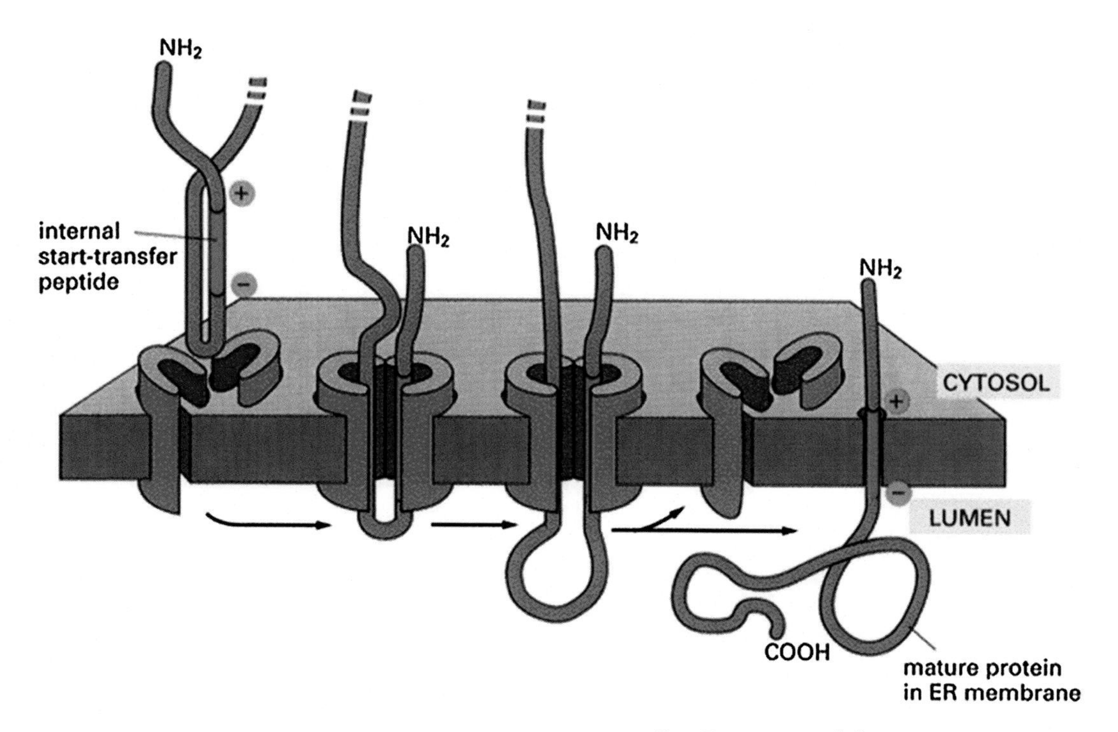

FIGURE 16.4 Synthesis of the single-span trans-membrane asialoglycoprotein receptor ($N_{cytosol}/C_{lumen}$). http://web.uconn.edu/mcb380/lecture04.html.

becomes a transmembrane α-helix and is not cleaved. The chain continues to grow while in the translocon and the nascent protein C-terminal eventually emerges into the ER lumen.

The orientation of transmembrane proteins is complicated. It is affected by the charge on the protein segments flanking the internal signal-anchor. A positive charge in this segment prefers an $N_{cytosol}/C_{lumen}$ orientation. An $N_{luminal}/C_{cytosol}$ orientation is favored by internal signal-anchor segments that are long (>20 amino acids) and hydrophobic. Shorter, less hydrophobic internal signal-anchor segments prefer an $N_{cytosol}/C_{lumen}$ orientation.

5. SYNTHESIS OF MULTI-SPAN PROTEINS

Many transmembrane (integral) proteins span the membrane more than once via α-helical segments (Chapter 6). Included in this multispan (also referred to as multipass or polytopic) category are the very large and important class of proteins with seven membrane spanning α-helices. Due to their importance, this class of proteins is often referred to as the "magnificent seven" (Chapter 6) [8]. Examples of proteins containing the seven α-helix motif have been previously discussed (Chapter 6) and include the large family of G-protein-coupled receptors (GPCR) (Chapter 18), a variety of channels and ion pumps, aquaporin, hormone receptors and the glucose transporter GLUT1 (Chapter 19). GPCRs account for 5% of the total human genes! Clearly, having multiple antiparallel α-helices in a single protein is a very common motif that presents a challenge in understanding transmembrane protein synthesis and orientation.

Multi transmembrane α-helices tend to bundle together with their orientation approximately perpendicular to the lipid bilayer [9]. Each α-helix has its most polar surface facing the bundle interior and its least polar surface facing the hydrophobic bilayer acyl chains. The transmembrane α-helices are enriched in nonpolar amino acids and the hydrophilic loops that connect the α-helices follow the "positive inside" rule [10]. This rule states that nontranslocated (extra membrane) loops facing the cytosol or "inside" with respect to their final membrane location in the cell, are enriched in positively charged amino acids compared to translocated loops. While a few signal protein sequences are now well known, undoubtedly many others remain undiscovered [11].

Complicating the basic understanding of synthesizing multispan transmembrane proteins is that each α-helix acts as a topogenic sequence functioning in the order in which they are synthesized. The example of a multipass transmembrane protein discussed here is the glucose transporter (GLUT1) [12,13]. The first α-helix segment initiates insertion of the nascent protein into the ER membrane (Fig. 16.5). It functions as an internal, uncleaved signal-anchor sequence similar to that of the single-span transmembrane insulin receptor ($N_{lumen}/C_{cytosol}$) depicted in Fig. 16.3. In both examples, the SRP and its receptor help assemble the nascent chain—ribosome—translocon complex. Continued insertion produces a hairpin turn in the chain facing the lumen with the N-terminal facing the cytoplasm. The nascent chain continues to elongate in the translocon until a second α-helix is formed out in the cytosol. The second α-helix functions as a stop-transfer membrane-anchor sequence. Thus, after synthesis of the first two α-helices both the C- and N-terminals are in the cytosol.

Continued extrusion of the nascent chain through the translocon ceases and the first two α-helices then leave the translocon and enter into the ER bilayer, keeping the hairpin

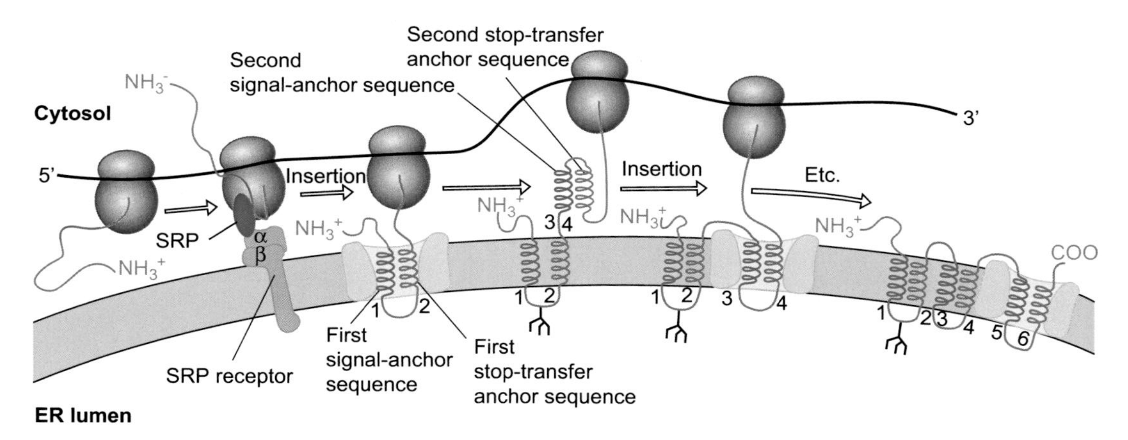

FIGURE 16.5 Synthesis and insertion of the multispan transmembrane glucose transporter (GLUT1) into the endoplasmic reticulum (ER) membrane. http://cc.scu.edu.cn/G2S/Template/View.aspx?courseType=1&courseId=17&topMenuId=113302&menuType=1&action=view&type=&name=&linkpageID=113473.

conformation linking the two α-helices. The C-terminus of the nascent chain continues to grow in the cytosol until subsequent α-helix pairs are generated and inserted into the ER bilayer. This, however, occurs without the assistance of SRP and the SRP receptor. The third α-helix is another internal uncleaved signal-anchor sequence and the fourth α-helix is another stop-transfer membrane-anchor sequence.

Experiments with recombinant multispan proteins confirm that it is the order of the α-helices relative to each other in the growing chain that determines whether a given helix functions as a signal-anchor sequence or as a stop-transfer membrane-anchor sequence. The first and all subsequent odd-numbered α-helices are signal-anchor sequences while all even-numbered α-helices are stop-transfer membrane-anchor sequences. The process continues until the GLUT1 chain is completed (Fig. 16.5).

6. SYNTHESIS OF GLYCOSYLPHOSPHATIDYLINOSITOL-LINKED PROTEINS

An unusual family of integral proteins, known as the glycosyl-phosphatidylinositol (GPI)-linked proteins, was discussed in Chapter 6. These proteins are anchored to the membrane lipid bilayer through the acyl chains of the phospholipid, phosphatidylinositol (PI) [14,15]. They do not have a hydrophobic, membrane-spanning protein segment. Synthesis and insertion of GPI-linked proteins is highly unusual, even for the strange menagerie of membrane proteins (Fig. 16.6).

GPI-linked proteins are initially synthesized and anchored in the ER membrane exactly like the single-span transmembrane insulin receptor protein ($N_{lumen}/C_{cytosol}$) with a cleaved N-terminal signal sequence and an internal stop-transfer membrane-anchor sequence. So initially, synthesis and insertion seem quite normal. However, the GPI-pathway diverges from the insulin receptor pathway when a short sequence of amino acids

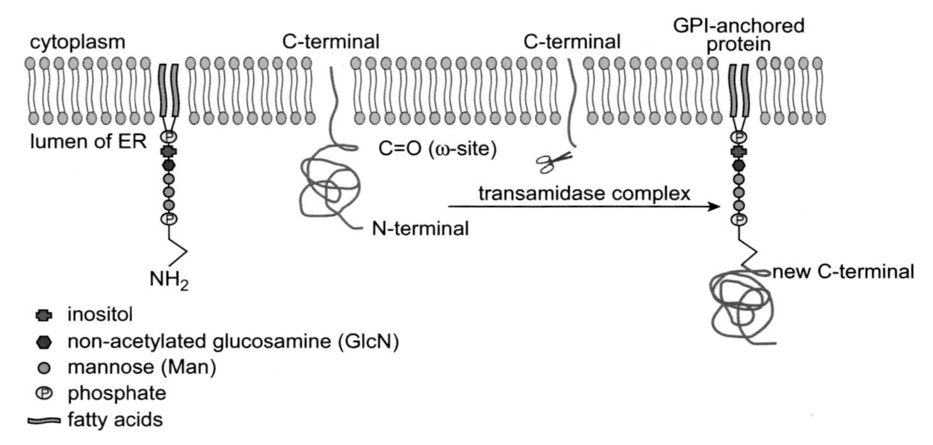

FIGURE 16.6 Synthesis and insertion of a glycosyl-phosphatidylinositol (GPI)-linked integral membrane protein into the endoplasmic reticulum (ER) membrane. http://scielo.sld.cu/scielo.php?pid=S1027-28522011000100002&script=sci_arttext.

adjacent to the membrane-spanning domain and sticking into the ER lumen is recognized by an endoprotease that simultaneously cleaves off the original stop-transfer membrane-anchor sequence and transfers the remainder of the protein to an awaiting, preformed GPI-anchor located in the ER membrane (Fig. 16.6). Therefore, for GPI-linked proteins the membrane-anchor abruptly changes from one based on protein to one based on lipid. There seems to be one significant advantage to what superficially appears to be a complex and unnecessary process. Upon switching to a GPI-link, the entire cytosolic exposure is eliminated, thus excluding any possible interaction with the cytoskeleton. Cytoskeletal involvement would greatly hinder lateral diffusion through the membrane. Once added to the plasma membrane, exposure of the GPI-linked protein is totally extracellular.

7. POSTTRANSLATIONAL MODIFICATIONS

Once translation of the protein in the ER is complete, there remains many additional modifications to the protein before it is fully functional. These steps are referred to as posttranslational modifications (PTMs) [16–18]. While genomics tells us that the human genome comprises 20,000 to 25,000 genes, proteomics estimates the number of proteins at over one million. Why is there such a large discrepancy? Clearly, single genes must encode multiple proteins. While some of the discrepancy can be attributed to alternative forms of nucleic acid recombination and splicing that generates different mRNA transcripts from a single gene, most of the proteins are likely the result of PTM. Therefore, PTM greatly increases complexity from the level of the genome to the level of the proteome, increasing life's functional diversity [16,17].

PTMs come in a bewildering array of formulations that defy logical categorization. As an example, it has been estimated that 5% of the human proteome comprises enzymes that perform more than 200 types of PTMs. These enzymes include kinases, phosphatases, transferases, and ligases. Also, PTM can occur at any step in the "life cycle" of a newly synthesized protein.

Only a brief subset of the many types of PTM will be considered here. Examples presented include:

1. Protein folding by molecular chaperones
2. Proteolytic cleavage
3. Formation of disulfide bonds
4. Lipidation
5. Ubiquitination
6. Amino acid modification: phosphate and friends
7. Addition of cofactors
8. Glycosylation
9. Dolichol

7.1 Protein Folding by Molecular Chaperones

Correct folding of a newly synthesized protein is an extremely complex and poorly understood process that is usually assisted by "molecular chaperones" [19–21]. Molecular chaperones are proteins that assist a protein to fold (or unfold) correctly through noncovalent interactions. Chaperones are not present when the protein has achieved its final, correct conformation and is functional. Chaperones are also involved with macromolecules other than proteins and often assist in putting together macromolecular complexes. In fact, the first discovered chaperone assists in the assembly of nucleosomes from histones and DNA [22]. The major functions of molecular chaperones are to assure the proper folding of a newly synthesized protein and to prevent unwanted aggregation of unrelated peptides. The first step for newly synthesized proteins on the ER is to be covered by molecular chaperones, protecting them from other proteins, giving them time to fold correctly. Fig. 16.7 depicts a protein about to enter a molecular chaperone complex where it will be stretched enough to allow for a new chance to correctly refold. If the misfolded protein cannot be saved, it is tagged for destruction by ubiquitin (Section 7.5).

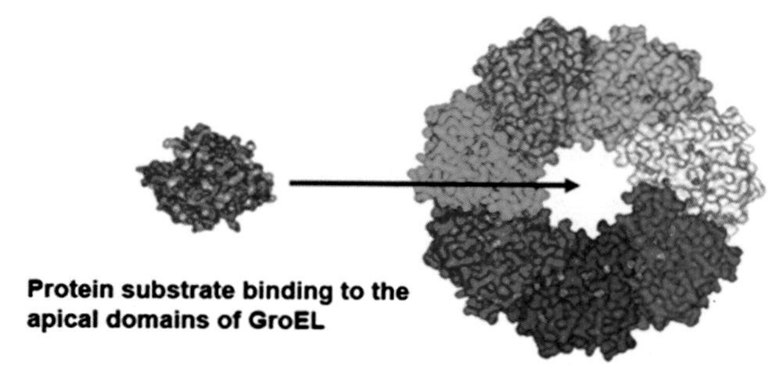

FIGURE 16.7 Protein about to enter the molecular chaperone GroEL. http://blogs.yahoo.co.jp/crazy_tombo/47613981.html?from%3DrelatedCat.

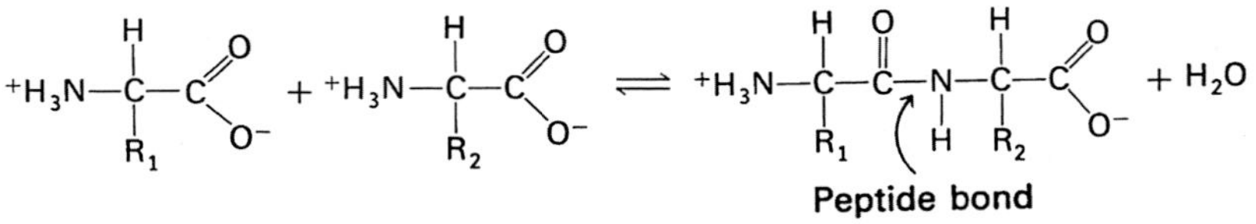

$$^{+}H_3N-\overset{H}{\underset{R_1}{C}}-COO^- + {}^{+}H_3N-\overset{H}{\underset{R_2}{C}}-COO^- \rightleftharpoons {}^{+}H_3N-\overset{H}{\underset{R_1}{C}}-\overset{O}{C}-\underset{H}{N}-\overset{H}{\underset{R_2}{C}}-COO^- + H_2O$$

Peptide bond

FIGURE 16.8 Reversible hydrolysis (right to left) of a peptide bond. http://cmgm.stanford.edu/biochem201/Slides/Protein%20Structure/.

7.2 Proteolytic Cleavage

While many types of PTM are reversible, proteolytic cleavage (proteolysis) is not. Proteolysis involves hydrolytic cleavage of a peptide bond, a thermodynamically favorable reaction that permanently breaks the peptide bond (Fig. 16.8, right to left).

Since peptide bonds are very stable under physiological conditions, they require enzymatic assistance to be cleaved. These enzymes are generally referred to as "proteases". Proteases comprise an enormous family of more than 11,000 different members, suggesting they perform many diverse and essential functions in living organisms [23]. With respect to protein biosynthesis, proteases cleave signal peptides from nascent proteins, degrade misfolded proteins, and convert inactive zymogens (precursors) to active enzymes by cleaving the peptide at specific locations on the protein chain.

A simple example involves the "start" codon, the first codon on the mRNA chain. This codon is usually AUG that in eukaryotes not only initiates translation but also codes for methionine. Therefore the first (N-terminal) amino acid in a nascent chain is initially a methionine that is normally proteolytically cleaved off during PTM.

In addition to the simple cleavage of a single amino acid off the N-terminal, additional cleavages at peptide bonds that are somewhere in the middle of the chain may also occur. An excellent example of this can be found in the peptide hormone insulin, a secretion product. During its synthesis, insulin is cleaved on two separate occasions. Insulin is produced by beta cells, which are pancreatic cells located in the islets of Langerhans. The insulin gene is translated into a primary protein, preproinsulin which is 110 amino acids long. The first 24 amino acids on the N-terminal comprise the signal peptide which binds to an SRP, the complex then migrates to the ER. At the ER, a signal peptide peptidase cleaves off the signal peptide, generating the 84 amino acid proinsulin. Proinsulin has three domains: the alpha-chain (A) domain, the beta-chain (B) domain, and the C-peptide domain. In the ER, endopeptidases (including proprotein convertase 1) cleave the 86 amino acid proinsulin chain into a 51 amino acid active insulin and a 35 amino acid C-chain that is degraded (Fig. 16.9). Active insulin is comprised of one alpha-chain (A) linked to one beta-chain (B) through disulfide bonds.

7.3 Formation of Disulfide Bonds

Once the protein is folded correctly, it must be locked in place to prevent denaturation. This is accomplished through disulfide bridges formed by oxidation of two cysteine side chains that are adjacent to one another in three dimensional space; not sequential in the peptide chain (Fig. 16.10).

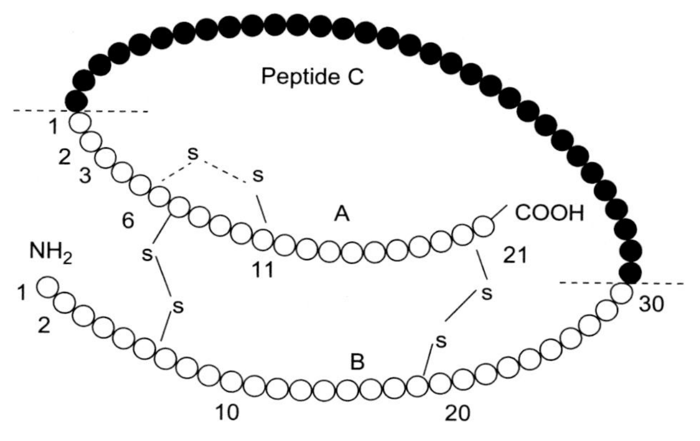

FIGURE 16.9 Cleavage of proinsulin into active insulin (A) one alpha-chain attached to (B) one beta-chain through disulfide bonds) and one C-peptide that is degraded. http://www.pharmacorama.com/en/Sections/Insulin_1.php.

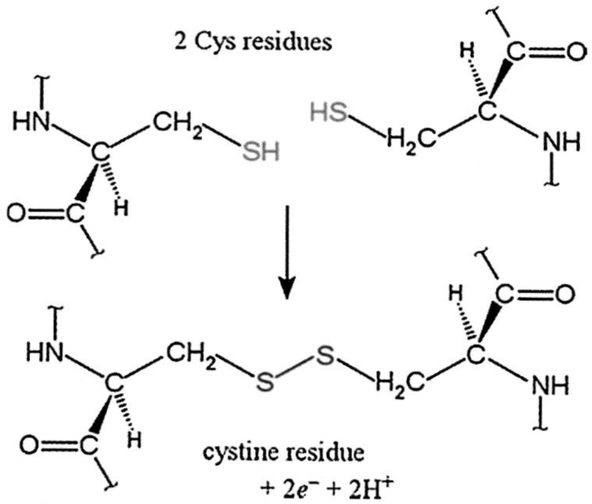

2 Cys residues

cystine residue
$+ 2e^- + 2H^+$

FIGURE 16.10 Disulfide bond formed by the oxidation of two adjacent cysteines to form one reduced cystine. http://guweb2.gonzaga.edu/faculty/cronk/biochem/D-index.cfm?definition=disulfide_bond.

7.4 Lipidation

Many membrane proteins are at least partially anchored to the membrane by long-chain hydrophobic lipids (Chapter 6) [24]. These lipids are covalently attached to the protein and so are examples of PTMs. The major function of lipidation is to increase protein hydrophobicity, leading to enhanced attraction to and insertion into a target membrane. Various combinations of anchoring lipids are employed to help target the protein to the different membranes. There are four major types of membrane protein lipidations: GPI-anchored proteins; myristoylation; palmitoylation; and prenylation. Protein lipidation was introduced in Chapter 6. The different types of lipidations are not mutually exclusive as several lipid types can simultaneously attach to a single protein. Lipidations anchor proteins to membranes only weakly.

7.4.1 Glycosyl-phosphatidylinositol-Anchored Proteins

GPI-anchored protein structure was discussed in Chapter 6 (Fig. 6.13) [14,15], and their biosynthesis was described in Section 6 of this Chapter (Chapter 16). GPI is involved in anchoring some 45 human cell surface proteins to the outer leaflet of the plasma membrane. The other three types of lipidation, myristoylation, palmitoylation, and prenylation, anchor proteins to the membrane inner leaflet. One interesting aspect of GPI-anchored proteins is that they often localize together with cholesterol and sphingolipids in lipid rafts, membrane domains that are important cell signaling platforms (Chapter 8). Being sequestered in lipid rafts greatly reduces the otherwise free diffusion that would be expected and helps keep the GPI-proteins where they belong. The protein−GPI anchor is reversible as the protein can be released by phosphoinositol-specific phospholipase C.

7.4.2 Myristoylated Proteins

Myristic acid is a 14-carbon, unsaturated fatty acid (14:0) that is found in human cellular membranes at much lower levels than its longer chain length, saturated cousins palmitic acid (16:0) and stearic acid (18:0) (Chapter 4). Myristic acid is sufficiently hydrophobic to provide membrane-anchorage for the attached protein [25], but not hydrophobic enough for the binding to be permanent. For this reason, myristoylation is often the lipid handle of choice to attach signal proteins, such as the Src-family kinases, to the plasma membrane.

Myristic acid is attached to the protein's N-terminal glycine via an amide linkage (Fig. 6.10). The reaction is catalyzed by N-myristoyl-transferase which uses myristoyl-CoA as the substrate. As with all newly synthesized eukaryotic proteins, the initial N-terminal amino acid is methionine that must be clipped off by PTM before a myristoyl is attached (a second PTM). Subsequent PTM events may also occur on this single protein. An example of an N-myrisoylated protein that has previously been discussed is rhodopsin, a transmembrane protein that features seven α-helices (Chapter 6).

7.4.3 Palmitoylated Proteins

Palmitic acid (16:0) is covalently added to a membrane protein through a thiolate side chain of cysteine residues (Fig. 6.11) [26]. The reaction is catalyzed by palmitoyl acyl transferases which uses palmitoyl-CoA as the substrate and generates a reversible thio-ester. Therefore, the protein is S-palmitoylated. Because palmitic acid is two carbons longer than myristic acid, it is more hydrophobic and thus its anchor to the membrane is considerably stronger and, without enzymatic intervention, is almost permanent. S-palmitoylation is often used to strengthen other types of lipidations, such as myristoylation or prenylation, and selectively concentrates in lipid rafts. The thio-ester attachment is readily cleaved by thioesterases. Palmitoylation reversibility is only caused by acyl-transferases and thio-esterases. This stands in sharp contrast to myristoylation and prenylation that are weakly bound but not hydrolysable.

7.4.4 Prenylated Proteins

The last major type of lipid anchor is through prenylated lipids (Chapter 6) [27,28]. Prenylated lipids are repeats of the 5-carbon, branched molecule isoprene (3-methyl-2-buten-1-yl) (Chapter 4 and Fig. 6.12). Three isoprenes (15 carbons) are referred to as a farnesyl chain whereas 4 isoprenes (20 carbons) comprise a geranylgeranyl chain. The prenyl chains are covalently added to specific cysteines located within five amino acids of the protein's C-terminal. Unlike S-palmitoylation, S-prenylation is through a non-hydrolyzible thioether and so is stable. Prenylation is a common protein PTM, affecting ~2% of all proteins. Although the major function of prenylation is to help anchor a protein to the membrane, prenylations are also important in promoting specific protein to protein binding and in steering the modified protein to specific membrane domains that are often involved in cell signaling.

7.5 Ubiquitination

Misfolded proteins present serious problems for the cell and must be dealt with promptly. This is accomplished by ubiquitin, a small 8 kDa, 76 amino acid protein (Fig. 16.11) that attaches to an ε-amine of lysine on the target, misfolded protein [29–31]. An isopeptide (amide) bond connects the C-terminal glycine of ubiquitin to the ε-amine of lysine on the target protein. Following the initial monoubiquitin event, additional ubiquitins are added as long chains, creating a polyubiquitinated protein that is recognized by the 26S proteasome for degradation of the ubiquitinated protein and recycling of ubiquitin. Ubiquitin was discovered by Gideon Goldstein et al. in 1975 [32] and since that time functions in addition to protein degradation have been ascribed. Also, ubiquitin has seven lysines that can be involved in putting together the polyubiquitin chains. The lysine attachment site for polyubiquitination induces different functions. For example, lysine 48-linked polyubiquitin chains signal the protein to be destroyed by the proteasome while lysine 63-linked chains regulate processes such as endocytic trafficking, inflammation, translation, and DNA repair (Fig. 16.11).

7.6 Amino Acid Modification: Phosphate and Friends

At first glance it seems that almost every loose, small molecule Nature could find was given some role in PTM. In addition, many amino acids with chemically reactive side chains (Chapter 6) can be posttranslationally modified as well. Even small additions and

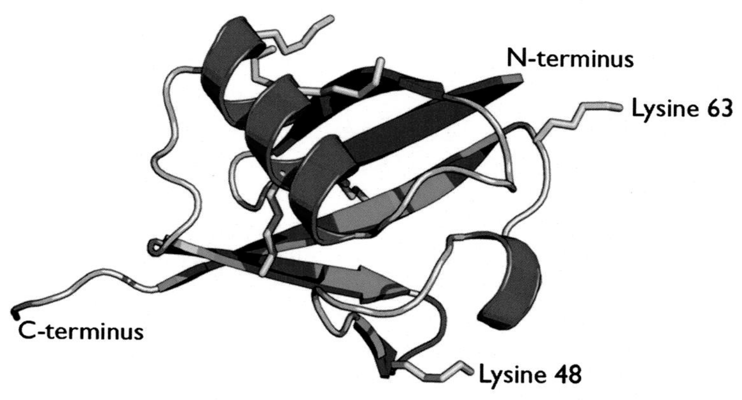

FIGURE 16.11 Structure of ubiquitin showing the N-terminus, C-terminus and the lysine 48 and lysine 63 binding sites for attachment of polyubiquitins. http://en.wikipedia.org/wiki/Ubiquitin.

TABLE 16.1 Posttranslational modification of proteins

1. Small chemicals added to the protein posttranslationally:
   a. Alkylation: Addition of methyl or ethyl at lysine or arginine
   b. Amide Link: To C- or N-terminal. Arginylation, polyglutamylation
   c. Polyglycylation: 1–40 glycine to tubulin C-terminal
   d. Butyrylation
   e. Gamma-carboxylation
   f. Malonylation
   g. Iodination
   h. Nucleotide addition
   i. Adenylation: O-linked to tyrosine, N-linked to histidine or lysine
   j. Propionylation
   k. S-glutathionylation
   l. S-nitrosylation
   m. Succinylation (to lysine)
   n. Sulfation (to tyrosine)
   o. Sumoylation
   p. Glycation: Addition of a single sugar without use of an enzyme
   q. Biotinylation
   r. Pegylation
   s. Addition of Peptides
   t. Nitrosylation: Add nitric oxide to a cysteine.
2. Posttranslational chemical modification of amino acids
   a. Deimination: Convert arginine to citrulline
   b. Deamination: Convert glutamine to glutamate and asparigine to aspartate
   c. Carbamylation: Convert lysine to homocitrulline
   d. Elimination: beta-elimination of phosphothreonine and phosphoserine
   e. Dehydration: threonine and serine
   f. Decarboxylation of cysteine
   g. Hydroxylation: Conversion of proline to 4-hydroxyproline
   h. Oxidation: Conversion of lysine to 5-hydroxylysine
   i. Carboxylation: Oxidation of glutamate to g-carboxyglutamate

modifications can profoundly affect protein structure, altering enzyme activity and cell function. Table 16.1 lists a few examples of post translational modification of amino acids comprising the target protein.

7.6.1 Phosphorylation

One of the most important and best studied examples of PTM involves reversible phosphorylation/dephosphorylation of enzymes [33]. Phosphorylation plays critical roles in regulating the activity of pivotal enzymes in many cellular processes including cell cycling, growth, apoptosis, and signal transduction pathways (Chapter 18). Phosphorylation occurs primarily on the side-chain alcohol of serine, threonine or tyrosine (O-linked), but occasionally on the imidazole nitrogen of histidine (N-linked) [34]. Phosphorylation/dephosphorylation is catalyzed by kinases and phosphatases, respectively (Fig. 16.12).

In most, but not all, cases, the phosphorylated form of the enzyme is the active one. However, there are examples where the dephosphorylated form is active. Since phosphorylation/dephosphorylation interconversions are under enzyme control, the process is rapid and generates essential amplification cascades (Chapter 18).

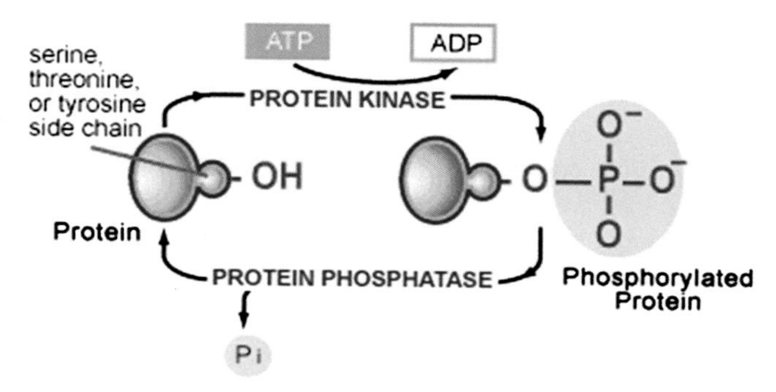

FIGURE 16.12 The phosphorylation/dephosphorylation cycle for protein regulation. http://www.scq.ubc.ca/protein-phosphorylation-a-global-regulator-of-cellular-activity/.

7.7 Addition of Cofactors

Often an enzyme requires an additional nonprotein component to function. This component is generally referred to as a cofactor and must be incorporated into the enzyme posttranslationally. The complete enzyme (holoenzyme) is, therefore, composed of the protein (apoenzyme) and the cofactor:

$$\text{Holoenzyme} = \text{Apoenzyme} + \text{Cofactor}.$$

Cofactors come in a wide variety of structures that participate in the particular catalytic function associated with the enzyme [35]. Some cofactors are loosely bound to the enzyme. If organic they are called coenzymes. Tightly bound cofactors are referred to as prosthetic groups, an example being heme. One group of cofactors have adenosine monophosphate (AMP) included in their structure (eg, adenosine triphosphate, coenzyme A, FMN, FAD, nicotinamide adenine dinucleotide), while others are also small organic molecules, but do not contain AMP (eg, lipoate and retinaldehyde). Inorganic cofactors including Mg^{2+}, Cu^{2+}, Mg^{2+}, Mn^{2+}, and iron–sulfur complexes, also participate in some reactions. Many of these cofactors are very well known to anyone with even a basic knowledge of biochemistry, cell biology, or cell physiology. Complex enzymes often employ multiple cofactors in their mechanism of action.

7.8 Glycosylation

Carbohydrates hold a special place in PTM. They significantly affect protein folding, conformation, charge, distribution, trafficking, stability, and activity. Sugars are added to proteins cotranslationally or posttranslationally via five types of linkages: N-linked to asparigine or arginine; O-linked to serine, tyrosine, threonine, hydroxylysine, or hydroxyproline; phospho-glycans linked through the phosphate of a phosphoserine; C-linked glycans; and glypiation where the GPI anchor is linked to a protein [36–38].

Glycosylation is so important in PTM because of the enormous amount of potential diversity involved in carbohydrate structures. This was discussed in Chapter 7 in terms of the information-carrying potential of sugars (the "carbohydrate code") which by far, exceeds nucleic acids and proteins. Carbohydrate diversity resides in the: types of sugars that comprise the glycan; sites of glycosidic bonds linking the various sugars together; glycan structure being unbranched or branched at every sugar; and length of the oligosaccharide chain. Glycosylation is the most complex of the PTM processes because of the large number of enzymatic steps that are involved. N-linked glycosylations occur in the ER lumen and are the most common of the linkage types (Chapter 7). O-linked glycosylations occur in the Golgi and are also common. The only other type of glycosylation that is common involves the GPI-linked proteins discussed in Section 6 of this chapter (Chapter 16) and in Chapter 6. C-linked sugars are extremely rare and are basically considered to be curiosities. A completely different type of sugar linkage to proteins is glycation which is believed to be nonenzymatic. It is evident that sugars linked to membrane proteins exhibit extraordinary diversity.

7.9 Dolichol

Another unusual, but essential, aspect of glycosylation in PTM involves the very hydrophobic molecule dolichol [39]. The structure of dolichol is depicted in Fig. 16.13. This rendition of dolichol was chosen to emphasize how long the hydrophobic component of this molecule actually is.

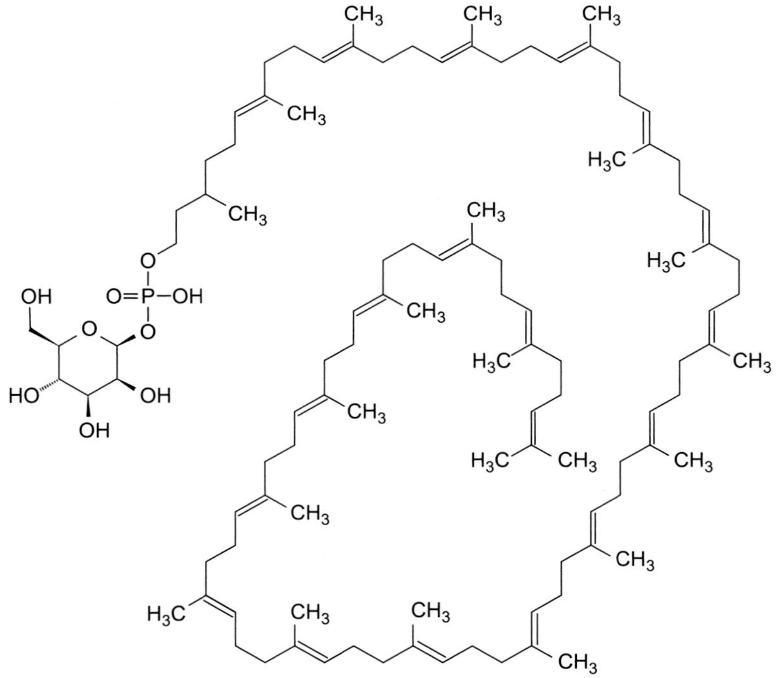

FIGURE 16.13 Structure of mono-glycosylated dolichol phosphate. http://en.wikipedia.org/wiki/Dolichol_monophosphate_mannose.

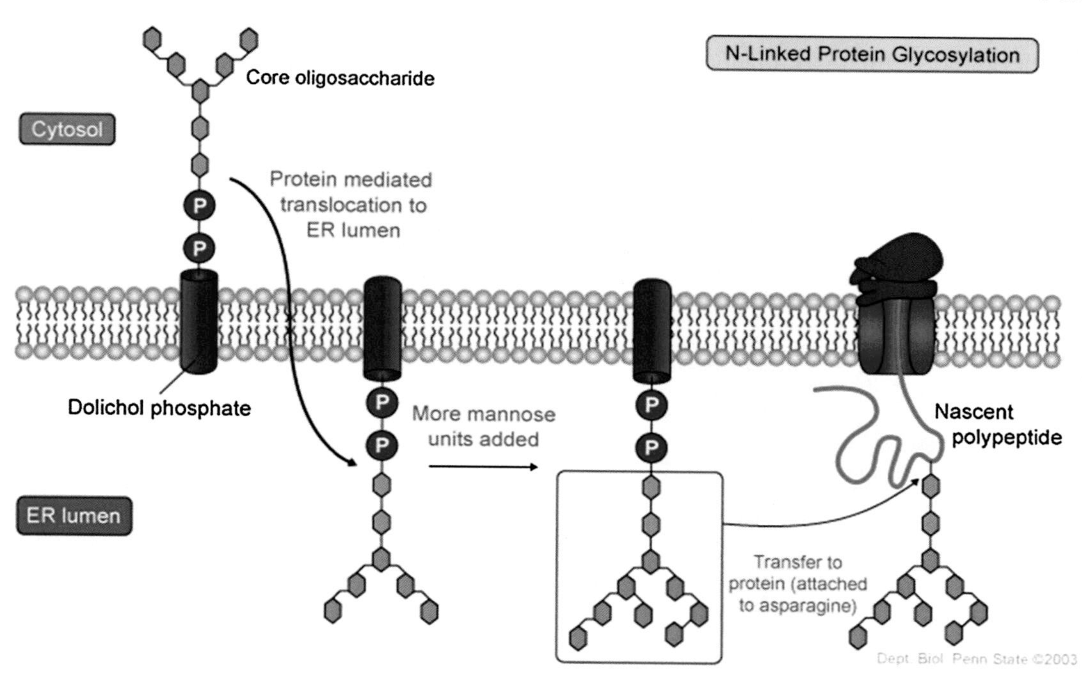

FIGURE 16.14 Involvement of dolichol phosphate in the transfer of oligosaccharides posttranslationally to an endoplasmic reticulum (ER) protein. https://wikispaces.psu.edu/pages/viewpage.action?pageId=112526377&navigatingVersions=true.

Dolichols are a group of very long hydrophobic chains that are composed of varying numbers of isoprene units terminating in a polar alcohol ($-CH_2-OH$) group. When its terminal alcohol is phosphorylated, dolichol can become glycosylated where it is involved in the cotranslational modification of proteins through N-glycosylation. Dolicol functions as a membrane-anchor that is the location on which the oligosaccharides are constructed. The initial oligosaccharide has 3 glucose, 9 mannose and 2 *N*-acetylglucosamine units. The individual sugar structures are discussed in Chapter 7. This oligosaccharide is transferred cotranslationally from dolicol onto certain asparagines on the nascent polypeptide. Also, complex, branched oligosaccharides attached to dolichol can be posttranslationally transferred to a protein in the ER (Fig. 16.14). Dolichol can also be used to attach single sugars as well.

8. SUMMARY

Since proteins are responsible for all biochemical activity, their biosynthesis and proper orientation into the membrane is essential. The vast array of membrane protein structures would imply that their biosynthesis might be hopelessly complex. Fortunately, all membrane protein biosynthesis is built around a central theme known as the "Signal Hypothesis".

Newly synthesized (nascent) proteins have intrinsic signals that govern their biosynthesis, transport, and ultimate localization in the cell. As a result, the biosynthesis of secretory proteins, single-span proteins ($N_{lumen}/C_{cytosol}$), single-span proteins ($N_{cytosol}/C_{lumen}$), multispan proteins, and GPI-linked proteins have a lot in common. Once initial translation of the protein in the ER is complete, there remains many additional modifications to the protein that assist in translocation through the endomembrane system before becoming fully functional. These steps are referred to as PTMs (post transational modifications) and come in a bewildering array of formulations. PTMs include protein folding by molecular chaperones. cleavage, phosphorylation, lipidation, addition of cofactors, and glycosylation.

Chapter 17 will discuss how various components are moved through the cell by membrane trafficking.

References

[1] Blobel G, Dobberstein B. Transfer of proteins across membranes. I. Presence of proteolytically processed and unprocessed nascent immunoglobulin light chains on membrane-bound ribosomes of murine myeloma. J Cell Biol 1975;67(3):835–51.
[2] Lodish H, Berk A, Zipursky SL, Matsudaira P, Baltimore D, Darnell J. Section 17.3. Overview of the secretory pathway. Molecular cell biology. 4th ed. New York: W.H. Freeman; 2000.
[3] Rapoport T. Protein translocation across the eukaryotic endoplasmic reticulum and bacterial plasma membranes. Nature 2007;450(7170):663–9.
[4] Do H, Falcone D, Lin J, Andrews DW, Johnson AE. The cotranslational integration of membrane proteins into the phospholipid bilayer is a multistep process. Cell 1996;85(3):369–78.
[5] Mothes W, Heinrich SU, Graf R, Nilsson I, von Heijne G, Brunner J, et al. Molecular mechanism of membrane protein integration into the endoplasmic reticulum. Cell 1997;89(4):523–33.
[6] Spiess M, Lodish HF. An internal signal sequence: the asialoglycoprotein receptor membrane anchor. Cell 1986;44(1):177–85.
[7] Shaw AS, Rottier PJ, Rose JK. Evidence for the loop model of signal-sequence insertion into the endoplasmic reticulum. Proc Natl Acad Sci USA 1988;85(20):7592–6.
[8] Sakmar TP. Twenty years of the magnificent seven. The Scientist 2005;19:22–5.
[9] Von Heijne G. Membrane-protein topology. Nat Rev Mol Cell Biol 2006;7:909–18.
[10] Von Heijne G. Membrane protein structure prediction hydrophobicity analysis and the positive-inside rule..I. Mol Biol 1992;225:487–94.
[11] van Geest M, Lolkema JS. Membrane topology and insertion of membrane proteins: search for topogenic signals (Review) Microbiol Mol Biol Rev 2000;64(1):13–33.
[12] Wessels HP, Spiess M. Insertion of a multispanning membrane protein occurs sequentially and requires only one signal sequence. Cell 1988;55(1):61–70.
[13] Mueckler M, Caruso C, Baldwin SA, Panico M, Blench I, Morris HR, et al. Sequence and structure of a human glucose transporter. Science 1985;229(4717):941–5.
[14] Ferguson MAJ, Williams AF. Cell-surface anchoring of proteins *via* glycosyl-phosphatidylinositol structures. Annu Rev Biochem 1988;57:285–320.
[15] Ikezawa H. Glycosylphosphatidylinositol (GPI)-Anchored proteins. Biol Pharm Bull 2002;25:409.
[16] Prabakaran S, Lippens G, Steen H, Gunawardena J. Post-translational modification: nature's escape from genetic imprisonment and the basis for dynamic information encoding. Wiley interdisciplinary reviews Syst Biol Med 2012;4(6):565–83.
[17] Walsh C. Posttranslational modification of proteins : expanding nature's inventory, xxi. Englewood, Colo: Roberts and Co. Publishers; 2006. p. 490.
[18] Han KK, Martinage A. Post-translational chemical modification(s) of proteins. Int J Biochem 1992;24(1):19–28.
[19] Ellis RJ, van der Vies SM. Molecular chaperones. Annu Rev Biochem 1991;60:321–47.
[20] Hartl FU, Bracher A, Hayer-Hartl M. Molecular chaperones in protein folding and proteostasis. Nature 2011;475:324–32.

[21] Kim YE, Hipp MS, Bracher A, Hayer-Hartl M, Hartl FU. Molecular chaperone functions in protein folding and proteostasis. Annu Rev Biochem 2013;82:323–55.
[22] Laskey RA, Honda BM, Mills AD, Finch JT. Nucleosomes are assembled by an acidic protein that binds histones and transfers them to DNA. Nature 1978;275(5679):416–20.
[23] Barrett AJ, Rawlings ND, Salvesen GD, editors. Handbook of proteolytic enzymes. 3rd ed. New York: Academic Press; 2013. p. 4094.
[24] Nadolski MJ, Linde ME. Protein lipidation. FEBS J 2007;274(20):5202–10.
[25] Farazi TA, Waksman G, Gordon JI. The biology and enzymology of protein N-myristoylation. J Biol Chem 2001;276:39501–4.
[26] Bijlmakers M-J, Marsh M. The on–off story of protein palmitoylation. Trends Cell Biol 2003;13:32–42.
[27] Gelb MH, Scholten JD, Sebolt-Leopold JS. Protein prenylation: from discovery to prospects for cancer treatment. Curr Opin Chem Biol 1998;2:40–8.
[28] Zhang FL, Casey PJ. Protein prenylation: molecular mechanisms and functional consequences. Annu Rev Biochem 1996;65:241–69.
[29] Hirsch C, Gauss R, Horn SC, Neuber O, Sommer T. The ubiquitylation machinery of the endoplasmic reticulum. Nature 2009;458:453–60.
[30] Cesari F. Protein degradation: catching ubiquitin. Nat Rev Mol Cell Biol 2008;9:498–9.
[31] Kerscher O, Felberbaum R, Hochstrasser M. Modification of proteins by ubiquitin and ubiquitin-like proteins. Annu Rev Cell Dev Biol 2006;22:159–80.
[32] Goldstein G, Scheid M, Hammerling U, Schlesinger DH, Niall HD, Boyse EA. Isolation of a polypeptide that has lymphocyte-differentiating properties and is probably represented universally in living cells. Proc Natl Acad Sci USA 1975;72(1):11–5.
[33] Krebs EG, Beavo JA. Phosphorylation-dephosphorylation of enzymes. Annu Rev Biochem 1979;48:923–59.
[34] Ubersax JA, Ferrell Jr JE. Mechanisms of specificity in protein phosphorylation. Nat Rev Mol Cell Biol 2007;8:530–41.
[35] Sauke DJ, Metzler DE, Metzler CM. Biochemistry: the chemical reactions of living cells. 2nd ed. San Diego: Harcourt/Academic Press; 2001.
[36] Kornfeld R, Kornfeld S. Comparative aspects of glycoprotein structure. Annu Rev Biochem 1976;45:217–38.
[37] Gamblin DP, Scanlan EM, Davis BG. Glycoprotein synthesis an update. Chem Rev 2009;109:131–63.
[38] Wood H. Hunting glycoproteins. Nat Rev Neurosci 2005;6:94.
[39] Kean EL. Studies on the activation by dolichol-P-mannose of the biosynthesis of GlcNAc-P-P-dolichol and the topography of the GlcNAc-transferases concerned with the synthesis of GlcNAc-P-P-dolichol and $(GlcNAc)_2$-P-P-dolichol: a review. Biochem Cell Biol 1992;70(6):413–21.

CHAPTER

17

Moving Components Through the Cell: Membrane Trafficking

OUTLINE

1. INTRODUCTION

If one were able to sit inside a cell and look around, one would be witness to a bewildering array of every type of life component (proteins, lipids, nucleic acids, membrane vesicles, etc.) zipping around in all directions, heading to specific cellular locations. It would first appear that everything is in total chaos, but on closer examination, well-designed patterns would emerge [1,2]. Movement of membrane-derived vesicles, referred to as "membrane trafficking," is of particular interest as it forms the heart of the endo-membrane concept. Three major patterns will be discussed here: endo-membrane flow (for secretion), receptor-mediated endocytosis (RME), and intracellular carrier proteins.

An Introduction to Biological Membranes
http://dx.doi.org/10.1016/B978-0-444-63772-7.00017-8

2. ENDO-MEMBRANE FLOW (SECRETION)

As discussed in Chapter 16, there are two basic types of protein synthesis: one made on cytosolic ribosomes for water-soluble cytosolic proteins and the other made on identical ribosomes attached to the outer (cytosolic) surface of the endoplasmic reticulum (ER) for generating transmembrane proteins, proteins to be delivered to the inside of cellular organelles and for proteins to be secreted from the cell. The process of movement through the cell from the nucleus to the plasma membrane with stops at different intracellular organelles is known as "endo-membrane flow" (Fig. 17.1). The first proposal that intracellular membranes form a single interconnected system (the endo-membrane concept) was by Morre and Mollenhauer in 1974 [3]. Here, steps will be followed for secretion of proteins initially synthesized in the lumen of the rough ER (RER) [4].

Steps involved in protein synthesis on an ER-bound ribosome are discussed in Chapter 16. Once inside the RER lumen, the newly synthesized protein begins its modification process [5]. The first step involves chaperone proteins that help fold the newly synthesized protein. If the initial folding is incorrect, a second attempt is made. If this too fails, the misfolded protein is

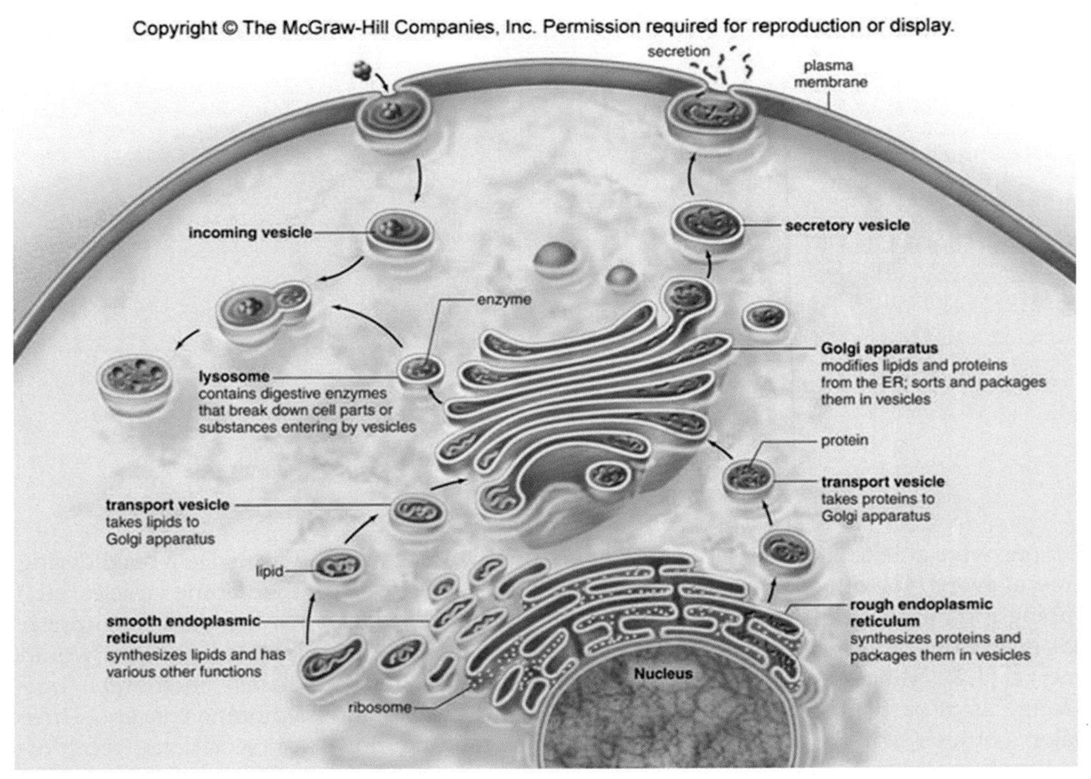

FIGURE 17.1 Endo-membrane flow from the nucleus to the plasma membrane. *Copyright McGraw Hill. http://www.studyblue.com/notes/note/n/test-2-ch-12-/deck/5782652.*

exported to the cytosol and labeled for destruction (see ubiquitin in Chapter 16). For the correctly folded protein, carbohydrates are then added to convert the new protein into a glycoprotein. The sugars are required to help further correctly fold the protein, targeting the glycoprotein to a final destination and for proper functioning of the protein once secreted. The modified proteins are encapsulated into transport vesicles that bud off from the ER. These vesicles are covered by a coating protein COPII (coat protein complex II) and are moved to the *cis* face of the Golgi (the Golgi is discussed in detail later in Chapter 17, Section 4). In passing through the Golgi, the protein is further modified before being dispatched from the *trans* face of the Golgi in a different type of uncoated transport vesicle. These vesicles are carried by the cytoskeleton to the plasma membrane for fusion and release of its contents to the extra-cellular solution (secretion). The transport (secretory) vesicles have surface components that recognize, and bind to receptors on the cytoplasmic side of the plasma membrane [6].

The process of secreting a protein follows the same basic steps that cells use to excrete waste and other large molecules from the cytoplasm to the cell exterior [7] and therefore is the opposite of endocytosis (discussed in Section 3). Fusion of the transport vesicles to the plasma membrane not only releases their aqueous sequestered contents to the outside but at the same time also adds vesicular membrane hydrophobic components (mostly lipids and proteins) to the plasma membrane. The process is depicted in Fig. 17.2 [8]. Steady state composition of the plasma membrane results from a balance between endocytosis and

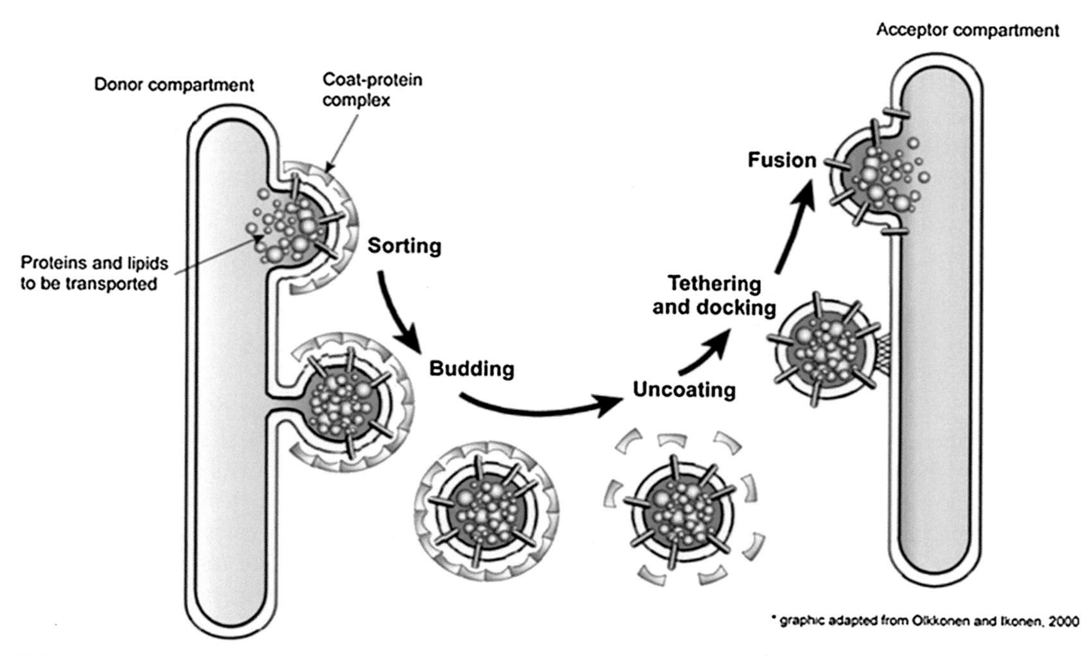

FIGURE 17.2 Mechanism of formation of transport (secretory) vesicles, their fusion to the plasma membrane, and release of sequestered material. *From The Rye Laboratory, Department of Biochemistry and Biophysics. Texas A&M University. http://ryeserv1.tamu.edu/secretory-vesicle-formation.html.*

exocytosis. The resultant process of plasma membrane recycling is amazingly fast. For example, pancreatic secretory cells recycle an amount of membrane equal to the whole surface of the cell in about 90 min. Even faster are macrophages that can recycle contents of the plasma membrane in only 30 min.

The transport vesicles must first dock with the plasma membrane, a process that keeps the two membranes separated by less than 5–10 nm. During docking, complex molecular rearrangements occur to prepare the membranes for fusion where separation becomes less than 1.5 nm. The process of vesicle fusion and release of aqueous compartment components is driven by SNARE proteins (see Chapter 10) [9]. Joining the vesicle SNARE (v-SNARE) to the plasma membrane SNARE (t-SNARE) is checked and locked by the regulatory GTP-binding protein Rab. Therefore, the process of exocytosis results in:

- The surface of the plasma membrane increasing by the size of the fused vesicular membrane. This is particularly important if the cell is growing.
- The material sequestered inside the vesicle is released to the cell exterior. Included in the vesicle contents may be waste products, intracellular toxins, and signaling molecules like hormones and neurotransmitters.
- Proteins imbedded in the vesicular membrane become part of the plasma membrane. This makes correct protein orientation in the vesicular membrane absolutely essential. The side of the protein facing the inside of the vesicle before fusion faces the outside of the plasma membrane after fusion.
- Some variations in the standard vesicle fusion process exist. Presynaptic vesicles carrying neurotransmitters do not fuse immediately but instead must await a fusion initiation signal (Chapter 18). A special type of exocytosis called "kiss-and-run" occurs in synapses [10]. The vesicles only make very brief contact with the plasma membrane whereupon they release their contents (neurotransmitters) to the outside (synaptic gap) and immediately return empty to the neuron cytoplasm. Since true fusion does not occur, the vesicle membrane is not incorporated into the plasma membrane.

For their work on the structure and function of cellular vesicles, James Rothman, Randy Schekman, and Thomas Sudhof shared the 2013 Nobel Prize in Physiology or Medicine.

3. RECEPTOR-MEDIATED ENDOCYTOSIS

RME [11,12] is also known as clathrin-dependent endocytosis because of involvement of the membrane-associated protein clathrin in forming membrane vesicles that become internalized into the cell. Clathrin plays a major role in formation of clathrin-coated pits and coated vesicles. Since clathrin was first isolated and named by Barbara Pearse in 1975 [13], it has become clear that clathrin and other coat proteins play essential roles in cell biology. Clathrin is an essential component in building small vesicles for uptake (endocytosis) and export (exocytosis) of many molecules. While the methods of membrane transport, discussed in Chapter 19, involved small solutes, RME is the primary mechanism for the specific internalization of most macromolecules by eukaryotic cells.

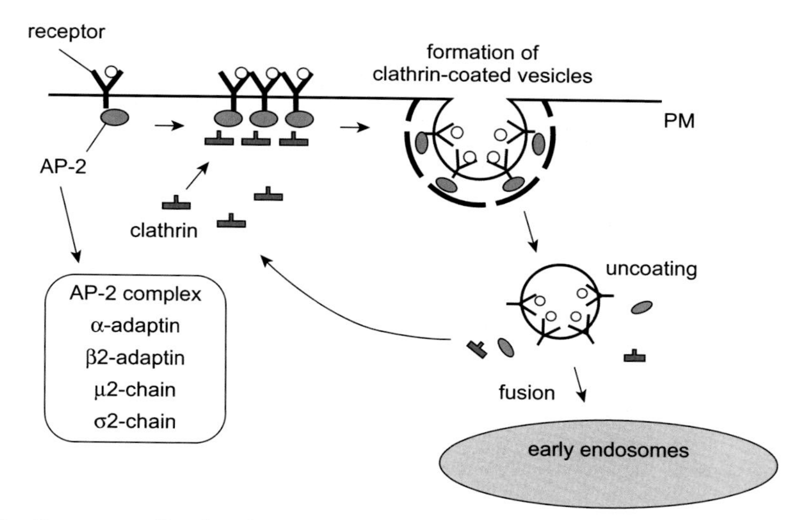

FIGURE 17.3 Receptor mediated endocytosis (also known as clathrin-dependent endocytosis). *Reference Grant BD, Sato M. http://www.wormbook.org/chapters/www_intracellulartrafficking/intracellulartrafficking.html; 2006.*

RME begins with an external ligand binding to a specific receptor that spans the plasma membrane (Fig. 17.3, [14,15]). Examples of these ligands include hormones, growth factors, enzymes, serum proteins, low-density lipoprotein (LDL) (with attached cholesterol), transferrin (with attached iron), antibodies, some viruses, and even bacterial toxins. After receptor binding, the complex diffuses laterally in the plasma membrane until it encounters a specialized patch of membrane called a coated pit. The receptor–ligand complexes accumulate in these patches as do other proteins including clathrin, adaptor protein, and dynamin. Since coated pits occupy about 20% of the plasma membrane surface area, they are not minor membrane features. The collection of these proteins starts to curve the adjacent section of the membrane that eventually pinches off to form an internalized coated vesicle. Clathrin and dynamin then recycle back to the plasma membrane, leaving an uncoated vesicle that is free to fuse with an early endosome. After the early endosomes mature into late endosomes, they then go to the lysosome for digestion. RME is a very fast process. Invagination and vesicle formation take about 1 min. One single cultured fibloblast cell can produce 2500 coated pits per minute.

One example of RME has received a great deal of attention because of its essential role in human health, namely maintaining the proper level of cholesterol in the body. Malfunctions in the RME process for uptake of cholesterol-carrying LDL (see Chapter 14) leads to hypercholesterolemia and cardiovascular disease [11,16]. RME and its role in cholesterol metabolism was discovered by Michael Brown and Joseph Goldstein of The University of Texas Health Science Center in Dallas (now the UT Southwestern Medical Center), who received the 1985 Nobel Prize in Physiology and Medicine for their iconic work.

4. THE GOLGI

Endo-membrane flow, RME, and many other essential cell functions use the Golgi as a central component [17]. The Golgi is often referred to as the "Post Office of the Cell" because its major functions involve packaging and labeling items to be shipped to various destinations around the cell. The Golgi is an intracellular organelle whose large size facilitated its early discovery and morphological description by Italian physiologist Camillo Golgi (Fig. 17.4) in 1897 (see also Chapter 18). After Golgi announced his discovery at a meeting of the Medical Society of Pavia on April 19, 1898, the organelle was named after him. To this day, the Golgi is the only organelle named after a scientist. Discovery of the organelle was facilitated by Golgi's development of silver as a stain for nerve cells. For his work, Camillo Golgi was awarded the 1906 Nobel Prize in Physiology or Medicine, one of the earliest Nobel Prizes. (In 1901, the first Nobel Prize in Physiology or Medicine went to the German physiologist Emil Adolf van Behring for his work on serum therapy and development of a vaccine against diphtheria.) Golgi's Italian birth town is now named Corteno Golgi in his honor.

The number and size of Golgi vary from cell to cell. Golgi tend to be larger and more numerous in cells that have a large secretion function. Morphologically, the Golgi appears to be composed of membrane-bound stacked structures called cisternae [18]. A "typical" Golgi is depicted in Fig. 17.5. A single mammalian cell usually contains between 40 and 100 Golgi stacks, each of which is composed of 4 to 8 cisternae. Each cisternae is a flat, membrane-enclosed disc that houses the enzymes that conduct the Golgi's business. The central spaces of the stacked cisternae are contiguous, allowing for processing of cargo as it passes through the Golgi [17]. Each cisternae is composed of four functional regions or networks referred to as the *cis*-Golgi, medial-Golgi, endo-Golgi, and *trans*-Golgi. Transport vesicles originating from the ER fuse with the *cis*-Golgi, releasing their cargo into the lumen of the cisternae. The cargo is modified as it passes through the Golgi, eventually reaching the *trans*-Golgi, where it is sorted, packaged, and sent to its final destination. Each region of the Golgi contains characteristic enzymes that selectively modify the cargo.

FIGURE 17.4 Camillo Golgi, 1843–1926. *http://www.vetmed.vt.edu/education/curriculum/vm8304/lab_companion/histo-path/vm8054/labs/Lab3/Notes/golginot.htm.*

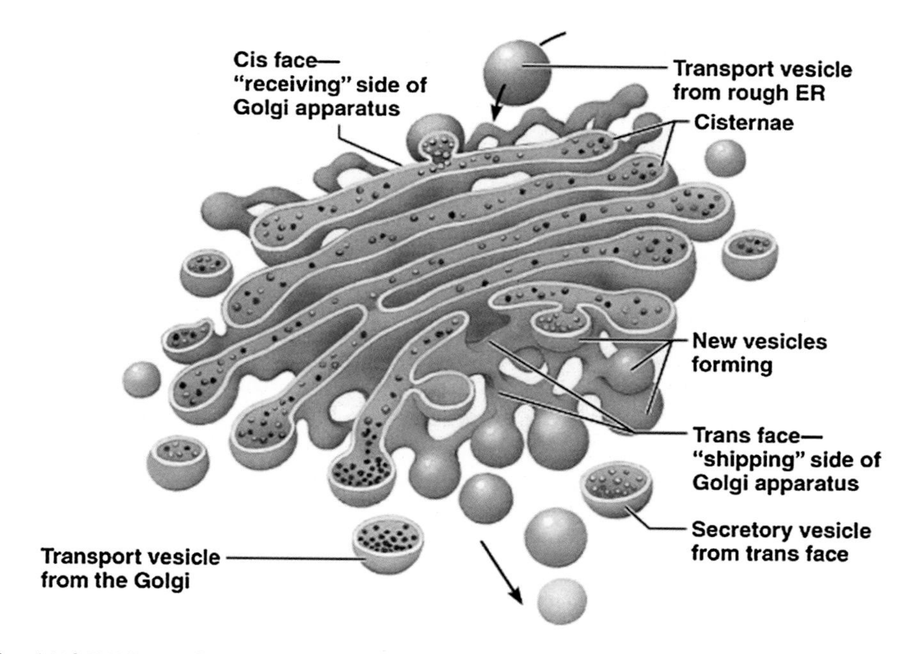

FIGURE 17.5 Diagram of a "typical" Golgi. *http://www.glogster.com/johnv4297/john-v-s-glog/g-6ltb1bblffg5cerho4nuna0.*

As the cargo proteins pass through the connected Golgi lumen, many modifications are made to ready the proteins for sorting, packaging, and eventual shipment to their appropriate destinations. Most modifications involve glycosylations, primarily to existing sugar chains, although phosphorylations and sulfations are also common. Some of the newly attached sugars may function as a type of signal, directing the protein to the appropriate transport vesicle forming in the *trans*-Golgi. One modification that has been known for decades involves attaching mannose-6-phosphate, targeting the protein to the lysosome. Sulfation is generally performed by the enzyme sulfotransferase in the *trans*-Golgi. Added sulfates give the cargo protein a net negative charge that helps in sorting. Cargo proteins destined for secretion or a particular organelle, depart the *trans*-Golgi in destination-specific transport vesicles and deposit their cargo at the target membrane via fusion.

5. INTRACELLULAR LIPID TRANSPORT

While the trafficking of membrane proteins through a cell can usually be followed precisely, the trafficking of membrane lipids is far more troublesome. To begin with for every protein there are 50 or more lipids found in its immediate environment, and these lipids are composed of many different molecular species (see Chapter 10). So, how are membrane lipids transported throughout a cell? While details remain elusive, several overlapping mechanisms are possible. Two of these mechanisms, serum albumin and lipoproteins, were discussed in Chapter 14. Other mechanisms include:

1. Vesicle transport
2. Lipid transfer proteins (LTPs)
3. Lipid lateral diffusion through membranes
4. Free diffusion through the cytosol
5. Membrane to membrane contact
6. Lipid flip-flop

5.1 Lipid Vesicles

A living cell is chock-full of lipid vesicles. In fact, there are so many vesicles that for many years cell biologists thought the vesicles were merely artifacts of electron microscopy procedures. A prime example is the secretory vesicles discussed earlier that are involved in membrane trafficking, endo-membrane flow, secretion, and RME. When one of these vesicles fuses with a cell membrane, an enormous number of vesicular membrane lipids are mixed with the cell membrane lipids, increasing the membrane size and altering the membrane lipid composition. This mechanism of lipid transfer likely moves mere lipids than does the other mechanisms. However, vesicle fusion does not have the ability to accurately control the type and amount of lipid transferred.

5.2 Lipid Transfer Proteins

LTPs are essential for the movement of lipids both intercellularly and intracellularly and help develop and maintain lipid compositions characteristic of organelles and membrane domains. LTPs were discovered by D.B. Zilversmidt in 1968 as soluble factors that accelerated the transfer of lipids between membranes [19]. We have already encountered an early example of an LTP called a phospholipid exchange protein (PLEP, see Chapter 9). PLEPs have been extensively used to measure transmembrane lipid asymmetry and flip-flop. Since their initial discovery, LTPs have been found in all eukaryotes and bacteria. They come in a wide variety of types, accounting for their ability to transfer many structurally diverse membrane lipids [20,21].

In general, LTPs mediate monomeric lipid exchange in which a single lipid molecule is transported through the aqueous phase sequestered in a hydrophobic pocket of the LTP. Most LTPs are low molecular weight structures that are dominated by β-sheet motifs (Fig. 17.6). A "lid" covers the opening to the hydrophobic pocket and acts as the gate for lipid uptake and release. Thus, LTPs exist in two distinct conformations, a closed conformation that transports the lipid through the aqueous phase and an open conformation where the LTP picks up its lipid cargo at the outer leaflet of a donor membrane and releases it to the outer leaflet of an acceptor membrane.

Sterol carrier proteins (also known as nonspecific lipid transfer proteins [NSLTPs]) are a family of LTPs that can transfer a variety of lipids from one intracellular membrane to another [22,23]. They are referred to as "nonspecific" since they have been reported to carry sterols, glycolipids, all common phospholipids, and gangliosides. The sterol carrier proteins

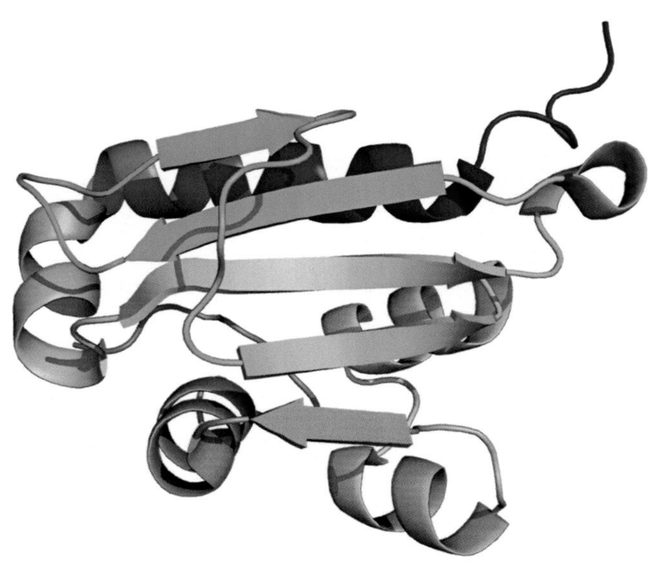

FIGURE 17.6 Structure of the lipid transfer protein, the sterol carrier protein-2 (SCP-2). *http://upload.wikimedia.org/wikipedia/commons/c/cc/Protein_SCP2_PDB_1c44.png.*

are actually two separate proteins: sterol carrier protein X (SCPx, 46 kDa) and the much smaller (13 kDa) sterol carrier protein-2 (SCP-2 is depicted in Fig. 17.6). The SCPs have been suggested to play a role in Zellweger syndrome, a peroxisome disorder.

Many other examples of LTPs exist for specific types of lipids: ceramide-transfer protein (CERT), sphingolipid-transfer proteins (CERT and FAPP2), phosphatidylcholine-transfer protein (PCTP), phosphatidylinositol-transfer protein (PITP), retinoid binding proteins (RBPs), and α-tocopherol transfer protein. In mammals, PCTPs, PITPs, and NSLTPs are the three major classes of phospholipid transfer proteins.

5.3 Lipid Lateral Diffusion Through Membranes

Although vesicle transfer and LTPs are the most appreciated mechanisms for lipid movement, other significant processes exist. One of the most important of these is lateral diffusion through the membrane plane. Lateral diffusion is responsible for lipid movement through endo-membrane flow. As discussed in Chapter 9, lipids can diffuse rapidly, moving completely around a liver cell in about 1 min. After transport vesicles fuse with a membrane, the newly added lipids rapidly spread out from the site of fusion via lateral diffusion. Lipid lateral diffusion rates are affected by many membrane properties, including lipid bilayer phase, type of lipid micro-domain, lipid heterogeneity, membrane lateral pressure, lipid–lipid affinities, lipid–protein affinities, membrane protein crowding, lipid interdigitation, and lipid–cytoskeleton interactions (Chapter 10). Lipid vesicle formation is also affected by the membrane lipid content, particularly the acyl chain double bond content, and levels of PS (phosphatidylserine), PA (phosphatidic acid), and DAG (diacylglycerol).

5.4 Free Diffusion Through the Cytosol

Initially, it would seem feasible that membrane lipids could simply be ejected from the bilayer and diffuse across the aqueous compartment to a distant membrane. While theoretically possible, the extremely low solubility of membrane lipids in water would preclude this mechanism from being significant (see Chapter 13, Detergents).

5.5 Membrane-to-Membrane Contact

Lipid transfer also occurs at points of contact between distant sites on a membrane.

5.6 Lipid Flip-Flop

Finally, lipid transfer is not only associated with diffusion laterally through the membrane plane but also transversely across the membrane in a process known as flip-flop (see Chapter 9). Inherently, lipid flip-flop is a slow event that can be greatly accelerated by protein–lipid interfaces and by enzymes known as flipases, flopases, and scrambleases (see Chapter 9).

Many examples of transfer of a specific lipid from one membrane to another for enzymatic modification have been reported [20]. For example, ER PS is transferred to the mitochondria, where it is decarboxylated to PE before being sent back to the ER. Surprisingly, this bidirectional, nonvesicular lipid transfer accounts for most PE synthesis in mammalian cells. Ceramide is also synthesized in the ER and is primarily transported by CERT to the Golgi, where it is converted into sphingomyelin (SM) by SM synthase.

It is clear that in cells, membrane lipids are constantly on the move via a variety of mechanisms.

6. SUMMARY

A characteristic of all living cells is the continuous movement of material from place to place within the cell. Translocation of membrane-associated constituents is referred to as "membrane trafficking" and forms the heart of the endo-membrane concept. Discussed here are methods commonly used in translocation: secretion (exocytosis), RME, and intracellular carrier proteins. A central component of the endo-membrane flow theory is the Golgi apparatus, a type of molecular processing center that takes in materials from various parts of the cell, modifies and repackages them, and sends them to new cellular locations. Both exocytosis (using SNARE fusion proteins) and endocytosis (through clathrin-coated pits) are largely responsible for the enormous number of intracellular vesicles. Intracellular vesicle fusion to membranes, and a large family of specific and nonspecific LTPs (lipid transfer proteins) are primarily responsible for the rapid movement of polar lipids from one cellular membrane to another.

Chapter 18 will discuss involvement of membranes in the essential cell functions of: anesthetic action, G protein–coupled reactions, membrane attack complex, nerve conduction, and electron transport/oxidative phosphorylation.

FIGURE 18.2 Hans Horst Meyer, 1853–1939. http://www.historiadelamedicina.org/meyer.html.

correlation for anesthetics [8]. This correlation formed the basis for the "nonspecific" mechanism of anesthetics (Section 1.3, Chapter 18).

Meyer quantified anesthetic potency in tadpoles and compared it with the anesthetic's olive oil–water partition coefficient [7]. He found an almost perfect linear correlation for many types of anesthetics – the more olive oil-soluble the anesthetic, the higher was its potency. The solubility correlation held over four to five orders of magnitude. Later, the correlation was found to be even better in *n*-octanol (a solvent that closely resembles the membrane hydrophobic interior).

1.3 "Nonspecific" Mechanism

From the earliest reports of anesthesia in the mid 19th century, it was believed that general anesthetics produce their effects by modulating activity in neuronal membranes. After all, it was assumed that local anesthetics must prevent transmission of nerve impulses to function. As more and more anesthetics were discovered, it became obvious that there was no clear relationship between anesthetic function and molecular structure, only the very general Meyer–Overton correlation [9]. The structures of several common anesthetics are shown in Fig. 18.3. It has also been shown that general anesthetics have only very weak affinities for their potential binding targets in neuronal membranes. The anesthetic concentrations required for function are at least an order of magnitude higher than those measured for most other types of drugs that exhibit specific binding. This observation questions the existence of any *specific* binding of an anesthetic to a protein. So the question remains as to where exactly anesthetics function in membranes [10].

The Meyer–Overton correlation was so convincing that a mechanism, the "nonspecific" mechanism of general anesthetic action, was proposed. By this hypothesis anesthetics are

GENERAL ANESTHETICS

DIETHYL ETHER

CHLOROFORM

HALOTHANE

PROPOFOL

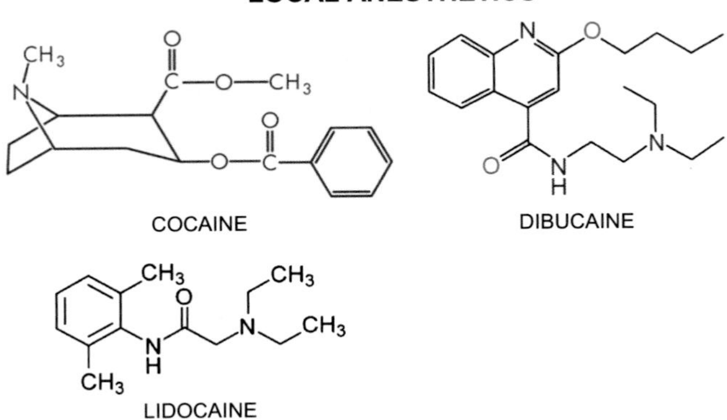

FIGURE 18.3 Structures of some common general (top) and local (bottom) anesthetics. Note, the names of local anesthetics end with −aine.

Diethylether ether: http://scores.espn.go.com/ncb/scoreboard?date = 20131116&confId = 50
Chloroform: http://simple.wikipedia.org/wiki/Chloroform
Halothane: http://chemistry.about.com/od/factsstructures/ig/Chemical-Structures—H/Halothane.htm
Propofol: http://scienceblogs.com/terrasig/2009/07/01/michael-jackson-cherilyn-lee-d/
Cocaine: http://scores.espn.go.com/ncb/scoreboard?date = 20131116&confId = 50
Dibucaine: http://www.selleckchem.com/products/dibucaine-cinchocaine-hcl.html
Lidocaine: http://www.chemistry-reference.com/q_compounds.asp?CAS=137-58-6.

small, volatile, lipid-soluble molecules that partition into membrane lipid bilayers, altering their basic physical properties, thereby affecting activity of proteins involved in nerve conduction. For about 70 years thereafter, many kinds of biophysical measurements on lipid monolayers and bilayers provided further support for the "nonspecific" mechanism of anesthetics. Included in these basic bilayer physical properties (discussed in Chapters 9–11) were membrane

- lipid packing (packing free volume)
- order parameter ("fluidity")
- thickness (hydrophobic match)

- phase separation (domains)
- lateral surface pressure
- curvature and elasticity
- protein activity

Finally, it was demonstrated that anesthetic action can be reversed by simply increasing atmospheric pressure, a process that increases lipid packing, driving the anesthetic out of the membrane. There was no doubt that anesthetics readily enter the lipid bilayer component of membranes where they can severely alter important membrane properties in a general way that is not dependent on an anesthetic's molecular structure. In short, there is no obvious structure–activity relationship for anesthetics.

As more was understood about the physical properties of the membrane lipid bilayer, the basic "nonspecific" mechanism of general anesthetic action had to be tweaked. In 1973, Miller et al. [11] proposed a "critical volume hypothesis" (also called the "lipid bilayer expansion hypothesis") where an anesthetic's molecular volume, and not its actual structure, was most important for activity – the greater the volume, the greater is the activity. A neuronal membrane bilayer would accumulate an anesthetic, causing membrane distortion. When a critical amount of anesthetic accumulated, ion channels would reversibly loose functionality, causing the anesthetic effect.

In 1977, Trudell [12] proposed an anesthetic model based on membrane lateral phase separation. By this hypothesis, anesthetics would accumulate, "fluidizing" (homogenizing) the neuronal membrane and destroying essential lipid domains. These domains are proposed to normally support ion channel activity, synaptic transmitter release, and transmitter–receptor binding.

Indeed, the "nonspecific" mechanism of general anesthetic action, based on weak anesthetic–lipid interactions, has decades of theoretical and experimental support. Unfortunately, even the strongest theories often meet their demise in the face of hash reality. Such is the case for the "nonspecific" mechanism of general anesthetic action [8,9,13]. Problems with this hypothesis can be grouped into five basic categories [14]:

1. **Action of Stereoisomers**: Most anesthetics are mixtures of different stereoisomers (enantiomers) that differ in the way they rotate polarized light. Since they are mirror images of one another, their physical properties are very similar and so their lipid–water partition coefficients are identical. Differences between stereoisomers should only be noticeable on interaction with agents that can distinguish between the stereoisomers. Such agents include biological, optically active (chiral) systems like proteins. If the "nonspecific" mechanism of general anesthetic action was correct, different stereoisomers of an anesthetic should partition identically into the achiral lipid bilayer component of neuronal membranes and so would be expected to have very similar anesthetic effects [15–17]. However, what is usually observed is that while one stereoisomer is a potent anesthetic, the other demonstrates little or no activity.
2. **Nonanesthetic Lipids**: Not all compounds that look like they should be anesthetics and partition into lipid bilayers are anesthetics. In fact, some of these compounds (called nonimmobilizers) do not follow the Meyer–Overton correlation and cause amnesia and not anesthesia [18]. This implies that normal anesthetics and nonimmobilizers must affect different targets and not just neurons.

3. **Temperature Effects**: At very low (clinical) concentrations, anesthetics cause small increases in membrane "fluidity" equivalent to that caused by physiological increases in body temperature (ie, fever). However, anesthesia is not associated with an increase in body temperature. Large increases in "fluidity" are only found with high (nonclinical) levels of anesthetic [8].
4. **Anesthetic Concentrations**: Many of the experiments supporting the "nonspecific" mechanism of general anesthetic action came from biophysical measurements on synthetic lipid monolayer and bilayer membranes. These experiments generally used at least 10 times more anesthetic than clinical levels. Just because a drug can cause an effect at very high concentrations does not mean it works similarly at low, clinical levels [8,9,13].
5. **Straight-Chain Alcohols or Alkanes**: Many straight-chain alcohols and alkanes are anesthetics. They also demonstrate a linear increase in lipid–water partition coefficient with increasing chain length. Therefore, the Meyer–Ocerton correlation would predict an increase in anesthetic potency with increasing chain length. However, this is not what was observed. The anesthetic effect was shown to disappear after carbon length of 13 for *n*-alcohols and after chain lengths 6 to 10 for *n*-alkanes [19,20]. This was called a "cutoff effect."

1.4 Lipid–Protein Hypothesis

The idea that anesthetics partition into the lipid bilayer of neuronal membranes ("nonspecific" mechanism of general anesthetic action) is necessary but not sufficient to explain the mechanism of anesthetic action. Something else must be involved – and evidence often points to proteins, particularly the Na^+ channel. However, the exact binding location and mechanism of action of anesthetics is far from being proven [8].

The experiments comparing anesthetic optical isomers to their activity [15–17] not only question validity of the "nonspecific" mechanism of general anesthetic action but also strongly imply the existence of a specific binding site (eg, a protein) that can distinguish between the optical isomers.

The "cutoff effect" measured for the homologous series of *n*-alcohol and *n*-alkane anesthetics [19,20] provided further evidence that anesthetic action is not entirely due to the Meyer–Overton correlation but rather in part probably occurs by inserting into hydrophobic pockets of well-defined volumes in proteins. As the anesthetic chain length grows the hydrophobic pocket fills up, increasing anesthetic–protein affinity. When the hydrophobic pocket is full, additional chain length cannot be accommodated and the oversized anesthetic is then excluded accounting for the "cutoff." Thus the volume of the *n*-alkanol chain at the "cutoff" length has been used to provide an estimate of the binding site volume [19,20].

While the Meyer–Overton correlation implied that only a lipid bilayer was necessary to produce an anesthetic effect, in the mid 1980s Franks and Lieb [21,22] reported they could reproduce the Meyer–Overton correlation using lipid bilayer–free, water-soluble proteins! Two classes of water-soluble proteins were inactivated by clinical levels of anesthetics: luciferase and cytochrome P450. In addition, anesthetic inhibition of luciferase exhibited a long-chain alcohol "cutoff" that was related to the size of the anesthetic-binding pocket. Anesthetics have also been shown to affect many other cytoplasmic (water-soluble) signaling proteins including protein kinase C.

An additional theory combines both the lipid bilayer and protein aspects of anesthetic action. This theory focuses on the membrane physical property of lateral pressure [23]. Lateral pressure was chosen because it is known to affect the conformation of ion channels and is affected by low, clinical levels of anesthetics. Anesthetics were shown to induce a much larger increase in lateral pressure at the bilayer–aqueous interface with a decreasing influence observed down to the bilayer center. The larger lateral pressure at the aqueous interface induced by anesthetics makes it more difficult for the ion channels to remain open, shifting the equilibrium back to the closed state. Thus, nerve impulses open ion channels that are subsequently closed by anesthetics. By this theory, anesthetics do not *directly* affect protein activity but instead perturb the membrane bilayer accommodating the ion channel. The lipid bilayer acts simply as a mediator for anesthesia.

Besides ion channels, several other proteins have been suggested to be possible targets for general anesthetics. At the top of this list is the Cys-loop family of receptors [21]. Anesthetics are proposed to affect these loops that connect transmembrane bundles of α-helices. They include γ–aminobutyric acid $(GABA)_A$, $GABA_C$, glycine, acetylcholine, and 5-hydroxytryptamine $(HT)_3$ receptors. While it was originally proposed that general anesthetics directly bind to protein channels by the common lock and key mechanism, it now seems more likely that the interaction is indirect through the lipid bilayer [24].

So the questions remain – What exactly is the anesthetic binding site? Is it in neuronal membranes? Is it the lipid bilayer? Is it a membrane protein or possibly even a cytosolic protein? Or is it some combination of both lipids and proteins?

2. G PROTEIN–COUPLED REACTIONS

2.1 Introduction

In Chapter 6, a large, diverse, and very important structural class of membrane proteins was briefly introduced. These integral proteins are all orientated in the membrane by seven transmembrane α-helices. Because of their biochemical importance, these proteins have been referred to as the "magnificent seven." Examples of proteins containing the seven α-helix motif include the large family of G protein–coupled receptors (GPCRs) that bind most important hormones and signaling molecules to the plasma membrane cell surface [25,26]. Already approximately 800 different GPCRs have been identified as being encoded in the human genome!

The concept of G proteins was introduced in 1970 by Martin Rodbell (Fig. 18.4) [27], work for which he shared the 1994 Nobel Prize in Physiology or Medicine with Alfred G. Gilman. Rodbell studied the effects of glucagon on rat liver plasma membrane receptors. He reported that guanosine triphosphate (GTP) disassociated glucagon from its receptor producing a GTP protein (G protein) that profoundly influenced the cell's metabolism. Thus he identified the G protein as a glucagon signal transducer.

FIGURE 18.4 Martin Rodbell, 1925–1998. *In NNDB http://www.nndb.com/people/181/000133779/.*

2.2 Phosphorylase Amplification Cascade

We will begin this discussion with the glycogen phosphorylase amplification cascade, a signal transduction-induced amplification cascade linked through cyclic adenosine monophosphate (cAMP) [28]. This pathway is detailed in Fig. 18.5. Signal transduction occurs when an extracellular signaling molecule (first messenger ligand, eg, epinephrine) binds to and activates a cell surface receptor (eg, GPCR) that causes a transmembrane conformational change, activating an enzyme (eg, adenylate cyclase) attached to the inner plasma membrane leaflet. The activated enzyme generates a second messenger (eg, cAMP), triggering an amplification cascade that in a final step activates an enzyme (eg, glycogen phosphorylase), producing a large amount of a final product (eg, glucose-1-phosphate).

A key component of this signaling pathway, and many others, is the second messenger cAMP (Fig. 18.6). In 1945, Earl Sutherland (Fig. 18.7) began an investigation of how epinephrine induces glycogen breakdown in the liver. The key finding occurred in 1956 when Sutherland discovered the first "second messenger," cAMP, and outlined the basic steps summarized in Fig. 18.5 and in the previous paragraph linking epinephrine to glycogen breakdown. The discovery of cAMP initiated the large field of intracellular signaling pathways. For this work, Sutherland was awarded the 1971 Nobel Prize in Physiology or Medicine.

The process begins with the extracellular first messenger ligand, epinephrine (also known as adrenaline) (Fig. 18.8) [29]. Epinephrine is one of a group of monoamines called catecholamines. However, epinephrine is a prime example of a biologically important

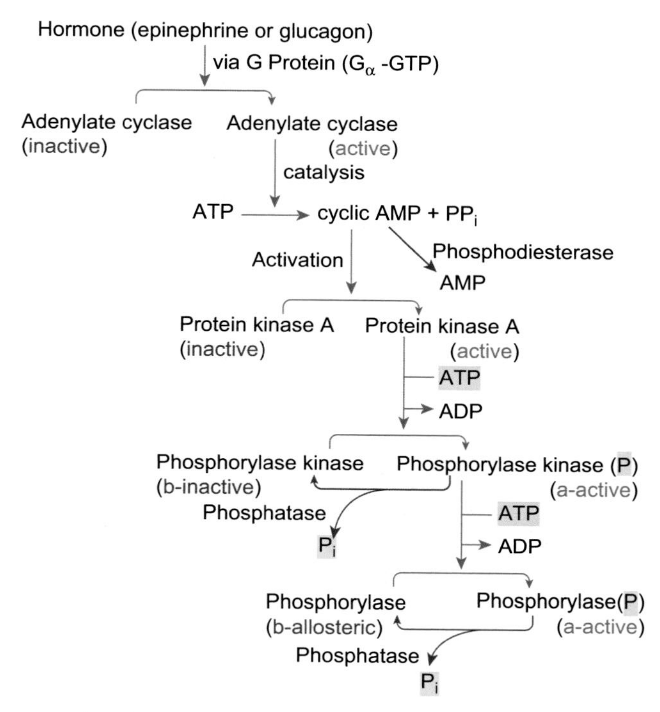

FIGURE 18.5 Glycogen phosphorylase amplification cascade, a signal transduction-induced amplification cascade linked through cyclic-AMP (cAMP). http://www.rpi.edu/dept/bcbp/molbiochem/MBWeb/mb1/part2/glycogen.htm.

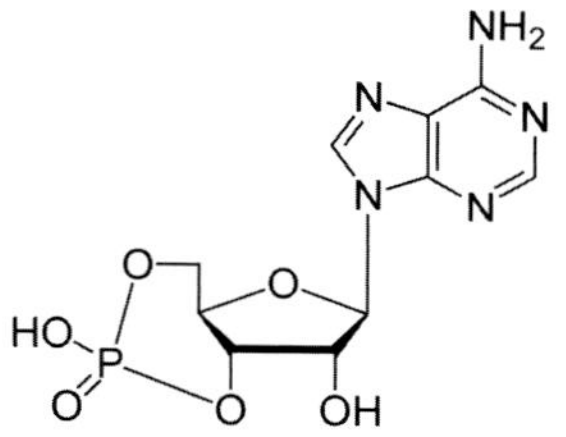

FIGURE 18.6 Cyclic AMP (cAMP, 3′,5′-cyclic adenosine monophosphate). http://en.wikipedia.org/wiki/Cyclic_nucleotide.

molecule that can easily fit into more than one distinct class. Epinephrine is a neurotransmitter when secreted by the central nervous system but a hormone when secreted by the adrenal medulla.

Epinephrine recognizes and binds to the outer face of its specific receptor, a β-adreneric GPCR. Epinephrine does not pass through the membrane itself but instead generates a signal across the membrane through a conformational change in the receptor. Epinephrine binding on the outside stimulates a series of events inside the cell. Different types of receptors

FIGURE 18.7 Earl W. Sutherland, 1915–1974. *In NNDB http://www.nndb.com/people/980/000130590/.*

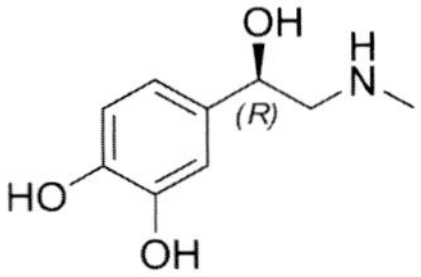

FIGURE 18.8 Epinephrine (also known as adrenaline). http://chemistry.about.com/od/factsstructures/ig/Chemical-Structures—E/Epinephrine.htm.

stimulate different responses, but the observed response is specific for the bound ligand. The conformational change is felt throughout the receptor, resulting in either the direct activation of an enzymatic activity already attached to the intracellular part of the receptor or, as with epinephrine, the exposure of a binding site for other intracellular signaling proteins within the cell.

On the inner leaflet of the plasma membrane, signal transduction begins with an inactive G protein coupled to the GPCR (Fig. 18.9). At this stage, the G protein exists as a heterotrimer consisting of the subunits Gα, Gβ, and Gγ. The inactive Gα subunit is bound to GDP. After the ligand (epinephrine) binds to the GPCR, the conformational change causes Gα to replace GDP with a molecule of GTP and dissociate from the other two G protein subunits, Gβ and

$$\mathrm{ATP} \rightarrow \mathrm{cAMP} + \mathrm{PP_i}.$$

FIGURE 18.9 G protein activation of adenylate cyclase producing the second messenger cAMP. http://employees.csbsju.edu/hjakubowski/classes/ch331/signaltrans/olsignalkinases.html.

Gγ. The detached Gβ and Gγ subunits initiate signaling from many downstream effector proteins such as phospholipases and ion channels and are not technically part of the amplification cascade for glycogen breakdown under discussion here. The Gα subunit with bound GTP diffuses through the plasma membrane inner leaflet until it binds to an adenylate cyclase enzyme, activating it to convert:

$$ATP \rightarrow cAMP + PP_i.$$

Each enzymatic step in the cascade of events leading to production of the final product, glucose-1-phosphate, from the first messenger, epinephrine, is dramatically amplified. The amplification enzymes in order are:

1. Adenylate cyclase
2. Protein kinase A
3. Phosphorylase kinase
4. Glycogen phosphorylase

For example, the turnover number for adenylate cyclase (number of ATP molecules converted to cAMP by a single adenylate cyclase per minute) is about 720 min^{-1} [18–30]. Therefore, several hundred molecules of cAMP can be made during the lifetime of a single Gα. And each subsequent enzymatic step further amplifies the amount of glucose-1-phosphate produced by glycogen phosphorylase. The glucose-1-phosphate rapidly enters glycolysis generating substantial quantities of biochemical energy in the form of ATP:

G protein activation of adenylate cyclase

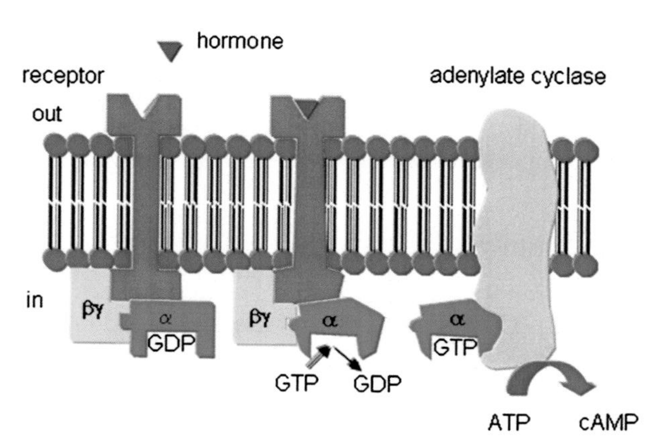

$$\text{glycogen (glucose}_n) \rightarrow \text{glycogen (glucose}_{n-1}) + \text{glucose-1-phosphate}$$

$$\downarrow$$

$$\text{ATP production} \leftarrow \text{GLYCOLYSIS}$$

Thus, one initial signaling molecule can result in the production of countless final products. This process is generally referred to as an "amplification cascade."

3. MEMBRANE ATTACK COMPLEX

3.1 Introduction

All organisms are continuously bombarded by a never-ending series of toxins and pathogens against which they must defend themselves. Leading the human defense arsenal are members of the immune system. Dictionary.com defines the immune system as "a diffuse, complex network of interacting cells, cell products, and cell-forming tissues that protects the body from pathogens and other foreign substances, destroys infected and malignant cells, and removes cellular debris." An immune system must not only be able to detect many types of foreign invaders; it must also distinguish between foreign and self and destroy potentially pathogenic organisms. Even single-celled bacteria have a primitive kind of immune system to ward off bacteriophages. One member of the human immune system that is closely associated with membrane structure and function is the complement-derived membrane attack complex (MAC) [31].

3.2 Complement

The complement system is an example of an amplification cascade mechanism (see G proteins, Chapter 18, Section 2) that accounts for the very fast speed of the process. The cascade begins with the attachment of complement to antibodies on the outer surface of a foreign cell (pathogen) and ends with destruction of that cell by the MAC [32,33]. The complement system is composed of about 25 different proteins that assist or "complement" antibodies in killing pathogens. The discovery of complement had its murky beginnings in the late 19th century when Hans Ernst August Buchner (Fig. 18.10) found that blood serum contains a "factor" capable of killing bacteria. Buchner's younger brother, Eduard, won the 1907 Nobel Prize in Chemistry for his work on fermentation. In 1896, Jules Bordet (Fig. 18.11) reported that this "factor" was composed of two parts, one that is heat stable and the other that is heat labile. The heat-labile component contains what is now known as complement.

3.3 Membrane Attack Complex

The MAC forms trans–plasma membrane channels on the surface of pathogenic bacteria, causing cell lysis and death [34]. The bacteria are first identified and tagged by antibodies that attach to specific bacteria-surface carbohydrates. This response stimulates the complement cascade leading to generation of the MAC.

The complement system consists of about 25 proteins that are synthesized in the liver and transported in the blood, mostly as inactive precursors. The system consists of complement proteins as well as serum proteins, serosal proteins, membrane receptors, and protein fragments. They comprise about 5% of the total serum globular proteins. Only five of the complement system proteins eventually form subunits of the MAC: one unit each of complements C5b, C6, C7, and C8 and several units of complement C9. Importantly, none of these five complement proteins have enzymatic activity, proteolytic or lipolytic, that could directly digest the bacterial membrane. Instead, another mechanism, a trans-membrane hole through the MAC, is implicated.

FIGURE 18.10 Hans E.A. Buchner, 1870–1961. http://en.wikipedia.org/wiki/Hans_Ernst_August_Buchner.

FIGURE 18.11 Jules Bordet, 1850–1902. *Journal of Medicinal Biography, http://jmb.sagepub.com/content/17/4/217/F1. expansion.*

Assembly of the MAC (Fig. 18.12) is initiated when the complement protease C5 convertase cleaves C5 into C5a and C5b. Complement protein C6 then binds to C5b, and this complex binds to complement C7 forming a larger complex. This complex undergoes a conformation change that exposes a hydrophobic site on C7 and draws the complex into the hydrophobic lipid bilayer portion of the pathogen's plasma membrane. Subsequent binding of the C5b–C6–C7 complex to C8 exposes another hydrophobic site that further anchors the 5b–8 complex into the pathogen's lipid bilayer. Complement C8 is composed of two proteins, C8β and C8αγ. C8αγ has a hydrophobic site that draws it into the growing complex housed in the pathogen lipid bilayer and also initiates polymerization of 10–16 molecules of complement C9. This completes the transmembrane pore known as the MAC (Fig. 18.13). The MAC has a hydrophobic external face that is stabilized by interaction with the membrane bilayer interior and a hydrophilic internal face that allows rapid passage of water and solutes. While it had been suspected for many years, the first clear evidence for complement-induced membrane pores was reported in 1959 by Green et al. [35].

A rare but interesting human disease is associated with a malfunction of the MAC. MAC synthesis is inhibited by CD59 (also known as MAC-inhibitory protein, or protectin) found on the surface of normal human cells. CD59 is an example of a glycosyl phosphatidylinositol (GPI)–linked protein (see Chapter 6). Its function is to protect normal human cells from being accidentally destroyed by their own antibacterial MAC. CD59 prevents polymerization of C9 by the complex C5b–C6–C7–C8, thus preventing synthesis of MAC on normal

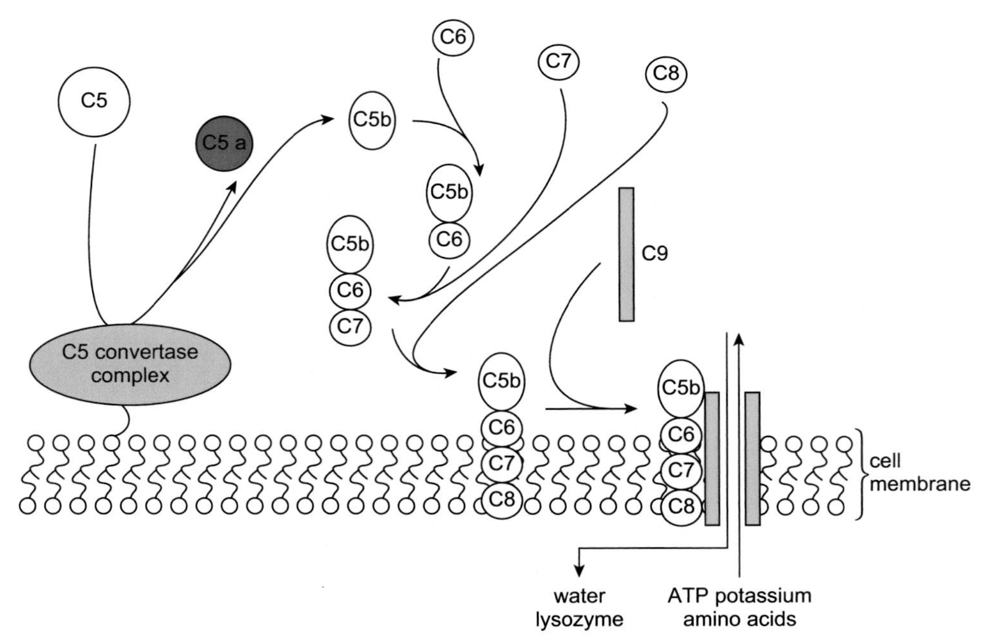

FIGURE 18.12 Complement pathway for generation of the membrane attack complex (MAC) on the surface of a pathogen membrane. http://en.wikipedia.org/wiki/Complement_membrane_attack_complex.

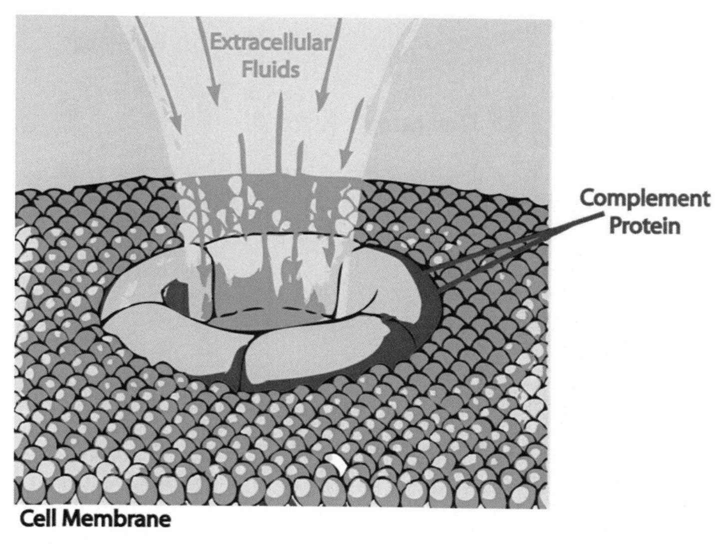

FIGURE 18.13 The membrane attack complex (MAC) punches a hole through the plasma membrane of the target cell, killing the pathogen. *Found in Membrane Attack complex in Wikipedia.*

cells. In a rare human condition known as paroxysmal nocturnal hemoglobinuria, erythrocytes lack CD59 and so can be lysed by MAC.

4. NERVE CONDUCTION

4.1 Introduction

A characteristic of all living cells is the development and maintenance of trans–plasma membrane gradients of water and every conceivable solute. As a result of unequal distribution of all solutes inside and outside the cell, many of which are charged, every plasma membrane exhibits a membrane electrical potential. In most cases this potential is about −70 mV, negative inside. A major contributor to this potential is the Na^+,K^+-ATPase, whose mechanism is discussed in Chapter 19 [36]. This enzyme is electrogenic, using energy in the form of ATP to drive 2 K^+ inside the cell against its gradient while simultaneously driving 3 Na^+ out of the cell against its gradient. As a result, cells normally maintain a large internal concentration of K^+ and a low internal concentration of Na^+ and a resting potential of about −70 mV. However, certain types of animal cells, called excitable cells (neurons, muscle and endocrine cells), are electrically active in the sense that their membrane potential can fluctuate rapidly over short spans of time. The function of this up-and-down fluctuation in membrane potential is called an action potential [37]: in neurons, it is cell-to-cell communication; in muscle cells, it is to stimulate intracellular events leading to muscle contraction; and in beta cells of the pancreas, it is to release insulin. A transmembrane potential of about −70 mV may not sound like much, but when generated across a very thin

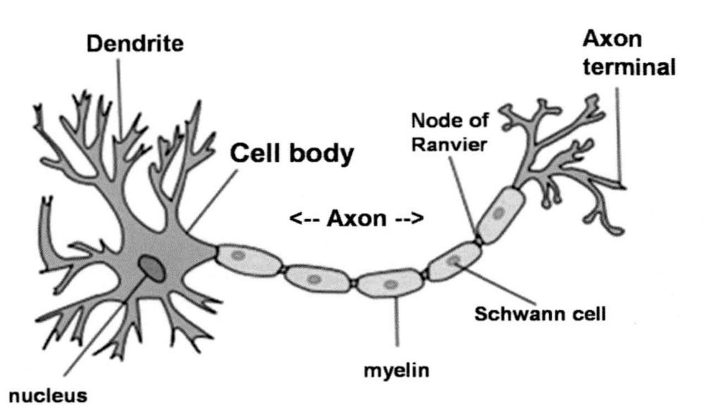

FIGURE 18.14 Schematic drawing of a "typical" neuron. http://users.tamuk.edu/kfjab02/Biology/Animal Physiology/B3408%20Systems/BIOL%203408%20Chapter%2011%20Neurons.htm.

membrane (<10 nm, see Chapter 9), very strong electric forces are produced. The very rapid up-and-down voltage cycles, called action potentials, are responsible for nerve conduction. In most neurons, a single action potential takes less than a millisecond (0.001 s) and neurons typically emit action potentials at rates up to 10–100 per second.

4.2 Neurons

The most studied, and excitable, of the cells that conduct nerve impulses are known as neurons. A "typical" neuron is depicted in Fig. 18.14. By even a cursory glance, the extreme asymmetry of a neuron is obvious. The major structural features of a neuron include a single soma or cell body, one or more dendrites, a single axon and one or more axon terminals. The cell nucleus is housed in the soma and regulates the neuron.

Signals enter the soma via dendrites that appear as branches on the soma surface. Electrical signals from an adjacent, upstream neuron enter the depicted neuron (Fig. 18.14) through synapses (discussed below in Section 4.4). A neuron has two types of synapses: one found on the dendrites and the other found on the axon terminals. While dendrites are populated by ligand-activated channels, the soma surface is populated by voltage-activated ion channels. Emerging from the soma is the axon hillock, a specialized part of the soma that is highly enriched in voltage-activated Na^+ channels [38]. It is here that the membrane potentials, propogated from the synaptic inputs, are accumulated before being transmitted to the axon. Therefore, axon hillocks concentrate and amplify the signals and initiate the action potentials (discussed below, Section 4.3) that pass down the axon away from the soma. The axon is a long, thin tube-like structure that is surrounded by a myelin sheath [39]. The myelin sheath is considered to be the least dynamic of all membranes, having only about 20% protein by weight (see Chapter 1). Myelin wraps around the axon like a jelly roll, essentially encasing it in layers of fat. Its only function seems to be nerve insulation where it prevents ions from entering or escaping from the axon. This insulation greatly reduces signal decay and increases signal transmission speed. However, the axon could not function if it was completely encased in myelin. Therefore, regularly spaced down the axon are small gaps where no insulation exists. These gaps, known as "nodes of Ranvier," function like "mini axon hillocks," boosting the signal and preventing its decay. At its distal end, the axon sheds its myelin sheath and branches into several axon terminals that contain the second type of synapse known as axon terminal buttons.

4.3 Action Potential

The link between electricity and nerve function has a very long history. In 1771, the Italian physician Luigi Galvani (Fig. 18.15), an early pioneer of electricity, discovered that the muscles of dead frog legs twitched when struck by an electrical spark. To put this discovery into historical perspective, this is at the same time that Benjamin Franklin was conducting his early experiments on lipid monolayers (oil on water, see Chapter 2). In about 1865, the German scientist Julius Bernstein (Fig. 18.16) was able to record the first time course of an action potential. Later Bernstein greatly advanced the field of electrophysiology with his highly regarded "membrane hypothesis," correctly explaining the origin of the resting and action potentials. He was the first to appreciate the selective transembrane permeability to K^+.

A characteristic of all living cells is the existence of trans–plasma membrane electrical gradients (membrane potential, negative interior). Establishment of the membrane potential is the result of complex interactions between many proteins, primarily ion pumps and ion channels, and the electrical insulating properties of the membrane hydrophobic lipid bilayer interior. Different parts of a neuron (dendrites, axons, and soma) have different types and concentrations of pumps and channels. In fact, some parts of a neuron are not even

FIGURE 18.15 Luigi Galvani, 1737–1798. http://www.probertencyclopaedia.com/cgi-bin/res.pl?keyword=Luigi+Galvani&offset=0.

FIGURE 18.16 Julius Bernstein, 1839–1917. http://de.wikipedia.org/wiki/Datei:Julius_bernstein.jpg.

excitable! There are two important levels of membrane potential: the resting (unperturbed) potential and the higher threshold potential. At the axon hillock the resting potential is about −70 mV and the threshold potential is about −55 mV. In neurons, the sequence of action potentials is referred to as a "nerve impulse" or "spike." The resting potential is largely established by the plasma membrane-bound Na^+,K^+-ATPase, while the action potentials are generated by voltage-gated ion channels also found in the cell's plasma membrane. Fig. 18.17 depicts a typical action potential [40].

The voltage-gated ion channels are shut when the membrane potential is close to the cell's resting potential, but they begin to open rapidly if the membrane potential increases to the threshold potential (the membrane becomes depolarized). Correlated opening and closing of Na^+ and K^+ channels are responsible for the action potential. The action potential begins with opening of the Na^+ channels and movement of Na^+ down its electrochemical gradient into the axon, causing depolarization. As more Na^+ moves into the axon, the membrane potential continues to rise, causing more channels to open. The process proceeds explosively until all of the available Na^+ channels are open. The large inward flow of Na^+ reverses the membrane polarity and, at some point, rapidly closes the Na^+ channels. These events activate K^+ channels, allowing K^+ to escape the axon down its electrochemical gradient. Internal Na^+ is then actively pumped from the cell as K^+ is pumped back in by the Na^+,K^+-ATPase, returning the membrane potential to its resting state. The action potential travels down the axon in one direction only to the axon terminal where the signal passes through a synapse to the next downstream (postsynaptic) neuron. In animal cells, there are two primary types of action potentials; one type is generated by voltage-gated

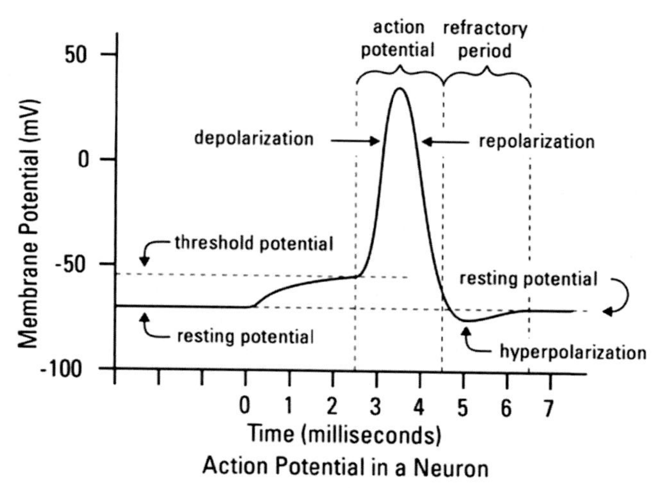

FIGURE 18.17 A typical action potential in a neuron. http://mikeclaffey.com/psyc170/notes/notes-neurons.html.

Na^+ channels that last for <1 ms, while the other type is generated by voltage-gated calcium channels that exist for >100 ms.

4.4 Neuronal Synapse

Since sequential neurons are not physically connected to one another, the nervous signal must pass through an extraneuronal space, referred to as a synapse [41]. A synapse is therefore a very narrow gap or cleft maintained by specialized structures found between a signal-passing (presynaptic) neuron and a signal-accepting (postsynaptic) neuron. Its purpose is to allow signals to pass from one neuron to another or from a neuron to a target. Synapses are highly complex structures and are an example of a common general cell process known as "membrane trafficking" (see Chapter 17).

On arriving at the axon terminus, the action potential triggers the influx of Ca^{2+} by opening plasma membrane–bound voltage-gated Ca^{2+} channels. The increased Ca^{2+} concentration is used later in the process to stimulate release of neurotransmitters by synaptic vesicle exocytosis into the synapse. Neurotransmitters [42] are a very large group of simple and very common biochemicals that are readily available through the diet or by very simple conversions from other common compounds. Countless neurotransmitters have already been discovered. Among the simplest neurotransmitters are the amino acids glycine, GABA, glutamate, and aspartate and the monoamines epinephrine, norepinephrine, histamine, dopamine, and serotonin. Even ATP and the gas nitric oxide are neurotransmitters. It is clear from this very abbreviated list that neurotransmitters are small, everyday biochemicals that function as moonlighting molecules (they have more than one function). Neurotransmitters are the molecules that carry the signal across the synapse.

The synapse was first observed as a 20- to 40-nm gap between neurons by Spanish pathologist Ramon y Cajal (Fig. 18.18), who is often considered to be the "father of modern

FIGURE 18.18 Ramon y Cajal, 1852–1934. *Nobelprize.org, http://www.nobelprize.org/nobel_prizes/medicine/laureates/1906/cajal-bio.html.*

neuroscience." In 1894, Cajal published the full details of his "neuron doctrine" in a lecture at the Royal Society of London. It was this presentation that revolutionized thought in neurology. It resulted in him receiving the 1906 Nobel Prize in Physiology or Medicine together with his arch foe, Camillo Golgi (Fig. 17.4). In fact, Cajal and Golgi only met once – at the 1906 Nobel Prize presentation. However, it was Cajal's ideas on neuron structure, not Golgi's, that lasted the test of time. Cajal was an unusual character with many examples of erratic behavior. For example, he was imprisoned at the age of 11 for destroying his neighbor's yard gate with a homemade cannon! Coincidentally, Cajal and Golgi shared an unusual aspect of their scientific careers: both conducted their research in homemade laboratories housed in their kitchens! In 1921, German pharmacologist Otto Loewi (Fig. 18.19) confirmed that neurons can communicate with one another by releasing chemicals known as neurotransmitters. It was Loewi who is credited with discovering the first neurotransmitter, acetylcholine (Fig. 18.20), for which he received the 1936 Nobel Prize in Physiology or Medicine.

A central feature of synapse function is the "synaptic vesicle cycle" (Fig. 18.21) [43]. Synaptic vesicles are small membrane-enclosed spherical bags clustered beneath the axon terminal plasma membrane on the presynaptic side of a synapse. This cycle is an example of "membrane trafficking" (see Chapter 17) and prepares the presynaptic nerve terminal for neurotransmitter release across the synapse.

Since this is a cycle, there is no formal beginning or end, just a continuum. Let us start with an empty synaptic vesicle that must be loaded with a neurotransmitter residing in

FIGURE 18.19 Otto Loewi, 1873–1961. http://en.wikipedia.org/wiki/Otto_Loewi.

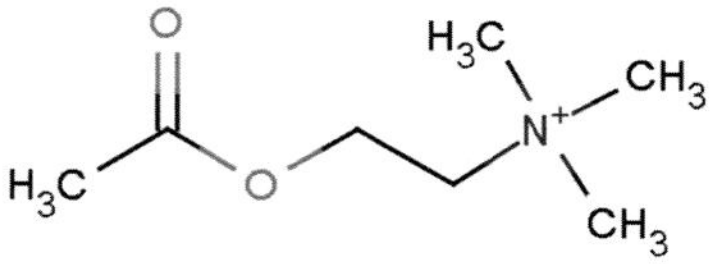

FIGURE 18.20 Acetylcholine, the first neurotransmitter. http://www.insupplements.com/54-acetylcholine-and-its-function-on-memory.

the cytoplasm (Fig. 18.21, steps 6 → 1 → 2). Neurotransmitter uptake is driven by a specialized transporter housed in the synaptic vesicle membrane. The transporter is a proton pump that creates an electrochemical gradient and so neurotransmitter uptake is an example of secondary active transport (see Chapter 19). After loading, the synaptic vesicles are moved to the "active zone" near the plasma membrane at the synapse (Fig. 18.21, step 3). It is estimated that there are about 200–500 synaptic vesicles per terminal. While most synaptic vesicles are found free in the cytosol of the neuron terminal (they are background synaptic vesicles), no more than 10–30 are found bound to the active zone membrane of a synapse. The synaptic vesicles are attached by "docking proteins" and are then ready to release the sequestered neurotransmitters into the synaptic cleft by a process known as exocytosis (see Chapter 19). Neurotransmitter release involves a Ca^{2+}- and ATP-dependent process that initiates a membrane fusion event (Fig. 18.21, step 4). The entire fusion occurs very quickly, in about 100 ms. After exocytosis, empty synaptic vesicles are rapidly reformed in the axon terminus cytosol via clathrin-dependent coated pit endocytosis (Fig. 18.21, step 5).

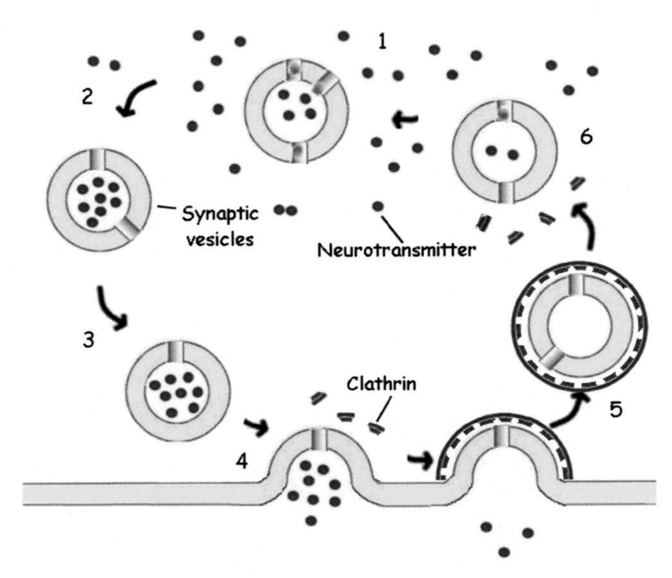

FIGURE 18.21 The synaptic vesicle cycle. **Steps 6 → 1 → 2**: Loading the synaptic vesicle with neurotransmitter. Neurotransmitter uptake is driven by a synaptic vesicle membrane-bound, electrochemical-dependent transporter. **Step 3**: The neurotransmitter-loaded synaptic vesicles are moved to the "active zone" near the plasma membrane at the synapse. **Step 4**: Exocytosis. Neurotransmitter release involves a Ca^{2+}- and ATP-dependent process that initiates a membrane fusion event. **Step 5**: Vesicle recycling. The empty synaptic vesicles are recycled via clathrin-dependent coated pit endocytos. http://origins.swau.edu/papers/complexity/trilo/gifs/vesicles.html.

The entire synaptic vesicle cycle takes approximately 60 s to complete. Ca^{2+}-triggered fusion tales <1 ms, docking takes 10–20 ms, and endocytosis takes a few seconds. Therefore, the majority of time is spent in neurotransmitter uptake into the synaptic vesicles and accounts for most synaptic vesicles being found unbound in the cytoplasm. One advantage of having the synaptic vesicle cycle located in the axon terminus, far removed from the neuron nucleus, is that the cycle is independent of the cell nucleus, allowing for more rapid responses. A question can be raised as to why Nature has chosen such a complex process as the "synaptic vesicle cycle" when it would appear that a simple neurotransmitter channel would suffice. The synaptic vesicle exocytosis process has an advantage in being a "cascade" where each synaptic vesicle can rapidly release a very large number of neurotransmitters at once.

4.5 Postsynaptic Neuron Receptors

Neurotransmitters released from the presynaptic vesicles cross the synaptic cleft and bind to receptors on the plasma membrane of the postsynaptic neuron (Fig. 18.22). Signal transduction through receptors is inherently complex, producing many possible outcomes on the postsynaptic cells. One important outcome is triggering an electrical signal by regulating the activity of ion channels. Thus, an electrical signal can be passed from one neuron to another. There are two basic types of neurotransmitter receptors: ligand-gated (or ionotropic)

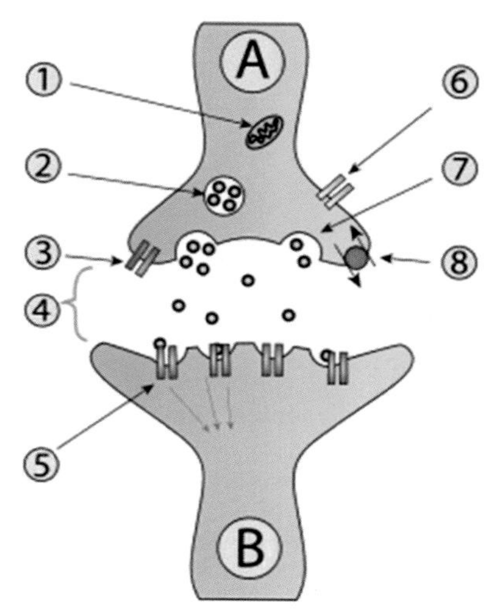

FIGURE 18.22 Diagram showing the diffusion of neurotransmitters from the presynaptic neuron (A) vesicles to receptors on the postsynaptic neuron (B). http://www.biologycorner.com/anatomy/nervous/notes_ch9a.html.

receptors and G protein–coupled (or metabotropic, GPCRs) receptors. Ligand-gated receptors can be directly excited or inhibited by different neurotransmitters. In contrast, GPCRs are neither directly excitatory nor inhibitory, but instead modulate the actions of excitatory and inhibitory neurotransmitters on the receptors. Most neurotransmitter receptors are G protein coupled.

4.6 Voltage-Gated Sodium Channels

A central participant in the mechanism of nerve conduction is the transmembrane voltage-gated Na^+ channel [44,45]. This channel (technically just the α subunit) is a large (>1800 amino acids), multiple transmembrane spanning protein. The membrane-spanning portion of the protein consists of bundles of α-helices that are tied together with short, nonhelical cytoplasmic chains (see the similar structure of bacteriorhodopsin discussed in Chapter 6). Within a transmembrane domain, the hydrophobic side chains of the α-helices face outward into the lipid bilayer hydrophobic interior while the polar peptide bonds face inward forming the Na^+ channel. For the voltage-gated Na^+ channel, there are 24 α-helices (Fig. 18.23). This figure depicts just the spread-out structure and not the actual folded three-dimensional protein structure as it would exist in the membrane. The 24 α-helices are divided into four packets (domains) of six α-helices per packet. At the center of the four packets is the Na^+ channel.

The voltage-gated Na^+ channel can be divided into three functional parts: the channel, the gate with voltage sensors, and the inactivation gate. The three functions are schematically depicted in Fig. 18.24.

The voltage-gated Na^+ channel can exist in three states: deactivated (closed), activated (open), and inactivated (closed). The channel pore is highly selective for Na^+ with one

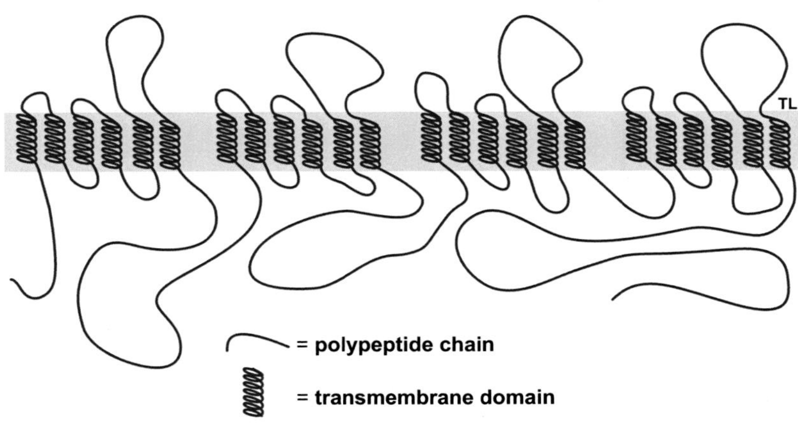

FIGURE 18.23 Basic structure of a voltage-gated sodium channel. Note, the 24 α-helices are divided into four packets (domains) of six α-helices per packet. http://courses.washington.edu/conj/membrane/nachan.htm.

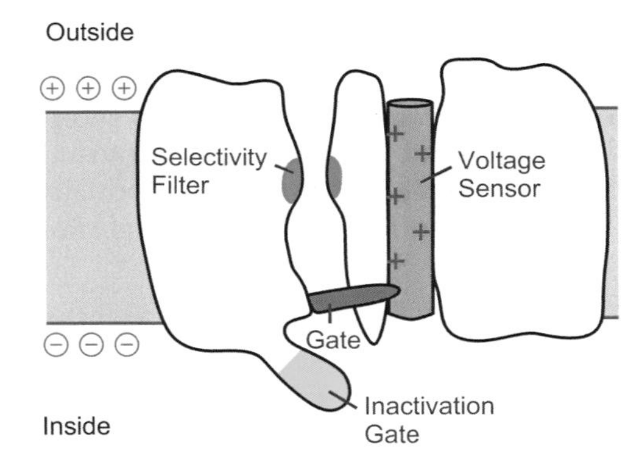

FIGURE 18.24 The three functional parts of the voltage-gated sodium channel: the channel, the gate with voltage sensors, and the inactivation gate. http://courses.washington.edu/conj/membrane/nachan.htm.

attached water of hydration. Even K^+ is too large to fit through the channel. Selectivity is provided by a negatively charged "selectivity filter" that attracts Na^+ while repulsing anions. In addition, the channel narrows to a passage of only 0.3 nm by 0.5 nm wide, too small to allow K^+ or other solutes larger than Na^+ through. In the resting state, before the action potential begins, the axonal membrane is at its resting potential (about −70 mV) and the Na^+ channel is in the deactivated (closed) state where entrance to the channel is blocked on the extracellular side by an activation gate. Opening or closing the gate is controlled by voltage sensors that respond to the membrane potential. The voltage sensor is a transmembrane domain with a positively charged amino acid found at every third position resulting in a total of four to eight positive charges per transmembrane domain.

At a typical resting potential of −70 mV, the voltage-gated Na^+ channel is closed. On arrival of an action potential, the membrane is depolarized to about −50 mV and the voltage sensor moves outward, opening the gate and allowing Na^+ to enter. The final functional component is an inactivation gate that limits the period of time the channel remains open despite continual stimulation.

5. ELECTRON TRANSPORT/OXIDATIVE PHOSPHORYLATION

The best example of a complex biological process being tightly associated with membrane structure is mitochondrial electron transport and oxidative phosphorylation – the method whereby eukaryotic cells generate most of their ATP. Because of its critical role in bioenergetics, mitochondria are often referred to as the "powerhouse of the cell."

5.1 Mitochondria

A mitochondrion (singular) is a membrane-surrounded organelle that resides inside eukaryotic cells. Mitochondria were first observed as small, ill-defined components of a cell in the 1840s [46]. The term "mitochondria" was introduced in 1898 by Carl Benda at a time when mitochondrial function was just beginning to be understood. A major breakthrough occurred in 1925 when David Keilin (Fig. 18.25) identified the cytochromes that form the foundation of electron transport. Later, new tissue fractionation technologies made it possible to isolate intact mitochondria and study their function independent of

FIGURE 18.25 David Keilin, 1887–1963. *Hartree EF. Obituary notice: David Keilin. Biochem J 1963;89(1–0). http://www.biochemj.org/bj/089/0001/bj0890001_b2_browsefigs.htm.*

FIGURE 18.26 Peter D. Mitchell, 1920–1992. *The Nobel Prize in Chemistry 1978, Peter Mitchell. http://www.nobelprize.org/nobel_prizes/chemistry/laureates/1978/mitchell-bio.html.*

other cellular components. By 1952, high-resolution electron microscopic images allowed for the first visualization of mitochondrial membranes. In 1957, Philip Siekevitz introduced the now commonly used descriptor of mitochrondria as "the powerhouse of the cell." Finally, in 1961, the chemiosmotic hypothesis for oxidative phosphorylation was proposed by Peter Mitchell (Fig. 18.26), for which he won the 1978 Nobel Prize in Chemistry.

It is not a coincidence that mitochondria are about the same size (0.5–1.0 μm in length) as bacteria. It is believed that in the very distant past mitochondria were free-living bacteria that, due to changing environmental conditions, were forced to take up refuge inside a much larger, single-celled prokaryote, thus creating the first eukaryote. A similar process is believed to have created the first chloroplast. The initial symbiotic event occurred at least 1.6–2.1 billion years ago, and this concept is now known as "endosymbiont theory" [47].

Through the aeons, as the mitochondria evolved, the original full-size bacterial genome was tremendously reduced and most genes were transferred to the nucleus [48]. The mitochondria now codes for only 13 proteins. Coding for the rest of the mitochondrial proteins (about 615 total proteins in human cardiac mitochondria) has been transferred to the nucleus. All of the mitochondria-coded proteins are components of electron transport and oxidative phosphorylation (cytochrome b, three subunits of cytochrome oxidase, one of the subunits of ATPase, and seven subunits for associated proteins of NADH dehydrogenase). Having some proteins coded by mitochondrial DNA while most proteins are coded in the nuclear DNA presents an interesting dilemma with regard to coordinating mitochondrial biogenesis. How are the complete proteins assembled? Mitochondria cannot be made from scratch and gene products must be made on separate ribosomes, one a cytoplasmic type and the other a mitochondrial type. Mitochondria have been defined as "semiautonomous, self-replicating organelles," meaning they grow and replicate *independent* of the cell in which they are housed. Instead, mitochondria multiply when the energy needs of a cell arise.

The number of mitochondria found in a eukaryotic cell varies over a large range depending on the cell's energy requirement. While sluggish yeast may have only a single mitochondrion, insect flight muscle may have as many as 10,000! A human liver cell has about 1000 mitochondria.

A schematic drawing of a "typical" mitochondrion is depicted in Fig. 18.27, left panel, while an electron micrograph of a mitochondrion is shown in Fig. 18.27, right panel. The basic structural components of a mitochondrion include: A surrounding outer mitochondrial membrane, a small intermembrane space, an inner mitochondrial membrane that in places is highly folded into what is known as the cristae (note it is not the cristae *membrane*), and, finally, the central aqueous space known as the matrix [49]. The outer and inner mitochondrial membranes are as different as two membranes can be. The outer mitochondrial membrane is full of large pores that are extremely permeable, even to small proteins like cytochrome c. As discussed in Chapter 24, release of cytochrome c into the cytoplasm is an important component of apoptosis. In sharp contrast, the inner mitochondrial membrane must be impermeable to protons in order to function (see below, Chapter 18, Section 5.3). The outer membrane also does not directly participate in electron transport and oxidative phosphorylation and in fact does not house functions that rise to the level of the much more complex inner membrane.

The processes of electron transport and oxidative phosphorylation are housed in the highly folded mitochondrial inner membrane (the cristae). The mitochondrial aqueous interior chamber, called the matrix, houses most of the enzymes involved in the Krebs cycle (terminal steps in sugar oxidation) and β-oxidation (fatty acid oxidation). These metabolic pathways provide the electrons (reduction potential) that drive electron transport and oxidative phosphorylation for production of ATP.

5.2 Redox Reactions

Oxidation/reduction (redox) reactions are the energetic driving force behind ATP production through oxidative phosphorylation [50]. Consider the following general reaction:

$$A_{red} + B_{ox} \rightarrow A_{ox} + B_{red} + \text{Energy}(\Delta G)$$

where A_{red} is the reduced form of compound A, A_{ox} is the oxidized form of compound A, B_{red} is the reduced form of compound B, and B_{ox} is the oxidized form of compound B.

In order to quantify the various redox reactions, the ability of many molecules to be reduced has been determined experimentally and listed in tables of "half cell reduction potentials." For a very abbreviated list of half-cell reduction potentials, see Table 18.1. Reduction potential is a quantitative measure of the ability of a molecule to acquire electrons and thereby be reduced. Reduction potentials are usually reported in volts. The more positive a reduction potential, the greater is the molecule's affinity for electrons and hence its ability to be reduced.

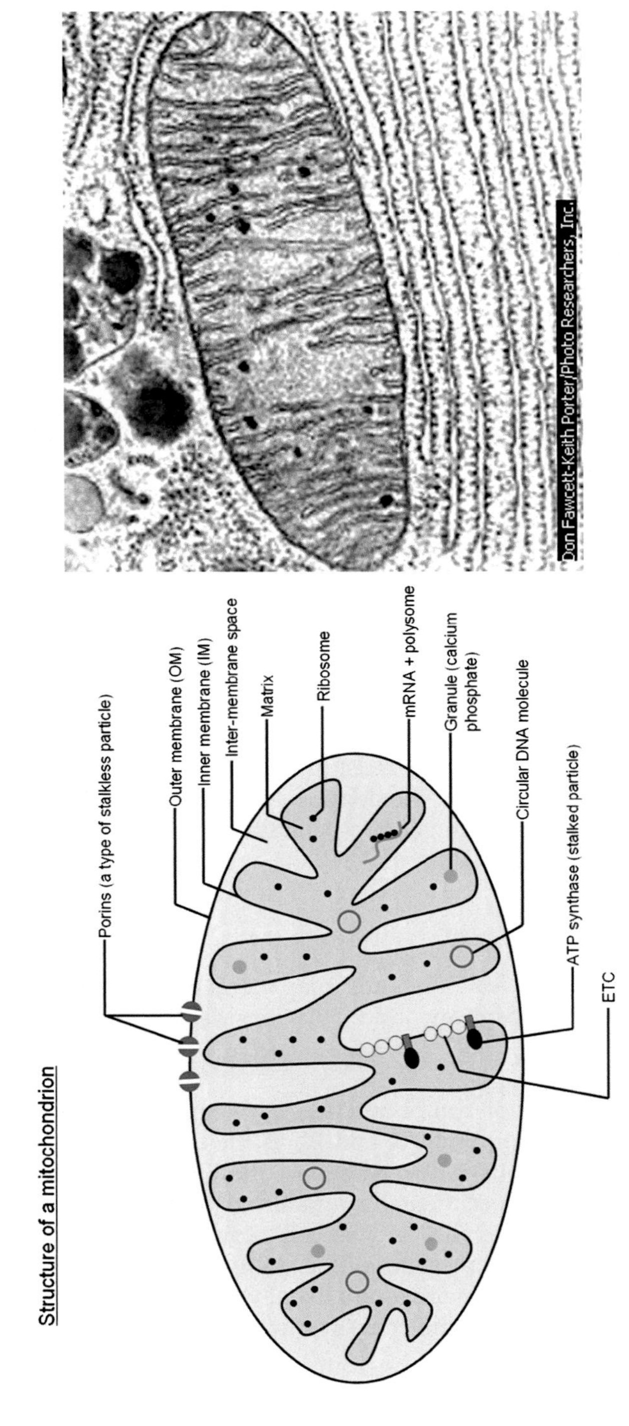

FIGURE 18.27 Left: schematic drawing of a "typical" mitochondrion. Right: an electron micrograph of a mitochondrion. *Left: http://www.mitoq.com/blog/the-mysterious-mitochondria/; Right: http://themagicschoolbus.blogspot.com/2006/05/cell-structure.html.*

TABLE 18.1 Selected Standard Half-Cell Reduction Potentials

Half-cell reactions			E'_o (V)
Succinate + CO_2 + $2e^-$	→	α-Ketoglutarate	−0.67
$2H^+$ + $2e^-$	→	H_2	−0.421
NAD^+ + $2H^+$ + $2e^-$	→	NADH	−0.32
FAD + $2H^+$ + $2e^-$	→	$FADH_2$	−0.22
Pyruvate + $2H^+$ + $2e^-$	→	Lactate	−0.185
Fumarate + $2H^+$ + $2e^-$	→	Succinate	+0.03
Ubiquinone + $2H^+$ + $2e^-$	→	Ubiquinol	+0.045
Cytochrome c (Fe^{3+}) + e^-	→	Cytochrome c (Fe^{2+})	+0.22
$^1/_2$ O_2 + $2H^+$ + $2e^-$	→	H_2O	+0.82

Note that the hydrogen electrode, $2H^+ + 2e^- \rightarrow H_2$ (where $[H^+]$ is 1 mol/L, $T = 298$ K, $P = 1$ atm, Pt electrode) is assigned a standard reduction potential, $E_o = 0$ V. However, in the biochemical convention, the standard state is $[H^+] = 10^{-7}$ mol/L (pH 7), $T = 298$ K, $P = 1$ atm, Pt electrodes, $E'_o = -0.421$ V.

Note that the general reaction above has coupled the oxidation of one molecule (A_{red}) to the reduction of another (B_{ox}). All redox reactions must couple an oxidation to a reduction. The larger the difference in reduction potentials between the two redox-coupled compounds, the larger is the ΔG of the reaction:

$$\Delta G = -nF\Delta E$$

where ΔG is the Gibbs free energy change of the reaction, n is the number of electrons transferred from the reduced to the oxidized molecule (n is either 1 or 2), F is the Faraday constant, and ΔE is the difference in reduction potentials between the two participants in the redox reaction.

It is the basic objective of redox couples in the electron transport chain to have a larger $-\Delta G$ than ATP hydrolysis ($\Delta G = -7.3$ kcal). Sufficiently large redox couples will generate enough energy to possibly drive ATP production by oxidative phosphorylation:

$$\text{ADP} + \text{Pi} \leftrightarrows \text{ATP} + H_2O \quad (\Delta G = 7.3\ \text{kcal})$$

Hydrolysis of ATP (the reverse reaction) releases 7.3 kcal of energy ($\Delta G = -7.3$ kcal). An abbreviated list of reduction potentials important in bioenergetics (ATP production) is listed in Table 18.1.

The more positive a reduction potential is, the stronger is the tendency of that compound to gain electrons (become reduced) and, conversely, the less positive a reduction potential is, the stronger is the tendency of that compound to loose electrons (become oxidized). In other words, in a redox-coupled reaction electrons flow from the compound that has the lower reduction potential to the compound with the higher reduction potential. Note the significance of O_2, which has a very high reduction potential of +0.82 V. Oxygen is, therefore, a very good "oxidizing agent."

In the mitochondrial matrix, glucose oxidation through the Krebs cycle (also called the TCA cycle or citric acid cycle) and β-oxidation of fatty acids produce high energy reducing electrons in the form of $FADH_2$ (flavin adenine dinucleotide) and NADH (nicotinamide adenine dinucleotide). It is the movement of electrons from $FADH_2$ or NADH to O_2 through the electron transport system that supplies the energy for ATP production (oxidative phosphorylation). Let us look at the energetics for each of these reactions.

$$\mathrm{FAD} + 2\mathrm{H}^+ + 2\mathrm{e}^- \rightarrow \mathrm{FADH_2} \quad -0.22$$

$$\frac{1}{2}\mathrm{O_2} + 2\mathrm{H}^+ + 2\mathrm{e}^- \rightarrow \mathrm{H_2O} \quad +0.82$$

$$\Delta G = -nF\Delta E$$

where $n = 2$, $F =$ the Faraday constant (23.061 kcal per volt gram equivalent), $\Delta E =$ electron acceptor—electron donor: $+0.82 - (-0.22) = +1.04$ V, $\Delta G = -2(23)(1.04) = -47.8$ kcal.

$$\mathrm{NAD}^+ + 2\mathrm{H}^+ + 2\mathrm{e}^- \rightarrow \mathrm{NADH} + \mathrm{H}^+ \quad -0.32$$

$$\frac{1}{2}\mathrm{O_2} + 2\mathrm{H}^+ + 2\mathrm{e}^- \rightarrow \mathrm{H_2O} \quad +0.82$$

$$\Delta G = -nF\Delta E$$

where $n = 2$, $F =$ the Faraday constant (23.061 kcal per volt gram equivalent), $\Delta E =$ electron acceptor—electron donor: $+0.82 - (-0.32) = +1.14$ V, $\Delta G = -2(23)(1.14) = -53.44$ kcal.

Since the ΔG of hydrolysis of ATP is about -7.3 kcal, the oxidation of both $FADH_2$ and NADH, when coupled to O_2, releases enough redox energy to theoretically drive production of several ATPs, depending on efficiency of the process. As the ΔG for NADH is significantly larger than the ΔG for $FADH_2$, it should produce more ATPs than $FADH_2$.

So far, we have just considered the simple case of $\Delta G = -nF\Delta E$ where the concentration of all participating molecular species is the same, and thus there is no concentration dependency in the process. An extension of this simple equation was presented by Walther Herman Nernst (Fig. 18.28) in an equation that now bares his name – the Nernst equation. Nernst was a renowned, pioneering German physical chemist who made numerous important contributions in electrochemistry, thermodynamics, solid state chemistry, and photochemistry. For his work on chemical affinities as embodied in the Third Law of Thermodynamics, Nernst was awarded the 1920 Nobel Prize in Chemistry.

Of importance to this discussion was his development of the Nernst equation. The simple equation, $\Delta G = -nF\Delta E$, indicates that you could completely predict the direction and magnitude of a redox reaction through half-cell reduction potential tables alone. However, Nernst showed that unfavorable reactions as predicted by the reduction potentials can be reversed by changing the concentrations of the redox molecular participants. In addition to the standard reduction potential E_o, a second part of the equation relating to concentration had to be added, resulting in the Nernst equation:

$$E = E_o - (RT/nF)\ln([\mathrm{Red}]/[\mathrm{OX}])$$

FIGURE 18.28 Walther Herman Nernst, 1864–1941. http://en.wikipedia.org/wiki/Walther_Nernst.

where E is the actual reduction potential, E_o is the standard reduction potential, R is the gas constant, T is the temperature in K, n is the number of electrons transferred (one or two), F is the Faraday constant, [Red] is the concentration of reduced compound A, [OX] is the concentration of oxidized compound A.

$$A_{red} + B_{ox} \rightarrow A_{ox} + B_{red} + \text{Energy}(\Delta G)$$

A similar calculation must be done for redox component B and the net ΔE (actual) calculated and this is used to calculate the actual $\Delta G = -nF\Delta E$.

While the source of the redox energy became clear, there was still a significant obstacle to understanding the *mechanism* of oxidative phosphorylation that baffled scientists for decades. Movement of electrons from a reduced molecule to O_2 releases *electrical* energy while ATP production involves *chemical* energy. What is the link between the two seemingly unrelated processes? The answer to this conundrum was based on membrane structure.

5.3 Electron Transport Chain

As electrons move down their reduction potential gradient from NADH or $FADH_2$ to O_2, energy (ΔG) for ATP production is generated. If this process occurred in one large step, most, perhaps all, of the energy would be wasted as heat. To better control the process and enhance the capture of useful chemical energy, the process is broken down into several, smaller steps

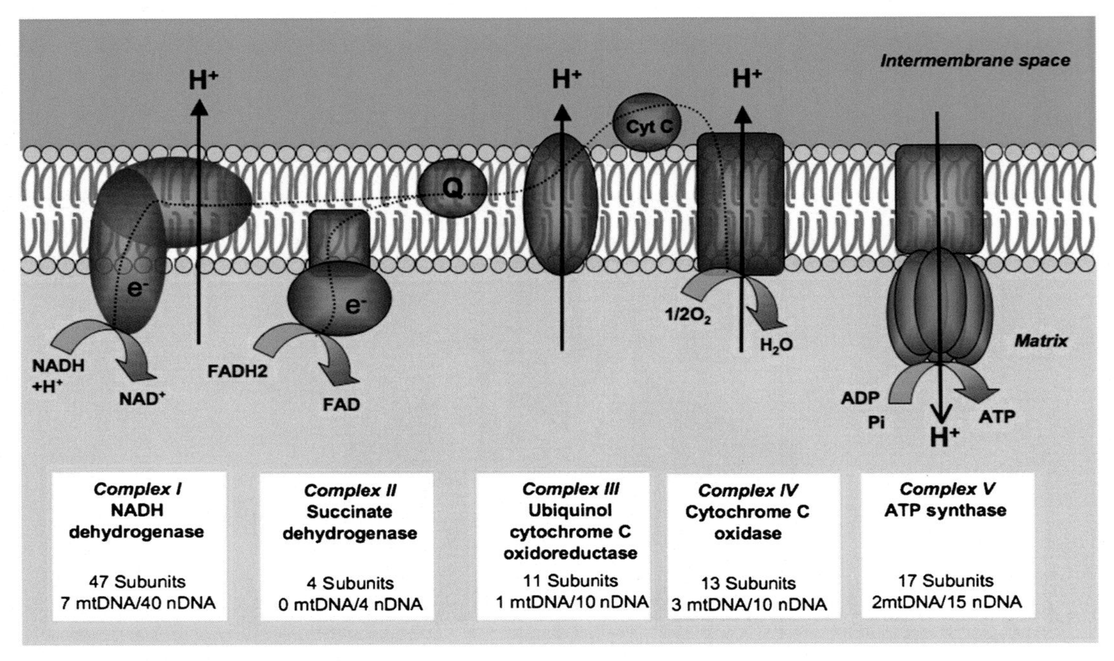

FIGURE 18.29 The five complexes of mitochondrial electron transport and oxidative phosphorylation. *Frontiers in Bioscience January 1, 2009;14:4015–34. http://www.bioscience.org/2009/v14/af/3509/fulltext.php?bframe=figures.htm.*

in a type of "bucket brigade" known as the electron transport chain. In the early 1960s, the basic concept of mitochondrial electron transport chain structure was being revamped. Prior to this time, the electron transport chain was believed to consist of a linear sequence of connected monomeric redox proteins called cytochromes:

$$\begin{array}{l}\text{NADH} \rightarrow \text{Cytochrome b} \rightarrow \text{Cytochrome } c_1 \rightarrow \text{cytochrome c} \rightarrow \text{Cytochrome } a,a_3 \rightarrow O_2\\ \qquad\qquad\qquad\qquad\qquad\quad \uparrow \\ \qquad\qquad\qquad\qquad\quad \text{FADH}_2\end{array}$$

Early work by Hatefi and co-workers [51,52], however, demonstrated that structure of the electron transport chain was far more complex than this. These workers, and others, were able to physically separate electron transport into four basic complexes referred to as complex I, complex II, complex III, and complex IV. A fifth complex, complex V, was also found. Complex V proved to be the mitochondrial F_1F_O-ATP synthase and did not participate directly in electron transport (Fig. 18.29) [53]:

$$\begin{array}{l}\text{NADH} \rightarrow \text{Complex I} \rightarrow \text{Q} \rightarrow \text{Complex III} \rightarrow \text{cytochrome c} \rightarrow \text{Complex IV} \rightarrow O_2\\ \qquad\qquad\qquad\qquad\quad \uparrow \\ \qquad\qquad\qquad\quad \text{Complex II} \\ \qquad\qquad\qquad\qquad\quad \uparrow \qquad\qquad\qquad\qquad\qquad\qquad \text{Complex V} \\ \qquad\qquad\qquad\quad \text{FADH}_2 \qquad\qquad\qquad\qquad\qquad (F_1F_O\text{-ATP synthase})\end{array}$$

Note that these complexes replaced the single component redox cytochromes but remain in the same sequence. These complexes are small islands of membrane composed of the electron transport protein (cytochrome), other proteins (including nonheme irons), and polar lipids (including phospholipids and coenzyme Q). The complexes are membrane microdomains discovered some three decades earlier than lipid rafts (discussed in Chapter 8)! The complexes communicate through "mobile elements," coenzyme Q and cytochrome c. Each isolated complex was shown to be affected by the classic electron transport inhibitors - rotenone (complex I), amytal (complex I), thenoyltrifluoroacetone (TTFA, complex II), antimycin A, (complex III), azide (complex IV), and cyanide (complex IV), and their effect on each complex is indistinguishable from that observed for the native mitochondrial chain. Discovery of the complexes led David Green (see Fig. 8.5) to propose his 1970 "protein crystal" model for membrane structure (see Chapter 8) [54]. Furthermore, the complexes could be isolated into separate tubes and, when simply mixed together, could reconstitute electron transport and oxidative phosphorylation. Not surprisingly, these powerful mitochondrial fractionation studies received considerable acclaim.

5.4 Chemiosmotic Theory

With NADH as the source of reducing electrons, three of these steps (referred to as complex I, complex III, and complex IV in Fig. 18.29) have enough energy to drive ATP production, but for $FADH_2$ only two steps (complex III and complex IV) can generate ATP. Fig. 18.29 depicts the basic components of the electron transport chain starting with NADH and terminating with O_2. Complex I contains the redox enzyme NADH dehydrogenase and is often called NADH–coenzyme Q reductase. Complex I has the lowest reduction potential of the complexes. Complex III is referred to as coenzyme Q–cytochrome c reductase, and complex IV, cytochrome c oxidase, has the highest reduction potential of the complexes. Complex II (succinate–CoQ reductase) contains the redox enzyme succinate dehydrogenase but does not generate ATP, and complex V is the mitochondrial F_1F_o-ATP synthase responsible for making ATP. Complexes I, III, and IV pump protons from the matrix to the intermembrane space, generating a transmembrane pH gradient. Complexes II is not a proton pump. An early 1977 review on electron transport and oxidative phosphorylation written by the six giants in the field while working in their prime is highly recommended reading [55].

Not all redox steps in the electron transport chain release equal amounts of energy ($\Delta G = -nF\Delta E$). Fig. 18.30 shows the half-cell reduction potentials of electron transport components from $NAD^+/NADH + H^+$ (−0.32 V) to $^1/_2\ O_2 + 2H^+/H_2O$ (+0.82 V). Three sections of the electron transport chain have sufficient ΔE to drive ATP synthesis. They are complex I (site 1 ATP synthesis), complex III (site 2 ATP synthesis), and complex IV (site 3 ATP synthesis).

While the basic information involving electron transport components, inhibitors, reduction potentials and sites for ATP production were known prior to 1961, it was still unknown how electrical energy (ΔE) could generate chemical energy in the form of ATP. This conundrum was solved by British biochemist Peter Mitchell (Fig. 18.26) with his highly creative "chemiosmotic hypothesis" [56]. As with many ingenious theories, it took several years for

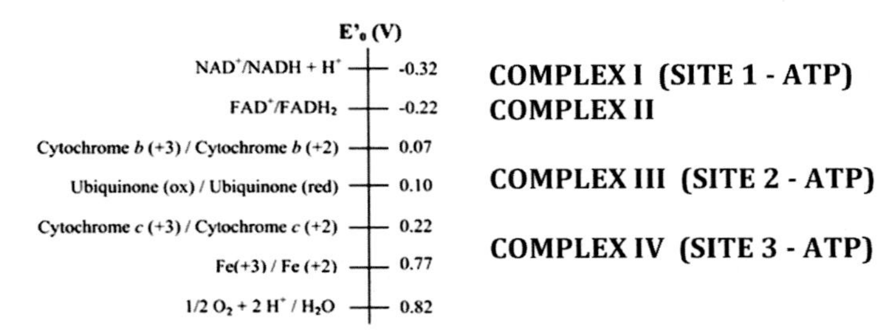

FIGURE 18.30 Half-cell reduction potentials of components of the electron transport chain and the three sites that have sufficient energy (ΔE) to drive ATP synthesis. http://pubs.rsc.org/en/content/articlelanding/2008/ee/b810642h/unauth#!divAbstract.

the chemiosmotic hypothesis to be understood. Mitchell had worked in the area of "vectorial metabolism" which was based on reactions occurring across biological membranes. While a water-soluble, cytoplasmic enzyme usually functions in any spatial orientation, transmembrane proteins bind a substrate on one side of the membrane and release products to the opposite side. Therefore, transmembrane-associated reactions are vectorial (directional). Mitchell proposed that the elusive link between ΔE and ATP synthesis was vectorial proton pumps associated with the electron transport chain in the cristae. The mitochondrial proton pumps move protons from the matrix to the intermembrane space, thus simultaneously creating a pH gradient (acidic outside) and an electrochemical gradient (negative inside). Transmembrane protons are generated at complexes I, III, and IV as electrons pass from the electron donor (NADH or $FADH_2$) to the terminal electron acceptor (O_2). Electron transport thus stores some of the redox electrical energy as a transmembrane pH gradient that is later discharged through a channel in the F_1F_o-ATP synthase (complex V) driving ATP synthesis in the process of oxidative phosphorylation:

$$ADP + Pi \leftrightarrows ATP + H_2O \quad (\Delta G = 7.3 \text{ kcal})$$

An electron transport component that appears often in mitochondrial membranes where it is found in three complexes is ubiquinone (also known as coenzyme Q (CoQ), Fig. 18.31, top). Ubiquinone is a major electron and proton carrying component of the mitochondrial electron transport system [57]. Ubiquinone has been described as being one of two "mobile carriers" (the other is cytochrome c) that is isolated as a component of complexes I, II, and III and therefore carries electrons and protons between these complexes. Ubiquinone is the most abundant of all the mitochondrial electron transport carriers, being about 10 times more abundant than any other carrier. Ubiquinone is also the most versatile of all of the electron transport components, having the ability to carry either one or two electrons and one or two protons, as depicted in Fig. 18.31, bottom. In complex III, ubiquinone is involved in cycling protons and electrons in what is known as the "Q-cycle." Ubiquinone was discovered in 1957 by Frederick Crane from an extract of beef heart. The molecule has a water-soluble quinone anchored to the hydrophobic membrane interior by a very long, methyl branched (isoprenoid) chain. One of the most common ubiquinone has a 50-carbon (10 isoprene units) anchoring chain. This molecule is often referred to as

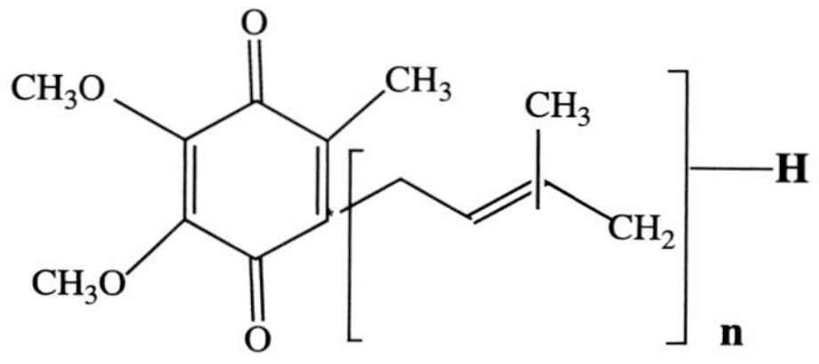

UBIQUINONE (COENZYME Q)

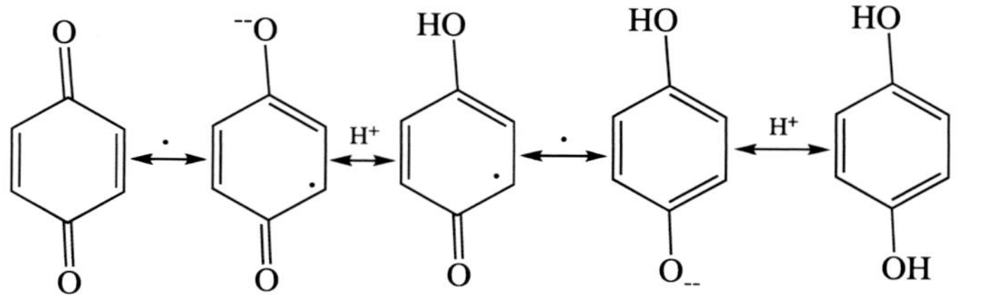

fully oxidized quinone | quinone anion radical | semi-quinone radical | quinone anion | fully reduced dihydroquinone

FIGURE 18.31 Structure (top) and function (bottom) of coenzyme Q (ubiquinone). CoQ is versatile as it can carry one or two electrons and one or two protons and is the major component of the mitochondrial electron transport chain.

coenzyme Q10 or just CoQ10. CoQ10 has recently been shown to be effective against Parkinson's disease [58].

5.5 F_1F_o-ATP Synthase

ATP is synthesized from ADP plus Pi in complex V (F_1F_o-ATP synthase), and reaction energy is derived by a trans-cristae electrochemical gradient. The gradient is established as electrons flow down the electron transport chain from low reduction potential NADH or $FADH_2$ to high reduction potential O_2. During electron transport, protons are pumped from the matrix (inner) side of the cristae to the inter-membrane space (outer) side. Since protons are charged, an electrochemical gradient (negative inside, low pH outside) is established. Therefore, there is both a chemical and electrical gradient to force the protons back through a channel in the center of the F_1F_o-ATP synthase. The force on the protons is referred to as the "proton motive force" (PMF) and consists of two parts: a proton chemical (ΔpH) gradient and a transmembrane electrical potential ($\Delta\psi$). By simple manipulation of the Gibbs free energy ($\Delta G = -nF\Delta E$) and Nernst equation and replacing the proton gradient with ΔpH and ΔE with $\Delta\psi$, the following equation for the proton motive force can be obtained:

$$\text{PMF} = -F\Delta\psi + 2.3RT\Delta\text{pH}$$

TABLE 18.2 PMF Values for the Bioenergetic Membranes, Mitochondria, and Chloroplasts

Membrane	$\Delta\psi$ (mV)	ΔpH	ΔG (kJ/mol)	H^+/ATP
Mitochondria	170	0.5	66	3.4
Chloroplasts	0	3.3	60	3.1

It is the *sum* of these two components that provide the energy for ATP synthesis. It is possible to synthesize ATP even if there is no pH gradient at all—if the $\Delta\psi$ is sufficiently large. The proton motive force must total about 50 kJ/mol to make ATP, where 1 kJ/mol = 10.4 mV. Table 18.2 lists the PMF values for the bioenergetic membranes, mitochondria and chloroplasts. Chloroplast chemiosmotic function is inside-out compared to mitochondria. In mitochondria, the electron transport−generated pH gradient is acidic outside (low pH), while in chloroplasts, the low pH is inside. In mitochondria, most of the PMF is the electrical component ($\Delta\psi$), while in chloroplasts that also make ATP through a PMF, the force is almost exclusively due to the pH gradient (ΔpH).

While the energetics and nature of the proton channel are basically understood, the mechanism by which Pi is coupled to ADP remains mostly a mystery. The F_1F_o-ATP synthase is composed of many subunits that can be divided into two functional domains known as the F_o and F_1 domains (Fig. 18.32) [59]. The F_o domain is an integral protein that forms the transmembrane proton channel. The F_1 domain is found on the matrix side of the membrane and is responsible for ATP synthesis. Together these domains form a rotary motor that couples proton movement to ATP production.

5.6 Proof of the Chemiosmotic Hypothesis

Mitchell's chemiosmotic hypothesis was so creative that it took several years to be understood and appreciated [60]. Fortunately, Mitchell's hypothesis made a number of predictions that could be validated experimentally.

1. Mitochondria with physically damaged (leaky) inner membranes should not be able to generate and support the required transmembrane proton gradient and so should not be able to make ATP. Indeed, physically damaged mitochondria have never been able to make ATP.
2. If the chemiosmotic hypothesis is correct, an artificially established pH gradient across the mitochondrial inner membrane should be able to drive ATP synthesis independent of electron transport. This possibility was tested in 1966 by Cornell University plant physiologist Andre Jagendorf in his famous "acid bath experiment" [61]. The acid bath experiment used chloroplasts as the test bioenergetic organelle. Chloroplasts function in the reverse direction of mitochondria, pumping protons into the chloroplast interior. Jagendorf found that when he incubated chloroplasts in a pH 4 buffer (hence, "acid bath") and subsequently placed them into a pH 8 buffer containing ADP and Pi, ATP was produced. Later, Peter Mitchell tested the same basic experiment with

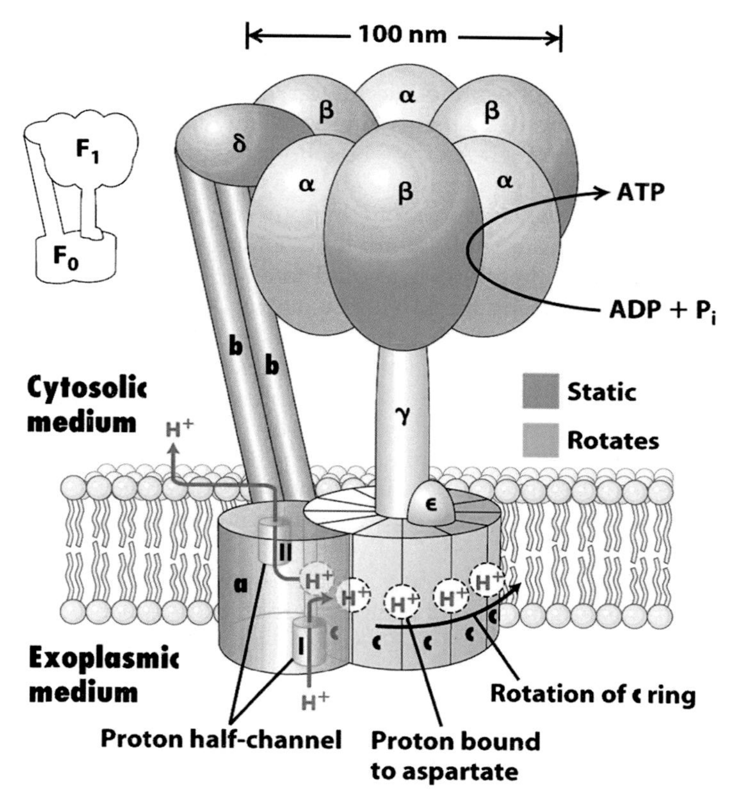

FIGURE 18.32 Schematic drawing showing the mechanism of action of the mitochondrial F_1F_o-ATP synthase. *From Molecular Cell Biology. 6th ed. WH Freeman and Company; 2008, Fig. 12-24. http://biochemist01.files.wordpress.com/2013/04/atpase.png*

mitochondria, however, reversing the direction of the pH gradient. Once again ATP was produced.

3. Another corner stone of the chemiosmotic hypothesis is that electron transport from NADH or $FADH_2$ to O_2, generates a transmembrane pH gradient that should be measurable. Do electron transport component proton pumps actually decrease the pH outside the mitochondria? When intact mitochondria were suspended in a lightly buffered, O_2-free solution containing an oxidizable substrate (eg, succinate, the substrate for complex II), addition of O_2 resulted in acidification of the suspending media.
4. The chemiosmotic hypothesis predicts a precise topography for all of the electron transport components. For decades many types of experiments have confirmed that mitochondrial inner membrane topography agrees with chemiosmotic predictions. This is particularly impressive considering Mitchell proposed his theory a full decade before the advent of the fluid mosaic model for membrane structure.
5. As a side-product, the chemiosmotic hypothesis proposed a new mechanism of action for uncouplers of oxidative phosphorylation. Uncouplers prevent the connection

between electron transport and ATP production through oxidative phosphorylation. Prior to the chemiosmotic hypothesis, it was believed that uncouplers inhibit the production of a high-energy, phosphorylated intermediate responsible for ATP production. Despite intense effort, no high-energy intermediate was ever found. By the chemiosmotic hypothesis, the high-energy phosphorylated intermediate does not exist because electron transport energy is instead trapped as a transmembrane proton (electrochemical) gradient. The chemiosmotic hypothesis predicted that the proton gradient is dissipated by uncouplers that are lipid-soluble proton ionophores. The mechanism of an uncoupler, therefore, is to bind to an H^+, which dissolves into the membrane lipid bilayer, returning the H^+ to the matrix before it has a chance of entering the F_1F_o-ATP synthase. Albert Lehninger tested this hypothesis using a phospholipid planar bimolecular lipid membrane (BLM, see Chapter 9) [62]. Lehninger found that indeed the classic uncoupler, 2,4-dinitrophenol, increased BLM conductivity sufficiently to account for uncoupling in mitochondria. While artificial uncouplers can be toxic, even lethal, the natural uncoupling protein, thermogenin, has an important biological role. In brown adipose tissue, mitochondria of hibernating animals dissipate the proton gradient resulting in heat production (thermogenesis) rather than ATP production.

6. Perhaps the strongest support for the chemiosmotic hypothesis comes from reconstitution experiments. Any source (mitochondrial, chloroplast, or bacterial) of electron transport and any source of F-ATPase can be mixed and successfully generate ATP, proving the mechanism's universality.

6. SUMMARY

Membranes play essential roles in virtually all cellular processes. Chapter 18 discusses involvement of membranes in the seemingly unrelated cell functions of (1) anesthetic action, (2) G protein–coupled reactions, (3) MAC (membrane attack complex), (4) nerve conduction, and (5) electron transport/oxidative phosphorylation.

1. The molecular mechanism of anesthetic action has remained a highly investigated mystery for more than a century. Starting in 1899, the Meyer–Overton correlation suggested the lipid bilayer component of nervous tissues was the anesthetic target. Targeting the lipid bilayer in a "nonspecific mechanism" proved to be necessary, but not sufficient, as a protein, likely the Na^+ channel, must also be involved.
2. A large class of plasma membrane surface receptors, known as G proteins, share the same seven transmembrane α-helix motif. They function by binding an external ligand (a first messenger, usually a hormone or signaling molecule) and then undergo a conformational change that transduces a signal across the plasma membrane and causes the synthesis of an intracellular second messenger. The second messenger triggers an amplification cascade resulting in the intracellular production of countless final products.
3. The immune system is responsible for defense against pathogens. One important component in the defense arsenal is the complement-derived MAC. The MAC is a

cascade that begins with attachment of complement to antibodies attached to the outer surface of a pathogen and ends with destruction of that cell by the complement-generated MAC. MAC functions by punching a hole through the pathogen plasma membrane causing cell lysis and death.

4. There is perhaps no biological process that is more completely dependent on membrane structure than nerve conductance. All plasma membranes develop and maintain trans-membrane electrochemical gradients. Nerve conduction is dependent on these gradients—Na^+ (low inside) and K^+ (high inside) with a resting potential of about −70 mV, established by the plasma membrane-bound Na^+,K^+-ATPase. An action potential (nerve impulse) occurs when the membrane potential fluctuates rapidly over a very short time span, the result of complex interactions between many proteins, primarily ion pumps and ion channels. The action potential is passed down the nerve axon until it reaches the terminal synapse where the nervous (electrical) signal is converted to a chemical signal in the form of a neurotransmitter. The neurotransmitter crosses a synapse extraneuronal space before entering the next neuron in the sequence at its postsynaptic neuron receptor. Therefore, an electrical signal in one neuron is transmitted across a synapse as a chemical signal to a second neuron where it converted back to an electrical signal.

5. Electron transport and oxidative phosphorylation is an excellent example of a highly complex function that is membrane dependent. This process describes how chemical energy originally trapped in the bonds of biochemical substrates is first converted to electrical energy as reduction potentials in the electron transport chain and this energy is subsequently stored as transmembrane electrochemical pH gradients which are finally converted back to chemical energy in the form of ATP. The mechanism for this process is known as the chemiosmotic hypothesis.

Chapter 19 will discuss how material can cross a membrane. Methods discussed include simple passive and facilitated diffusion, active transport, ionophores, and gap junctions.

References

[1] Franks NP. General anaesthesia: from molecular targets to neuronal pathways of sleep and arousal. Nat Rev Neurosci 2008;9:370–86.

[2] Hendrickx JF, Eger EI, Sonner JM, Shafer SL. Is synergy the rule? A review of anesthetic interactions producing hypnosis and immobility. Anesth Analg 2008;107(2):494–506.

[3] Hall BA, Chantigian RC. Anesthesia: a comprehensive review: expert consult. 4th ed. Mosby/Elsevier Amsterdam; 2010.

[4] Milam SB. General anesthetics: a comparative review of pharmacodynamics. Anesth Prog 1984;31(3):116–23.

[5] Boland FK. The first anesthetic: the story of Crawford Long. Athens, Georgia: University of Georgia Press; 2009. ISBN:13-978-0-8203-3436-3.

[6] Meyer HH. Zur Theorie der Alkoholnarkose. Arch Exp Pathol Pharmacol 1899;42(2–4):109–18.

[7] Overton CE. Studien über die Narkose zugleich ein Beitrag zur allgemeinen Pharmakologie. Jena, Switzerland: Gustav Fischer; 1901.

[8] Franks NP, Lieb WR. Where do general anaesthetics act? Nature 1978;274(5669):339–42.

[9] Ueda I. The window that is opened by optical isomers. Anesthesiology 1999;90(1):336.

[10] Franks NP. Molecular targets underlying general anaesthesia. Br J Pharmacol 2006;147(1):72–81.

[11] Miller KW, Paton WD, Smith RA, Smith EB. The pressure reversal of general anesthesia and the critical volume hypothesis. Mol Pharmacol 1973;9(2):131–43.
[12] Trudell JR. A unitary theory of anesthesia based on lateral phase separations in nerve membranes. Anesthesiology 1977;46(1):5–10.
[13] Janoff AS, Miller KW. A critical assessment of the lipid theories of general anaesthetic action. Biol Membr 1982;4(1):417–76.
[14] Cameron JW. The molecular mechanisms of general anaesthesia: dissecting the $GABA_A$ receptor. Contin Educ Anaesth Crit Care Pain 2006;6(2):49–53.
[15] Janoff AS, Pringle MJ, Miller KW. Correlation of general anesthetic potency with solubility in membranes. Biochim Biophys Acta 1981;649(1):125–8.
[16] Nau C, Strichartz GR. Drug chirality in anesthesia. Anesthesiology 2002;97(2):497–502.
[17] Franks NP, Lieb WR. Stereospecific effects of inhalational general anesthetic optical isomers on nerve ion channels. Science 1991;254(5030):427–30.
[18] Johansson JS, Zou H. Nonanesthetics (nonimmobilizers) and anesthetics display different microenvironment preferences. Anesthesiology 2001;95(2):558–61.
[19] Liu J, Laster MJ, Taheri S, Eger EI, Koblin DD, Halsey MJ. Is there a cutoff in anesthetic potency for the normal alkanes? Anesth Analg 1993;77(1):12–8.
[20] Pringle MJ, Brown KB, Miller KW. Can the lipid theories of anesthesia account for the cutoff in anesthetic potency in homologous series of alcohols? Mol Pharmacol 1981;19(1):49–55.
[21] Franks NP, Lieb WR. Do general anaesthetics act by competitive binding to specific receptors? Nature 1984;310(16):599–601.
[22] Franks NP, Lieb WR. Mapping of general anesthetic target sites provides a molecular basis for cutoff effects. Nature 1985;316(6026):349–51.
[23] Cantor RS. The lateral pressure profile in membranes: a physical mechanism of general anesthesia. Biochemistry 1997;36(9):2339–44.
[24] Liu R, Loll PJ, Eckenhoff RG. Structural basis for high-affinity volatile anesthetic binding in a natural 4-helix bundle protein. FASEB J 2005;19(6):567–76.
[25] Filmore D. It's a GPCR world. Modern drug discovery. American Chemical Society; November 24–28, 2004.
[26] Krauss G. Biochemistry of signal transduction and regulation. Wiley-VCH; 2008. p. 15.
[27] Rodbell M. The role of hormone receptors and GTP-regulatory proteins in membrane transduction. Nature 1980;284(5751):17–22.
[28] Berg JM, Tymoczko JL, Stryer L. Section 21.3 epinephrine and glucagon signal the need for glycogen breakdown. Biochemistry. 5th ed. New York: W.H. Freeman; 2002.
[29] Ronnett GV, Moon C. G proteins and olfactory signal transduction. Annu Rev Physiol 2002;64(1):189–222.
[30] Dessauer CW, Gilman AG. Purification and characterization of a soluble form of mammalian adenylyl cyclase. J Biol Chem 1996;271:16967–74.
[31] Muller-Eberhard HJ. The membrane attack complex of complement. Annu Rev Immunol 1986;4:503–28.
[32] Peitsch MC, Tschopp J. Assembly of macromolecular pores by immune defense systems. Curr Opin Cell Biol 1991;3(4):710–6.
[33] Stanley KK, Luzio JP, Tschopp J, Kocher HP, Jackson P. The sequence and topology of human complement component of C9. EMBO J 1985;4(2):375–82.
[34] Tschopp J, Masson D, Stanley KK. Structural/functional similarity between proteins involved in complement- and cytotoxic T-lymphocyte-mediated cytolysis. Nature 1986;322(6082):831–4.
[35] Green H, Fleischer RA, Barrow P, Goldberg B. The cytotoxic action of immune gamma globulin and complement on Krebs ascites tumor cells. II. Chemical studies. J Exp Med 1959;109:511.
[36] Lopina OD. Na^+, K^+ ATPase: structure, mechanism, and regulation. Membr Cell Biol 2000;13:721–44.
[37] Barnett MW, Larkman PM. The action potential. Pract Neurol June 2007;7(3):192–7.
[38] Wollner D, Catterall WA. Localization of sodium channels in axon hillocks and initial segments of retinal ganglion cells. Proc Natl Acad Sci USA 1986;83(21):8424–8.
[39] Dominique D, Campanac E, Andrzej B, Carlier E, Alcaraz G. Axon physiology. Physiol Rev 2011;91(2):555–602.
[40] BOUNDLESS. Biology. The nervous system. How neurons communicate. Nerve impulse transmission within a neuron. https://www.boundless.com/biology/the-nervous-system/how-neurons-communicate/nerve-impulse-transmission-within-a-neuron-action-potential/.

[41] Schacter DL, Gilbert DT, Wegner DM. Psychology. 2nd ed. New York: Worth Publishers; 2011. p. 80.
[42] Neuroscience. In: Purves D, Augustine GJ, Fitzpatrick D, et al., editors. What defines a neurotransmitter?. 2nd ed. Sunderland (MA): Sinauer Associates; 2001. http://www.ncbi.nlm.nih.gov/books/NBK10957/.
[43] Sudhof TC. The synaptic vesicle cycle. Annu Rev Neurosci 2004;27:509–47.
[44] Jessell TM, Kandel ER, Schwartz JH. Principles of neural science. 4th ed. New York: McGraw-Hill; 2000. p. 154–69.
[45] Hillel B. Ion channels of excitable membranes. 3rd ed. Sunderland (Mass): Sinauer; 2001. p. 73–7.
[46] Ernster L, Schatz G. Mitochondria: a historical review. J Cell Biol 1981;91(3 Pt 2):227s–55s.
[47] Miller FP, Vandome AF, McBrewster J. Endosymbiont theory. VDM Publishing House Ltd; 2010. p. 74.
[48] Scheffler IE. Mitochondria. 2nd ed. John Wiley & Sons; 2007. 484 pp.
[49] Caprette DR. Structure of mitochondria. Experimental biosciences. Resources For Introductory and Intermediate Level Laboratory Courses; 1996. caprette@rice.edu.
[50] Testa B. Biochemistry of redox reactions. Metabolism of drugs and other xenobiotics. New York: Academic Press; 1994. 471 pp.
[51] Hatefi Y, Haavik AC, Griffiths DE. Studies on the electron transfer system: XL. Preparation and properties of mitochondrial DPNH-coenzyme Q reductase. J Biol Chem 1962;237:1676–80.
[52] Hatefi Y, Haavik AC, Griffiths DE. Studies on the electron transfer system: XLI. Reduced coenzyme Q (QH_2)-cytochrome c reductase. J Biol Chem 1962;237:1681–5.
[53] Green DE, Silman I. Structure of the mitochondrial electron transfer chain. Annu Rev Plant Physiol 1967;18:147–78.
[54] Green DE, Vanderkooi G. Biological membrane structure, I. The protein crystal model for membranes. Proc Natl Acad Sci USA 1970;66:615–21.
[55] Boyer PD, Chance B, Ernster L, Mitchell P, Racker E, Slater EC. Oxidative phosphorylation and photophosphorylation. Annu Rev Biochem 1977;46:955–66.
[56] Mitchell P. Coupling of phosphorylation to electron and hydrogen transfer by a chemi-osmotic type of mechanism. Nature 1961;191(4784):144–8.
[57] Ernster L, Dallner G. Biochemical, physiological and medical aspects of ubiquinone function. Biochim Biophys Acta 1995;1271:105–204.
[58] Liu J, Wang L, Zhan SY, Xia Y. Coenzyme Q10 for Parkinson's disease. Cochrane Database Syst Rev December 7, 2011;12.
[59] Amzel LM, Bianchet MA, Leyva JA. Understanding ATP synthesis: structure and mechanism of the F1-ATPase (Review) Mol Membr Biol 2003;20(1):27–33.
[60] Saier MH. Peter mitchell and his chemiosmotic theories. ASM News November 1, 1997;63:13–21.
[61] Jagendorf AT, Uribe AT. ATP formation caused by acid-base transition of spinach chloroplasts. Proc Natl Acad Sci USA 1966;55:170–7.
[62] Bielawski J, Thompson TE, Lehninger AL. The effect of 2,4-dinitrophenol on the electrical resistance of phospholipid bilayer membranes. Biochem Biophys Res Comm 1966;24(6):94.

CHAPTER

19

Membrane Transport

OUTLINE

1. INTRODUCTION

Life depends on a membrane's ability to precisely control the level of solutes in the aqueous compartments, inside and outside, bathing the membrane. The membrane determines what solutes enter and leave a cell. Transmembrane transport is controlled by

An Introduction to Biological Membranes
http://dx.doi.org/10.1016/B978-0-444-63772-7.00019-1

complex interactions between membrane lipids, proteins, and carbohydrates. How the membrane accomplishes these tasks is the topic of Chapter 19.

A biological membrane is semipermeable, meaning it is permeable to some molecules, most notably water, while being very impermeable to most solutes (various biochemicals and salts) found in the bathing solution. This very important concept of unequal transmembrane distribution and, hence, permeability between water and other solutes came out of the pioneering work of Charles Overton in the 1890s (see Chapter 2). How does a biological membrane accomplish semipermeability? The barrier to solute movement is largely provided by the membrane's hydrophobic core, a very thin (~40 Å thick), oily layer. The inherent permeability of this core varies from membrane to membrane. Generally, the more tightly packed the lipids comprising the bilayer, the lower its permeability will be. Lipid bilayers are very impermeable to most solutes because of their tight packing. Fig. 19.1 depicts the membrane permeability of a variety of common solutes [1]. Note the data are presented as a log scale of solute permeability (P in cm/s) and ranges from $Na^+ = 10^{-12}$ cm/s to water $= 0.2 \times 10^{-2}$ cm/s, spanning almost 10 orders of magnitude!

Lipid bilayer permeability is not a constant but instead is affected by environmental factors. For example, LUVs (large unilamellar veicles) made from DPPC (16:0, 16:0 PC) have a sharp phase transition temperature, T_m, of 41.3°C. At temperatures well below T_m, the LUVs are in the tightly packed gel state and permeability is extremely low. At temperatures well above T_m, the LUVs are in the loosely packed liquid disordered state (l_d, also called the liquid crystalline state) and permeability is high. However, maximum permeability is not found in the l_d state, but rather at the T_m [2]. As the LUVs are heated from the gel state and approach the T_m, domains of l_d start to form in the gel state. Solutes can then pass more readily through the newly formed l_d domains than the gel domains resulting in an increase in permeability. At T_m there is a maximum amount of coexisting gel and l_d state domains that exhibit extremely porous domain boundaries. It is through these boundaries that most permeability occurs. As the temperature is further increased, the LUVs pass into the l_d state and the interface boundaries disappear, reducing permeability to that observed for the single-component l_d state. Thus, maximum permeability is observed at the T_m.

1.1 Fick's First Law

The tendency for solutes to move from a region of higher concentration to one of lower concentration was first defined in 1855 by the physiologist Adolf Fick (Fig. 19.2). His work is summarized in what is now the very well-known Fick's Laws of Diffusion [3]. The laws apply to both free solution and diffusion across membranes. Fick developed his laws by measuring concentrations and fluxes of salt diffusing between two reservoirs through connecting tubes of water.

Fick's First Law describes diffusion as:

$$\text{Diffusion rate} = -DA\frac{dc}{dx}$$

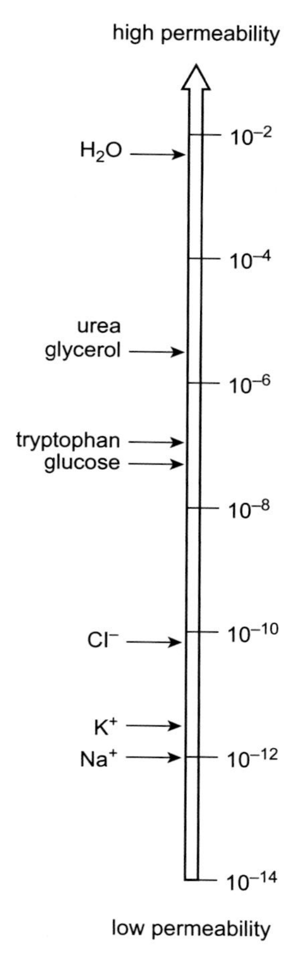

FIGURE 19.1 Log of the permeability (P in cm/s) across lipid bilayer membranes for common solutes ranging from Na^+ (10^{-12} cm/s) to water (0.2×10^{-2} cm/s). This range spans almost 10 orders of magnitude [1].

Where D = diffusion coefficient (bigger molecules have lower Ds); A = cross-sectional area over which diffusion occurs; dc/dx is the solute concentration gradient (diffusion occurs from a region of higher concentration to one of lower concentration).

The relationship between a solute's molecular weight and its diffusion coefficient is shown in Table 19.1. Large solutes have low diffusion coefficients and therefore diffuse more slowly than small solutes. The diffusion rate for a particular solute under physiological conditions is a constant and cannot be increased. This defines the theoretical limit for an enzymatic reaction rate and also limits the size of a cell. If a solute starts at the center of a bacterial cell, it takes about 10^{-3} s to diffuse to the plasma membrane. For this reason, typical cells are microscopic (see Chapter 1). At about 3.3 pounds and the size of a cantaloupe, the largest cell on Earth today is the ostrich egg. However a fossilized dinosaur egg in the American Museum of Natural History in New York is about

FIGURE 19.2 Adolf Fick, 1829–1901.

TABLE 19.1 Relationship Between a Solute's Molecular Weight and Its Diffusion Coefficient, *D*

Compound	O_2	Acetyl choline	Sucrose	Serum albumin
D ($cm^2/s \times 10^6$)	19.8	5.6	2.4	0.7
Molecular weight	32	182	342	69,000

the size of basketball. Since an egg's only function is to store nutrients for a developing embryo, its size is many orders of magnitude larger than a normal cell.

1.2 Osmosis

Osmosis is a special type of diffusion, namely the diffusion of water across a semipermeable membrane. Water readily crosses a membrane down its potential gradient from high to low potential (Fig. 19.3) [4]. Osmotic pressure is the force required to prevent water movement across the semipermeable membrane. Net water movement continues until its potential reaches zero. An early application of the basic principles of osmosis came from the pioneering work on hemolysis of red blood cells by William Hewson in the 1770s (see Chapter 2). It has also been discussed that MLVs (multilamellar vesicles, liposomes) behave as almost perfect osmometers, swelling in hypotonic solutions and shrinking in hypertonic solutions (see Chapter 3) [5,6]. Liposome swelling and shrinking can be easily followed by changes in absorbance due to light scattering using a simple spectrophotometer. Therefore, osmosis has been investigated for many years using common and inexpensive methodologies and a lot is known about the process.

Membranes are rarely, if ever, perfectly semipermeable. Deviation from ideality is defined by a reflection coefficient (σ). For an ideal semipermeable membrane where a solute is totally

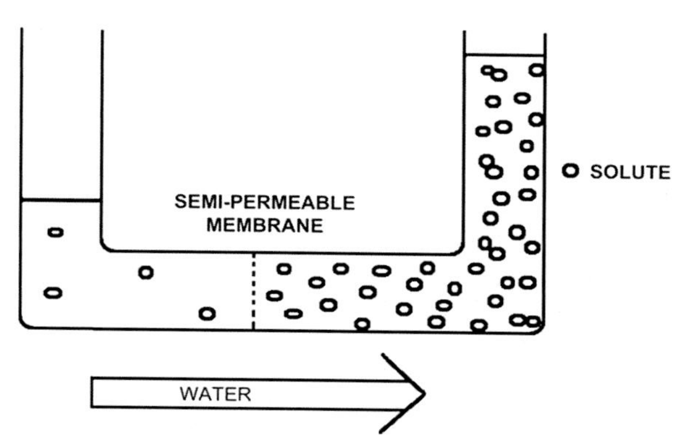

FIGURE 19.3 Osmosis and osmotic pressure. Water is placed in a U-shaped tube where each of the tube arms is separated by a semipermeable membrane with pores of a size that water can easily pass through but a solute cannot. Upon addition of the solute to the tube's right arm, water diffuses from left to right (high water potential to low). The column of water in the tube's right arm (the one containing the solute) rises until the extra weight of the column equals the osmotic pressure caused by the solute. A pump could then be used to counter the osmotic pressure whereupon the solution columns in the right and left arms of the tube are made the same. The pump pressure required to equalize the height of the two columns is the osmotic pressure [4]. Note a small amount of the solute leaks from right to left since no filter is perfect.

impermeable, $\sigma = 1$. If a solute is totally permeable (its permeability is equal to water), $\sigma = 0$. Biological membranes are excellent semipermeable barriers with $\sigma = 0.75$ to 1.0.

2. SIMPLE PASSIVE DIFFUSION

Movement of solutes across membranes can be divided into two basic types: passive diffusion and active transport [7]. Passive diffusion requires no additional energy source other than what is found in the solute's electrochemical (concentration) gradient and results in the solute reaching equilibrium across the membrane. Passive diffusion can be either simple passive diffusion where the solute crosses the membrane anywhere by simply dissolving into and diffusing through the lipid bilayer, or facilitated passive diffusion where the solute crosses the membrane at specific locations where diffusion is assisted by solute-specific facilitators or carriers. Active transport requires additional energy, often in the form of ATP, and results in a nonequilibrium, net accumulation (uptake) of the solute on one side of the membrane. The basic types of membrane transport, simple passive diffusion, facilitated diffusion (by channels and carriers) and active transport are summarized in Fig. 19.4 [8]. There are countless different examples of each type of membrane transport process [7]. Only a few representative examples will be discussed here.

Even simple passive diffusion requires energy to cross a bilayer membrane. In order to cross a membrane, the solute must first lose its waters of hydration, diffuse across the membrane, and then regain its waters on the opposite side. The limiting step involves the energy required to lose the waters of hydration. Table 19.2 shows the relationship between the waters of hydration (proportional to the number of –OH groups on a homologous series

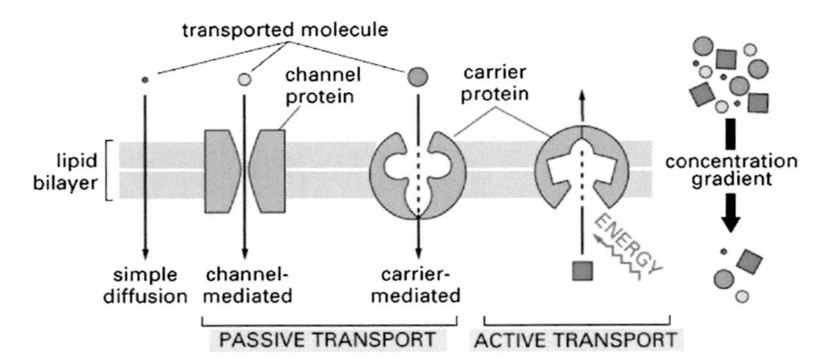

FIGURE 19.4 Basic types of membrane transport, simple passive diffusion, facilitated diffusion (by channels and carriers), and active transport [8].

TABLE 19.2 Relationship Between the Waters of Hydration (Number of –OH Groups on a Homologous Series of Solutes) and the Activation Energy for Transmembrane Diffusion

Solute	Activation energy (kJ/mol)
Glycol ($HO–CH_2–CH_2–OH$)	60
Glycerol ($HO–CH_2–CH(OH)–CH_2–OH$)	77
Erythritol ($HO–CH_2–CH(OH)–CH(OH)–CH_2–OH$)	87

of solutes) and the activation energy for transmembrane diffusion. As the number of waters of hydration increases from glycol < glycerol < erythritol, the activation energy for diffusion also increases. The activation energy compares very well with the energy of hydration.

However, water diffusion does not fit this model. Water permeability is just too high. Several possibilities have been suggested to account for the abnormally high membrane permeability of water:

1. Water is very small and so it just dissolves in bilayers better than larger solutes.
2. Due to its size, water can readily enter very small statistical pores (~4.2 Å in diameter). Statistical pores result from the simultaneous lateral movement of adjacent membrane phospholipids in opposite directions. Statistical pores have only a fleeting existence and cannot be isolated or imaged.
3. Passage down water chains.
4. Water can be carried down kinks in acyl chains that result from acyl chain melting (see lipid melting in Chapter 9).
5. Water may rapidly cross membranes through nonlamellar regions (eg, micelles, cubic or H_{II} phase—see Chapter 10).
6. High water permeability will occur at regions of packing defect (eg, surface of integral membrane proteins, boundary between membrane domains).
7. Through pores or channels used to conduct ions.
8. Through specific water channels known as aquaporins (see below, Chapter 19, Section 3.5).

The only molecules that can cross a membrane by simple passive diffusion are water, small noncharged solutes, and gasses. Charged or large solutes are virtually excluded from membranes and so require more than just simple passive diffusion to cross a membrane.

3. FACILITATED DIFFUSION

Facilitated diffusion (also known as carrier-mediated diffusion) is, like simple passive diffusion, dependent on the inherent energy in a solute gradient. No additional energy is required to transport the solute and the final solute distribution reaches equilibrium across the membrane. Facilitated diffusion, unlike simple passive diffusion, requires a highly specific transmembrane integral protein or carrier to assist in the solute's membrane passage. Facilitators come in two basic types: carriers and gated channels. Facilitated diffusion exhibits Michaelis-Menton saturation kinetics (Fig. 19.5, Part A, right),

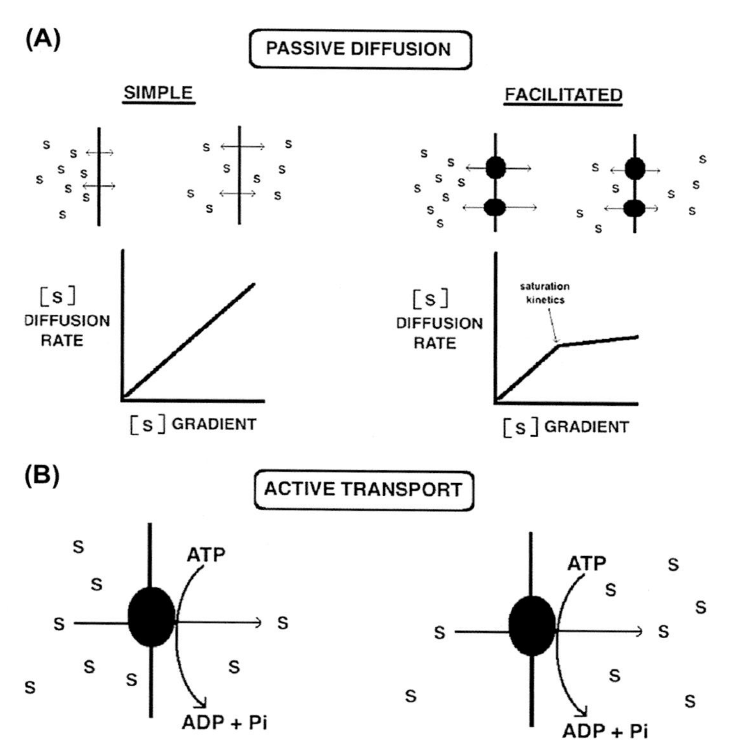

FIGURE 19.5 (A) Simple passive diffusion (top, left) and facilitated passive diffusion (top, right) both result in a final equilibrium distribution of a solute across the membrane. For a noncharged solute, the final distribution of the solute would find equal amounts of S on both sides of the membrane. Facilitated diffusion employs a specific transporter and exhibits Michaelis–Menten saturation kinetics. (A, center right) Active transport (bottom) utilizes energy, often in the form of ATP, to drive solute uptake against its gradient resulting in a net accumulation of the solute.

indicating the carrier has an enzyme-like active site. Like enzymes, facilitated diffusion carriers exhibit saturation kinetics and recognize their solute with exquisite precision, easily distinguishing chemically similar isomers like D-glucose from L-glucose. Fig. 19.5 (Part A) compares simple passive diffusion to facilitated diffusion. The figure is not to scale, however, as facilitated diffusion is orders of magnitude faster than simple passive diffusion.

3.1 Glucose Transporter

A well-studied example of a facilitated diffusion carrier is the glucose transporter, or GLUT [9]. From the activation energies for transmembrane simple passive diffusion of glycol, glycerol and erythritol presented in Table 19.2, it can be estimated that the activation energy for glucose should be well over 100 kJ/mol, but instead it is only 16 kJ/mol. This large discrepancy is attributed to the presence of a glucose-facilitated diffusion carrier. Fig. 19.6 demonstrates the mode of action of one of these transporters, GLUT-1, from the erythrocyte [10]. GLUTs occur in nearly all cells and are particularly abundant in cells lining the small intestine. GLUTs are but one example in a superfamily of transport facilitators. GLUTs are integral membrane proteins whose membrane-spanning region is composed of 12 α-helices. GLUTs function through a typical membrane transport mechanism [10]. Glucose binds to the membrane outer surface site causing a conformational change associated with transport across the membrane. At the inner side of the membrane, glucose is released into the internal aqueous solution (Fig. 19.6).

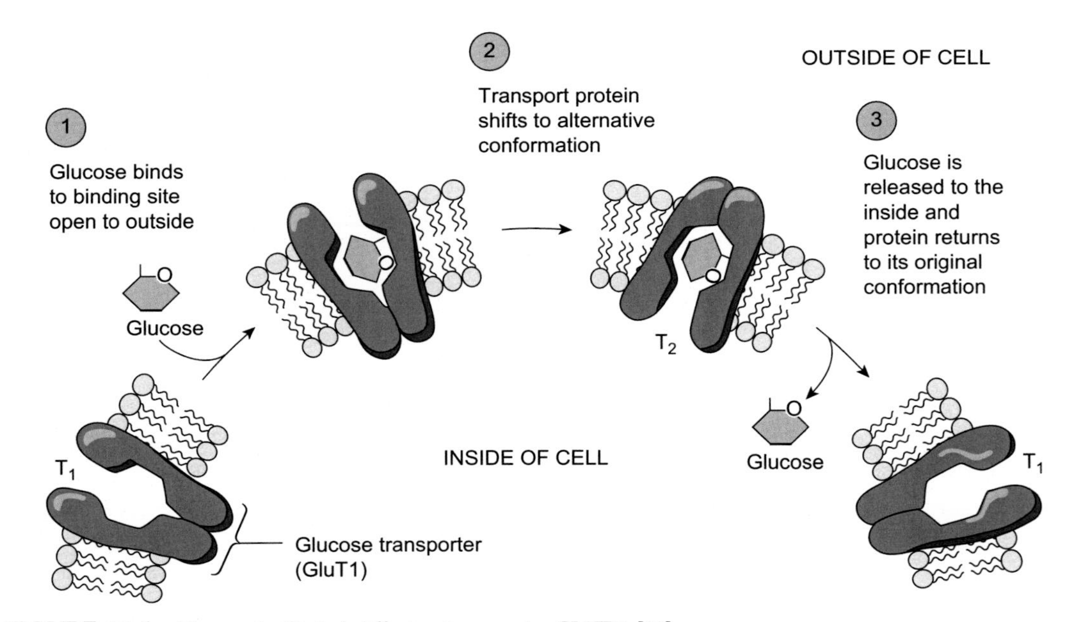

FIGURE 19.6 Glucose-facilitated diffusion transporter GLUT-1 [10].

3.2 Potassium Channels

In virtually all organisms there exists a wide variety of ion channels, the most widely distributed being potassium channels [11]. There are four basic classes of potassium channels, all of which provide essential membrane-associated functions including setting and shaping action potentials and hormone secretion:

1. Calcium-activated potassium channel
2. Inwardly rectifying potassium channel
3. Tandem pore domain potassium channel
4. Voltage-gated potassium channel

Potassium channels are composed of four protein subunits that can be the same (homotetramer) or closely related (heterotetramer). All potassium channel subunits have a distinctive pore-loop structure that sits at the top of the channel and is responsible for potassium selectivity [12]. This is often referred to as a selectivity or filter loop. The selectivity filter strips the waters of hydration from the potassium ion, allowing it into the channel. Farther down the structure is a 10-Å-diameter, transmembrane, water-filled central channel that conducts potassium across the membrane. Elucidating the three-dimensional structure of this important integral membrane protein by X-ray crystallography (Fig. 19.7) [12] was a seminal accomplishment in the field of membrane biophysics. For this

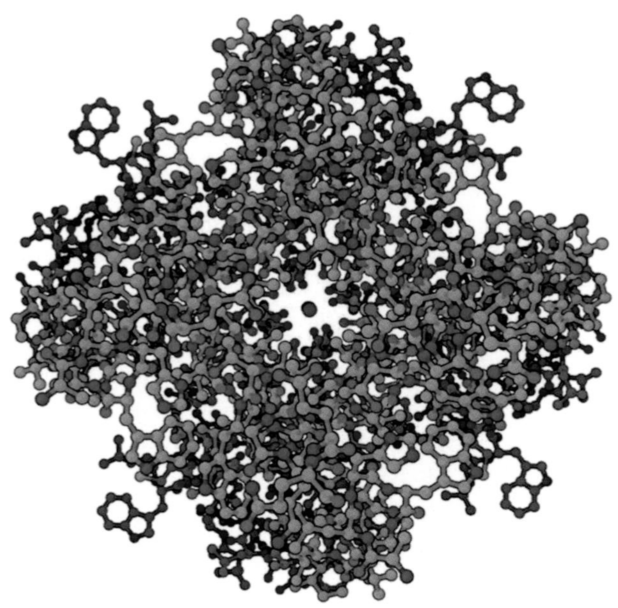

FIGURE 19.7 Three-dimensional structure of the potassium channel [12]. The channel itself is the clear opening in the center of the structure and a single K^+ is shown in the center of the channel.

FIGURE 19.8 Rod MacKinnon, 1956–.

work from 1998, Rod MacKinnon (Fig. 19.8) of Rockefeller University was awarded the 2003 Nobel Prize in Chemistry. Until the potassium channel work, just obtaining the structure of non–water-soluble proteins was next to impossible. MacKinnon's work elucidated not only the structure of the potassium channel but also its molecular mechanism. It has served as a blueprint for determining the structure of other membrane proteins and has greatly stimulated interest in the field.

3.3 Sodium Channel

In some ways, Na^+ channels [13] parallel the action of K^+ channels. They are both facilitated diffusion carriers that conduct the cation down the ion's electrochemical gradient. In excitable cells such as neurons, myocytes, and some glia, Na^+ channels are responsible for the rising phase of action potentials (see Chapter 18). Therefore agents that block Na^+ channels also block nerve conduction and so are deadly neurotoxins. There are two basic types of Na^+ channels: voltage-gated and ligand-gated. The opening of a Na^+ channel has a selectivity filter that attracts Na^+. From there the Na^+ ions flow into a constricted part of the channel that is about 3–5 Å wide. This is just large enough to allow the passage of a single Na^+ with one attached water. Since the larger K^+ cannot squeeze through, the channel is selective for Na^+. Of particular interest are two extremely potent biological toxins, tetrodotoxin (TTX) and saxitoxin (STX) (Fig. 19.9, [14]), that, in seafood, have killed and injured many humans. Both toxins shut down Na^+ channels by binding from the extracellular side.

TTX is encountered primarily in puffer fish but also in porcupine fish, ocean sunfish, and triggerfish. TTX (Fig. 19.9, left) is a potent neurotoxin that blocks Na^+ channels while having no effect on K^+ channels. Puffer fish is the second most poisonous vertebrate in the world, trailing only the Golden Poison Frog that is endemic to the rain forests on the Pacific Coast of Colombia. In some parts of the world puffer fish are considered to be a delicacy but

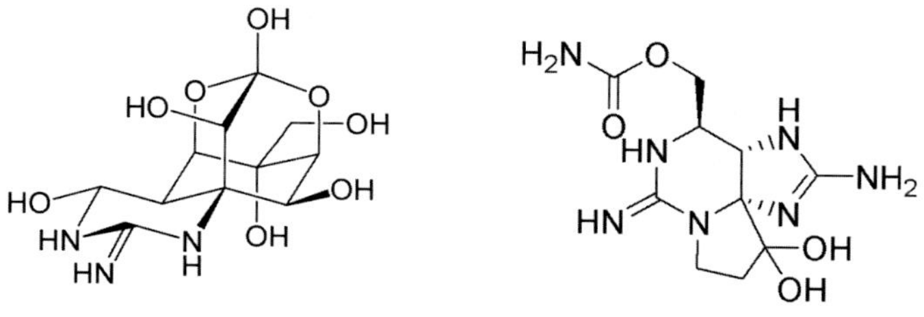

Tetrodotoxin (TTX) **Saxitoxin (STX)**

FIGURE 19.9 Structures of the extremely potent neurotoxins, tetrodotoxin (TTX) and saxitoxin (STX). Both neurotoxins function by blocking the Na^+ channel.

must be prepared by chefs who really know their business, as a slight error can be fatal. Puffer poisoning usually results from consumption of incorrectly prepared puffer soup, and TTX has no known antidote!

Saxitoxin (STX, Fig. 19.9, right) is a Na^+ channel–blocking neurotoxin produced by some marine dinoflagellates that can accumulate in shellfish during toxic algal blooms known as Red Tide. Saxitoxin is one of the most potent natural toxins, and it has been estimated that a single contaminated mussel has enough STX to kill 50 humans! STX's toxicity has not escaped the keen eye of the United States military, which has weaponized the toxin and given it the designation TZ.

3.4 Solute Equilibrium

The driving force for transmembrane solute movement by simple or passive diffusion is determined by the free energy change, ΔG.

$$\Delta G = RT \ln \left[s'_o\right]/[s_o] + ZF\Delta\Psi$$

Where ΔG is the free energy change; $\left[s'_o\right]$ is the solute concentration on the right side of a membrane; $[s_o]$ is the solute concentration on the left side of a membrane; R is the gas constant; T is the temperature in K; Z is the charge of the solute; F is the Faraday; $\Delta\Psi$ is the transmembrane electrical potential.

Solute movement will continue until $\Delta G = 0$. If ΔG is negative, solute movement is left to right (it is favorable as drawn). If ΔG is positive, solute movement is right to left (it is unfavorable in the left-to-right direction) or energy must be added for the solute to go from left to right. The equation has two parts; a transmembrane chemical gradient $\left(\left[s'_o\right]/[s_o]\right)$ and a transmembrane electrical gradient ($\Delta\Psi$). The net movement of a solute is therefore determined by a combination of the solute's chemical gradient and an electrical gradient inherent to the cell. If the solute has no charge, $Z = 0$ (as is the case for glucose) and the right hand part of the equation ($ZF\Delta\Psi$) drops out. Therefore, the final equilibrium distribution of glucose across the membrane will have the internal glucose concentration equal to the external glucose concentration and is independent of $\Delta\Psi$, the electrical potential. At equilibrium for a noncharged solute, $\Delta G = RT \ln \left[s'_o\right]/[s_o]$ and ΔG can only be = zero if $\left[s'_o\right] = [s_o]$.

The situation for a charged solute like K^+ is more complicated. The net ΔG is determined by both the chemical gradient $\left([s'_o]/[s_o]\right)$ and electrical gradient ($\Delta\Psi$). The $\Delta\Psi$ results from the sum of all charged solutes on both sides of the membrane, not just K^+. Therefore even if the K^+ concentration is higher inside the cell than outside (the chemical gradient is unfavorable for K^+ uptake), the $\Delta\Psi$ may be in the correct direction (negative interior) and of sufficient magnitude to drive K^+ uptake against its chemical gradient.

3.5 Aquaporins

Aquaporins are also known as water channels and are considered to be "the plumbing system for cells" [15,16]. For decades it was assumed that water simply leaked through biological membranes by numerous processes described above (Chapter 19, Section 2). However, these methods of water permeability could not come close to explaining the rapid movement of water across some cells. Although it had been predicted that water pores must exist in very leaky cells, it was not until 1992 that Peter Agre (Fig. 19.10) at Johns Hopkins University identified a specific transmembrane water pore that was later called aquaporin-1. For this accomplishment Agre shared the 2003 Nobel Prize in Chemistry with Rod MacKinnon for his work on the potassium channel. Aquaporins are usually specific for water permeability and exclude the passage of other solutes. A type of aquaporin known as aqua-glyceroporins can also conduct some very small uncharged solutes such as glycerol, CO_2, ammonia, and urea across the membrane. However, all aquaporins are impermeable to charged solutes. Water molecules traverse the aquaporin channel in single file (Fig. 19.11) [17].

4. ACTIVE TRANSPORT

A characteristic of all living membranes is the formation and maintenance of transmembrane gradients of all solutes including salts, biochemicals, macromolecules, and even water. In living cells, large gradients of Na^+ and K^+ are particularly important. Typical cell concentrations are:

Cell interior:	400 mmol/L K^+, 50 mmol/L Na^+
Cell exterior:	20 mmol/L K^+, 440 mmol/L Na^+

FIGURE 19.10 Peter Agre, 1949–.

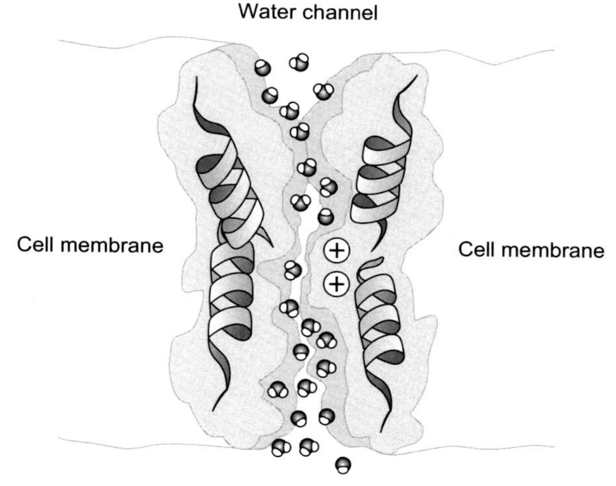

FIGURE 19.11 Aquaporin. Water molecules pass through the aquaporin channel in single file.

Living cells will also have a $\Delta\Psi$ from −30 to −200 mV (negative interior) resulting from the uneven distribution of all ionic solutes including Na^+ and K^+. The chemical and electrical gradients are maintained far from equilibrium by a multitude of active transport systems. Active transport requires a form of energy (often ATP) to drive the movement of solutes against their electrochemical gradient, resulting in a nonequilibrium distribution of the solute across the membrane. A number of nonexclusive and overlapping terms are commonly used to describe the different types of active transport. Some of these are depicted in Fig. 19.12 [18].

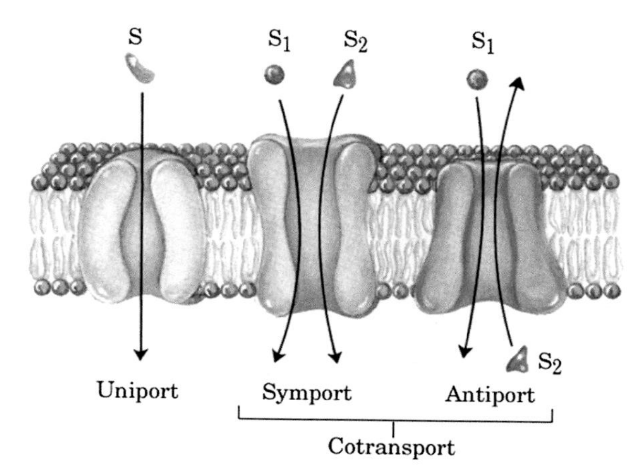

FIGURE 19.12 Basic types of active transport [18].

4.1 Primary Active Transport

Primary active transport is also called direct active transport or uniport. It involves using energy (usually ATP) to directly pump a solute across a membrane against its electrochemical gradient.

The most studied example of primary active transport is the plasma membrane Na^+,K^+-ATPase discussed below (Chapter 19, Section 4.2). Other familiar examples of primary active transport are the redox H^+-gradient generating system of mitochondria (see Chapter 18), the light-driven H^+-gradient generating system of photosynthetic thylakoid membranes, and the ATP-driven acid (H^+) pump found in the epithelial lining of the stomach. There are four basic types of ATP-utilizing primary active transport systems (Table 19.3).

4.2 Na^+,K^+-ATPase

Arguably the most important active transport protein is the plasma membrane-bound Na^+,K^+-ATPase. This single enzyme accounts for one-third of human energy expenditure and is often referred to as the "pacemaker for metabolism." As a result the Na^+,K^+-ATPase has been extensively studied for more than 50 years. The enzyme was discovered in 1957 by Jens Skou (Fig. 19.13) who, 40 years later, was awarded the 1997 Nobel Prize in Chemistry.

As is often the case in biochemistry, a serendipitous discovery of a natural product from the jungles of Africa has been instrumental in unraveling the enzyme's mechanism of action. The compound is ouabain (Fig. 19.14), a cardiac glycoside first discovered in a poison added to the tip of Somali tribesmen's hunting arrows. In fact the name ouabain comes from the

TABLE 19.3 Four Types of ATP-Using Primary Active Transport Systems

ATP-using primary active transport systems	Example
P-type	Na^+,K^+-ATPase Ca^{2+} pump H^+ acid pump
F-type	Mitochondrial ATP synthase Chloroplast ATP synthase
V-type	Vacuolar ATPase
ABC (ATP binding cassette transporter)	Many

FIGURE 19.13 Jens Skou, 1918–.

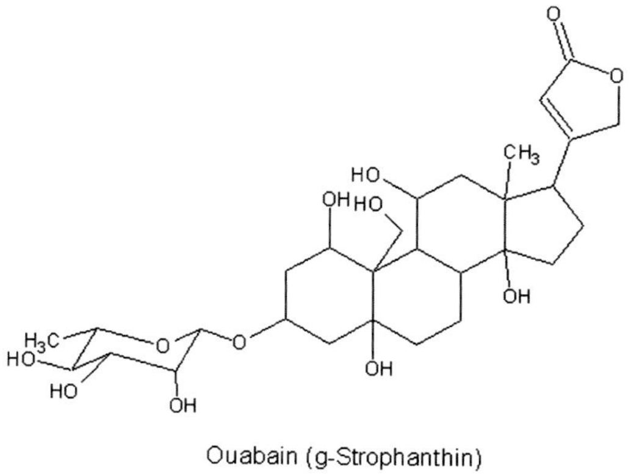

Ouabain (g-Strophanthin)

FIGURE 19.14 Structure of ouabain.

Somali word *waabaayo* that means "arrow poison." The sources of ouabain are ripe seeds and bark of certain African plants and ouabain is potent enough to kill a hippopotamus with a single arrow. For decades after its discovery, ouabain was routinely used to treat atrial fibrillation and congestive heart failure in humans. More recently, ouabain has been replaced by digoxin, a structurally related, but more lipophilic cardiac glycoside.

There are several important observations about Na^+,K^+-ATPase that had to be factored in before a mechanism of action could be proposed. These include:

1. Na^+,K^+-ATPase is an example of active antiport and primary active transport.
2. Na^+,K^+-ATPase is inhibited by ouabain, a cardiac glycoside.
3. Ouabain binds to the outer surface of Na^+,K^+-ATPase and blocks K^+ transport into the cell.
4. Na^+ binds better from the inside.
5. K^+ binds better from the outside.
6. ATP phosphorylates an aspartic acid on the enzyme from the inside.
7. Phosphorylation is related to Na^+ transport.
8. Dephosphorylation is related to K^+ transport.
9. Dephosphorylation is inhibited by ouabain.
10. Three Na^+ ions are pumped out of the cell as two K^+ ions are pumped in, driven by hydrolysis of one ATP.
11. Na^+,K^+-ATPase is electrogenic.

Mechanism of Na^+,K^+-ATPase [19] is based on toggling back and forth between two conformational states of the enzyme, ENZ-1 and ENZ-2 (Fig. 19.15). Three Na^+s bind from the inside to Na^+,K^+-ATPase in one conformation (ENZ-1). This becomes phosphorylated by ATP causing a conformation change producing ENZ-2~P. ENZ-2~P does not bind Na^+, but does bind two K^+ ions. Therefore, three Na^+ ions are released to the outside and two K^+ ions are bound from the outside, generating ENZ-2~P ($2K^+$). Upon hydrolysis of ~P, Na^+,K^+-ATPase (ENZ II) reverts back to the original ENZ-1 conformation that releases two K^+ ions and binds three Na^+ ions from the inside. Ouabain blocks the dephosphorylation step.

ENZ-1 ($3Na^+$) + ATP → ENZ-2~P + $3Na^+$ released outside
(inside) (outside)

↓

ENZ-2~P ($2K^+$) ← ENZ-2~P + 2 K^+
(outside) (outside)

↓

ENZ-1 ($3Na^+$) + Pi + $2K^+$s released inside
(inside)

FIGURE 19.15 Mechanism of the Na^+,K^+-ATPase.

4.3 Secondary Active Transport

Secondary active transport (also known as cotransport) systems are composed of two separate functions. The energy-dependent movement of an ion (eg, H^+, Na^+, or K^+) generates an electrochemical gradient of the ion across the membrane. This ion gradient is coupled to the movement of a solute in either the same direction (symport) or in the opposite direction (antiport, see Fig. 19.12, [18]). Movement of the pumped ion down its electrochemical gradient is by facilitated diffusion. The purpose of both types of co-transport is to use the energy in an electrochemical gradient to drive the movement of another solute against its gradient. An example of symport is the SGLT1 (sodium-glucose transport protein-1) in the intestinal epithelium [20]. SGLT1 uses the energy in a downhill transmembrane movement of Na^+ to transport glucose across the apical membrane against an uphill glucose gradient so that the sugar can be transported into the bloodstream.

4.4 Bacterial Lactose Transport

The secondary active symport system for lactose uptake in *Escherichia coli* is shown in Fig. 19.16 [21]. Lactose uptake is driven through a channel by a H^+ gradient generated by the bacterial electron transport system [22]. The free energy equation for transport described above $\left(\Delta G = RT \ln \left[s'_o\right]/\left[s_o\right] + ZF\Delta\Psi\right)$ can be rearranges for cases employing H^+ gradients (see Chapter 18) to:

$$\Delta\mu_{H^+} = \Delta\Psi - RT/n\mathrm{F}\ \Delta\mathrm{pH}$$

Where $\Delta\mu_{H^+}$ is the proton motive force; $\Delta\Psi$ is the transmembrane electrical potential; R is the gas constant; T is the temperature in °K; n is the solute charge (+1 for protons); F is the Faraday; ΔpH is the transmembrane pH gradient.

It is the *force* on an H^+ (called the proton motive force) that drives lactose uptake. Note that the ability to take up lactose is a combination of the electrical gradient and the pH gradient. Although lactose uptake is directly coupled to H^+ transmembrane movement, it is possible to take up lactose even if the pH gradient is zero (ie, if the $\Delta\Psi$ is sufficiently large).

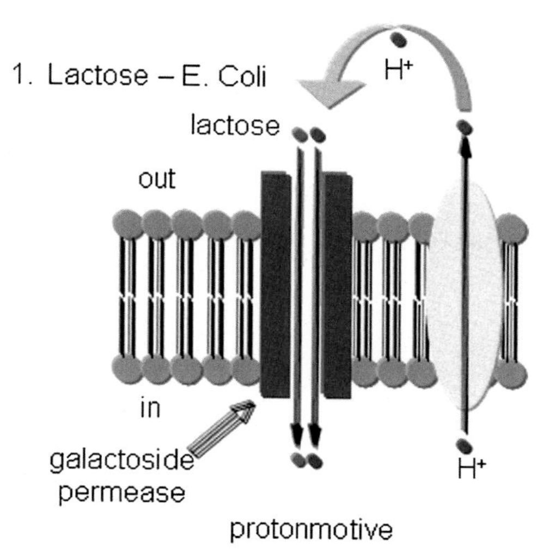

FIGURE 19.16 Lactose transport system in *Escherichia coli* [21]. Uptake of lactose is coupled to the movement of an H^+ down its electrochemical gradient. This is an example of active transport, co-transport, and active.

4.5 Vectorial Metabolism, Group Translocation

Over 50 years ago, Peter Mitchell (see Chapter 18, Fig. 18.26) recognized the importance of what he termed "vectorial metabolism" [23,24]. Water-soluble enzymes convert substrate to product without any directionality. Mitchell proposed that many enzymes are integral membrane proteins that have a specific transmembrane orientation. When these enzymes convert substrate to product they do so in one direction only. This enzymatic conversion is therefore unidirectional, or "vectorial." Mitchell expanded this basic concept into his now famous "chemiosmotic hypothesis" for ATP synthesis in oxidative phosphorylation (Chapter 18) [25,26]. For this revolutionary idea Mitchell was awarded the 1997 Nobel Prize in Chemistry.

Vectorial metabolism has been used to describe the mechanism for several membrane transport systems. For example, it has been reported in some cases the uptake of glucose into a cell may be faster if the external source of glucose is sucrose rather than free glucose. Through a vectorial transmembrane reaction, membrane-bound sucrase may convert external sucrose into internal glucose plus fructose more rapidly than the direct transport of free glucose through its transport system.

Mitchell defined one type of vectorial transport as group translocation, the best example being the PTS (phosphotransferase system) discovered by Saul Roseman in 1964.

PTS is a multicomponent active transport system that uses the energy of intracellular phosphoenol pyruvate (PEP) to take up extracellular sugars in bacteria. Transported sugars include glucose, mannose, fructose, and cellobiose. Components of the system include both plasma membrane and cytosolic enzymes. PEP is a high-energy phosphorylated compound (ΔG of hydrolysis is -61.9 kJ/mol) that drives the system. The high-energy phosphoryl group is transferred through an enzyme bucket brigade from PEP to

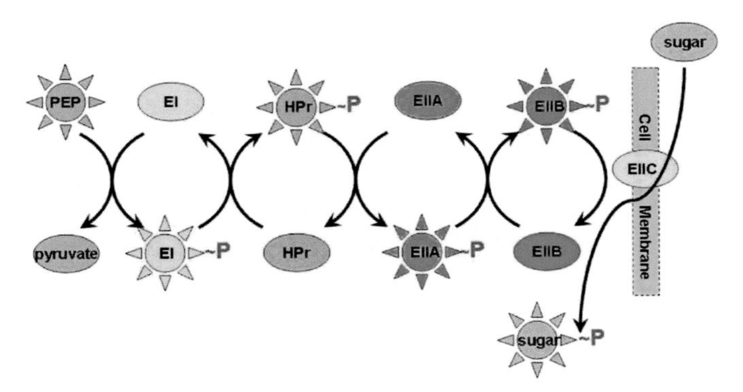

FIGURE 19.17 The bacterial PTS system for glucose transport [27].

glucose producing glucose-6-phosphate in several steps (PEP → EI → HPr → EIIA → EIIB → EIIC → glucose-6-phosphate). The sequence is depicted in more detail in Fig. 19.17 [27]. HPr stands for heat-stable protein that carries the high-energy ~P from EI (enzyme-I) to EIIA. EIIA is specific for glucose and transfers ~P to EIIB that sits next to the membrane where it takes glucose from the transmembrane EIIC and phosphorylates it producing glucose-6-phosphate. Although it is glucose that is being transported across the membrane, it never actually appears inside the cell as free glucose but rather as glucose-6-phosphate. Free glucose could leak back out of the cell via a glucose transporter, but glucose-6-phosphate is trapped inside the cell where it can rapidly be metabolized through glycolysis. Group translocation is defined by a transported solute appearing in a different form immediately after crossing the membrane.

5. IONOPHORES

The term ionophore means "ion bearer." Ionophores are small, lipid-soluble molecules, usually of microbial origin, whose function is to conduct ions across membranes [28,29]. They are facilitated diffusion carriers that transport ions down their electrochemical gradient. Ionophores can be divided into two basic classes: channel formers and mobile carriers (Fig. 19.18) [30]. Channel formers are long lasting, stationary structures that allow many ions at a time to rapidly flow across a membrane. Mobile carriers bind to an ion on one side of a membrane, dissolve in and cross the membrane bilayer and release the ion on the other side. They can only carry one ion at a time. Four representative ionophores will be discussed: the K^+ ionophore valinomycin, the proton ionophore 2,4-dinitrophenol, synthetic crown ethers, and the channel-forming ionophore nystatin (Fig. 19.19).

5.1 Valinomycin

Superficially valinomycin resembles a cyclic peptide (Fig. 19.19). However, upon closer examination the ionophore is actually a 12-unit (dodeca) *depsi*peptide where amino acid peptide bonds are alternated with amino alcohol ester bonds. Therefore the linkages that

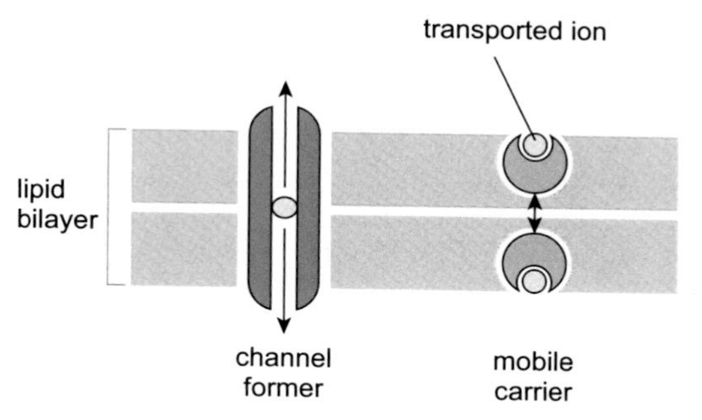

FIGURE 19.18 Two basic types of ionophores: channel formers (left) and mobile carriers (right) [30].

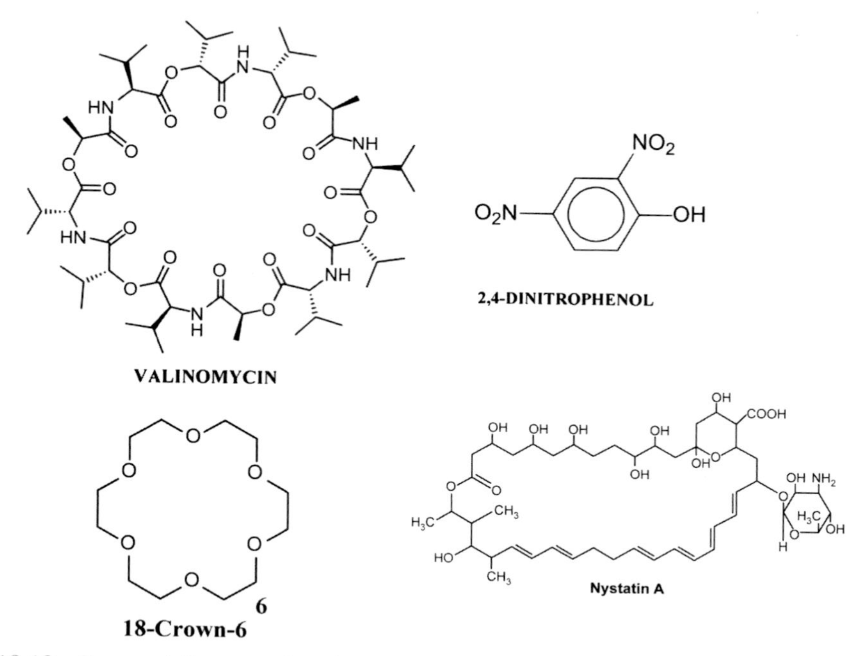

FIGURE 19.19 Representative examples of ionophores: the K^+ ionophore valinomycin, the proton ionophore 2,4-dinitrophenol, the synthetic crown ether 18-crown-6, and the channel forming ionophore nystatin.

hold the molecule together alternate between nitrogen esters (peptide bonds) and oxygen esters. The units that comprise valinomycin are D- and L-valine (hence the name "valinomycin"), hydroxyvaleric acid and L-lactic acid. The circular structure is a macrocyclic molecule with the 12 carbonyl oxygens facing the inside of the ring where they chelate a single K^+. The outside surface of valinomycin is coated with nine hydrophobic side chains of D- and L-valine and L-hydroxyvaleric acid. The polar interior of valinomycin precisely fits one K^+. The binding constant for K^+-valinomycin is 10^6 while Na^+-valinomycin is only 10. This emphasizes the high selectivity valinomycin has for K^+ over Na^+. Valinomycin, therefore, has an oily surface that readily dissolves in a membrane lipid bilayer, carrying K^+ across the membrane down its electrochemical gradient.

Valinomycin was first recognized as a potassium ionophore by Bernard Pressman in the early 1960's [31,32]. He reported that valinomycin, a known antibiotic, stimulated K^+ uptake and H^+ efflux from mitochondria. Many studies showed that valinomycin dissipates essential transmembrane electrochemical gradients causing tremendous metabolic upheaval in many organisms including microorganisms. It is for this reason that valinomycin was recognized as an antibiotic long before it was identified as an ionophore. Currently several ionophores are added to animal feed as antibiotics and growth enhancing additives [33]. Recently valinomycin has been reported to be the most potent agent against SARS-CoV (severe acute respiratory-syndrome coronavirus), a severe form of pneumonia first identified in 2003 [34].

5.2 2,4-Dinitrophenol

2,4-Dinitrophenol (DNP, Fig. 19.19) is considered to be the classic uncoupler of oxidative phosphorylation (see Chapter 18). It is a synthetic lipid-soluble proton ionophore that dissipates proton gradients across bioenergetic membranes (mitochondrial inner, thylakoid, bacterial plasma). An uncoupler is therefore an H^+-facilitated diffusion carrier. Elucidating the role of DNP in uncoupling oxidative phosphorylation was an essential component in support of Peter Mitchell's chemiosmotic hypothesis [25]. Electron movement from NADH or $FADH_2$ to O_2 via the mitochondrial electron transport system generates a considerable amount of electrical energy that is partially captured as a transmembrane pH gradient (see Chapter 18). The movement of H^+s back across the membrane, driven by the electrochemical gradient, is through a channel in the F_1ATPase (an F-type primary active transport system discussed above, (Chapter 19, Section 4.1)) that is coupled to ATP synthesis. DNP short-circuits the H^+ gradient before it can pass through the F_1ATPase, thus uncoupling electron transport, the energy source for the H^+ gradient, from ATP synthesis. Therefore, in the presence of DNP, electron transport continues, even at an accelerated rate, but ATP production is diminished. The energy that should have been converted to chemical energy in the form of ATP is then released as excess heat.

This combination of properties led to the medical application of DNP to treat obesity from 1933 to 1938 [35]. Upon addition of DNP:

- The patient became weak due to low ATP levels.
- Breathing increased due to increased electron transport to rescue ATP production.
- Metabolic rate increased.
- Body temperature increased due to inability to trap electrical energy as chemical energy in the form of ATP, releasing heat.
- Body weight decreased due to increased respiration burning more stored fat.

DNP was indeed a successful weight loss drug. Two of the early proponents of DNP use as a diet drug, Cutting and Tainter at Stanford University, estimated that more than 100,000 people in the United States had tested the drug during its first year in use [35]. DNP, however, did have one disturbing side effect—death! Fatality was not caused by a lack of ATP, but rather by a dangerous increase in body temperature (hyperthermia). In humans, 20–50 mg/kg of DNP can be lethal. Although general use of DNP in the United States was discontinued in 1938, it is still employed in other countries and by bodybuilders to eliminate fat before competitions.

5.3 Crown Ethers

Crown ethers are a family of synthetic ionophores that are generally similar in function to the natural product valinomycin [36]. The first crown ether was synthesized by Charles Pederson (Fig. 19.20) while working at DuPont in 1967. For this work Pedersen was co-awarded the 1987 Nobel Prize in Chemistry. Crown ethers are cyclic compounds composed of several ether groups. The most common crown ethers are oligomers of ethylene oxide with repeating units of $(-CH_2CH_2O-)_n$ where $n = 4$ (tetramer), $n = 5$ (pentamer), or $n = 6$ (hexamer). Crown ethers are given structural names, *X*-crown-*Y*, where *X* is the total number of atoms in the ring and *Y* is the number of these atoms that are oxygen. Crown refers to the crown-like shape the molecule takes. Crown ether oxygens form complexes with specific cations that depend on the number of atoms in the ring. For example, 18-crown-6 (Fig. 19.19) has high affinity for K^+, 15-crown-5 for Na^+, and 12-crown-4 for Li^+. Like valinomycin, the exterior of the ring is hydrophobic, allowing crown ethers to dissolve in the membrane lipid bilayer while carrying the sequestered cation down its electrochemical gradient. It is now possible to tailor make crown ethers of different sizes that can encase a variety of catalysts for phase transfer into the bilayer hydrophobic interior where they can be used to catalyze reactions inside the membrane.

5.4 Nystatin

Nystatin (Fig. 19.19) is a channel-forming ionophore that creates a hydrophobic pore across a membrane [37,38]. Channel-forming ionophores allow for the rapid facilitated diffusion of various ions that depend on the dimensions of the pore. Nystatin, like other channel-forming ionophores (eg, amphotericin B and natamycin), is a commonly used antifungal agent. Finding medications that can selectively attack fungi in the presence of

FIGURE 19.20 Charles Pedersen, 1904–1989.

normal animal cells presents a difficult challenge since both cell types are eukaryotic. Bacteria, being prokaryotes, are sufficiently different to present a variety of anti-bacterial approaches not amenable to fungi. However, fungi do have an Achilles heel. Fungal plasma membranes have as their dominant sterol ergosterol, not the animal sterol cholesterol (see Chapter 5). Nystatin binds preferentially to ergosterol, thus targeting fungi in the presence of animal cells. When present at sufficient levels, nystatin complexes with ergosterol and forms transmembrane channels that lead to K^+ leakage and death of the fungus. Nystatin is a polyene antifungal ionophore that is effective against many molds and yeast including *Candida*. A major use of nystatin is as a prophylaxis for AIDS patients who are at risk for fungal infections.

6. GAP JUNCTIONS

Gap junctions are a common structural feature of many animal plasma membranes [39,40]. In plants similar structures are known as plasmodesmata. Gap junctions were introduced earlier in Chapter 11 (see Fig. 11.6). Gap junctions represent a primitive type of intercellular communication that allows transmembrane passage of small solutes like ions, sugars, amino acids, and nucleotides while preventing migration of organelles and large polymers like proteins and nucleic acids. Gap junctions connect the cytoplasms of two adjacent cells through nonselective channels. Connections through adjacent cells are at locations where the gap between cells is only 2–3 nm. This small gap is where the term "gap junction" originated. Gap junctions are normally clustered from a few to over a 1000 in select regions of a cell plasma membrane.

Early experiments involved injecting fluorescent dyes, initially fluorescein (molecular weight 300), into a cell and observing the dye movement into adjacent cells with a fluorescence microscope [41,42]. Currently Lucifer Yellow has become the fluorescent dye of choice for gap junction studies, replacing fluorescein. At first, the dye only appeared in the initially labeled cell. With time, however, the dye was observed to spread to adjacent cells through what appeared to be points on the plasma membrane. These points were later recognized as gap junctions. By varying the size of the fluorescent dye, it was shown that there was an upper size limit for dye diffusion. Solutes had to have a molecular weight of less than $\sim$1200 to cross from one cell to another [41].

Although gap junctions were obviously channels that connected the cytoplasms of adjacent cells, it was years before their structure, shown in Fig. 19.21, was determined [43,44]. Each channel in a gap junction is made up of 12 proteins called connexins. Six hexagonally arranged connexins are associated with each of the adjacent cell plasma membranes that the gap junction spans. Each set of six connexins is called a connexon and forms half of the gap junction channel. Therefore, one gap junction channel is composed of 2 aligned connexons and 12 connexins. Each connexin has a diameter of about 7 nm and the hollow center formed between the 6 connexins (the channel) is about 3 nm in diameter. Gap junctions allow adjacent cells to be in constant electrical and chemical communication with one another. Of particular importance is the rapid transmission of small second messengers, such as inositol triphosphate (IP_3) and Ca^{2+}.

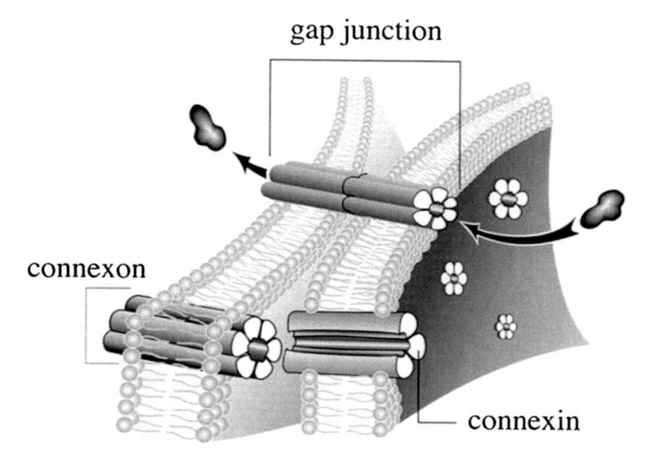

FIGURE 19.21 Gap junction [43]. Six connexins form a connexon and one connexon from each cell unite to form a gap junction.

It appears that all cells in the liver are interconnected through gap junctions. This presents a possible dilemma. If even a single cell is damaged, deleterious effects may be rapidly spread throughout the entire liver. Preventing this is one important function of Ca^{2+}. Extracellular Ca^{2+} is $\sim 10^{-3}$ mol/L while intracellular levels are maintained at $\sim 10^{-6}$ mol/L. If a cell is damaged, Ca^{2+} rushes in, dramatically increasing intracellular Ca^{2+}. Gap junction channels close if intracellular Ca^{2+} reaches 10^{-3} mol/L, thus preventing the spread of damage.

Gap junctions are particularly important in cardiac muscle as the electrical signals for contraction are passed efficiently through these channels [45]. As would be expected, malfunctions of gap junctions lead to a number of human disorders including demyelinating neurodegenerative diseases, skin disorders, cataracts, and even some types of deafness.

7. OTHER WAYS TO CROSS THE MEMBRANE

There are several other ways that solutes, including large macromolecules, can cross membranes. These methods include receptor-mediated endocytosis (RME, discussed in Chapter 17), phagocytosis, pinocytosis, exocytosis, and membrane blebbing. These methods involve large sections of a membrane containing many lipids and proteins.

Two similar transport processes that have been known for a long time are pinocytosis and phagocytosis [46]. Both involve nonspecific uptake (endocytosis) of many things from water and ions through to large macromolecules and, for phagocytosis, even whole cells. *Pinocytosis* is Greek for "cell drinking" and involves the plasma membrane invaginating a volume of extracellular fluid and anything it contains including water, salts, biochemicals and even soluble macromolecules. *Phagocytosis* is Greek for "cell eating" and involves the plasma membrane invaginating large insoluble solids.

7.1 Pinocytosis

Pinocytosis is a form of endocytosis involving fluids containing many solutes. In humans, this process occurs in cells lining the small intestine and is used primarily for absorption of fat droplets. In endocytosis the cell plasma membrane extends and folds around desired extracellular material, forming a pouch that pinches off creating an internalized vesicle (Fig. 19.22, [19–47]). The invaginated pinocytosis vesicles are much smaller than those generated by phagocytosis. The vesicles eventually fuse with the lysosome whereupon the vesicle contents are digested. Pinocytosis involves a considerable investment of cellular energy in the form of ATP and so is many 1000 times less efficient than RME (see Chapter 17). Also, in sharp contrast to RME, pinocytosis is nonspecific for the substances it accumulates. Pinocytosis is not a recent discovery as it was first observed decades before the other transport systems discussed in Chapter 19. Its discovery is attributed to Warren Lewis in 1929.

7.2 Phagocytosis

Phagocytosis is a type of endocytosis that involves uptake of large solid particles, often >0.5 mm [47]. The particles are aggregates of macromolecules, parts of other cells, and even whole microorganisms and, in contrast to pinocytosis (shown in Fig. 19.22), phagocytosis has surface proteins that specifically recognize and bind to the solid particles. Fig. 19.23 [48] depicts events in phagacytosis. Phagocytosis is a routine process that ameba and ciliated protozoa use to obtain food. In humans, phagocytosis is restricted to specialized cells called phagocytes that include white blood cell neutrophils and

Endocytosis – Pinocytosis

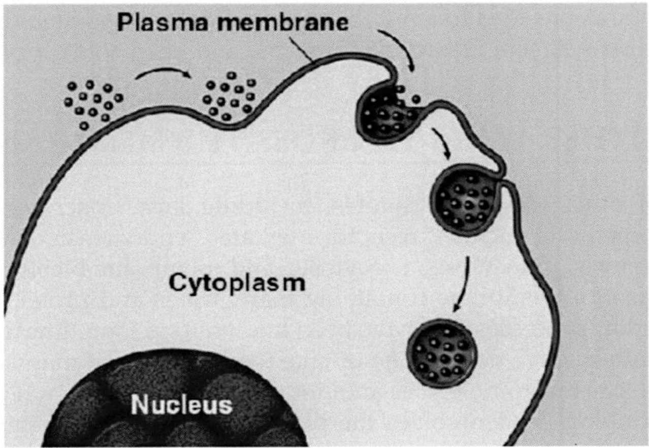

FIGURE 19.22 Pinocytosis, a type of endocytosis. An invagination of the plasma membrane encapsulates many water-soluble solutes ranging in size from salts to macromolecules.

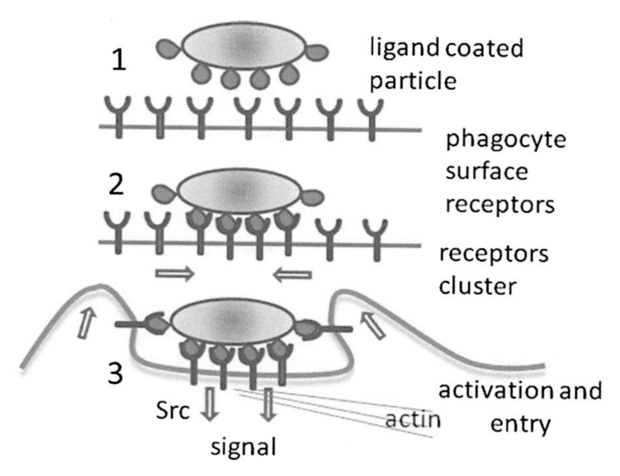

FIGURE 19.23 Phagocytosis, a type of endocytosis that involves uptake of large solid particles.

macrophages. As with pinocytosis, phagocytosis generates intracellular vesicles called phagosomes that have sequestered solid particles they transport to the lysosome for digestion. Phagocytosis is a major mechanism used by the immune system to remove pathogens and cell debris. In fact, very early studies of the immune system led Elie Metchnikoff to discover phagocytosis in 1882. For this work Metchnikoff shared the 1908 Nobel Prize in Medicine with Paul Ehrlich.

7.3 Exocytosis

Exocytosis is the process by which cells excrete waste and other large molecules from the cytoplasm to the cell exterior [49] and therefore is the opposite of endocytosis. Exocytosis generates vesicles referred to as secretory or transport vesicles (Chapter 17). In exocytosis, intracellular (secretory) vesicles fuse with the plasma membrane and release their aqueous sequestered contents to the outside at the same time that the vesicular membrane hydrophobic components (mostly lipids and proteins) are added to the plasma membrane (Fig. 19.24, [50]). Steady state composition of the plasma membrane results from a balance between endocytosis and exocytosis. The resultant process of plasma membrane recycling is amazingly fast. For example, pancreatic secretory cells recycles an amount of membrane equal to the whole surface of the cell in ~90 min. Even faster are macrophages that can recycle contents of their plasma membrane in only 30 min.

Before approaching the plasma membrane for fusion, exocytosis vesicles had a prior life that is considered in Chapter 17. The vesicles must first dock with the plasma membrane, a process that keeps the two membranes separated by <5–10 nm. During docking, complex molecular rearrangements occur to prepare the membranes for fusion. The process of vesicle fusion and release of aqueous compartment components is driven by SNARE proteins (see Chapters 10 and 17) [51,52].

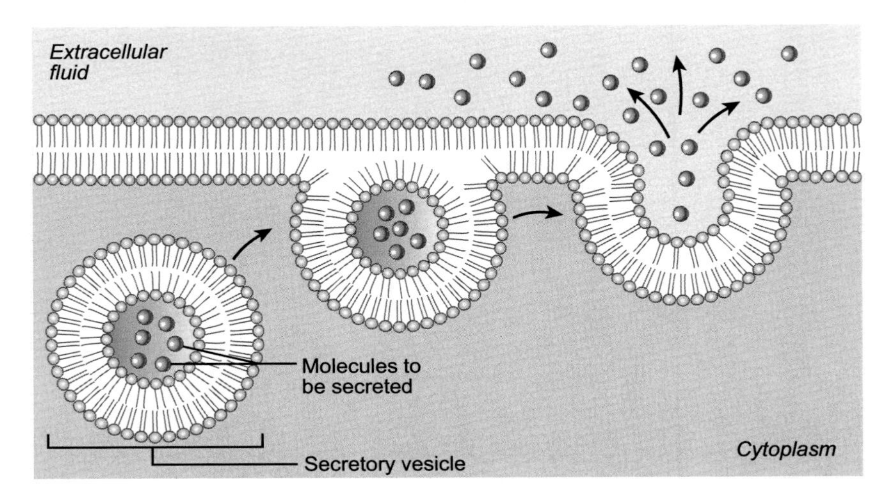

FIGURE 19.24 Exocytosis. Intracellular secretory vesicles fuse with the plasma membrane releasing their water-soluble contents to the outside and adding membrane material to the plasma membrane [50].

7.4 Blebbing

Blebbing of the plasma membrane is a morphological feature of cells undergoing late stage apoptosis (programmed cell death, see Chapter 24) [53]. A bleb is an irregular bulge in the plasma membrane of a cell caused by localized decoupling of the cytoskeleton from the plasma membrane. The bulge eventually blebs off from the parent plasma membrane taking part of the cytoplasm with it. It is clear in Fig. 19.25 [54] that the plasma membrane of an apoptotic cell is highly disintegrated and has lost the integrity required to maintain essential transmembrane gradients. Blebbing is also involved in some normal cell processes, including cell locomotion and cell division.

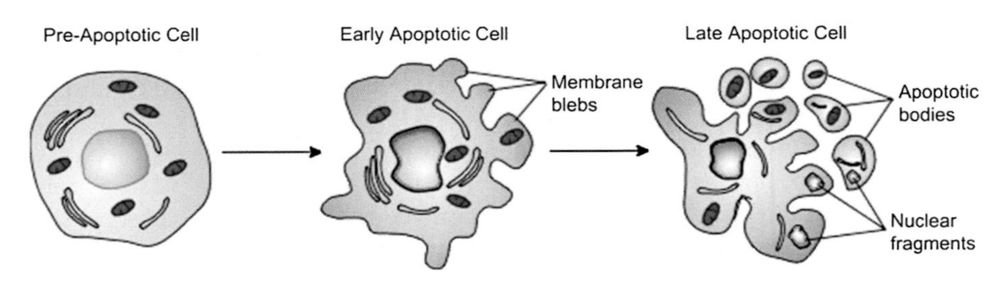

FIGURE 19.25 Membrane blebbing during apoptosis [54].

8. SUMMARY

Carefully controlled solute movement into and out of cells is an essential feature of life. There are many ways solutes are transported across the thin (~40 Å) membrane hydrophobic barrier. Transport is divided into passive diffusion and active transport. A biological membrane is semipermeable, being permeable to some molecules, most notably water (osmosis), while being very impermeable to most solutes that require some form of transporter. Passive diffusion (simple and facilitated) only requires the energy inherent in the solute's electrochemical gradient and results in its equilibrium across the membrane. In contrast, active transport requires additional energy (ie, ATP), and results in a nonequilibrium, net accumulation of the solute. Passive transport can involve simple diffusion or facilitated carriers including ionophores and channels. Active transport comes in many, often complex forms. Examples of active transport include primary active transport (uniport), secondary active transport (co-transport, antiport), and group translocation. Besides the multitude of transport systems, transport can be accomplished by gap junctions, receptor mediated endocytosis, phagocytosis, pinocytosis, exocytosis, and apoptotic membrane blebbing.

Chapter 20 will discuss bioactive lipids, highly specialized lipids that are functional at very low levels. Discussed bioactive lipids include ceramides, diacylglycerol, eicosanoids, steroid hormones, and phosphatidic acid.

References

[1] Alberts B, Bray D, Lewis J, Raff M, Roberts K, Watson JD. Principles of membrane transport. In: Molecular biology of the cell. 3rd ed. New York: Garland Science; 1994.

[2] Blok MC, van Deenen LLM, de Gier J. Effect of the gel to liquid crystalline phase transition on the osmotic behaviour of phosphatidylcholine liposomes. Biochim Biophys Acta 1976;433:1–12.

[3] Miller FP, Vandome AF, McBrewster J. Fick's laws of diffusion. VDM Publishing House Ltd; 2010. p. 76.

[4] Water systems: aqua technology for the 21st century. Introductory information on reverse osmosis.

[5] Bangham AD, de Gier J, Greville GD. Osmotic properties and water permeability of phospholipids liquid crystals. Chem Phys Lipids 1967;1:225–46.

[6] De Gier J. Osmotic behaviour and permeability properties of liposomes (review). Chem Phys Lipids 1993;64:187–96.

[7] Baldwin SA. Membrane transport: a practical approach. USA: Oxford University Press; 2000. p. 320.

[8] Alberts B, Bray D, Johnson A, Lewis J, Raff M, Roberts K, et al. Figure 12.4. Comparison of passive and active transport. In: Essential cell biology. 2nd ed. Garland Publishing: Taylor Francis Group; 2004.

[9] Huang S, Czech MP. The GLUT4 glucose transporter. Cell Metab 2007;5:237–52.

[10] Biologia Medica. Transporte de Glucosa: GLUT y SGLT Seminarios de Biología Celular y Molecular – USMP Filial Norte. 2010.

[11] Lippiat JD. Potassium channels. methods and protocols. In: Series in methods in molecular biology, vol. 491. New York (NY): Humana Press; 2009. p. 302.

[12] Doyle DA, Morais CJ, Pfuetzner RA, Kuo A, Gulbis JM, Cohen SL, et al. The structure of the potassium channel: molecular basis of K^+ conduction and selectivity. Science 1998;280:69–77.

[13] Goldin AL. Resurgence of sodium channel research. Ann Rev Physiol 2001;63:871–94.

[14] Penzotti JL, Fozzard HA, Lipkind GM, Dudley SC. Differences in saxitoxin and tetrodotoxin binding revealed by mutagenesis of the Na^+ channel outer vestibule. Biophys J 1998;75(6):2647–57.

[15] Noda Y, Sohara E, Ohta E, Sasaki S. Aquaporins in kidney pathophysiology. Nat Rev Nephrol 2010;6:168–78.

[16] Hill AE, Shachar-Hill B, Shachar-Hill Y. What are aquaporins for? J Membr Biol 2004;197:1–32.

[17] Kungl Vetenskapsakademien, The Royal Swedish Academy of Sciences. The Nobel prize in chemistry: Peter Agre, Roderick MacKinnon. 2003. Nobelprize.org. The official web site of the Nobel Prize.
[18] Nelson DL, Cox MM. Chapter 10. Biological membranes and transport. Solute transport across membranes. Figure 10-21. In: Lehninger: principles of biochemistry. Worth Publishers; 2006.
[19] Lopina OD. Na^+, K^+ ATPase: structure, mechanism, and regulation. Membr Cell Biol 2000;13:721–44.
[20] Wright EM. Renal Na^+-glucose cotransporters. Am J Physiol Renal Physiol 2001;280:F10–8.
[21] Jakubowski H. In: Chapter 9 – signal transduction. A. Energy transduction: uses of ATP; 2002. http://employees.csbsju.edu/hjakubowski/classes/ch331/signaltrans/olsignalenergy.html.
[22] Martin SA. Nutrient transport by ruminal bacteria: a review. J Anim Sci 1994;72:3019–31.
[23] Mitchell P, Moyle J. Group-translocation: a consequence of enzyme-catalysed group-transfer. Nature 1958;182:372–3.
[24] Mitchell P, Moyle J. Coupling of metabolism and transport by enzymic translocation of substrates through membranes. Proc R Phys Soc Edinb 1959;28:19–27.
[25] Mitchell P, Moyle J. Chemiosmotic hypothesis of oxidative phosphorylation. Nature 1967;213:137–9.
[26] Mitchell P. Proton current flow in mitochondrial systems. Nature 1967;214:1327–8.
[27] Herzberg O, Canner D, Harel M, Prilusky J, Hodis E. Enzyme I of the phosphoenolpyruvate: sugar phosphotransferase system. Proteopedia, Weizmann Institute of Science in Israel; 2011.
[28] Pressman BC. Biological applications of ionophores. Annu Rev Biochem 1976;45:501–30.
[29] Szabo G. Structural aspects of ionophore function. Fed Proc 1981;40:2196–201.
[30] Alberts B, Johnson A, Lewis J, Raff M, Roberts K, Walter P. Principles of membrane transport. Figure 11.5. In: Molecular biology of the cell. 4th ed. New York: Garland Scientific; 2002.
[31] Moore C, Pressman BC. Mechanism of action of valinomycin on mitochondria. Biochem Biophys Res Comm 1964;15:562–7.
[32] Pressman BC. Induced active transport of ions in mitochondria. Proc Natl Acad Sci USA 1965;53:1076–83.
[33] Page SW. The role of enteric antibiotics in livestock production. Canberry (Australia): Avcare Limited; 2003.
[34] Cheng YQ. Deciphering the biosynthetic codes for the potent anti-SARS-CoV cyclodepsipeptide valinomycin in *Streptomyces tsusimaensis* ATCC 15141. Chembiochem 2006;7:471–7.
[35] Tainter ML, Stockton AB, Cutting WC. Use of dinitrophenol in obesity and related conditions: a progress report. J Am Med Assoc 1933;101:1472–5.
[36] Huszthy P, Toth T. Synthesis and molecular recognition studies of crown ethers. Period Polytech 2007;51:45–51.
[37] Borgos SEF, Tsan P, Sletta H, Ellingsen TE, Lancelin J-M, Zotchev SB. Probing the structure–function relationship of polyene macrolides: engineered biosynthesis of soluble nystatin analogues. J Med Chem 2006;49:2431–9.
[38] Lopes S, Castanho MARBJ. Revealing the orientation of nystatin and amphotericin B in lipidic multilayers by UV-Vis linear dichroism. Phys Chem B 2002;106:7278–82.
[39] Revel JP, Karnovsky MJ. Hexagonal array of subunits in intracellular junctions of the mouse heart and liver. J Cell Biol 1967;33:C7–12.
[40] In: Peracchia C, editor. Gap junctions: molecular basis of cell communication in health and disease. Benos D, series editor. Current topics in membranes and transport, series ed. New York: Elsevier Publishing; 1999. 648 pp.
[41] Simpson I, Rose B, Loewenstein WR. Size limit of molecules permeating the junctional membrane channels. Science 1977;195:294–6.
[42] Imanaga I, Kameyama M, Irisawa H. Cell-to-cell diffusion of fluorescent dyes in paired ventricular cells. Am J Physiol Heart Circ Physiol 1987;252:H223–32.
[43] Echevarria W, Nathanson MH. Gap junctions in the liver. In: Chapter: channels and transporters. Molecular pathogenesis of cholestasis. Madame Curie Bioscience Database; 2003.
[44] Cao F, Eckert R, Elfgang C, Nitsche JM, Snyder SA, Hulsen DF, et al. A quantitative analysis of connexin-specific permeability differences of gap junctions expressed in HeLa transfectants and Xenopus oocytes. J Cell Sci 1998;111:31–43.
[45] Jongsma HJ, Wilders R. Gap junctions in cardiovascular disease. Circ Res 2000;86:1193–7.
[46] Aderem A, Underhill DM. Mechanisms of phagocytosis in macrophages. Annu Rev Immunol 1999;17:593–623.
[47] Underhill DM, Ozinsky A. Phagocytosis of microbes: complexity in action. Ann Rev Immunol 2002;20:825–52.
[48] Ernst JD, Stendahl O, editors. Phagocytosis of bacteria and bacterial pathogenicity. Cambridge University Press; 2006. p. 6.
[49] Li L, Chin L-S. The molecular machinery of synaptic vesicle exocytosis. Cell Mol Life Sci 2003;60:942–60.

[50] Benjamin Cummings an imprint of Addison Wesley Longman. 2001. Cell and cell structure. IV. Membrane transport processes, http://www.highlands.edu/academics/divisions/scipe/biology/faculty/harnden/2121/notes/cell.htm.

[51] Sudhof TC, Rothman JE. Membrane fusion: grappling with SNARE and SM proteins. Science 2009;323:474–7.

[52] Wightman RM, Haynes CL. Synaptic vesicles really do kiss and run. Nat Neurosci 2004;7:321–2.

[53] Coleman ML, Sahai EA, Yeo M, Bosch M, Dewar A, Olson MF. Membrane blebbing during apoptosis results from caspase-mediated activation of ROCK1. Nat Cell Biol 2001;3:339–45.

[54] O'Day D. 2011. Human Development, Bio 380F. As modified from: Walker NI, Harmon BV, Gobe, GC, Kerr JF. Patterns of cell death. Methods Achiev Exp Pathol 1988;13:18–54.

CHAPTER

20

Bioactive Lipids

OUTLINE

1. INTRODUCTION

Historically, lipids have been associated with two basic functions: as a structural component of membranes; and as a source of metabolic energy. More recently, a third, very different function, referred to as "bioactive lipids", has emerged and has had a tremendous impact on cell biology. In a review article in *Nature Reviews Molecular Cell Biology*, Hannun and

An Introduction to Biological Membranes
http://dx.doi.org/10.1016/B978-0-444-63772-7.00020-8

Obeid stated "It has become increasingly difficult to find an area of cell biology in which lipids do not have important, if not key, roles as signaling and regulatory molecules" [1]. Included in a long and ever increasing litany of functions supported by bioactive lipids are: regulation of cell growth, death, senescence, adhesion, migration, inflammation, angiogenesis, and intracellular trafficking. And this abbreviated list goes on and on. Hannun states that bioactive lipids can be "broadly defined as changes in lipid levels that result in functional consequences". The concept of bioactive lipids has its origins in the discovery that the lipid diacylglycerol (DAG) affects protein kinase C (PKC) activity thus influencing cell signaling.

There are several classes of bioactive lipids, often with overlapping targets affecting major, complex biochemical networks and pathways. Chapter 20 will briefly outline the mode of action and basic properties of the bioactive lipids: ceramide, DAG, eicosanoids, steroid hormones, and phosphatidic acid (PA).

2. CERAMIDES

2.1 Ceramide

Sphingolipids, discussed in Chapters 5 and 15, are a large and complex family of dynamic membrane lipids [2]. Sphingolipids are classified by possessing a C-18 amino alcohol backbone moiety known as sphingosine (Fig. 5.13). Free sphingosine however can only be found in trace amounts in living organisms due to its high toxicity. Sphingosine is rendered "safe" and given biological function after attaching polar moieties to its C-1 alcohol and esterifying various acyl chains to its C-2 nitrogen through an amide linkage. All functional sphingolipids are therefore amphipathic molecules, having both a water-soluble polar head and one or two water-insoluble apolar tails (see Chapter 5). At a basic level, their structure is similar to that of membrane phospholipids.

The simplest sphingolipids comprise the class known as ceramides [3]. Ceramides have only the underivatized (−OH) at the polar C-1 position. The esterified fatty acids attached at C-2 consist primarily of very long chain (C16 to C24 and occasionally even longer) saturated or mostly n-9 mono-unsaturated fatty acids. With the exception of some testicular cells, polyunsaturated fatty acids (PUFAs) are generally absent in most cell ceramides. More than 200 distinct ceramide molecular species have been characterized from mammalian cells. The structure of a "typical" ceramide (C16-ceramide) is shown in Fig. 20.1.

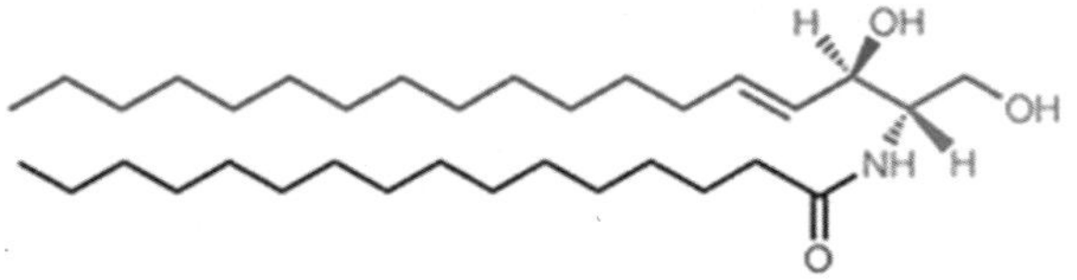

H OH
OH
NH H
O

FIGURE 20.1 Structure of a "typical" ceramide: C16-ceramide or palmitoylceramide. http://www.premierbiosoft.com/tech_notes/lipidomics.html.

For decades it was believed that the function of all sphingolipids was to provide basic structure to membranes [4]. This was based on the abundance of sphingomyelin (SM) in mammalian plasma membranes. While membrane structure is indeed an important role for many sphingolipids, it is now understood that some sphingolipids, including ceramides, also play essential roles as bioactive, signaling molecules. It has even been suggested that the bioactive and structural roles may be one and the same. Small amounts of ceramides are produced in cell membranes of all tissues. In sharp contrast, ceramides accumulate at very high levels as structural components of skin [5]. In fact, it has been reported that ceramides account for $\sim$50% of stratum corneum (skin) lipids! For this reason, ceramides are a major component of most skin care products. The role of ceramides in cell membranes and skin are very different and only the membrane functions will be considered here.

Ceramides, like other signaling lipids, are produced rapidly and transiently, keeping their cell membrane levels low [6]. They are produced as a response to specific stimuli and subsequently affect the activity of target proteins. Ceramide biosynthesis is highly complex (Chapter 15) and can be initiated by a variety of stress-induced activators. Activation can differ between cell compartments and between ceramide molecular species containing different acyl chains. Generally, ceramides can be synthesized de novo by attaching a long acyl chain to the sphingosine nitrogen via an amide linkage. Attachment is catalyzed by a variety of ceramide synthases. Alternately, ceramides can be produced as hydrolysis products of larger sphingolipids (eg, glucosides, cerebrosides, gangliosides etc., see Chapter 5) using many pathways. Since the degradation pathways are so much faster than de novo synthesis pathways, they are preferred for rapid ceramide-induced cell signaling events. Ceramide synthesis involves a bewildering array of at least 28 distinct enzymes, including six ceramide synthases and five sphingomyelinases (SMases) each of which produce distinctive ceramide molecular species. The enzymes are located on different intracellular membranes and their products shuttled throughout the cell by cytosolic ceramide transporters. Indeed, the biochemistry of ceramide metabolism is challenging.

Recently ceramides have been linked to a wide variety of physiological functions and as such, are classified as "bioactive lipids". Included in the list of important physiological events controlled in part by ceramides are cell growth and arrest, cell differentiation, migration, adhesion, senescence, and apoptosis [1,7]. Since ceramides lie at the heart of many essential cell processes, it is not surprising that they can be involved in many human pathologies including cancer, neurological disorders, and inflammation [8]. Understanding their molecular mode of action should provide many fruitful targets for future drug development.

The best characterized functions of ceramides are on protein kinase cascades and primarily as proapoptotic molecules. Apoptosis is an essential process in maintaining normal cellular homeostasis (Chapter 24). Ceramide accumulation has been shown to link various types of cellular stress (eg, ionizing radiation, UV light, heat, tumor necrosis factor-alpha, 1,25-dihydroxy-vitamin D_3, endotoxin, gamma-interferon, interleukins, nerve growth factor, and chemotherapeutic agents) to apoptosis [6]. Of particular interest to medicine is ceramide's potential to induce apoptosis in cancer cells. In this regard, ceramide has been termed the "tumor suppressor lipid". However, not all ceramide molecular species support apoptosis similarly. For example, C16 ceramide is especially important in apoptosis in nonneuronal tissues, while C18 ceramide is proapoptotic in some carcinomas. Although the precise mode of action is uncertain, it is clear that ceramide acts as an intermediary signal connecting external signals to internal cell metabolism.

The important question remains, how ceramides function at the molecular level. Ceramide low-water solubility implies their activity may be associated with membranes and their close link to cell signaling further indicates a possible link to membrane lipid rafts. Indeed the earliest studies on lipid domains (later to become lipid rafts) involved sphingolipids and cholesterol (Chapter 8). In the 1970s, evidence was starting to emerge that biological membranes were not homogeneous but instead existed in lipid patches known as domains [9]. The first observed lipid patches were liquid ordered domains (l_o) (Chapter 8) enriched in SM and cholesterol. In 1997, Kai Simon assigned these patches a catchy name, "lipid rafts", and an essential function in cell signaling [10]. Since that time, membrane studies have focused on lipid raft structure and function. A first place to investigate how ceramides might affect lipid raft structure and function should begin with the molecule's molecular properties.

Ceramides are composed of two distinct functional components, the sphingosine structure and the esterified fatty acyl chain. The sphingosine *trans*-double bond between carbons 4 and 5 facilitates tight hydrogen bonding near the aqueous interface of adjacent ceramides. This results in tight ceramide–ceramide packing. Membrane structure is also impacted by the nature of the attached fatty acyl chains. Ceramide-short chains induce positive curvature strain to the adjacent patch of membrane housing local resident proteins. In contrast, long chain ceramides induce negative curvature strain and increase order in the surrounding patch of membrane. These properties assure that, despite their relatively low concentration, ceramides have a significant influence on membrane structure and therefore function [11,12].

One type of lipid raft that is enriched in ceramide has been termed "ceramide-rich platforms" [8]. Like lipid rafts, they too are l_o microdomains, but they differ in a number of facets including basic composition. Platform ceramides often contain C24 fatty acids that actually displace cholesterol from rafts. Importantly, cholesterol is believed to be the molecular "glue" that holds traditional lipid rafts together (Chapter 8), yet is greatly reduced or missing from ceramide platforms. Ceramides can be generated in situ by the action of acid SMase on raft SM. As the lipid raft is modified by SMase, specific receptor and signaling proteins cluster, creating the ceramide-rich platforms that respond to stress signals. In contrast, the related bioactive sphingolipids, ceramide-1-phosphate, sphingosine, and sphingosine-1-phosphate (S1P), do not facilitate raft formation.

An approach to measure the affect of ceramides on raft structure involves addition of a variety of exogenous ceramides to either model membranes or living cells. In both systems, ceramides cause a significant displacement of cholesterol from lipid rafts. In Schwan cells, Yu et al. [12] reported that bacterial SMase increased ceramide levels several fold through conversion from SM, while decreasing the cholesterol content of detergent-insoluble caveolin-enriched membranes (lipid rafts) by 25–50%. Associated with this was a change in protein composition of the caveolin-enriched (lipid raft) membranes. This experiment clearly linked an increased ceramide level to reduced cholesterol, as a lipid raft's structure and function evolved into a ceramide-rich platform.

In model membrane studies, C16- and C18-ceramides, have been shown to affect membrane properties including permeability, fusion and fission, transmembrane flip-flop and lateral separation into ceramide-rich and phospholipid-rich domains [4]. In a 2012 report, Silva et al. [13] analyzed membranes isolated from a ceramide synthase 2 (CerS2)-null mouse, which is unable to synthesize very long acyl chain (C22 to C24) ceramides. These membranes displayed major alterations in their basic physical properties including fluidity, phase behavior, domain size and shape, intrinsic membrane curvature, vesicle adhesion, and fusion. Ceramides have also been suggested to organize large channels traversing the mitochondrial outer membrane, allowing for apoptosis by inducing cytochrome c to leak out of the inter-membrane space.

A striking feature of long chain ceramide structure is their large discrepancy in hydrophobic chain length, between the short, fixed sphingosine and the long, variable esterified acyl chain. This chain asymmetry feature is characteristic of molecules that interdigitate across lipid bilayers from one leaflet to the other (see Chapter 10). Interdigitation allows for inter-leaflet communication, an essential feature of lipid rafts [14]. In a 2008 report, Pinto et al. [15] employed several biophysical instrumentations to determine the affect of a very long chain ceramide, nervonoylceramide (NCer) (Fig. 20.2), where the acyl chain is $24{:}1^{\Delta 15}$. This is a very common ceramide found in the liver, kidney, and brain [16]. A binary model bilayer membrane was made with this ceramide and 1-palmitoyl-2-oleoyl-sn-glycero-3-phosphocholine (POPC) (16:0, 18:1 PC) as a "typical" fluid phase, bulk bilayer phospholipid. The ceramide was shown to phase separate as a gel from the fluid POPC. The ceramide-rich phase was shown to exist in a variety of interdigitated states, depending on the temperature. At 37°C, the partially interdigitated ceramide gel coexists with the DOPC fluid phase. The ceramide interdigitated gel phases also demonstrate major morphological changes including cochleate-type tubular structures integrated into the normal lamellar bilayer. Clearly, long chain ceramides can have a profound influence on membrane structure and hence function.

2.2 Sphingosine-1-Phosphate

Like ceramide, S1P (Fig. 20.3) is a bioactive lipid molecule with multiple actions, regulating many processes [17]. It is involved throughout the body in many aspects of cell growth, cell

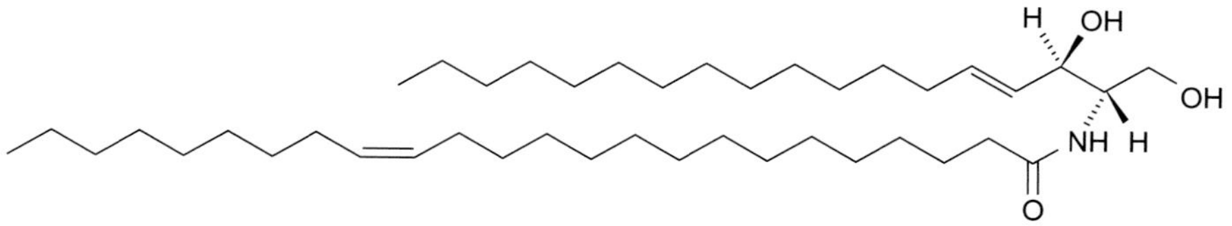

FIGURE 20.2 Structure of nervonoylceramide. http://web.ist.utl.pt/ist10881/Articles/%5B94%5D.PDF. *This is Fig. 1 in Pinto SN, Silva LC, de Almeida RFM, Prieto M. Membrane domain formation, interdigitation, and morphological alterations induced by the very long chain asymmetric C24:1. Ceramide Biophys J 2008;95(6):2867–79.*

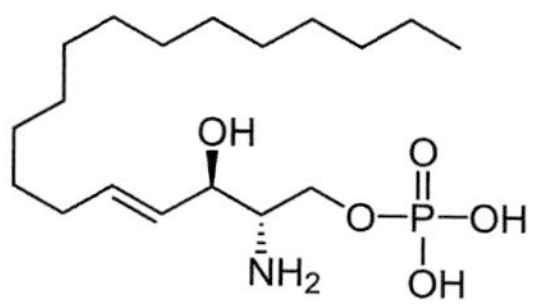

FIGURE 20.3 Structure of sphingosine-1-phosphate (S1P). *File: Sphingosine 1-phosphate.svg. From Wikimedia Commons, the free media repository http://upload.wikimedia.org/wikipedia/commons/2/2a/Sphingosine_1-phosphate.svg.*

CERAMIDE

↓ (ceraminidase)

SPHINGOSINE

(sphingosine phosphatase) ↑↓ (sphingosine kinase)

SPHINGOSINE – 1 – PHOSPHATE

(sphingosine ↓ phosphate lyase)

irreversible degradation

FIGURE 20.4 Metabolic production and degradation of sphingosine-1-phosphate.

signaling, and immune function. For example, in the immune system it has been demonstrated that S1P is a major regulator of T-cell and B-cell trafficking.

S1P emerged as a cell signaling bioactive lipid in 1991 when a report by Sarah Spiegel and coworkers linked it to cell growth [18]. All bioactive lipids must accurately maintain their cellular concentration by balancing their production with their degradation. S1P is primarily obtained by phosphorylation of sphingosine by sphingosine kinase (Fig. 20.4). S1P in turn can be dephosphorylated back to sphingosine by sphingosine phosphatase or can be degraded by sphingosine phosphate lyase (Fig. 20.4). S1P is carried via the blood primarily on high-density lipoprotein. S1P tissue levels are considerably lower than found in the blood, creating an S1P gradient that is thought to drive immune cell trafficking. In 1998 it was discovered that S1P, originally assumed to be an intracellular second messenger, was surprisingly, an *extracellular* ligand for the G protein-coupled receptor, S1PR1. It is now believed that S1P mediates cell signaling by binding to one of at least five kinds of outside-facing surface receptors [19]. Finally, whereas many sphingolipids are closely related to lipid raft structure and function, S1P, sphingosine, and ceramide-1-phosphate are not. S1P does not facilitate lipid raft formation.

3. DIACYLGLYCEROL

3.1 Diacylglycerol

Bioactive lipids are often simple molecules whose structure places them at the intersection of major metabolic pathways. The large family of DAGs (diacylglycerols or diglycerides) are such molecules and, like ceramides, reside in and influence membrane structure and function [20,21]. DAGs' membrane effects are a consequence of their amphipathic molecular structure, composed of one small polar (alcohol) head, and two esterified apolar acyl chains (Fig. 20.5). This motif orients DAG perpendicular to the plane of the membrane, similar to phospholipids, sphingolipids and cholesterol.

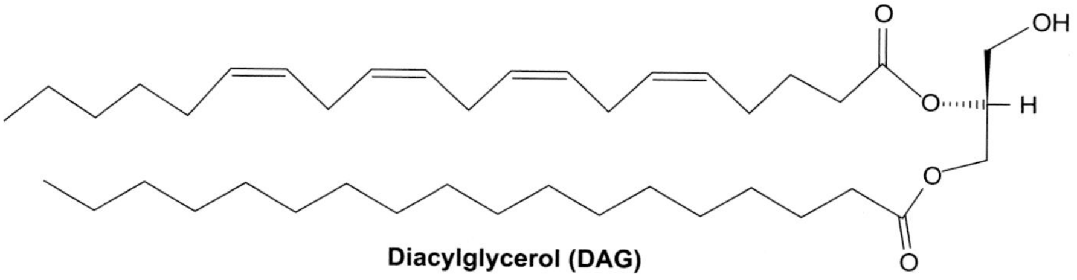

FIGURE 20.5 Structure of diacylglycerol (DAG). http://www.pharmacology2000.com/Autonomics/Adrenergics1/Adrenergic-4.htm.

A wide variety of acyl chains can be found at either glycerol location. DAGs can exist in three stereochemical forms, sn-1,2- and 2,3-DAGs are sometimes termed α,β-DAGs, while sn-1,3-DAGs may be designated α,α′-DAGs [20,21].

DAGs accumulate transiently in membranes, where they can cause changes in the physical properties of the membrane bilayer and can bind via strong hydrophobic interactions to target proteins [22]. As their polar head group is small (−OH) and their acyl chains wide, DAGs tend to prefer inverted micellar structure (HII). As previously discussed in Chapter 10, lipids with this structure introduce negative curvature stress to small adjacent membrane patches. These membrane instabilities may facilitate membrane fission or fusion and may also affect the conformation of adjacent resident proteins, thus altering their activity and hydrophobic interactions.

Although DAGs are minor components of most animal tissues, they: are a basic component of membrane structure; are a central intermediate in lipid metabolism; and play essential roles as second messengers in many cellular processes, modulating vital biochemical mechanisms [20,21]. The major interest in DAGs involves their physiological activation of a large, important family of signaling enzymes known as PKCs [23,24]. Other less appreciated functions of DAGs include: a source of prostaglandins; as a precursor of 2-arachidonylglycerol; and as an activator of a subfamily of transient receptor potential cation channels.

3.2 Protein Kinase C

Protein kinase C is a central component of many crucial cell signaling cascades [25]. Kinase is a general name given to any enzyme that transfers phosphate groups from high energy molecules (eg, adenosine triphosphate and phosphoenol pyruvate) to a substrate. These enzymes are therefore "phosphotransferases" (see Chapter 6). PKC is a specific type of kinase that transfers phosphate to precise substrate proteins. Some of PKC's most important functions include: modulating membrane structure, regulating transcription, mediating immune responses, regulating cell growth, and affecting receptor desensitization. In addition, PKC plays a pivotal role as a relay system in many Ca^{2+} dependent processes and serves as a promoter of lipid hydrolysis. With such a diversity of significant physiological functions, it is not surprising that there have already been some 15 human PKC isozymes with well over 100 specific substrates identified.

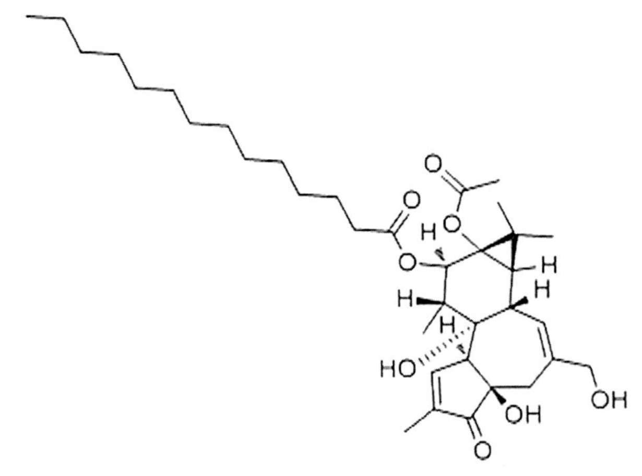

FIGURE 20.6 Structure of phorbol-12-myristate-13-acetate (PMA). http://www.biovision.com/phorbol-12-myristate-13-acetate-pma-232.html.

The large family of PKC enzymes function by controlling the short- and long-term activities of other proteins by phosphorylating specific serines or threonines in their structures. Protein phosphorylation increases intracellular Ca^{2+} levels which in turn stimulates translocation of protein kinase C to the inner leaflet of the plasma membrane. There, PKC encounters its protein kinase substrate. The importance of Ca^{2+} to the activation of PKC accounts for the C in the enzyme name. The 15 PKC isozymes are divided into three subfamilies based on their second messenger requirements: conventional (requires Ca^{2+}, DAG, and an anionic phospholipid, primarily PS (phosphatidylserine) for activation); novel (requires DAG but does not require Ca^{2+} for activation); and atypical (requires neither Ca^{2+} nor DAG for activation).

As we previously encountered with the fortuitous discovery of the arrow poison ouabain in elucidating the mechanism of action of the Na^+/K^+ ATPase (Chapter 19), a similar lucky discovery resulted in determining how DAGs function. The action of DAGs can be mimicked by the tumor-promoting compounds known as phorbol esters. Phorbol is a plant diterpene discovered in 1934 as a hydrolysis product of croton oil. Like DAG, phorbol esters are amphipathic molecules, being soluble in both polar organic solvents and water. Phorbol esters have been shown to exhibit important biological properties, particularly as tumor promoters, through activation of PKC. Therefore, phorbol esters resemble DAG and so can be used to replace DAG in many physiological pathways [26,27]. Phorbol esters have an important advantage over DAG, they are not rapidly degraded in the cell. As a result, phorbol esters produce a prolonged activation of protein kinase C. The most commonly used phorbol ester is phorbol-12-myristate-13-acetate (PMA) (Fig. 20.6), which is used as a biomedical research tool in models of carcinogenesis.

Because of DAGs' essential roles in many signaling cascades, their cellular concentration must be tightly controlled. Loss of DAG control can result in a plethora of human disease states, including a variety of cancers. In addition to their central role in signaling events, DAGs lie at the intersection of several basic metabolic pathways.

3.3 Metabolism of Diacylglycerol

Hydrolysis of phosphatidylinositol (PI) or polyphosphatidylinositides (eg, phosphatidylinositol-4,5-bisphosphate) by phospholipase C are the major source of DAG in cells. Phospholipase C constitutes a large family of at least 13 related enzymes that are subdivided into 4 subclasses. Phospholipase C was introduced in Chapter 5:

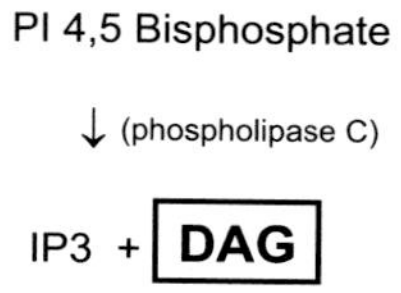

PI 4,5 Bisphosphate

↓ (phospholipase C)

IP3 + **DAG**

Synthesis of DAG begins with glycerol-3-phosphate, which is derived primarily from dihydroxyacetone phosphate, a product of glycolysis (Fig. 20.7). This usually occurs in the cytoplasm of liver or adipose tissue cells. Glycerol-3-phosphate is first acylated with

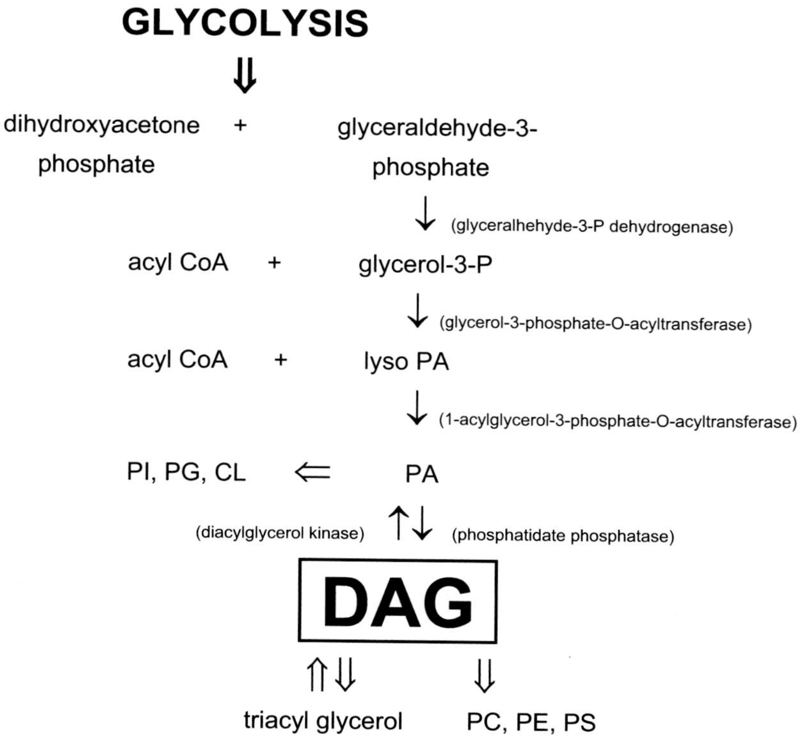

FIGURE 20.7 Synthesis of diacylglycerol from glyceraldehyde-3-phosphate.

acyl-coenzyme A (acyl-CoA) to form lysophosphatidic acid (LPA), which is further acylated with another molecule of acyl-CoA to yield PA. PA is then dephosphorylated to form DAG:

DAG levels can be decreased by draining off PA for synthesis of phosphatidylinositol (PI), phosphatidylglycerol (PG), or cardiolipin (CL), or by using DAG in the biosynthesis of phosphatidylcholine (PC), phosphatidylethanolamine (PE), or phosphatidylserine (PS) (Chapter 15). Most of the PA that serves as the precursor for higher phospholipids is generated as outlined in Fig. 20.7, but a second mechanism involves the action of a specific phospholipase D on PC:

$$\text{PC} \underset{\text{(phospholipase D)}}{\rightarrow} \text{PA} + \text{choline}$$

DAG are also formed as intermediates in the hydrolysis of triacylglycerols by pancreatic lipase:

$$\text{triacylglycerol} \rightarrow \text{diacylglycerol} + \text{fatty acid} \rightarrow \text{monoacylglycerol} + \text{fatty acid} \rightarrow \text{glycerol} + \text{fatty acid}$$

Although all triacylglycerols (fats and oils) are a potential source of DAGs, some sources are naturally highly enriched [28]. At one extreme, edible oils consisting of 80% 1,3-DAGs are marketed in Japan as nutritional supplements. All DAGs will isomerize slowly on standing in inert solvents or in the dry state, even at low temperatures. Due to their availability and amphipathic structure, DAGs are commonly used in food processing to emulsify oils and water. For these reasons, DAGs are a common component in the human diet.

DAG fatty acid compositions reflect their mode of synthesis. Fatty acid compositions mirror that of their parent phospholipids. For example, DAGs derived from PIs are highly enriched in molecular species containing stearic acid in position *sn*-1 and arachidonic acid (AA, 20:4) in position *sn*-2. There is accumulating evidence that DAGs in most cells and organelles must contain PUFAs to achieve their optimal function.

DAGs also bind to protein kinase D (PKD), a cytosolic enzyme that translocates to the *trans*-Golgi membrane network and regulates transport of proteins to the cell surface.

4. EICOSANOIDS

Eicosanoids are lipid-soluble signaling molecules made by enzyme-controlled oxidation of 20-carbon, PUFAs. The most studied of the eicosanoic fatty acids is AA ($20{:}4^{\Delta 5,8,11,14}$, omega 6) although eicosapentaenoic acid (EPA, $20{:}5^{\Delta 5,8,11,14,17}$, omega 3) and dihomo-γ-linolenic acid (DGLA, $20{:}\ 3^{\Delta 8,11,14}$, omega 6) have important physiological properties as well (Fig. 20.8). This chapter will confine comments to AA although similar statements could be made for EPA and DGLA.

4.1 Arachidonic Acid

The roles of AA in the human body are multifaceted and very complex [29]. AA has two very different, major biological functions; as an important structural component of membranes and as oxidized products known as eicosanoids.

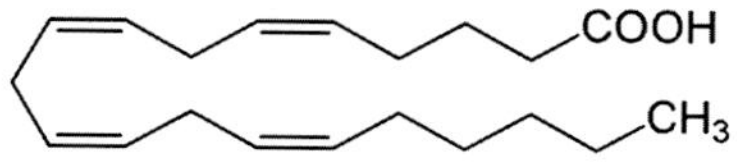

COOH

CH_3

Arachidonic Acid (AA, 20:4$^{\Delta 5,8,11,14}$, omega 6)

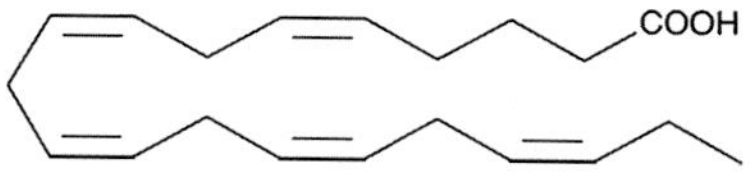

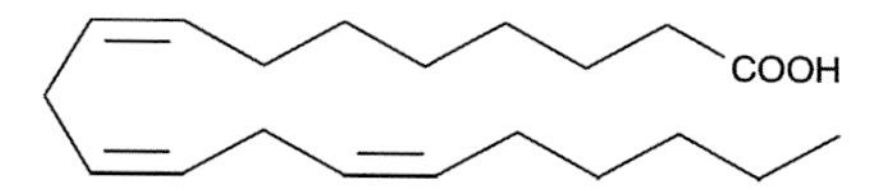

Eicosapentaenoic Acid (EPA, 20:5$^{\Delta 5,8,11,14,17}$, omega 3)

COOH

Dihomo-γ-linolenic Acid (DGLA, 20:3$^{\Delta 8,11,14}$, omega 6)

FIGURE 20.8 Structures of three common eicosanoic (20-carbon) fatty acids: arachidonic acid (AA, 20:4$^{\Delta 5,8,11,14}$, omega 6); eicosapentaenoic acid (EPA, 20:5$^{\Delta 5,8,11,14,17}$, omega 3); and dihomo-γ-linolenic acid (DGLA, 20:3$^{\Delta 8,11,14}$, omega 6). These three fatty acids are all polyunsaturated fatty acids (PUFAs) and are either omega-3 or omega-6. They are drawn in a bent, hairpin-structure as compared to the stick structures depicted in Chapter 4. This is to facilitate understanding their mechanism of oxidation into eicosanoids *The three structures come from different sources.* ***Arachidonic Acid*** *From Wikipedia Commons: File: Arachidonic acid structure.svg http://commons.wikimedia.org/wiki/File: Arachidonic_acid_structure.svg.* ***Eicosapentaenoic Acid*** *From Cayman Chemical https://www.caymanchem.com/app/ template/Product.vm/catalog/90110.* ***Dihomo-γ-linolenic Acid*** *From Cayman Chemical https://www.caymanchem.com/app/ template/Product.vm/catalog/90230.*

AA is a major fatty acid component of mammalian membranes (Chapter 4, Table 4.9). It is primarily found in the *sn*-2 position of the major structural phospholipids PE and PC and is often found at very high levels in PI. Brain, muscle, and liver are particularly enriched in AA. AA is not technically an "essential fatty acid" in humans as it can be produced from linoleic acid (18:2$^{\Delta 9,12}$. Omega-6), a common component of plants. Plants, however, are almost totally devoid of AA and so cannot be a *direct* dietary source.

Free AA is obtained by hydrolysis from membrane phospholipids with the enzyme phospholipase A_2 (PLA_2) (Chapter 5, Fig. 5.18) [30], but can also be obtained from DAG by hydrolysis with DAG lipase. As a component of membranes, AA supports a plethora of general human health functions [29,31]. For example, AA has often been positively linked with ameliorating heart disease and maintaining optimal metabolism and cell growth. AA is necessary for the growth and repair of skeletal muscle where it functions as a central nutrient controlling the anabolic—tissue-rebuilding response to weight training [32]. For this reason body builders have used AA supplements to enhance muscle growth [33]. AA, particularly when combined with the long chain omega-3 PUFA, docosahexaenoic acid (Chapter 23), is also involved in early neurological development. In one very interesting study from the National Institute of Child Health and Human Development, 18 month old infants were given AA supplements for 17 weeks [34]. These infants demonstrated improvements in intelligence as measured by the Mental Development Index! In adults, AA deficiency has been linked to neurological disorders including Alzheimer's Disease and Bipolar Disorders [35]. Therefore, AA is probably involved in maintaining optimal brain function.

The action of eicosanoids is considered to be among the most complex biological networks ever discovered. In this capacity, AA is involved in cellular signaling as a lipid second messenger helping to regulate many signaling enzymes including isoforms of the PKC and PLC (phospholipase C) families. AA oxidation products (eicosanoids) are also key inflammatory intermediates and are important vasodilators [36].

4.2 Eicosanoid Biosynthesis

The large number of known eicosanoids are divided into four distinct families: prostaglandins, prostacyclins, thromboxanes, and leukotrienes [37–39]. For each family there are two or three separate series derived from either omega-3 or omega-6 eicosanoic PUFAs. Generally, the omega-6 eicosanoids are more proinflammatory than are the omega-3 eicosanoids. In addition, AA-derived eicosanoid products can be either proinflammatory or antiinflammatory molecules. And AA is not the only eicosanoic acid capable of generating eicosanoids of unusual structure.

Synthesis of eicosanoids results from the enzymatically-controlled oxidation of 20-carbon PUFAs. The basic pathway for AA oxidation, known as the "arachidonic acid cascade", is depicted in Fig. 20.9. This figure follows the pathway from AA-linked to a phospholipid through to some of the final eicosanoid products (prostaglandins, prostacyclins, thromboxanes, and leukotrienes) [40]. The molecular mechanism for the cyclo-oxygenase

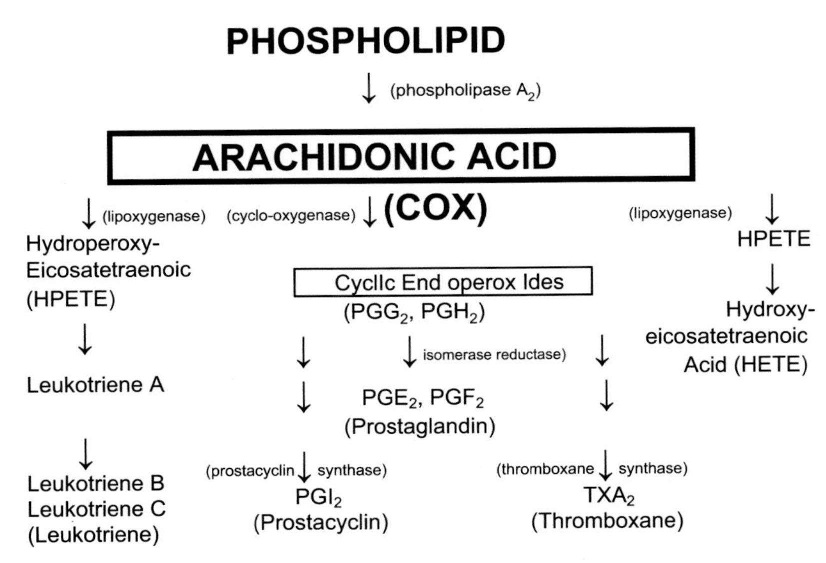

FIGURE 20.9 The arachidonic acid (AA) cascade for the production of eicosanoid hormones: prostaglandins, prostacyclins, thromboxanes and leukotrienes. The key step in the pathway is catalyzed by the enzyme cyclo-oxygenase (COX).

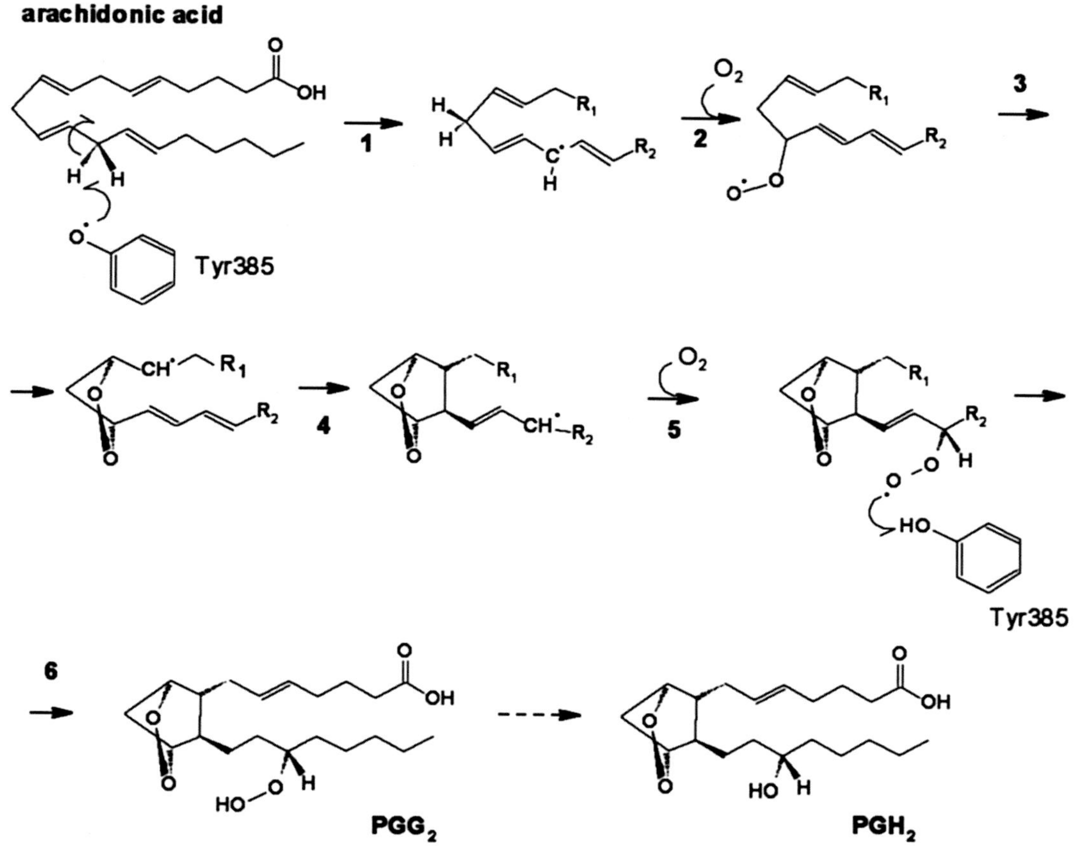

arachidonic acid

Tyr385

PGG$_2$

PGH$_2$

FIGURE 20.10 The cyclo-oxygenase (COX) reaction mechanism that oxidizes arachidonic acid (AA) to prostaglandins prostaglandin G_2 (PGG_2) and prostaglandin H_2 (PGH_2). It is this mechanism that is inhibited by aspirin. *This figure is from the reaction mechanism of cyclooxygenation of arachidonic acid by prostaglandin H. http://homepage.ufp.pt/pedros/science/cox/cox.htm.*

(COX)–catalyzed oxidation of AA is shown in Fig. 20.10. Note that the pathway between a PUFA and a final eicosanoid product passes through several cyclic intermediates that have connected different carbons of a single PUFA chain via intramolecular oxidations. A large number of starting PUFAs and numerous types of connecting oxidations accounts for the vast array of prostanoids found in nature. Clearly, even this abbreviated schematic for only one eicosanoic acid, AA, is complex.

4.3 Analgesics

The AA cascade is central to the mechanism of medicines that reduce inflammation, prevent fever, and relieve pain (analgesics). These compounds function by inhibiting the COX enzyme, thus shutting down the AA cascade and preventing eicosanoid hormone

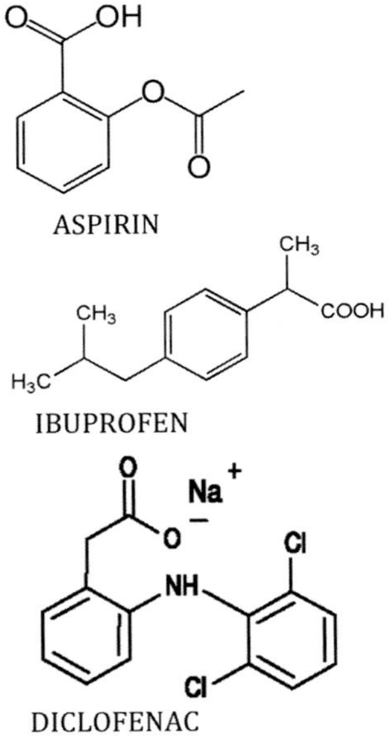

ASPIRIN

IBUPROFEN

DICLOFENAC

FIGURE 20.11 Structures of the common nonsteroidal antiinflammatory drugs (NSAIDs): Aspirin, Ibuprofen, and Diclofenac. *The 3 structures that comprise this figure all come from different sources.* ***Aspirin*** *http://drpinna.com/is-aspirin-good-for-you-5980.* ***Ibuprofen*** *http://vintageprintable.com/wordpress/vintage-printable-printed-material-ephemera-typography/vintage-printable-printed-material-ephemera-typography-2/printed-matter-5/science-chemistry-ibuprofen-structure-2-3/.* ***Diclofenac*** *http://totalpict.com/information%20on%20diclofenac%20sod.*

production. Actually there are two, very similar cyclo-oxygenases termed COX-1 and COX-2. Both forms of COX have similar molecular weights (70 and 72 kDa, respectively), have 65% amino acid sequence homology and near-identical catalytic sites.

A class of analgesics, referred to as nonsteroidal antiinflammatory drugs (NSAIDs) are among the most used medicines in the World. In the United States alone, 2010 NSAID sales topped 5.4 billion dollars. The best known NSAID is Aspirin (acetylsalicylic acid), although Ibuprofen and Diclofenac are also widely used (Fig. 20.11).

Aspirin is one of the oldest medicines recorded in human history. In the second millennium BC, Egyptians reported an early medicine from willow and other salicylate-rich plants, which inhibited fever and pain. Later, Hippocrates (400 BC) referred to the use of salicylic tea to reduce fevers. During their famous 1803–1806 expedition exploring the American West, Lewis and Clark used willow bark tea as a remedy for fever. Despite centuries-long use of crude aspirin-containing extracts, it was not until 1899 that aspirin was manufactured and marketed for medical use.

Aspirin inhibits the COX enzyme by irreversibly acylating a serine residue at the COX active site. Other common NSAIDS, ibuprofen and diclofenac, also inhibit the COX enzyme by affecting substrate binding at the active site, but do so noncovalently and reversibly.

5. STEROID HORMONES

5.1 Classes of Steroid Hormones

A well-known family of bioactive lipids is collectively known as steroid hormones. A hormone is a molecule produced in small amounts by a cell that is transported to a distant cell where it alters the target cell's metabolism. Therefore, a hormone is a chemical messenger that transports a signal from one cell to another. Steroids are common biochemicals that are characterized by having four attached nonconjugated (nonbenzene) rings. The major steroid in mammalian cells is cholesterol (Fig. 20.12) (Chapter 5).

Steroid hormones are divided into five classes, based on the type of receptor they bind to [41]: glucocorticoids, mineralocorticoids, androgens, estrogens, and progestogens. Vitamin D derivatives represent a sixth class of similar hormones, but since they are not steroids, will be discussed separately in Chapter 21, Lipid-Soluble Vitamins. In addition, each basic class of steroid hormone has several known molecular examples that are modifications of the parent steroid, further complicating their study. Steroid hormones affect a wide variety of physiological processes including: control of metabolism, inflammation, immune functions, salt and water balance, development of sexual characteristics, and the ability to withstand illness and injury [42]. Steroid hormones are all derivatives of the common membrane structural polar lipid, cholesterol. By a mechanism very different from that of the steroid hormones, cholesterol influences many of the same processes through its crucial role in affecting cell signaling through lipid rats (Chapter 8).

Common examples of the five classes of steroid hormones are listed in Table 20.1 and are discussed individually in Sections 5.2.–5.7.

5.2 Steroid Hormone Functions

Steroid hormones are synthesized from cholesterol in the gonads and adrenal glands. A quick glance at their structures (Figs. 20.13–20.17) shows how closely they resemble cholesterol (and one another) and therefore are lipids that are at home in membranes. They can readily move about laterally, undergo flip-flop and can pass through membranes until they encounter their appropriate steroid hormone receptor in the cytosol or nucleus

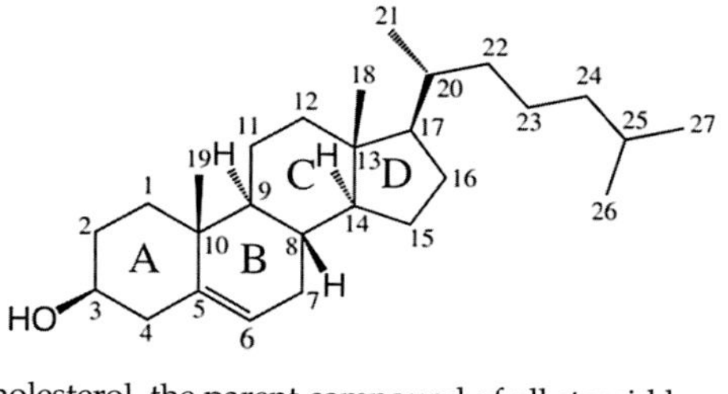

FIGURE 20.12 Structure of cholesterol, the parent compound of all steroid hormones. The basic steroid structure has four attached non-conjugated rings designated ring A, ring B, ring C, and ring D. http://andthatsscience.blogspot.com/2011/11/animals-on-steroids-steroids-on-oxygen.html.

TABLE 20.1 Representative Examples of Each of the Five Classes of Steroid Hormones and the Parent Steroid, Cholesterol

	Steroid hormone class	**Example**	
	Parent steroid	**Cholesterol**	**Moniker**
1.	Glucocorticoids	Cortisol	"Stress hormone"
2.	Mineralocorticoids	Aldosterone	"Blood pressure hormone"
3.	Androgens	Testosterone	"Male hormone"
4.	Estrogens	Estradiol	"Female hormone"
5.	Progestogens	Progesterone	"Pregnancy hormone"

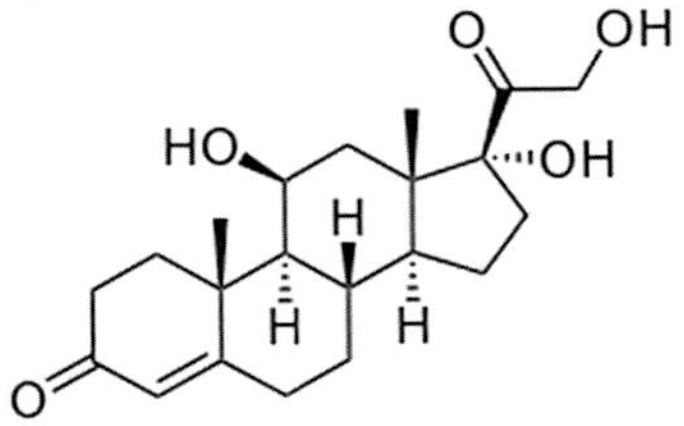

FIGURE 20.13 Cortisol. http://en.wikipedia.org/wiki/Cortisol.

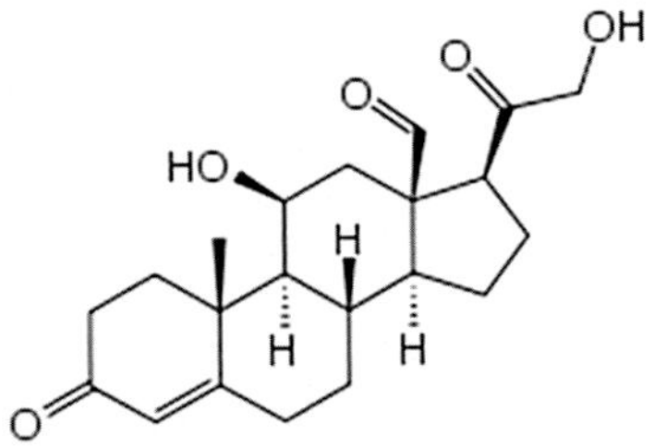

FIGURE 20.14 Aldosterone. http://www.worldofchemicals.com/chemicals/chemical-properties/aldosterone.html.

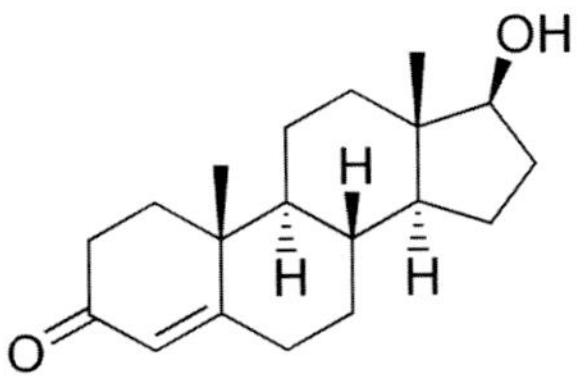

FIGURE 20.15 Testosterone. *From Wikipedia Commons. File: Testosterone. PNG http://commons.wikimedia.org/wiki/File:Testosterone.PNG.*

FIGURE 20.16 Estradiol. http://www.icgeb.org/~p450srv/ligand/estradiol.html.

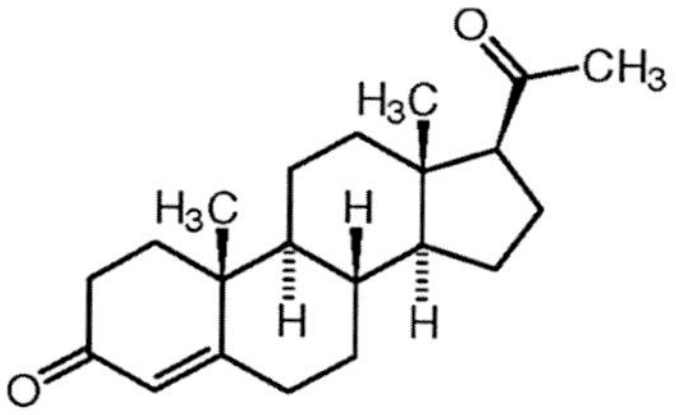

FIGURE 20.17 Progesterone. http://embryology.med.unsw.edu.au/embryology/index.php?title=File:Progesterone.jpg.

[41,43]. In the cytosol, the steroid hormone may or may not be further enzymatically modified and can then bind to its receptor. The steroid hormone–receptor complex can immediately alter intracellular metabolic events by rapid, nongenomic mechanisms [42]. Alternatively, some steroid hormones bind to nuclear receptors [44]. The nuclear steroid hormone–receptor complex then binds to specific DNA sequences, inducing transcription of its target genes. This instigates slow-acting genomic responses. In either case, the hormone–receptor complexes cause changes in the target cell's basic physiology. With limited water-solubility, steroid hormones must be transported through the blood bound to specific carrier proteins.

5.3 Cortisol

Cortisol (Fig. 20.13, also known as hydrocortisone) is a member of the class of steroid hormones known as glucocorticoids. It is produced by the zona fasciculate of the adrenal gland in response to stress and so is often referred to as the "stress hormone". Its primary functions are to: increase blood sugar levels through stimulation of gluconeogenesis [45], preserving glucose for brain function; suppress the immune system; and aid in fat, protein and carbohydrate metabolism [46]. In addition, cortisol is involved in countless other important physiological processes including counteracting insulin, stimulating gastric-acid secretion, reducing bone formation, functioning as a diuretic, and acting as a biomarker of psychological stress. Various synthetic forms of cortisol even have medical applications to treat a number of human afflictions. Cortisol's seemingly endless biological effects make the hormone a versatile tool but also presents a conundrum. Cortisol must affect many complex processes simultaneously making its single purpose use next to impossible. Adding to the confusion are many seemingly contradictory reports. For example, several studies have indicated that cortisol has a lipolytic (fat breakdown) function while others claim cortisol actually suppresses lipolysis. And elevated levels of cortisol may not be beneficial at all but instead may lead to proteolysis and muscle wasting. Cortisol is indeed a complex and confusing beast.

5.4 Aldosterone

Aldosterone (Fig. 20.14) is the best known member of the family of steroid hormones known as mineralocorticoids. Aldosterone is produced from cholesterol in the cortex of the adrenal gland. Its primary function involves maintaining salt and water balance in the body and so has a major influence on blood pressure and blood volume [47,48]. In the kidney, aldosterone supports active reabsorption of sodium with associated passive reabsorption of water and active secretion of potassium. Aldosterone's primary function therefore is retention of sodium, a mineral. Aldosterone accounts for about 2% of filtered sodium in the kidney, an amount that is about the same as found in the blood [47].

Like cortisol, aldosterone has both slow, genomic mechanisms that result from gene-specific transcription and rapid nongenomic mechanisms through membrane-associated receptors and signaling cascades [49].

5.5 Testosterone

Testosterone (Fig. 20.15) is the best known member of the class of steroid hormones referred to as androgens. This hormone controls development and maintenance of male characteristics and is often referred to as the "male hormone" [50]. In human males, testosterone is primarily produced in the testes although a lesser amount is also produced by the female ovaries. As with other steroid hormones, testosterone is synthesized from cholesterol and men produce about 20 times more than do women. Testosterone induces its many effects by first binding to androgenic receptors [51,52]. Testosterone effects are of two major types, anabolic and androgenic.

Anabolic effects: The American Heritage Dictionary defines anabolism as: "The phase of metabolism in which simple substances are synthesized into the complex materials of living tissue". Testosterone therefore stimulates protein synthesis and the growth of muscle, bone, and any tissues with androgenic receptors. It is for this reason that many athletes take anabolic steroids, primarily synthetic, to gain a competitive edge. In a related manner, there has been a recent surge in testosterone replacement therapy for aging males diagnosed with "low T". Unfortunately there have been many reports of increases in prostate and other cancers, cardiovascular disease, stroke, and even death with anabolic steroid supplementation. The risk/benefit analysis remains controversial [53] and the debate rages (actually "roid rages") on.

Androgenic effects: The American Heritage Dictionary defines androgenic as: "Controlling the development and maintenance of masculine characteristics". This aspect of testosterone function involves development of male reproductive tissues including the penis, scrotum, prostate, testes, and sperm. Also affected by testosterone are male secondary sex characteristics including deepening of the voice, beard and body hair growth, body odor, and acne.

Like most hormones, testosterone is synthesized in one location (primarily the testes) and transported to target tissues elsewhere in the body via the blood. A specific plasma protein, the sex hormone-binding globin, is responsible for testosterone transport.

5.6 Estradiol

Estradiol (Fig. 20.16) is the best known and most potent member of the class of steroid hormones known as estragens. Estradiol controls development and maintenance of female sex characteristics and is often referred to as the "female hormone" [54]. Actually, estradiol is the central member of a triad of structurally similar estragens. Estradiol, the most androgenic of the three, has two (−OH) groups while estrone has only one (−OH) and estriol has three (−OH) groups. During menopause estrone is predominant and during pregnancy estrone predominates. However, estradiol is the primary estrogen during reproductive years. Like all steroid hormones, estradiol is a cholesterol derivative and is mainly produced by granulose cells of the ovaries. Estradiol is carried from the ovaries to target cells in the blood where, like testosterone, it is primarily bound to sex hormone-binding globulin. Estradiol simply diffuses across the target cell plasma membrane and binds to a cytosolic estrogen receptor [55]. The estradiol−receptor complex then enters the nucleus where it binds to DNA, thus regulating gene transcription.

Through its affects on transcription, estradiol acts as a growth hormone for female reproductive organs including the vaginal lining, cervical glands, lining of the fallopian tubes, the endometrium, and the myometrium. Estradiol is responsible for maintaining the proper environment for oocytes in the ovary. Estradiol also plays a significant role in initiating and maintaining postpubescent female secondary sex characteristics including breast development, changes in body shape, and affecting bones and fat deposition. Estradiol has an important role in maintaining pregnancy. Estradiol is also employed in hormone contraception and hormone replacement therapy. On the down side, estradiol is suspected to activate certain oncogenes, most notably those for breast cancer [56].

5.7 Progesterone

Progesterone (Fig. 20.17) is the most important member of the class of steroid hormones known as progestogens. Since progesterone's primary functions involve the female menstrual cycle, pregnancy, and embryogenesis, it is often referred to as the "hormone of pregnancy". Progesterone is also a cholesterol derivative and so structurally resembles the other steroid hormones discussed previously (Sections 5.3−5.6). It is primarily produced in the corpus luteum of the ovaries, the adrenal gland, and the placenta during pregnancy. The hormone's major role involves development of the fetus in females [57]. However, progesterone also provides significant benefits to the male component of reproduction where, via nongenomic signaling, sperm is properly channeled through the female tract before fertilization occurs [58]. Progesterone accomplishes this by modulating the activity of a voltage-gated Ca^{2+}

channel on the sperm plasma membrane. In addition, it has been postulated that progesterone released from the egg may serve as a chemotaxis homing signal directing the sperm to the egg.

As is characteristic of all steroid hormones, progesterone is involved in many other processes that are seemingly unrelated to its better described, major roles in pregnancy and fetal development. For example, progesterone is an antiinflammatory agent that helps regulate the immune response [59], reduces gall bladder activity, normalizes blood clotting, controls zinc and copper levels, controls stored fat utilization, and has an important role in the signaling of insulin release. Recently there has been considerable medical interest in applying progesterone to recovery from traumatic brain injury. While the mechanism for this application is unclear, it may be related to progesterone's antiinflammatory properties [59].

6. PHOSPHATIDIC ACID

6.1 Structure of Phosphatidic Acid and Lysophosphatidic Acid

PA (phosphatidic acid, Fig. 20.18) is an example of a simple molecule for which Nature has found a variety of uses. In Chapter 5, PA's role as an anionic membrane structural lipid was introduced. PA is also the simplest phospholipid that possesses two acyl chains (it does not have an attached alcohol head group) and so serves as a metabolic precursor for the more complex membrane phospholipids (Chapter 15). While the membrane structural and metabolic roles have been understood for decades, a third function, as a bioactive lipid, is just now being appreciated.

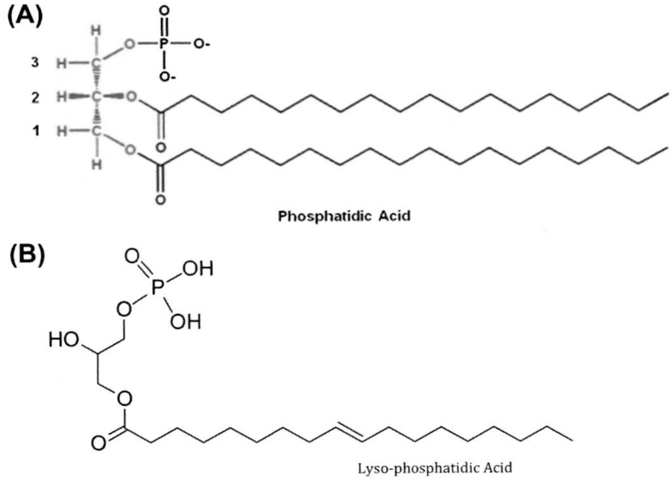

(A)

Phosphatidic Acid

(B)

Lyso-phosphatidic Acid

FIGURE 20.18 (A) Structure of phosphatidic acid (PA) with the glycerol positions indicated (*sn*-1, *sn*-2, and *sn*-3). (B) Structure of lysophosphatidic acid (LPA). *The structures come from two different sources. A: Structure of phosphatidic acid http://www.mikeblaber.org/oldwine/BCH4053/Lecture13/Lecture13.htm. B: Structure of lyso-phosphatidic acid. From Wikipedia Commons http://commons.wikimedia.org/wiki/File:Lysophosphatidic_acid.PNG.*

At first glance, PA (Fig. 20.18 (A)) appears to be basically similar to the other phospholipids. It is an amphipathic lipid composed of a polar head group and two long chain, hydrophobic acyl tails (Chapter 5). PA associates with water to produce lipid bilayers (Chapter 8). It mixes with the other phospholipids to produce multicomponent bilayers with usual physical characteristics. So why is it that PA supports a large number of nonmembrane, physiological processes, while other phospholipids do not? In other words, why is PA a bioactive lipid [60,61]?

Both in vitro and in vivo studies have linked PA and its more potent hydrolysis product, LPA (lysophosphatidic acid) (Fig. 20.18 (B)) to countless critical biochemical reactions, pathways, and signaling networks [60,61]. In recent years there has been a growing interest in LPA [60–62]. Since LPA can be readily derived from PA and is found at much lower levels than PA, it is possible that many of the cellular functions originally attributed to PA are actually caused by LPA. LPA differs from PA by having only one acyl chain. In its place is a glycerol (–OH). As a result LPA is the simplest possible glycerophospholipid.

6.2 Bioactive Lipid Roles for Phosphatidic Acid and Lysophosphatidic Acid

Partial regulation of many cellular functions have been attributed to PA and/or LPA and most of these effects parallel those promoted by S1P. Table 20.2 lists some of the reputed roles for PA and LPA.

6.3 Electrostatic–Hydrogen Bond Switch Model for PA/LPA

It is clear from the lists in Table 20.2 that PA and LPA are not "ordinary phospholipids". Indeed, they are more important than their well-known functions as structural membrane lipids and as basic metabolic intermediates in the synthesis of more complex phospholipids, would indicate.

So what makes PA and LPA so special? A first approach to this problem would involve investigating the unusual nature of the PA and LPA polar phosphate head group which is attached to the glycerol backbone by a phosphomonoester. PA has two dissociable phosphate (–OH) groups while all other phospholipids are phosphodiesters and so have only one dissociable phosphate (–OH). The PA phosphate monoester head group lies very close to the hydrophobic interior of the lipid bilayer and so has a large influence on membrane structure [61,63]. At neutral pH, PA is cone-shaped (Chapter 10) and imparts a negative curvature strain to its adjacent membrane neighbors, while LPA has an inverted cone-shape that imparts the most positive curvature strain of any membrane lipid measured to date. Head group charge is partly responsible for both phospholipid molecular shape, surrounding membrane strain, and ability to selectively bind to and affect so many different types of proteins [63]. Since all phospholipids have the potential to be cone- or inverted cone-shaped (Chapter 10), this property is clearly not what makes PA or LPA so unusual.

Edgar Kooijman and coworkers have pursued ionization (proton dissociation) of the phosphate head group for PA and LPA to determine if this is related to the bioactive roles

TABLE 20.2 Some of the Reputed Roles for PA and LPA

A. A partial list of processes reputed to be affected by PA and LPA:
- Cell proliferation and differentiation
- Regulation of fusion and fission events
- Cell transformation, tumor progression
- Cell signaling
- Regulation of membrane trafficking
- Defense mechanism against infection
- Vesicle formation and transport
- Vesicle trafficking, secretion and endocytosis
- Intercellular lipid mediator
- Growth factor-like activities
- Release of AA
- Mammalian reproductive system
- Oocyte maturation and spermatogenesis
- Proinflammatory events
- Wound healing

B. A partial list of proteins reputed to be affected by PA and LPA:
- Protein kinases
- Protein phosphatases
- G-proteins
- Nicotinamide adenine dinucleotide phosphate hydrogen oxidase
- Phospholipase A_2
- Phospholipase C
- Adenylyl cyclase

C. A partial list of potential medical applications for PA and LPA:
- Ovarian cancer
- Other cancers
- Biomarker for cardiovascular disease
- Topical applications for skin wounds
- Rheumatoid arthritis
- Multiple sclerosis

for these special phospholipids [61,63,64]. These investigators employed magic angle spinning ^{31}P nuclear magnetic resonance to monitor ionization of PA and LPA in mixed phospholipid bilayers. Several important conclusions were drawn from their experiments. Despite having identical phosphomonoester head groups, in mixed PC bilayers, LPA carries a more negative charge than PA (ie, its pKa is substantially lower). Upon addition of the primary amine lipid PE to the PC bilayer, ionization of PA and LPA was substantially enhanced, implying that hydrogen bonding to a nearby amine further destabilizes the phosphomonoester group.

These ionization studies have led to a novel proposal of the mechanism for specific binding of PA or LPA to their receptors. This model is referred to as the "electrostatic–hydrogen bond switch model" [63,64]. By this model, the phosphomonoester head group of PA or LPA hydrogen bonds to a basic amine on the target protein thus leading to increased ionization of the head group, and increased negative charge. This enables additional and stronger hydrogen bonding resulting in tight docking to the target protein.

Although PA and LPA are relatively small, simple molecules they do have the ability to carry a high content of information, accounting for their large number of diverse target proteins. The phosphomonoester head group is dynamic and can have its basic physical properties changed by the presence of adjacent basic groups. Additional information can be associated with the nature of the attached acyl chains (ie, acyl chain lengths, number and positional locations of any acyl chain double bonds, presence of ester and/or ether acyl chains, and the acyl chain positional location on the glycerol backbone) (Chapter 5).

It is not clear how many, if any, of the bioactive lipid functions attributed to PA are actually the result of minor PA metabolic or degradation products. For example it is easy to hydrolyze one of PA's acyl chains producing low levels of the more potent LPA. Also, PA can be metabolized to DAG, a stimulator of many signaling processes (Section 3). LPA can be further metabolized to cyclic LPA (with a cyclic phosphate bridging the *sn*-2 and *sn*-3 positions on the glycerol) and LPA can be further phosphorylated to LPA-pyrophosphate. It is not even clear if some of the functions attributed to PA can be accomplished by other anionic phospholipids (eg, PS, PI, PG, and CL).

7. SUMMARY

Bioactive lipids are defined by changes in their concentration affecting cell function, often signaling and regulation. Five examples discussed in this chapter are: ceramide, DAG (diacylglycerol), eicosanoids, steroid hormones, and PA.

Ceramides are sphingolipids that perhaps are best known for their major structural role in skin. Less obvious, but more significant, are their bioactive lipid roles linking external signals to internal cell metabolism, affecting cell growth and arrest, cell differentiation, migration, adhesion, senescence, and apoptosis. Ceramide has been termed the "tumor suppressor lipid" for its affect on cancers. Recently, ceramides' close link to cell signaling has led to proposals where the lipid affects the structure and function of "ceramide-rich platforms", a type of lipid raft.

Diacylglycerols (DAGs) lie at the intersection of several crucial metabolic pathways where they reside in and influence membrane structure and function. DAGs prefer inverted micellar structure and thereby introduce negative curvature stress to small adjacent membrane patches. This may affect the conformation (and hence activity) of adjacent membrane resident proteins and may induce membrane fission or fusion. DAGs are best known for activating the major regulatory protein PKC thereby altering many crucial cell signaling cascades. In addition, PKC plays a pivotal role as a relay system in many Ca^{2+}-dependent processes and serves as a promoter of lipid hydrolysis.

Eicosanoids are lipid-soluble signaling molecules made by oxidation of the 20-C PUFAs, AA, EPA, and DGLA. The action of eicosanoids is considered to be among the most complex biological networks ever discovered. AA is the vanguard of the family of 20-C fatty acids as more is known about it than the others, but all eicosanoids have a lot in common. They are made by cyclic oxidation into prostaglandins, prostacyclins, thromboxanes, and leukotrienes.

AA (and EPA and DGLA) has two very different, biological functions: as an important structural component of membranes, and as a source of eicosanoids. AA supports a plethora of general human health functions including: ameliorating heart disease and inflammation, maintaining optimal metabolism and cell growth, repairing skeletal muscle, early neurological development, and prevention of neurological disorders including Alzheimer's Disease and Bipolar Disorders. Eicosanoids are synthesized through the "arachidonic acid cascade" with its central enzyme (COX) that is inhibited by a class of analgesics, referred to as NSAIDs. Aspirin is the best known analgesic.

Steroid hormones are chemical messengers that transports a signal from one cell to another. The five classes of steroid hormones are synthesized from the same parent sterol, cholesterol, and so are structurally similar. The classes are: glucocorticoids (eg, cortisol), mineralocorticoids (eg, aldosterone), androgens (eg, testosterone), estrogens (eg, estradiol), and progestogens (progesterone). Cortisol, the "stress hormone", has a primary function of increasing blood sugar levels through stimulation of gluconeogenesis. Aldosterone, the "blood pressure hormone", maintains salt and water balance in the body. Testosterone, the "male hormone", controls development and maintenance of male characteristics, stimulates protein synthesis, and stimulates the growth of muscle and bone. Estradiol, the "female hormone", controls development and maintenance of female reproductive organs and sex characteristics. Progesterone, the "pregnancy hormone", controls the female menstrual cycle, pregnancy, and embryogenesis.

PA and LPA. PA's role as an anionic membrane structural lipid, activator of enzymes, and metabolic precursor for more complex membrane phospholipids have previously been discussed (see Chapter 5). Chapter 20 outlines potential roles for PA and LPA in critical biochemical reactions, pathways, and signaling networks. What makes these membrane lipids different than other phospholipids is their head groups that possess a phosphate monoester rather than a phosphate diester. This affects PA and LPA molecular shape, curvature stress, ionization, and affinity for cations. These properties have led to proposal of the "electrostatic–hydrogen bond switch model" for PA and LPA activity.

Chapter 21 will continue on the theme presented in Chapter 20. Instead of "Bioactive Lipids", Chapter 21 will discuss the lipid-soluble vitamins: vitamin A, Vitamin D, vitamin E, and vitamin K.

References

[1] Hannun YA, Obeid LM. Principles of bioactive lipid signalling: lessons from sphingolipids. Nat Rev Mol Cell Biol 2008;9:139–50.
[2] Christie W, Sphingolipids. An introduction to sphingolipids and membrane rafts. AOCS. The lipid library; 2015.
[3] Christie W, Ceramides. Chemistry, occurrence, biology and analysis. AOCS. The lipid library; 2014.
[4] van Blitterswijk WJ, van der Luit AH, Veldman RJ, Verheij M, Borst J. Ceramide: second messenger or modulator of membrane structure and dynamics? (Review). Biochem J 2003;369:199–211.
[5] Coderch L, Lopez O, de la Maza A, Parra JL. Ceramides and skin function. Am J Clin Dermatol 2003;4(2):107–29.
[6] Zheng W, Kollmeyer J, Symolon H, Momin A, Munter E, Wang E, et al. Ceramides and other bioactive sphingolipid backbones in health and disease: lipidomic analysis, metabolism and roles in membrane structure, dynamics, signaling and autophagy. Biochim Biophys Acta 2006;1758(12):1864–84.
[7] Ohanian J, Ohanian V. Sphingolipids in mammialian cell signaling. Cell Mol Life Sci 2001;58:2053–68.

[8] Schenck M, Carpintessu A, Grassme H, Lang F, Gulbins F. Ceramide physiological and pathological aspects. Arch Biochem Biophys 2007;462:171–5.
[9] Karnovsky MJ, Kleinfeld AM, Hoover RL, Klausner RD. The concept of lipid domains in membranes. J Cell Biol 1982;94:1–6.
[10] Simons K, Ikonen E. Functional rafts in cell membranes. Nature 1997;387:569–72.
[11] Edidin M. The state of lipid rafts: from model membranes to cells. Annu Rev Biophys Biomol Struct 2003;32:257–83.
[12] Yu C, Alterman M, Dobrowsky RT. Ceramide displaces cholesterol from lipid rafts and decreases the association of the cholesterol binding protein caveolin-1. J Lipid Res 2005;46(8):1678–91.
[13] Silva LC, Ben David O, Pewzner-Jung Y, Laviad EL, Stiban J, Bandyopadhyay S, et al. Ablation of ceramide synthase 2 strongly affects biophysical properties of membranes. J Lipid Res 2012;53(3):430–6.
[14] Putzel GG, Schick M. Phase behavior of model bilayer membrane with coupled leaves. Biophys J 2008;94:869–77.
[15] Pinto SN, Silva LC, de Almeida RFM, Prieto M. Membrane domain formation, interdigitation, and morphological alterations induced by the very long chain asymmetric C24:1 ceramide. Biophys J 2008;95(6):2867–79.
[16] Laviad EL, Albee L, Pankova-Kholmyansky I, Epstein S, Park H, Merrill Jr AH, et al. Characterization of ceramide synthase 2: tissue distribution, structure specificity, and inhibition by sphingosine 1-phosphate. J Biol Chem 2008;283(9):5677–84.
[17] Spiegel S, Milstein S. Sphingosine 1-Phosphate, a key cell signaling molecule. J Biol Chem 2002;277:25851–4.
[18] Zhang H, Desai NN, Olivera A, Seki T, Brooker G, Spiegel S. Sphingosine-1-Phosphate, a novel lipid, involved in cellular proliferation. J Cell Biol 1991;114:155–67.
[19] Spiegel S, Milstien S. Sphingosine-1-phosphate: an enigmatic signalling lipid. Nat Rev Mol Cell Biol 2003;4:397–407.
[20] Christie W. Diacylglycerols: structure, composition, function and analysis. AOCS. The Lipid Library; 2013.
[21] Becker KP, Hannun YA. Diacylglycerols. In: Nicolaou A, Kokotos G, editors. Bioactive lipids. Bridgwater: The Oily Press; 2004. p. 37–61.
[22] Goñi FM, Alonso A. Structure and functional properties of diacylglycerols in membranes. Prog Lipid Res 1999;38:1–48.
[23] Newton AC. Protein kinase C family. In: Lennarz WJ, Lane MD, editors. The Encyclopedia of biological chemistry, 3. Waltham, MA: Academic Press; 2013. p. 637–40.
[24] Newton AC. Diacylglycerol's affair with protein kinase C turns 25. Trends Pharmacol Sci 2004;25(4):175–7.
[25] Newton AC. Protein kinase C: poised to signal. Am J Physiol Endocrinol Metab 2010;298:E395–402.
[26] Newton AC. Protein kinase C: structure, function, and regulation. J Biol Chem 1995;270(48):28495–8.
[27] Ron D, Kazanietz MG. New insights into the regulation of protein kinase C and novel phorbol ester receptors. FASEB J 1999;13(13):1658–76.
[28] Matsuo N. Diacylglycerol oil: an edible oil with less accumulation of body fat. Lipid Technol 2001;13:129–33.
[29] Bayness JW, Dominiczak MH. Medical biochemistry. 2nd ed. Elsevier Mosby; 2005. p. 555.
[30] Gelb MH, Lambeau G. PLA_2: A short phospholipase review. Cayman chemical issue 14. 2003.
[31] Piomelli D. Arachidonic acid. In: Neuropsychopharmacology – 5th generation of progress. American College of Neuropsychopharmacology; 2000.
[32] Ordway RW, Walsh JV, Singer JJ. Arachidonic acid and other fatty acids directly activate potassium channels in smooth muscle cells. Science 1989;244(4909):1176–9.
[33] Roberts MD, Iosia M, Kerksick CM, Taylor LW, Campbell B, Wilborn CD, et al. Effect of arachidonic acid supplementation on training adaptations in resistance-trained males. J Intern Soc Sports Nutr 2007;4:21.
[34] Birch EE, Garfield S, Hoffman DR, Uauy R, Birch DG. A randomized controlled trial of early dietary supply of long-chain polyunsaturated fatty acids and mental development in term infants. Dev Med Child Neurol 2007;42(3):174.
[35] Amtul Z, Uhrig M, Wang L, Rozmahel RF, Beyreuther K. Detrimental effects of arachidonic acid and its metabolites in cellular and mouse models of Alzheimer's disease: structural insight. Neurobiol Aging 2012;33(4):831.
[36] Li B, Birdwell C, Whelan J. Antithetic relationship of dietary arachidonic acid and eicosapentaenoic acid on eicosanoid production in vivo. J Lipid Res 1994;35(10):1869–77.

[37] Harizi H, Carcuff JB, Guagde N. Arachidonic-acid-derived eicosanoids: roles in biology and immunopathology. Trends Mol Med 2008;14(10):461–9.
[38] Curtis-Prior P. The eicosanoids. John Wiley & Sons; 2004. 654 pp.
[39] Marks F, Furstenberger G, editors. Prostaglandins, leukotrienes, and other eicosanoids: From Biogenesis to Clinical application. Wiley; 2008. 408 pp.
[40] Samuelsson B. An elucidation of the arachidonic acid cascade. Discovery of prostaglandins, thromboxane and leukotrienes. Drugs 1987;33(Suppl. 1):2–9.
[41] Norman AW, Mizwicki MT, Norman DPG. Steroid-hormone rapid actions, membrane receptors and a conformational ensemble model. Nat Rev Drug Discov 2004;3:27–41.
[42] Falkenstein E, Tillmann H-C, Christ M, Feuring M, Wehling M. Multiple actions of steroid hormones – a focus on rapid, nongenomic effects. Pharmacol Rev 2000;52(4):513–56.
[43] DeSombre ER, Jensen EV. Chapter 52. Steroid hormone binding and hormone receptors. In: Bast Jr RC, Kufe DW, Pollock RE, Weichselbaum RR, Holland JF, Frei III E, editors. Holland-Frei Cancer medicine. 5th ed. Hamilton, ON: BC Decker; 2000.
[44] Lu NZ, Wardell SE, Burnstein KL, Defranco D, Fuller PJ, Giguere V, et al. International Union of Pharmacology. LXV. The pharmacology and classification of the nuclear receptor superfamily: glucocorticoid, mineralocorticoid, progesterone, and androgen receptors. Pharmacol Rev 2006;58(4):782–97.
[45] Coderre L, Srivastava AK, Chiasson JL. Role of glucocorticoid in the regulation of glycogen metabolism in skeletal muscle. Am J Physiol 1991;260(6 Pt 1):E927–32.
[46] Hoehn K, Marieb EN. Human anatomy & physiology. San Francisco: Benjamin Cummings; 2010.
[47] Sherwood L. Human Physiology: From cells to systems. Pacific Grove, CA: Brooks/Cole; 2001.
[48] Booth RE, Johnson JP, Stockand JD. Aldosterone. Adv Physiol Educ 2002;26(1–4):8–20.
[49] Thomas W, Harvey BJ. Mechanisms underlying rapid aldosterone effects in the kidney. Annu Rev Physiol 2011;73:335–57.
[50] Kohn FM. Testosterone and body functions. Aging Male 2006;9(4):183–8.
[51] Hiipakka RA, Liao S. Molecular mechanism of androgen action. Trends Endocrinol Metab 1998;9(8):317–24.
[52] McPhaul MJ, Young M. Complexities of androgen action. J Am Acad Dermatol 2001;45(Suppl. 3):S87–94.
[53] Spitzer M, Huang G, Basaria S, Travison TG, Bhasin S. Risks and benefits of testosterone therapy in older men. Nat Rev Endocrinol 2013;9:414–24.
[54] Zelinski-Wooten MB, Chaffin CL, Duffy DM, Schwinof K, Stouffer RL. The role of estradiol in primate ovarian function. Reprod Med Rev 2000;8(1):3–23.
[55] Dahlman-Wright K, Cavailles V, Fuqua SA, Jordan VC, Katzenellenbogen JA, Korach KS, et al. International union of pharmacology. LXIV. Estrogen receptors. Pharmacol Rev 2006;58(4):773–81.
[56] Lyytinen H, Pukkala E, Ylikorkala O. Breast cancer risk in postmenopausal women using estradiol-progestogen therapy. Obstet Gynecol 2009;113(1):65–73.
[57] Boomsma D, Paoletti J. A review of current research on the effects of progesterone. Inter J Pharmac Compd 2002;6(4):245–9.
[58] Correia JN, Conner SJ, Kirkman-Brown JC. Non-genomic steroid actions in human spermatozoa. Persistent tickling from a laden environment. Semin Reprod Med 2007;25(3):208–19.
[59] Pan DS, Liu WG, Yang XF, Cao F. Inhibitory effect of progesterone on inflammatory factors after experimental traumatic brain injury. Biomed Environ Sci 2007;20(5):432–8.
[60] Christie WW. 2. Phosphatidic acid – Biological functions in animals. The AOCS Lipids Library; 2013.
[61] Kooijman EE, Carter KM, van Laar EG, Chupin V, Burger KNJ, et al. What makes the bioactive lipids phosphatidic acid and lysophosphatidic acid so special? Biochemistry 2005;44:17007–15 [PA].
[62] Moolenaar WH, van Meeteren LA, Giepmans BNG. The ins and outs of lysophosphatidic acid signaling. BioEssays 2004;26(8):870–81.
[63] Kooijman EE, Tieleman DP, Testerink C, Munnik T, Rijkers DT, Burger KN, et al. An electrostatic/hydrogen bond switch as the basis for the specific interaction of phosphatidic acid with proteins. J Biol Chem 2007;282:11356–64.
[64] Kooijman EE, Burger KN. Biophysics and function of phosphatidic acid: a molecular perspective. (Review). Biochim Biophys Acta 2009;1791(9):881–8.

CHAPTER

21

Lipid-Soluble Vitamins

OUTLINE

1. INTRODUCTION

Vitamins are a group of organic molecules that, in minute quantities, are essential for normal growth and several body processes [1]. They often act as coenzymes or precursors of coenzymes in the regulation of metabolic processes but do not directly provide energy or serve as building blocks of polymers. Mostly, they are not produced in the body and so must be obtained through the diet (the exception is vitamin D). Vitamins are classified as

An Introduction to Biological Membranes
http://dx.doi.org/10.1016/B978-0-444-63772-7.00021-X

being either water-soluble or lipid-soluble. In humans, there are four lipid-soluble (A, D, E, and K) and nine water-soluble (eight B vitamins and vitamin C) vitamins. Here, we are interested in the lipid-soluble vitamins due to their favorable partitioning into membranes [2].

2. VITAMIN A

The family of vitamin A's [3,4] exist in three oxidation states: alcohol (retinol), aldehyde (retinal), and acid (retinoic acid) (Fig. 21.1). The chemical parent of the vitamin A family is β-carotene, which is essentially two retinyl groups attached tail to tail. All members of the vitamin A family have the same basic structure, a β-ionone ring attached to an isoprene chain.

Each vitamin A type supports different functions, but since the oxidation states in many cases can be interconverted (Fig. 21.2), one vitamin A type may appear to support several functions. Vitamin A functions include vision, gene transcription, immune function, embryonic development, reproduction, bone metabolism, hematopoiesis, skin health, cell health, and membrane antioxidant activity [4,5].

2.1 β-Carotene

There are two major sources of vitamin A in the human diet: animal and plant. Dietary animal-derived vitamin A comes in the form of retinyl esters. Pure retinol is unstable and

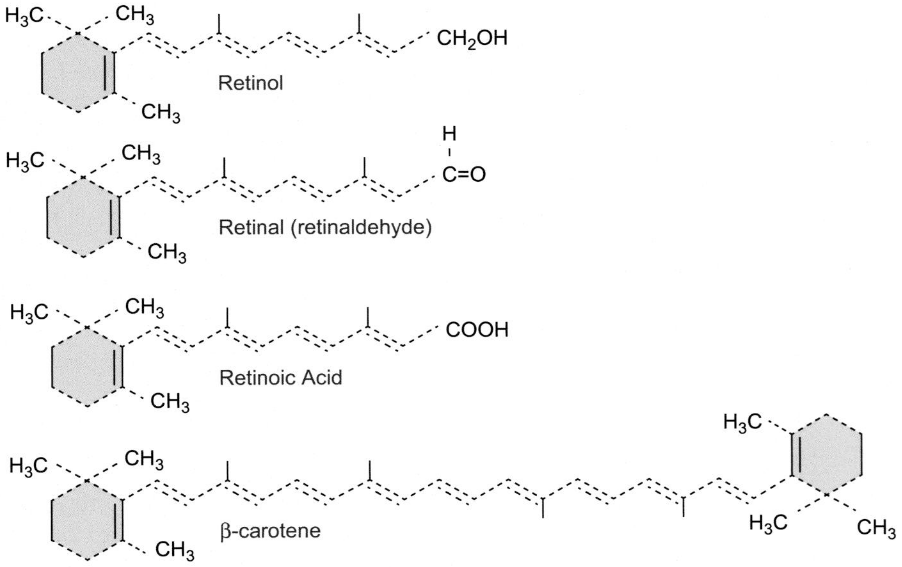

H_3C CH_3 CH_2OH CH_3 Retinol

H_3C CH_3 H C=O CH_3 Retinal (retinaldehyde)

H_3C CH_3 COOH CH_3 Retinoic Acid

H_3C CH_3 CH_3 β-carotene H_3C H_3C CH_3

FIGURE 21.1 Chemical structures for the family of vitamin A's: Carotene, retinol, retinal, and retinoic acid. The β-ionone rings are colored in *yellow*. *Bates CJ, 1995. Vitamin A (review). Lancet 345:31–5. http://www.vivo.colostate.edu/hbooks/pathphys/misc_topics/vitamina.html.*

β-CAROTENE
(reduced)

↓ (β-carotene 15,15'-monooxygenase)

(alchol) (aldehyde) (acid)
RETINOL ⟷ RETINAL → RETINOIC ACID
(retinol dehydrogenase) (retinal dehydrogenase)

FIGURE 21.2 Interconversion of oxidation states for the vitamin A family.

so is stored as retinyl esters (primarily palmitoyl ester). The major plant source of vitamin A is β-carotene (Fig. 21.1), the orange color in carrots. However, two additional carotenoids, α-carotene and γ-carotene, are also found in plants but have much lower vitamin A activity. In contrast to humans, absolute carnivores (eg, cats) lack the enzyme β-carotene 15,15′-monooxygenase (Fig. 21.3) and so cannot cleave any carotenoid. This means that for these animals, none of the carotenoids can be a source of vitamin A. Plant sources highly enriched in β-carotene include dandelion greens, carrots, broccoli, sweet potatoes, and spinach.

2.2 Retinol

The major forms of vitamin A found In the human body are retinol and retinal, which are readily interconverted (Fig. 21.2). Retinol is directly produced from hydrolysis of animal retinyl esters or from reduction of retinal derived from β-carotene. Retinol esters are the major form of vitamin A in the human body where they serve as a source of retinal (involved in vision) and retinoic acid (involved in growth and cell differentiation). The first recognized vitamin A, retinol, was discovered around 1913 by Elmer McCollum and Marguerite Davis in butterfat and cod liver oil. The vitamin was shown to prevent and reverse night blindness. The period between 1910 and 1920 was an active one for the discovery of vitamins. During this time, a water-soluble bioactive component was discovered and named "water-soluble factor B," or vitamin B. A newly discovered water-insoluble component was therefore named "fat-soluble factor A," or vitamin A. Retinol esters (usually of acetate or palmitate) are now added to skin creams, where they are advertised to prevent acne. Retinol also functions as a membrane antioxidant since it is readily oxidized to retinal.

2.3 Retinal

Retinal, the aldehyde form of vitamin A (Fig. 21.1), is the light-absorbing molecule necessary for both low-light or night vision (scotopic vision) and color vision. Retinal can be obtained directly from the enzymatic cleavage of β-carotene (Fig. 21.3) or by oxidation

$$\beta\text{-carotene} + O_2 \rightarrow 2\ \text{RETINAL}$$

FIGURE 21.3 Reaction catalyzed by the enzyme β-carotene 15,15′-monooxygenase.

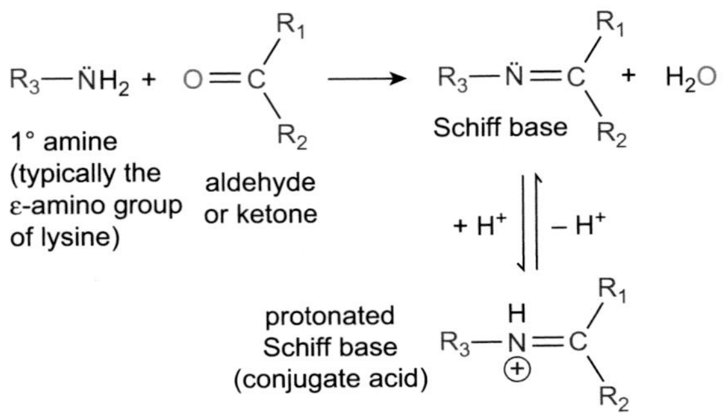

FIGURE 21.4 Formation of a Schiff base between an aldehyde (or ketone) and a primary amine. In vision, the aldehyde is retinal and the primary amine on the side chain of lysine on the protein opsin. The protonated Schiff base shifts light absorption to a longer wavelength, giving the complex its purple color and accounting for color vision. *Biochemistry Dictionary, http://guweb2.gonzaga.edu/faculty/cronk/biochem/S-index.cfm?definition=Schiff_base.*

of retinol. Retinal can be further oxidized (irreversibly) to retinoic acid (Fig. 21.2). In the eye, retinal attaches via a reversible Schiff base (Fig. 21.4) to a lysine on the protein opsin. The retinal/opsin conjugate, known as rhodopsin or visual purple, is responsible for color vision.

The "visual cycle" [6] is based on rapid interconversion of retinal isomers. In the rods, 11-*cis*-retinal binds via a Schiff base to a specific lysine on opsin, creating rhodopsin. A parallel conjugation involving 11-*cis*-retinal binding to the protein iodopsin occurs in the cones. The visual cycle begins when light enters the eye, striking the rhodopsin complex. The 11-*cis*-retinal is isomerized to all-*trans*-retinal (light reaction) that dissociates from the opsin in a series of steps called photobleaching (Fig. 21.5). The isomerization induces a nervous signal that travels down the optic nerve to the visual center of the brain. The all-*trans*-retinal is then converted back to 11-*cis*-retinal (dark reaction) where the cycle can begin again.

Some of the all-*trans*-retinal can be reduced to all-*trans*-retinol where it is stabilized as retinyl esters. These esters serve as storage forms of all-*trans*-retinol.

2.4 Retinoic Acid

Vitamin A also functions in a very different role as an irreversibly oxidized form of retinal known as retinoic acid (Fig. 21.1). While retinoic acid can be readily made from retinal (and hence retinol, Fig. 21.2), neither retinal nor retinol can be made from retinoic

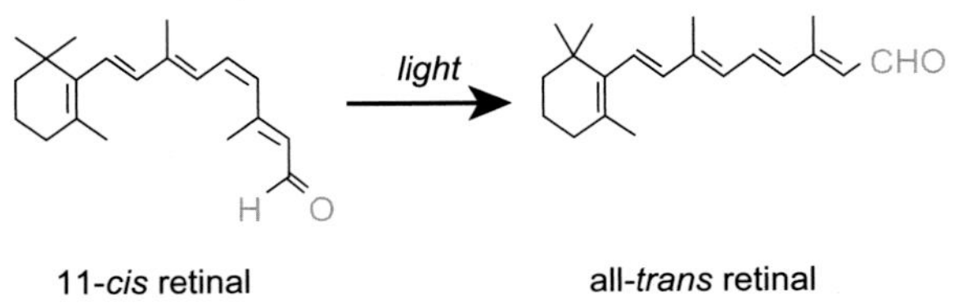

FIGURE 21.5 Isomerization between 11-*cis*-retinal and all-*trans*-retinal (light reaction). http://www.photobiology.info/Crouch.html.

acid. For this reason, retinoic acid has only partial vitamin A activity and has no function in the visual cycle. Instead, retinoic acid's major function is as an important hormone-like growth factor for epithelial and other cells resulting from its role as a regulator of gene transcription.

All-*trans*-retinoic acid plays an important role in regulating gene transcription through binding to nuclear receptors known as retinoic acid receptors (RARs). These receptors are bound to DNA as heterodimers with retinoid X receptors (RXRs). RAR/RXR receptors dimerize before binding to DNA.

For decades vitamin A, and particularly retinoic acid, has been successfully used in medical dermatology to alleviate a number of human skin conditions. Retinoic acid maintains normal skin health by switching on genes that control the differentiation of keratinocytes (immature skin cells) into mature epidermal cells. The mechanism of action is not clear. The most prescribed vitamin A product is 13-*cis*-retinoic acid, originally marketed as Accutane by Hoffman–La Roche. In 2009 Accutane was removed from the U.S. market due to it being linked to inflammatory bowel disease. Currently, 13-*cis*-retinoic acid, marketed under different names, is the standard of care for treatment of severe, scarring cystic acne. Another commercial use of vitamin A is skin creams containing retinyl palmitate. These creams are reputed to reduce acne through oxidization of retinol to retinoic acid.

As with all vitamins, too much or too little vitamin A can result in serious health problems. Deficiency of vitamin A results in the worldwide death of 670,000 children under the age of 5 annually. An additional 250,000 to 500,000 children go blind each year due to vitamin A deficiency. Not surprisingly, the major health problems linked to vitamin A deficiency in the United States are impaired vision, particularly night blindness.

Excess vitamin A (hypervitaminosis A) occurs when the vitamin A content of the body exceeds the carrying capacity of the liver [7]. Since vitamin A is lipid-soluble, it accumulates in fat and membrane bilayers until toxic levels are reached. Once accumulated, excess vitamin A is hard for the body to eliminate. However, most patients do fully recover.

Although hypervitaminosis A can manifest many symptoms, only two colorful and well-documented examples will be discussed here: polar bear liver and carrot poisoning.

2.5 Polar Bear Liver Poisoning

Animal liver is a common component of many human diets. One exception is the livers of polar bears and other arctic animals that are known to be toxic [8]. Inuits (Eskimos) often eat polar bear meat, but strictly avoid consuming the bear's liver. Polar bear liver toxicity was first reported by Europeans in 1597 when the Dutch explorer Gerrit de Veer wrote in his diary that while taking refuge during the winter in Nova Zemlya (an archipelago in the Arctic Sea in northern Russia) that he and his men became seriously ill after eating polar bear liver. Since that early report, other similar reports of arctic explorers becoming ill and even dying after consuming polar bear liver have appeared. So why is polar bear liver so toxic? For decades the finger has been pointed at vitamin A. A single polar bear liver (about 500 g) has an astonishing 9 million IU of vitamin A, and acute human toxicity occurs at about 300,000 IU! Long-term (chronic) toxicity can be achieved by ingesting 4000 IU/kg every day for 6–15 months. Therefore, a 90-kg (about 200-lb) man would have to ingest 360,000 IU every day to achieve toxicity. Since vitamin A supplements usually have about

10,000 IU/capsule, the man would have to consume about 36 vitamin A capsules daily for about 1 year. For this reason, there is little chance of an accidental vitamin A overdose from vitamin supplements. However, a single ounce of polar bear liver may be fatal.

Symptoms of polar bear liver poisoning depend on the amount of liver consumed but usually include weakness, vomiting, diarrhea, drowsiness, headaches, hair loss, bone pain, peeling skin, and blurred vision. At very high vitamin A levels, symptoms include full-body skin loss, hemorrhaging, coma, and death. One related set of symptoms led to a proposal that at high levels vitamin A may function as an uncoupler of oxidative phosphorylation. These symptoms include heavy breathing, increased body temperature, sweating, weight loss, weakness due to decreased ATP levels, and possible death. Symptoms are the same as those noted with the classic uncoupler 2,4-dinitrophenol discussed in Chapter 18.

The vitamin A hypothesis for polar bear liver poisoning has been challenged by observations made by a team of Swiss scientists in the late 1980s. They noted that polar bear livers adsorb very large amounts of the metal cadmium found in arctic waters. Among the symptoms of cadmium poisoning are dermatitis and skin defoliation. This hypothesis places the toxicity of polar bear liver on cadmium, not vitamin A.

2.6 Carrot Poisoning

A different type of vitamin A—related poisoning has been reported for β-carotene, primarily obtained from prodigious consumption of carrots. The extreme poisoning symptoms noted for polar bear liver are not observed for carrot toxicity. Excessive consumption of carrots can lead to carotenodermia, a condition that causes the skin to attain an orange hue.

Carrot poisoning has an unusual history that involves England during World War II [9]. Part of this story may actually be better classified as an "urban legend." During the Battle of Britain, British gunners and fighter pilots were enjoying remarkable success shooting down German aircraft, particularly at night. The reason for this was the secret development of radar and the use of red light in aircraft instruments, new technologies the British desperately wanted to keep out of German hands. To dissuade the Germans from pursuing similar technologies, the RAF circulated a story that their pilots' success was due to eating large quantities of carrots that allowed them to see in the dark. A "Dr. Carrot" advertising campaign further encouraged carrot consumption for rural Britons to help spot German aircraft in the dark (Fig. 21.6). Many Britons started to grow and consume carrots. By one report a woman living in northern England consumed 5 pounds of carrots per day for 5 years to help the British cause. She turned a brilliant orange!

3. VITAMIN D

3.1 Introduction

Vitamin D is an example of a biocheminal structure that falls between the cracks and so is hard to classify. On the one hand, it behaves like a typical vitamin, functioning at low levels

FIGURE 21.6 Example of a British World War II poster promoting carrot consumption through its spokesman, Doctor Carrot. http://www.telegraph.co.uk/foodanddrink/7222430/Waging-a-war-for-a-healthy-thrifty-diet.html.

to promote a variety of biological processes. However, a true vitamin cannot be synthesized by the human body and so must be obtained solely through the diet. Vitamin D, however, can be synthesized in small amounts by humans when exposed to UV light. For this reason, vitamin D has earned the moniker the "sunshine vitamin." In several respects, vitamin D resembles steroid hormones, a family of compounds discussed in Chapter 20 that, like vitamin D, are influential at low levels and are derived from cholesterol. Steroid hormones by definition retain the four intact sterol rings of cholesterol. Vitamin D is technically a secosteroid where one of the steroid rings (the B ring) is broken. In vitamin D, carbons C9 and C10 of the B ring are not joined, whereas in true steroids they are.

There are several forms of vitamin D (known as vitamers) that have been identified in the human body, the two major ones being vitamin D_2 (ergocalciferol) and vitamin D_3 (cholecalciferol) (Fig. 21.7) [10,11]. All vitamers of vitamin D are lipid-soluble. Many published reports just refer to "vitamin D" without identifying the actual vitamer. In these cases, "vitamin D" usually refers to either vitamin D_2, vitamin D_3, or a mixture of the two. Both vitamin D_2 and vitamin D_3 can be synthesized or obtained through the diet. The major function of vitamin D involves absorption of calcium and phosphate in the intestine.

3.2 Rickets and Other Diseases

The major symptom resulting from vitamin D deficiency is osteomalacia, a condition where the bones are softened due to defective mineralization of phosphorous and calcium. This condition in children is better known to as "rickets" [12]. In children, rickets leads to stunted growth and deformity of the long bones. Rickets was first described in detail in 1650 by the British physician Francis Glisson (Fig. 21.8) in his self-published (in Latin) "A Treatise of the Rickets." It was not until 1918 that Edward Mellanby (Fig. 21.9), experimenting with dogs, discovered vitamin D, and the role of the vitamin in preventing rickets soon followed (in 1919).

In recent years there has been an explosion of interest in possible human afflictions affected by vitamin D [10,11]. While the link of vitamin D and bone health is well

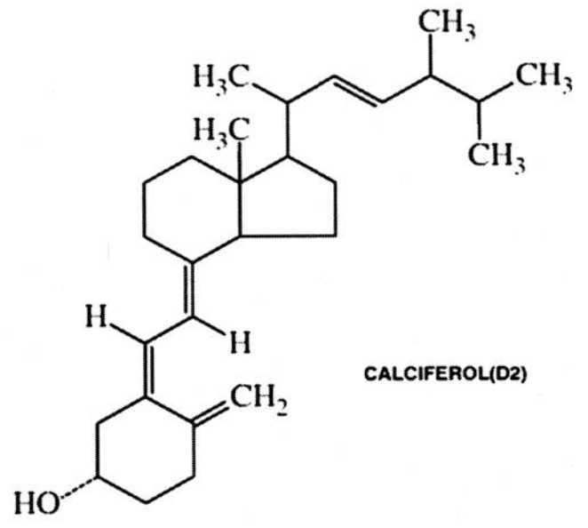

VITAMIN D_2 – ERGOCALCIFEROL

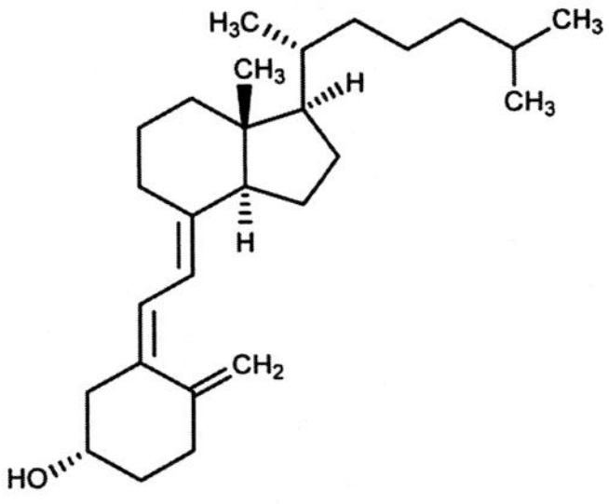

VITAMIN D_3 - CHOLECALCIFEROL

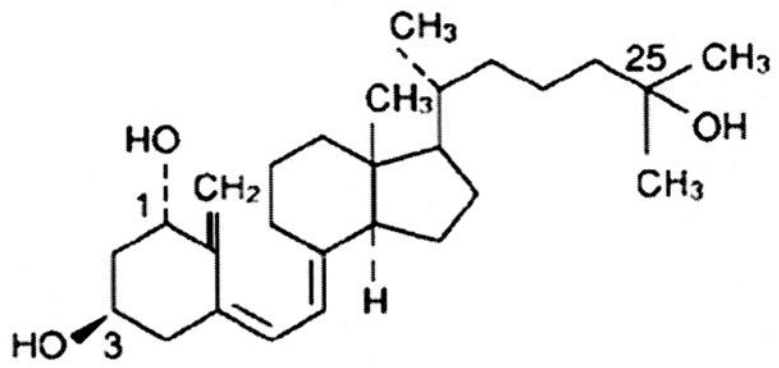

1,25-DIHYDROXY VITAMIN D - CALCITRIOL

FIGURE 21.7 Three important vitamers of vitamin D: vitamin D_2 (ergocalciferol), vitamin D_3 (cholecalciferol), and 1,25-dihydroxyvitamin D (calcitriol). *Vitamin D_2 (ergocalciferol): http://sixnutrition.com/blog/what-is-vitamin-d/; vitamin D_3 (cholecalciferol): http://chemistry.about.com/od/factsstructures/ig/Chemical-Structures—C/Cholecalciferol—Vitamin-D3.htm; 1,25-dihydroxy vitamin D (calcitriol): http://www.rxlist.com/rocaltrol-drug.htm.*

documented, many of the other possible roles for vitamin D are interesting but uncertain. Among the human afflictions suggested to be linked to vitamin D deficiency are multiple sclerosis, some cancers, immune function, influenza during winter months, tuberculosis, HIV infection, asthma, neuromuscular function, premature aging, and mortality in the elderly. At best, the evidence supporting these health effects is inconclusive. At a cell level, the active, intracellular form of vitamin D, calcitriol, has been suggested to enhance proliferation, differentiation, and apoptosis of cells. In model membrane systems, vitamin D has been shown to be a lipid antioxidant [13], similar to vitamins E and A.

FIGURE 21.8 Francis Glisson, 1597–1677. http://en.wikipedia.org/wiki/Francis_Glisson.

FIGURE 21.9 Edward Mellanby, 1884–1956. *CSIR-Central Drug Research Institute, http://www.cdriindia.org/museumpic6.htm.*

Although not common, hypervitaminosis D has been observed. Overt toxicity has been reported following 50,000 IU/day for several months. Among the symptoms of vitamin D poisoning are anorexia, nausea, vomiting, polyuria, polydipsia, weakness, insomnia, nervousness, and, ultimately, renal failure. Most symptoms can be readily reversed by eliminating dietary vitamin D. Renal failure, however, is usually irreversible.

3.3 Vitamin D → Calcitrol Metabolic Pathway

Fig. 21.10 summarizes the vitamin D pathway for the production of calcitrol, the biologically active form of vitamin D. In the skin, 7-deydrocholesterol is converted by

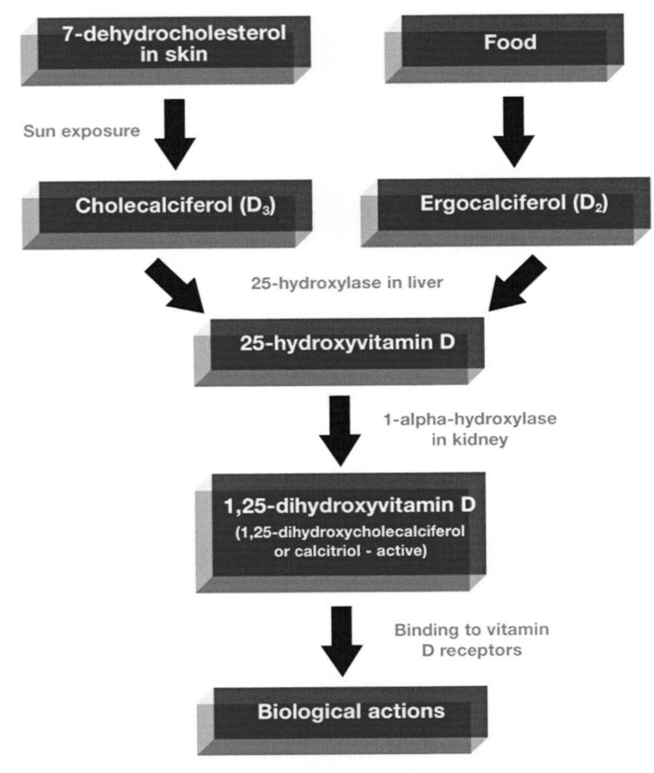

FIGURE 21.10 Pathway for the biosynthesis of 1,25-dihydroxyvitamin D (calcitriol), the biologically active form of vitamin D. *Clinuvel, Clinuvel's vitiligo program – Q&A September 2013, Tanning beds and the vitamin D debate, http://www.clinuvel.com/en/blog/lightandhealth/tanning-beds-and-the-vitamin-d-debate/.*

sunlight to vitamin D_3 (cholecalciterol). Both vitamin D_3 and vitamin D_2 (ergocalciferol) can be converted by liver 25-hydroxylase to 25-hydroxyvitamin D (calcidiol). Calcidiol is the vitamin D metabolite that is measured in serum to determine a person's vitamin D status. In the kidney, 1-α-hydroxylase converts part of the 25-hydroxyvitamin D to 1,25-dihydroxyvitamin D (calcitriol), which is the biologically active form of vitamin D. Calcitrol circulates in the blood, regulating calcium and phosphate levels. On binding of calcitriol to an appropriate vitamin D receptor (VDR), the biological action is initiated. VDRs are principally located in the nuclei of target cells.

4. VITAMIN E

4.1 Introduction

Vitamin E [14,15] was discovered in 1922 by Herbert M. Evans (Fig. 21.11) and Kathrine S. Bishop at the University of California, Berkeley. These investigators discovered that rats fed

FIGURE 21.11 Herbert Evans, 1882–1971. http://webpac.uvi.edu/imls/fb_baa/E/Evans_Melvin_Herbert/index.shtml.

diets with lard as the fat source failed to reproduce. However, when the lard diet was supplemented with wheat germ or lettuce, the problem was alleviated. The active component was originally called an "antisterility factor," but in 1925 Evans changed the name to vitamin E since the last vitamin to be discovered was vitamin D. Evans purified vitamin E from wheat germ in 1937.

Vitamin E is a general term covering a group of eight similar, lipid-soluble vitamins that are all antioxidants [14,15]. The eight vitamin Es (Fig. 21.12) are grouped into four tocopherols and four tocotrienols and are further identified by prefixes; alpha (α), beta (β), gamma (γ), and delta (δ). For example, α-tocopherol is the most studied of all eight of the vitamin Es and has the highest biological activity. Unless stated otherwise, many reports on "vitamin E" actually refer to α-tocopherol. γ-Tocopherol is the most abundant vitamin E in the North American diet, just ahead of α-tocopherol. Both tocopherols are common components of plant oils and nuts and so are readily available in the human diet making deficiency of vitamin E rare.

All four natural tocopherols occur only in the RRR-configuration (Fig. 21.13). In contrast, synthetic tocopherols are mixtures of the eight possible sterioisomers and so are called all-*rac*-tocopherol. The consumer of vitamin E supplements must be careful not to obtain a racemic mixture, since the unnatural isomers likely have no biological function.

4.2 Antioxidant Function

The family of vitamin Es have many health-related biological functions [16]. The best known and most important of these is its role as the major lipid antioxidant in membranes [17]. In fact, α-tocopherol resides almost solely in membrane lipid bilayers. Vitamin E stops the production of reactive oxygen species (ROS) formed upon lipid oxidation and in so doing

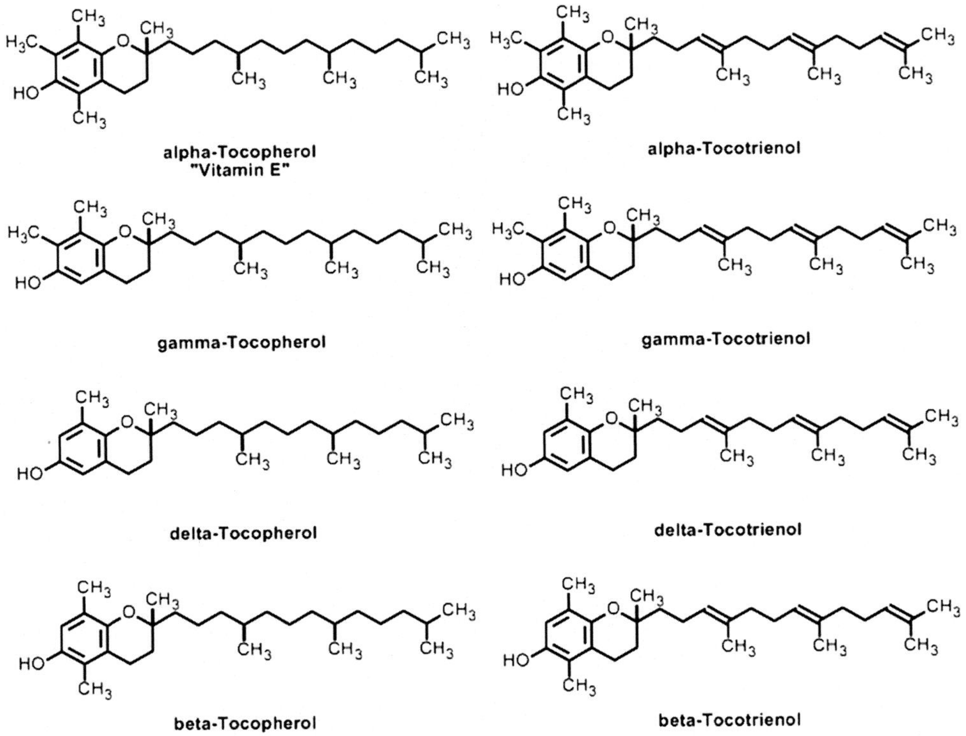

FIGURE 21.12 Structures of the eight types of vitamin E. http://www.vita-dose.com/structure-of-vitamin-e.html.

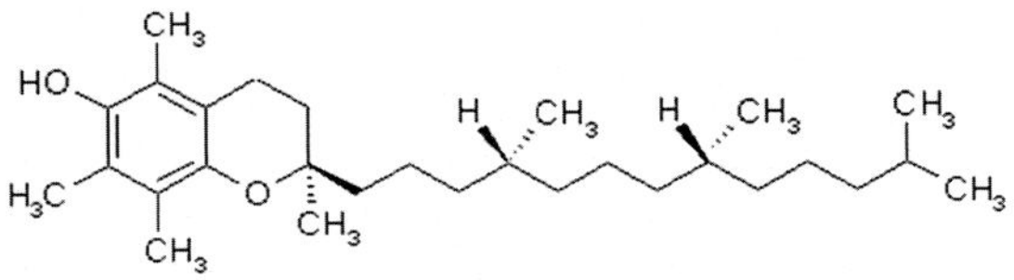

FIGURE 21.13 RRR-α-tocopherol. http://en.wikipedia.org/wiki/Tocopherol.

prevents the destruction of membranes. Particularly susceptible to oxidation are polyunsaturated fatty acyl chains (PUFAs) that are highly abundant in the membrane lipid bilayers of animals. As a general rule, the more acyl chain double bonds a membrane has, the more susceptible it is to oxidation. Therefore, membranes with substantial levels of arachidonic acid (AA, 20:4), eicosapentaenoic acid (EPA, 20:5), and docosahexaenoic acid (DHA, 22:6) should require larger levels of resident vitamin E for protection. The acyl chain locations that are most susceptible to oxidation are the ($-CH_2-$) locations, linking sequential double bonds ($-CH{=}CH-\mathbf{CH_2}-CH{=}CH-$). Fig. 21.14 demonstrates how α-tocopherol functions as a scavenger of peroxyl radicals, preventing the potentially explosive propagation

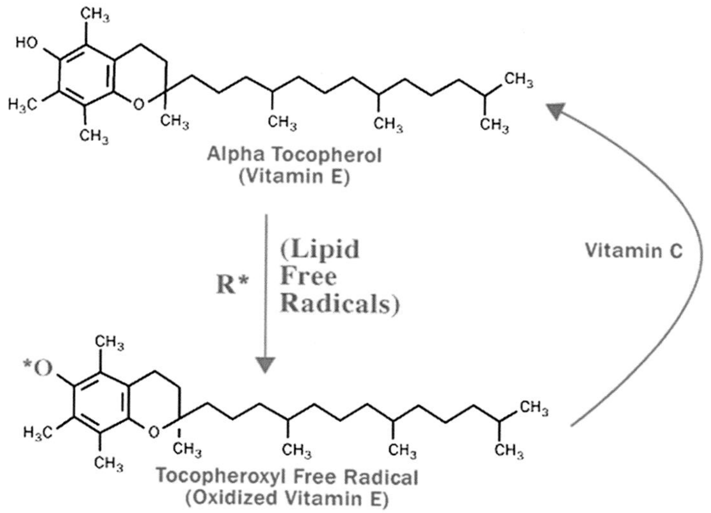

FIGURE 21.14 The "α-tocopherol cycle," demonstrating how α-tocopherol functions as an antioxidant. http://www.life-enhancement.com/magazine/article/2274-break-the-bonds-of-dementia.

of free radicals in membranes. This is known as the "α-tocopherol cycle." α-Tocopherol reacts with peroxyl radicals, reducing them and in turn becoming oxidized generating an α-tocopherol radical. The α-tocopherol radical is then reduced back to its original reduced state by hydrogen donors like vitamin C (ascorbate), retinol, or ubiquinol.

A major conundrum surrounding vitamin E concerns how low levels of the vitamin can protect such a vast excess of unsaturated acyl chains from oxidation [17]. One question that has often been asked is does α-tocopherol attract PUFAs or visa versa? In other words, is there a structural relationship between α-tocopherol and PUFAs that enhances antioxidant function? In the early 1970s, Diplock and Lucy [18] proposed on the basis of molecular models that the multiple *cis*-double bonds in arachidonic acid are perfectly spaced to create pockets that can exactly accommodate the 4′ and 8′ methyl groups on the isoprenoid side chain of α-tocopherol. This idea of PUFA/α-tocopherol complexes was supported by physical measurements of α-tocopherol/PUFA mixtures dissolved in organic solution. Unfortunately, when these experiments were extended to lipid bilayers, it became evident that such complexes probably do not exist in membranes. Recent advances in conformational dynamics of PUFAs have demonstrated that very rapid interconversion between tortional states, make long-lived α-tocopherol/PUFA complexes highly unlikely. However, preferential associations like those observed for cholesterol and sphingolipids (Chapter 8) are still feasible. Instead of the PUFA/α-tocopherol attraction being through the isoprenoid chain, involvement of the chromanol ring is now considered to be more likely.

To better understand how α-tocopherol may function as both an antioxidant and membrane structural component in PUFA-containing membranes, its bilayer depth, orientation, and conformational dynamics were determined [17,19]. Biophysical studies of α-tocopherol in lipid bilayers paralleled those previously done with cholesterol. At first glance, α-tocopherol bears a striking, but superficial, resemblance to cholesterol. Both molecules are amphipathic, being anchored to the aqueous interface by an (−OH) group attached to a rigid ring that constitutes the molecule's mid-structure. Finally, both molecules have a floppy, branched acyl chain protruding down into the bilayer interior. The orientation of cholesterol in lipid bilayer membranes was discussed in Chapter 10. Neutron scattering suggests that α-tocopherol stands upright in the bilayer with the hydroxyl group just above the depth of the first acyl chain carbon [17,19]. The α-tocopherol isoprenyl tail is highly disordered since the C9′ carbon is found in a broad range of locations.

4.3 PUFA/α-Tocopherol Domains

Cholesterol has been shown to associate strongly with SM and with saturated acyl chains (Chapter 10). This association is the "molecular glue" that holds lipid rafts together. Cholesterol has also been shown to strongly avoid contact with PUFAs [20]. It might be advantageous if PUFAs, excluded from traditional saturated acyl chain-enriched lipid rafts, then associated instead with α-tocopherol to form another type of as-yet-undiscovered lipid domain. In sharp contrast to SM/cholesterol lipid rafts, PUFA/α-tocopherol domains would be highly disordered (l_d state) and have a completely different set of resident proteins and hence functions.

There is some evidence supporting the association of α-tocopherol and PUFAs. A 1996 DSC study by Sanchez-Migallon et al. [21] showed that α-tocopherol broadened and lowered the phase transition temperature of PUFA-containing PCs, which was interpreted as being due to lateral phase separation of α-tocopherol−enriched domains [21]. Detergent extraction and DSC studies offered persuasive support for the concept that α-tocopherol co-localizes with DHA [22].

The difference in fatty acyl association behavior between cholesterol and saturated acyl chains versus α-tocopherol and PUFA acyl chains can be attributed to the size of the much larger rigid sterol ring of cholesterol compared to the smaller rigid chromanol ring of α-tocopherol. In fact, a chromanol ring is only about half the size of a sterol ring. For PUFAs, the chromanol ring of α-tocopherol can fit adjacent to the acyl chain from carbon one to the first double bond (position Δ4 for DHA, Δ5 for AA, and Δ5 for EPA). The sterol ring of cholesterol cannot fit into this small space but instead needs at least nine straight saturated carbons (positions 1−9) to attain a tight fit. For this reason, α-tocopherol associates with PUFAs while cholesterol excludes them. Additionally, hydrogen bonding of the α-tocopherol hydroxyl to the phospholipid ester carbonyl further strengthens the association. Co-localization of α-tocopherol and PUFA would produce a concentration amplification that optimizes the protection of membranes from destructive oxidation.

It is the phenomenal flexibility of PUFA acyl chains [23] that exposes peroxyl radicals from all oxidizable positions on the PUFA chains to the chromanol group of α-tocopherol that in turn can be rapidly regenerated at the nearby aqueous interface by hydrogen donors like vitamin C. PUFA isomerization is so rapid it can explore all conformational space within 50 ns!

It has been stated that: "α-Tocopherol remains a most unusual character; it is a function in search of a home." Perhaps this home is an unusual type of lipid raft driven by α-tocopherol's affinity for PUFA.

4.4 Other Functions

Vitamin E is involved in several important functions in addition to its role as an antioxidant. These functions include participation in enzymatic activities, gene expression, neurological functions, and cell signaling [24].

Enzymatic Activity: An example of vitamin E's role as a regulator of enzymatic activity can be found in smooth muscle growth. α-Tocopherol stimulates the dephosphorylation enzyme, protein phosphatase 2A, which cleaves phosphate from protein kinase C (PKC) leading to its deactivation, thus inhibiting smooth muscle growth.
Gene Expression: An example of α-tocopherol affecting gene expression has come from studies of atherosclerosis. In macrophages, α-tocopherol was found to downregulate expression of the CD36 scavenger receptor gene and the scavenger receptor class A (SR-A) and modulate expression of the connective tissue growth factor (CTGF). The *CTGF* gene is responsible for the repair of wounds and regeneration of the extracellular tissue that is lost or damaged during atherosclerosis.
Other: The effect of α-tocopherol on PKC activity through activation of protein phosphatase 2A indicates a likely link to cell signaling events. Vitamin E also plays a role in neurological functions and inhibition of platelet aggregation.

4.5 Vitamin E Deficiency

Vitamin E deficiency is rare and almost never related to a dietary deficiency [25]. In fact, vitamin E deficiency is only found in very low weight, premature infants or in individuals with either an inability to absorb dietary fat or rare disorders of fat metabolism.

Symptoms of vitamin E deficiency are often neurological or neuromuscular. However, deficiency can also result in anemia due to oxidative damage to red blood cells, retinopathy, immune response impairment, and male infertility.

4.6 Tocotrienols

Although tocotrienols comprise four of the eight types of vitamin Es, comparatively little is known about them [26]. Less than 1% of the total vitamin E articles on PubMed concern tocotrienols, but interest is growing. It now appears that tocotrienols may even be more potent antioxidants than tocopherols. Recent studies have suggested that tocotrienols protect neurons from damage and may affect cholesterol biosynthesis by inhibiting HMG-CoA reductase. On a medical note, oral tocotrienols are thought to protect against brain damage due to stroke. Clearly, tocotrienol studies are just in their infancy, but their future is bright.

5. VITAMIN K

5.1 Vitamins K_1 and K_2

Vitamin K is the "poor sister" of the lipid-soluble vitamins. In general, vitamin K is not widely appreciated. While vitamin A is essential for vision, vitamin D is required for bone

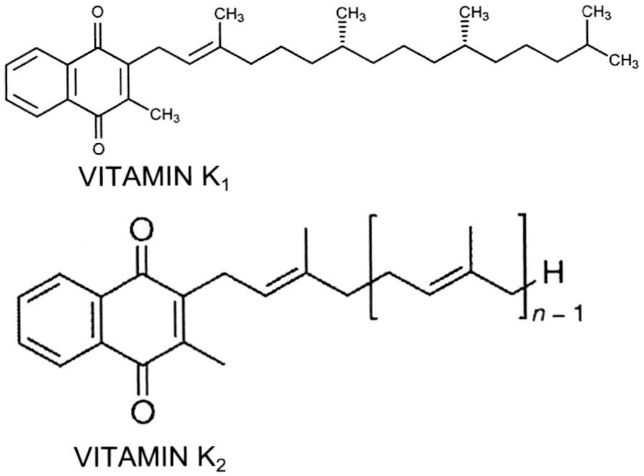

FIGURE 21.15 Structures of vitamin K_1 and vitamin K_2. *Vitamin K_1: About.com Chemistry, http://chemistry.about.com/od/imagesclipartstructures/ig/Vitamin-Chemical-Structures/Vitamin-K1.htm; Vitamin K_2: http://www.hiwtc.com/products/vitamin-k2-mk7-200212-32016.htm.*

health and vitamin E is the major membrane antioxidant, vitamin K does not have a marquee function. Part of this problem stems from both hypervitaminosis K and hypovitaminosis K being almost nonexistent in humans. Vitamin K's essential role in blood coagulation is important but not obvious.

Like vitamins A, D, and E, vitamin K is lipid-soluble and can be stored in fat deposits [27,28]. The family of vitamin K are all 2-methyl-1,4-naphthoquinone derivatives that have varying isopropyl side chains. The two major vitamin Ks are vitamin K_1 and vitamin K_2 (Fig. 21.15).

Vitamin K_1, also known as phylloquinone, phytomenadione or phytonadione, is the plant form of vitamin K. Vitamin K_1 is an abundant molecule in Nature since it is an important component of photosynthesis and is found in highest amounts in green leafy vegetables. Due to its abundance, vitamin K_1 is commonly found in both plants and animals that consume the plants. Animals can easily convert vitamin K_1 to vitamin K_2 (menaquinone), the animal form of vitamin K. Due to its ready availability and rapid recycling, vitamin K deficiency in humans is rare.

The family of vitamin K_2 are the active forms of vitamin K used in animals. All vitamin K_2s have the same naphthoquinone head group but vary as to the number of five-carbon isoprene units composing the long hydrophobic isoprenoid tail. Since it is generally agreed that the naphthoquinone head group is responsible for functionality, all vitamin Ks have similar mechanisms of action. Menaquinones are abbreviated MK-*n*, where M stands for menaquinone, K stands for vitamin K, and *n* represents the number of isoprenes in the side chain. The most prevalent of the vitamin K_2s in animals is abbreviated MK-4, indicating the presence of four isoprenes (20 carbons) in the side chain. MK-4 is the most easily converted vitamin K_2 since it is directly made by desaturation of plant vitamin K_1 that already has a 20-carbon side chain. It is this form of vitamin K_2 that is involved in the

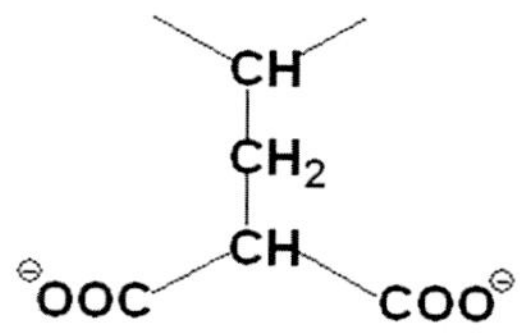

FIGURE 21.16 Gamma-carboxyglutamate (Gla). http://www.cram.com/flashcards/biochem-1-2913211.

production of blood clotting proteins [29]. Another common vitamin K_2, with a longer side chain than MK-4, is MK-7. This vitamin, however, is not produced by human tissue and must be obtained through the diet. MK-7 has been proposed to be involved in reducing the risk of bone fractures and cardiovascular disorders.

5.2 Gamma-Carboxyglutamate Proteins

In animals, vitamin K functions by carboxylating certain gultamate residues in target proteins to form gamma-carboxyglutamate (Gla, Fig. 21.16) [30] residues. The structure of Gla with its two carboxylates allows this amino acid to function as a Ca^{2+}-binding chelating agent. The Gla–Ca^{2+} complexes trigger the function or binding of Gla proteins such as the vitamin K–dependent clotting factors. This type of post-translational modification is uncommon and produces what is known as "Gla proteins."

5.3 Health Effects

There are currently about 15 known human Gla proteins that participate in the regulation of three major physiological processes:

1. Blood coagulation
2. Bone metabolism
3. Vascular biology

The vitamin Ks have been investigated for use in a number of medical applications. A major implementation involves using either vitamins K_1 or K_2 to reverse the powerful anticoagulant (blood-thinning) activity of the highly prescribed drug warfarin (brand name Coumadin) [29]. Since it is hard to accurately control the level of warfarin in the body, toxic levels may occur. Toxicity may be reversed by vitamin K. Warfarin works by blocking normal recycling of vitamin K resulting in depleted levels of vitamin K and hence diminished blood coagulation. Vitamin K is commonly used as an antidote for poisoning by anticoagulants including warfarin and the rat poison bromadione.

Vitamin K has been studied with respect to a number of human afflictions unrelated to blood coagulation. For example, a Nurses Health Study from 1998 reported an inverse relationship between dietary vitamin K and the risk of hip fracture [31]. Preliminary results have also correlated vitamin K levels with reduction of neuronal damage, Alzheimer's disease, osteoporosis, and coronary heart disease.

6. SUMMARY

Vitamins are a group of organic molecules that, in minute quantities, are essential for normal growth and several body processes. With the exception of vitamin D, vitamins are not produced in the human body and so must be obtained through the diet. Chapter 21 discusses only the lipid-soluble vitamins, A, D, E, and K, since they have a high affinity for membranes. *Vitamin A* is found in three oxidation states: alcohol (retinol), aldehyde (retinal), and acid (retinoic acid). Retinol esters are the major storage form of vitamin A in the body and can be readily converted to retinal and retinoic acid. Retinol is a membrane antioxidant. Retinal is essential for vision and retinoic acid is a hormone-like growth factor that is involved in growth and cellular differentiation through regulation of gene transcription. Retinoic acid has also been successfully employed in medical dermatology to alleviate a number of human skin conditions, including acne.

Vitamin D is not a true vitamin as it can be synthesized in small amounts by humans when exposed to UV light. Vitamin D has therefore earned the nickname the "sunshine vitamin," although in many ways it more closely resembles steroid hormones than vitamins. There are two major forms of vitamin D (known as vitamers): vitamin D_2 (ergocalciferol) and vitamin D_3 (cholecalciferol). The major function of vitamin D involves absorption of calcium and phosphate in the intestine and therefore plays a large role in bone growth, development and health. Vitamin D deficiency results in osteomalacia, a condition where the bones are softened due to defective mineralization of phosphorous and calcium. This condition in children is better known to as "rickets." Recently, vitamin D has been linked to a plethora of nonbone afflictions including multiple sclerosis, some cancers, immune function, influenza during winter months, tuberculosis, HIV, asthma, neuromuscular function, premature aging, and mortality in the elderly.

Vitamin E is a general term covering a group of eight similar, lipid-soluble vitamins that are all antioxidants. Four members of the vitamin E family are tocopherols and four are the less familiar tocotrienols. α-Tocopherol is the most studied of all eight of the vitamin Es and has the highest biological activity. Vitamin Es cannot be synthesized by humans, but fortunately are highly available in plant oils and nuts. Although vitamin Es are reputed to have many biological functions, by far the best known is as the major anti-oxidant in membranes. α-Tocopherol resides almost solely in membrane lipid bilayers where its task is to scavenge peroxyl radicals, preventing the potentially explosive propagation of free radicals in PUFA-rich-membranes. A major conundrum is explaining how low levels of the vitamin can protect such a vast excess of PUFA chains from oxidation. A search is now underway looking for a structural relationship between α-tocopherol and PUFAs that enhances antioxidant function.

Vitamin K is the "poor sister" of the lipid-soluble vitamins as it is not widely appreciated by the general public. Vitamin K however does have one essential function - blood clotting. Vitamin K comes in two major forms, Vitamin K_1 and vitamin K_2 (menaquinone), and many other lesser forms. Vitamin K_1 is abundant in plants as it is involved in photosynthesis and is therefore plentiful in the human diet where it is rapidly converted to vitamin K_2, the active form of vitamin K in humans. All vitamin K_2s have the same naphthoquinone head group, but vary as to the number of five-carbon isoprene units comprising the long hydrophobic tail. The most active form of menaquinone has four isoprene units (20 carbons) and functions by carboxylating certain gultamate residues on target proteins to form Gla, a Ca^{2+}-binding chelating agent. The Gla–Ca^{2+} complexes trigger formation of the vitamin K–dependent clotting factors.

Chapter 22 will next consider how alterations in membrane composition, structure, and function can lead to some of the most debilitating and lethal diseases and conditions that humans confront: cystic fibrosis, Duchenne muscular dystrophy, Alzheimer's disease, demyelination diseases, and aging.

References

[1] Lieberman S, Bruning N. The real vitamin & mineral book. New York: Avery Group; 1990.

[2] McDonald A, Natow A, Heslin J-A, Smith SM, editors. Complete book of vitamins and minerals. Publications International, Ltd; 2000.

[3] Higdon J, 2003. Updated by Drake VJ, 2007. Micronutrient Information Center. Vitamin A. Linus Pauling Institute Oregon State University.

[4] Bates CJ. Vitamin A (review). Lancet 1995;345:31–5.

[5] Combs Jr GF. The vitamins. Fundamental aspects in nutrition and health. 4th ed. New York: Academic Press; 2012. 600 pp.

[6] Kiser PD, Golczak M, Palczewski K. Chemistry of the retinoid (visual) cycle. Chem Rev 2014;114(1):194–232.

[7] Penniston KL, Tanumihardjo SA. The acute and chronic toxic effects of vitamin A. Am J Clin Nutr 2006;83(2):191–201.

[8] Rodahl K, Moore T. The vitamin A content and toxicity of bear and seal liver. Biochem J July 1943;37(2):166–8.

[9] Smith KA. A WWII propaganda campaign popularized the myth that carrots help you see in the dark. August 13, 2013. smithsonianmag.com.

[10] Norman AW. From vitamin D to hormone D: fundamentals of the vitamin D endocrine system essential for good health. Am J Clin Nutr 2008;88(2):491S–9S.

[11] Holick MF. High prevalence of vitamin D inadequacy and implications for health. Mayo Clin Proc March 2006;81(3):353–73.

[12] MedlinePlus Encyclopedia *Rickets*.

[13] Wiseman H. Vitamin D is a membrane antioxidant. Ability to inhibit iron-dependent lipid peroxidation in liposomes compared to cholesterol, ergosterol and tamoxifen and relevance to anticancer action. FEBS Lett 1993;326(1–3):285–8.

[14] Preedy VR, Watson RR. The Encyclopedia of vitamin E. 1st ed. CABI; 2007. 960 pages.

[15] Litwack G. Vitamin E. Vitamins and hormones [Hardcover], vol. 76. London: Academic Press; 2007. 616 pp.

[16] Papas A. The vitamin E factor: the miraculous antioxidant for the prevention and treatment of heart disease, cancer, and aging. New York: HarperCollins Publishers; 1999.

[17] Atkinson J, Epand RF, Epand RM. Tocopherols and tocotrienols in membranes: a critical review. Free Rad Biol Med 2008;44:739–64.

[18] Diplock AT, Lucy JA. The biochemical modes of action of vitamin E and selenium: a hypothesis. FEBS Lett 1973;1973(29):205–10.

[19] Atkinson J, Harroun T, Wassall SR, Stillwell W, Katsaras J. The location and behavior of α-tocopherol in membranes. Mol Nutr Food Res 2010;54:1–11.

[20] Wassall SR, Brzustowicz MR, Shaikh SR, Cherezov V, et al. Order from disorder, corralling cholesterol with chaotic lipids. The role of polyunsaturated lipids in membrane raft formation. Chem Phys Lipids 2004;132:79–88.

[21] Sanchez-Migallon MP, Aranda FJ, Gomez-Fernandez JC. Interaction between alpha-tocopherol and heteroacid phosphatidylcholines with different amounts of unsaturation. Biochim Biophys Acta 1996;1279:251–8.

[22] Stillwell W, LoCascio DS, Turker Gorgulu S, Wassall SR. Is α-tocopherol the "glue" that holds PUFA non-raft domains together? Biophys J 2008;94:396.

[23] Feller SE, Gawrisch K, MacKerrell Jr AD. Polyunsaturated fatty acids in lipid bilayers: intrinsic and environmental contributions to their unique physical properties. J Am Chem Soc 2002;124:318–26.
[24] Azzi A. Non-antioxidant activities of vitamin E. Curr Med Chem 2004;11(9):1113–33.
[25] Brigelius-Flohé R, Traber MG. Vitamin E: function and metabolism. FASEB J 1999;13(10):1145–55.
[26] Tan B, Watson RR, Preedy VR. Tocotrienols: vitamin E beyond tocopherols. 2nd ed. Boca Raton: CRC Press; 2013.
[27] Vitamin K overview. University of Maryland Medical Center; 2013. http://umm.edu/health/medical/altmed/supplement/vitamin-k.
[28] Higdon J. Vitamin K. Linus Pauling Institute, Oregon State University, Micronutrient Information Center; 2014. http://lpi.oregonstate.edu/infocenter/vitamins/vitaminK/.
[29] Ansell J, Hirsh J, Poller L, Bussey H, Jacobson A, Hylek E. The pharmacology and management of the vitamin K antagonists: the Seventh ACCP Conference on Antithrombotic and Thrombolytic Therapy. Chest 2004;126(Suppl. 3):204S–33S.
[30] Stenflo J, Suttie JW. Vitamin K-dependent formation of γ-carboxyglutamic acid. Annu Rev Biochem 1995;46:157–72.
[31] Booth SL, Tucker KL, Chen H, Hannan MT, Gagnon DR, Cupples LA, et al. Dietary vitamin K intakes are associated with hip fracture but not with bone mineral density in elderly men and women. Am J Clin Nutr 2000;71(5):1201–8.

CHAPTER

22

Membrane-Associated Diseases

OUTLINE

An Introduction to Biological Membranes
http://dx.doi.org/10.1016/B978-0-444-63772-7.00022-1

Since almost all biological processes are at some level affected by membranes, it is difficult to select just a few representative examples of the many membrane-associated human afflictions. The fundamental importance of membranes in life makes their malfunctions particularly life-threatening. A few examples of membrane-associated diseases have been selected to demonstrate the nature and severity of the problem. Examples to be discussed here include cystic fibrosis (CF), Duchenne muscular dystrophy (DMD), Alzheimer's disease (AD), demyelination diseases, and aging.

1. CYSTIC FIBROSIS

1.1 Health Effects

Cystic fibrosis (CF) is a prime example of a well-known membrane-associated disease [1]. CF is an autosomal recessive genetic disorder that is primarily known for its affect on lungs, resulting in fluid buildup, difficulty breathing, fatigue, and frequent lung infections. Besides the lung, CF is also known to exhibit deleterious affects on the pancreas, liver, and intestine. Effects are associated with improper mucus, sweat, saliva, tears, and digestive fluids. The primary problem in CF is with mucus, a slippery substance that lubricates linings of the lung, digestive system, reproductive system, and other tissues.

CF most commonly afflicts Caucasians. The gene manifests itself only as a homozygous recessive and, in Europe is found in 1 in 2000–3000 births. The ultimate outcome of CF is often a lung transplant or death. It has been estimated that CF first appeared about 3000 BC, just before building of the Great Pyramid of Giza (2560 BC). CF has been recognized in the medical literature for hundreds of years. 18th century German and Swiss literature warned "Woe to the child who tastes salty from a kiss on the brow, for he is cursed and soon must die." Salty sweat is currently employed as a diagnosis for CF. In 1938, Dorothy Hansine Andersen (Fig. 22.1) published a seminal paper tying together for the first time various aspects of CF [2]. In 1989, Tsui and coworkers at the Hospital For Sick Children in Toronto identified and sequenced the gene for CF [3].

1.2 Mechanism of Action

The protein responsible for CF, CF transmembrane conductance regulator (CFTR), is an anion channel, responsible for the flow of H_2O and particularly Cl^- in and out of cells [4]. Abnormal transport of Cl^- and associated Na^+ leads to thick, viscous secretions that are characteristic of CF. The most common mutation leading to CF results in a deletion of a phenylalanine at position 508 (ΔF508) on the CFTR protein. Although this mutation accounts for about 90% of CF in the United States, more than 1500 other CF mutations have been identified. CFTR is a glycoprotein composed of 1480 amino acids and is an adenosine triphosphate binding cassette (ABC) transporter-class ion channel whose primary job is to transport Cl^-. The protein has two ATP-hydrolyzing domains (nucleotide binding domains) and two transmembrane domains comprised of six alpha helices apiece that form the Cl^- channel (Fig. 22.2). A regulatory site on CFTR allows for activation by phosphorylation with a cyclic

FIGURE 22.1 Dorothy Hansine Andersen, 1901–1963. http://www.fountia.com/who-discovered-cystic-fibrosis/.

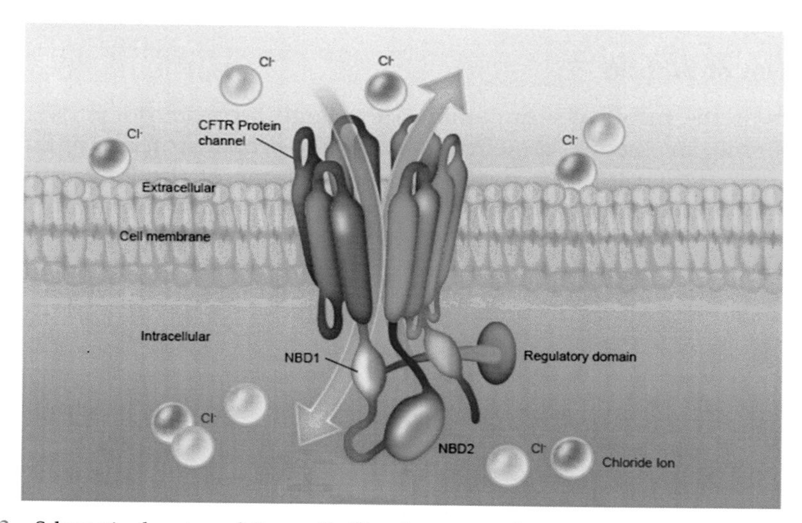

FIGURE 22.2 Schematic drawing of the cystic fibrosis transmembrane conductance regulator (CFTR) channel protein. CFTR INFO http://www.cftr.info/about-cf/role-of-cftr-in-cf/genetics-and-cell-biology-of-cftr/cftr-function-and-regulation/.

adenosine monophosphate (cAMP)-dependent protein kinase A. Finally, the carboxyl terminal of the protein is anchored to the cytoskeleton by a PDZ domain interaction. The ion channel is open only when the regulatory domain has been phosphorylated by protein kinase A and ATP is bound at the nucleotide-binding domain. The open channel is the normal state, and channel blockage leads to CF.

2. DUCHENNE MUSCULAR DYSTROPHY

2.1 Health Effects

Duchenne Muscular Dystrophy (DMD) [5] is another well-known example of a fatal genetic disorder whose mutated protein (dystrophin) is linked to a membrane (the muscle cell plasma membrane). DMD is a recessive X-chromosome-linked form of muscular dystrophy that affects 1 in 3600 boys, while having virtually no affect on girls. In boys, serious symptoms appear by age six and get progressively worse with time. Muscle mass loss and weakness is first observed in the legs but spreads to the arms, neck, and other areas. Most patients are wheelchair-bound by age 12 and die by age 25. For this reason, famous people with DMD are almost nonexistent. One exception was the race car designer Alfredo (Dino) Ferrari who designed the 1.5 L DOHC V6 engine shortly before his death from DMD in 1956 at the age of 24.

The Duchenne form of muscular dystrophy was first described by Giovanni Semmola in 1834. However, it was not until 1861 that the French neurologist Guillaume Benjamin Amand Duchenne (Fig. 22.3) described in detail the symptoms of a boy who had the disease. Subsequent descriptions and photos followed, resulting in his name being attached to the disease.

2.2 Mechanism of Action

DMD occurs as the result of a mutation in the large X-chromosome dystrophin gene, affecting dystrophin, an essential protein in maintaining proper structure of the muscle dystroglycan complex (DGC) [6]. The dystrophin gene is one of the longest in the human genome. It is composed of 2.5 million base pairs and is so long that it takes 16 h to transcribe and is produced with 79 exons [7]! Dystrophin is absent in DMD patients. In skeletal muscle the DGC serves as a transmembrane linkage between the extracellular matrix and

FIGURE 22.3 Guillaume Benjamin Amand Duchenne, 1806–1875. http://www.uic.edu/depts/mcne/founders/page0026.html.

intracellular actin attached to the cytoskeleton. A diagram of the multicomponent DGC is shown in Fig. 22.4 [8].

Dystrophin is a very large (427 kDa, 3685 amino acid), rod-shaped protein that connects actin of the cytoskeleton to the transmembrane β-subunit of dystroglycan (β-DG). Dystrophin is not a transmembrane protein and in fact is usually grouped with the cytoskeleton proteins.

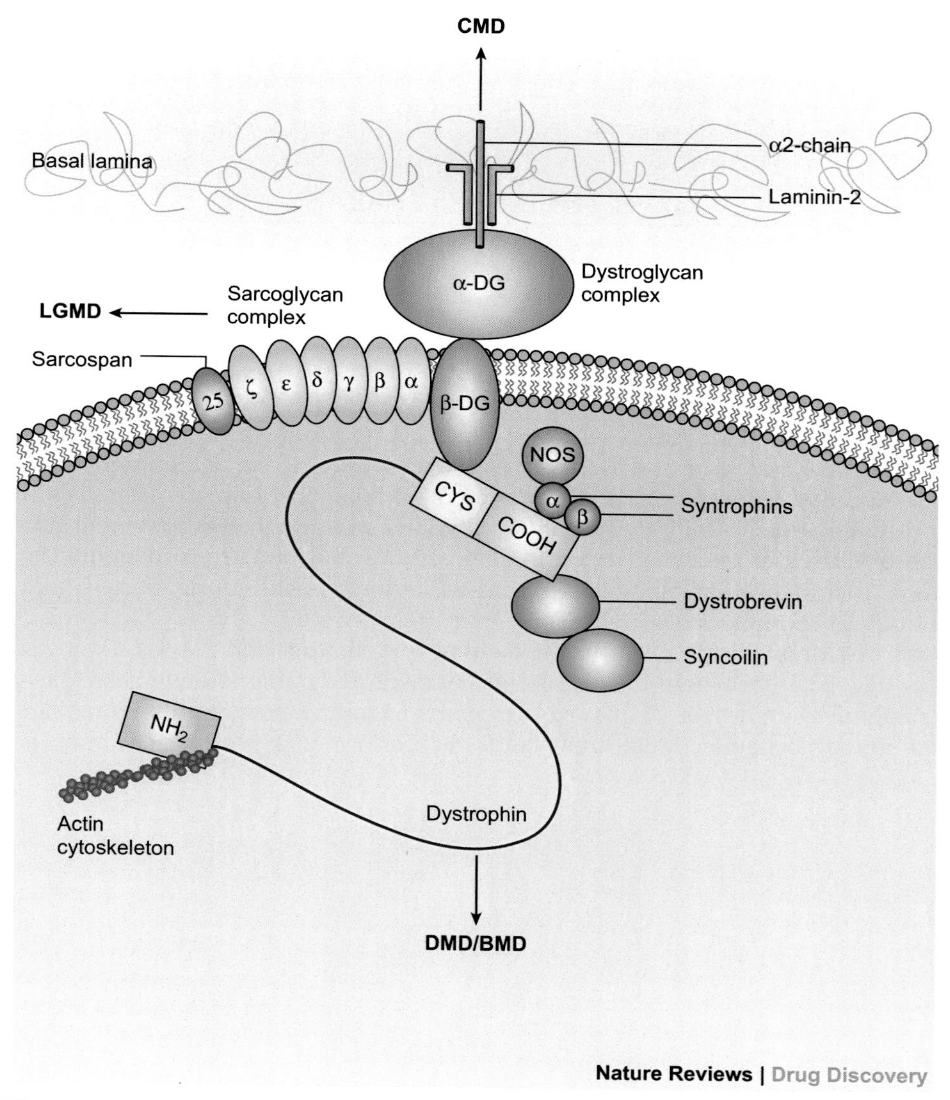

FIGURE 22.4 Diagram of the muscle dystroglycan complex (DGC). The transmembrane protein, dystrophin, links the extracellular matrix to the intracellular cytoskeletal actin. http://www.nature.com/nrd/journal/v2/n5/fig_tab/nrd1085_F1.html. *Khurana TS, Davies KE. Pharmacological strategies for muscular dystrophy. Nat Rev Drug Discov 2003;2:379–90.*

Dystrophin's major function involves providing essential muscle characteristics by stabilizing and strengthening the DGC, increasing muscle fiber strength, reducing muscle stiffness, and increasing sarcolemmal deformability.

Besides dystrophin, the other major component of the DGC is dystroglycan, that is translated as one large protein that is cleaved into two noncovalently associated subunits, α-dystroglycan and β-dystroglycan. α-Dystroglycan (the N-terminal peptide) is extracellular where it binds to alpha-2-laminin of the extracellular matrix basement membrane. β-Dystroglycan (the C-terminal peptide) is a transmembrane protein that binds to dystrophin inside the cell that in turn binds to actin in the cytoplasm. Thus the DGC links the extracellular matrix to intracellular actin.

3. ALZHEIMER'S DISEASE

3.1 Health Effects

There is perhaps no human affliction as confounding as Alzheimer's disease (AD). Alzheimer's is the most common form of dementia, affecting some 5.2 million Americans in 2013, and this number is expected to grow rapidly in the immediate future [9,10]. Worldwide there are five million new cases per year or one every 7 s. AD worsens as it progresses, is incurable, and invariably results in death. AD is currently the sixth leading cause of death in the United States. Most diseases, even the most severe, offer a glimmer of hope for a successful outcome. At present, Alzheimer's offers no hope, just despair. There is a tremendous worldwide effort to find a cure for AD. By 2012 more than 1000 potential AD drugs have been or currently are being tested in clinical trials, but no cure is in sight. The hope, of course, is that a cure for AD will be found while there is still an unaffected population smart enough to continue the search!

AD was first described in 1906 by the German psychiatrist and neuropathologist Alois Alzheimer (Fig. 22.5) for whom the disease was named. Alzheimer initially based his studies on a mentally ill woman who displayed symptoms including memory loss, language problems, and strange behavior. Most importantly, he performed a postmortem brain autopsy

FIGURE 22.5 Alois Alzheimer, 1864–1915. http://www.pyroenergen.com/articles12/alzheimers-disease-diabetes.htm.

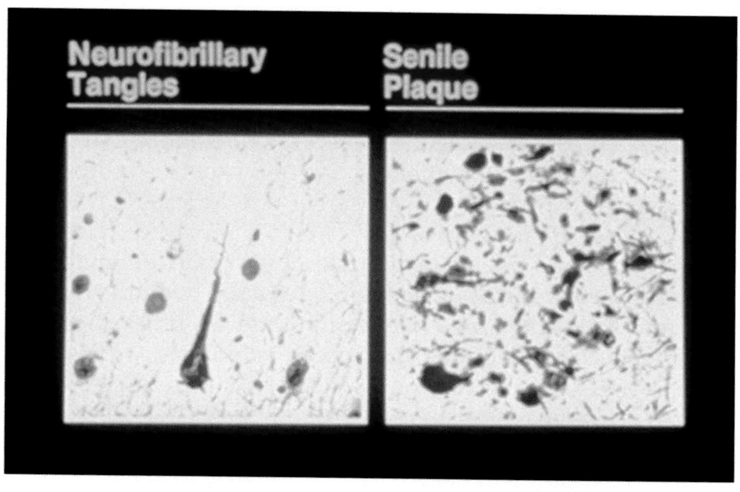

FIGURE 22.6 Neurofibrillary tangles (left) amyloid plaques(right). http://nihseniorhealth.gov/alzheimersdisease/symptomsanddiagnosis/tangles_popup.html.

where he found many abnormal clumps (now known as amyloid plaques) and tangled bundles of fibers (now known as neurofibrillary tangles) (Fig. 22.6). What is most remarkable about his observations is that after more than a century, the plaques and tangles he discovered remain the major characteristic to confirm AD [11].

3.2 Mode of Action

No one yet fully understands the origin and progression of AD. However, this has not hindered scientists from proposing a wide range of possibilities. One possibility, pertinent to this book, involves several aspects of membrane structure and function. AD develops due to a complex series of events in the brain that are stretched over many years. It appears unlikely that AD is the result of a single event or cause but likely is initiated by an unknown mix of genetic, environmental, and lifestyle factors. Briefly outlined below are several possible causes/mechanisms of AD that are currently being pursued or have been investigated in the past. All examples have their advantages, disadvantages, and overlaps, but none have led to success in preventing, halting, reversing, or curing AD. Possible causes/mechanisms of AD include:

3.2.1 Genetic Hypothesis

An obvious solution to the AD conundrum would be "it is all genetic." However, only 1–5% of all AD cases can be directly attributed to genetics and of these only 0.1% are autosomal dominant mutations. Most of these can be attributed to mutations in three genes that encode for the proteins amyloid precursor protein, presenilin 1, and presenilin 2. Mutations in these three genes increase the production of a small protein $A_{\beta}42$, which is the main component of plaques. Although, at best, genetics can directly account for only a small fraction of

AD cases, it has recently been suggested that "Genetic factors play a major role in determining a person's risk to develop Alzheimer's disease (AD)" [12].

3.2.2 *Virus Hypothesis*

Of course, there has been considerable interest in the possibility that viruses are in some way linked to AD. In one example that combines viral and genetic studies, the herpes simplex virus type 1 was proposed to play an AD causative role in people carrying a susceptible version of the apolipoprotein E (apoE) gene [13]. ApoE is involved in normal catabolism of triglyceride-rich lipoproteins and assists in cholesterol transport to neurons. One variant of apoE (E4) is the largest known risk factor for late-onset sporadic AD. Caucasian and Japanese carriers of two E4 alleles have between 10 and 30 times the risk of developing AD by age 75. While the mechanism of this interaction is unknown, a possible link to amyloid has been suggested. Apo(E4) remains a strong genetic risk factor for AD.

3.2.3 *Cholinergic Hypothesis*

One of the oldest AD hypothesis is based on the assumption that dementia (AD) is the result of diminished brain neurological activity that in turn is a reflection of reduced synthesis of the best known neurotransmitter, acetylcholine [14]. Unfortunately, medications used to treat acetylcholine deficiency have not been effective at treating AD.

3.2.4 *Amyloid Hypothesis*

From its first description by Alois Alzheimer, amyloid plaques have played a central, if poorly understood role in AD. It seems plausible that preventing amyloid plaque development may stop AD. In 1991, the Amyloid Hypothesis proposed that extracellular beta-amyloid (A_β) deposits are responsible for AD [15]. An experimental vaccine was even developed that cleared the amyloid plaques, but unfortunately had almost no affect on dementia [16]. The Amyloid Hypothesis has been extended several times since its initial proposal. Nonplaque oligomers of A_β have been suggested to be the pathogenic form of A_β. They bind to neuron surface receptors and change synapse structure. In 2009, the Amyloid Hypothesis was again modified to include a close relative of A_β that cleaves neuronal connections by a process that is triggered by aging related events [17].

3.2.5 *Myelin Breakdown Hypothesis*

Since AD is a neurological malfunction, a reasonable place to search for causation would be myelin, the lipid jelly roll that surrounds and insulates nerves (Chapter 18). By one hypothesis, AD may be caused by age-related myelin breakdown in the brain [18]. This process was proposed to release iron, further exasperating the damage through catalyzing lipid peroxidation. Age-related oxidative stress is complex and may generate many potential targets for causing AD.

3.2.6 *Degeneration of Locus Ceruleus Hypothesis*

Because of the possibility that AD may be in some way linked to stress, a logical location to search for a cause of AD would be the locus ceruleus, a location in the brain involved with physiological responses to stress and panic. Locus ceruleus cells, the principal site for brain synthesis of norepinephrine, is reduced by ~70% in AD patients [19]. From work done on mice, it has been suggested that degradation of the locus ceruleus might be responsible for increased $A_β$ deposition in AD brain.

3.2.7 *Hyperphosphorylation of* Tau *Hypothesis*

The two classical tests for AD are the presence of $A_β$ plaques and neurofibrillary tangles. The tangles are aggregates of the *tau* proteins that have become hyperphosphorylated. *Tau* are proteins that stabilize microtubules and are abundant in central nervous system neurons. The *tau* hypothesis proposes that hyperphosphorylated *tau* initiates AD [20].

3.2.8 *Protein Misfolding Hypothesis*

It has often been stated that AD is the result of misfolded $A_β$ and *tau* proteins, resulting in plaque and tangle accumulation in the brain. Recently, protein misfolding in AD has been linked to problems with chaperone complexes [21]. However, details of the link between misfolded protein accumulation and AD is not yet known.

3.2.9 *Membrane Hypothesis*

As the examples outlined previously indicate, many diverse approaches to AD have been attempted, and yet a final solution to the problem remains elusive. One last hypothesis concerning the development and progression of AD involves membrane structure and function, the theme of this book. Arispe et al. [22] have proposed a hypothesis in which $A_β$ induces neuron cytotoxicity through an $A_β$ channel that allows external Ca^{2+} to flood into the cell at toxic levels. Two crucial histidines (the diad, His^{13}–His^{14}) form the entrance of the channel and so are essential to control access to the channel. A closed channel prevents cytotoxicity. Two important experiments have helped establish the channel hypothesis. The $A_β$ (channel) peptide was incorporated into model planar bimolecular lipid membranes (BLMs) (Chapter 8). The peptide induced a large increase in electrical conductivity, thus establishing the presence of a channel. Also, it was shown that addition of fresh aggregates of $A_β$ to cell cultures generated a potentially toxic increase in intracellular Ca^{2+}. Finally, the investigators tested a number of ligands for their affect on channel conductivity in BLMs and their ability to stop cytotoxicity [23]. Various ligands that are known to associate with His (eg, Ni, imidazole, His, and a series of His-related compounds) block the $A_β$ channel at the characteristic His^{13}–His^{14} diad, decreasing electrical conductivity and stopping cytotoxicity. In all cases tested, molecules that blocked channels in BLMs also kept cells alive, and ligands that did not block the channels (eg, methyl-imidazole) resulted in cell death. While these experiments are interesting and do nibble at the edges of AD, they do not clearly link $A_β$ channel activity to AD.

Although the $A_β$ channel experiments identified a *specific* target to study, it is possible that AD is instead due to a *generalized* defect in plasma membrane structure–function [24]. The complexity of plasma membranes offers a plethora of possible targets to investigate for roles in AD. Experiments from 1984 [25] demonstrated that platelet plasma membranes from AD versus normal subjects showed "substantial differences" with regard to their microviscosity, determined by fluorescence polarization of 1,6-diphenylhexatriene (DPH) (Chapter 9). At that time it was known that the microviscosity of platelet and lymphocyte membranes increased with age. It was also generally believed that AD probably accelerated the aging process. If this assumption is correct, AD platelets should exhibit a *higher* microviscosity than normal platelets at all subject ages. However, what was actually observed was quite surprising. The AD platelets had a considerably *lower* microviscosity than the normal platelets, implying that AD is not the result of just an accelerated aging process.

A further advance of the general "membrane hypothesis" for AD (discussed above in Section 3.2.9) is incorporated into the "Two-Step Mechanism of Membrane Disruption" as reviewed by Sciacca et al. [26]. The process of amyloid formation through assembly of the A_β, leading ultimately to amyloid fibrils appears to be at the center of AD pathology. It is now believed that the mechanism for this is through a "Two Step Mechanism" [27]. The early stages of amyloidogenesis are affected by abnormal A_β peptide misfolding, the result of oxidative metabolites, including cholesterol-derived aldehydes [26,27]. This "First Step" involves Schiff Base formation and produces characteristic spherical aggregates that alter membrane structure—function (Fig. 22.7, left). The "Second Step" involves the formation of fibrillary aggregates (Fig. 22.7, right) which also involves cholesterol and damages membranes.

The initial membrane-destructive component (the first step spherical aggregates in Fig. 22.7) have been suggested to be ion-selective pores that are responsible for allowing Ca^{2+} to flow into the cell at lethal levels [26]. In fact, one of the major observations associated with AD is an elevation of cytoplasmic Ca^{2+}. These channels can be selectively stopped by zinc. In sharp contrast, the second step involves subsequent amyloid fiber formation and elongation on the membrane surface, leading to deleterious, but nonspecific, membrane fragmentation [26]. Elongation of existing fibrils adsorb on the membrane surface and remove lipids via a detergent-like mechanism. This causes general membrane disruption and cell death and is not affected by zinc.

Recent advances in understanding plasma membrane structure as a conglomeration of lipid rafts (Chapter 8) opens a new vista to study the cause of AD. All of the proteins involved in formation of plaques and tangles are associated with neuronal plasma membranes, and several of the metabolic processes involved in AD appear to be associated with lipid rafts. It may be predicted that lipid rafts will be at the center of the next generation of AD investigations.

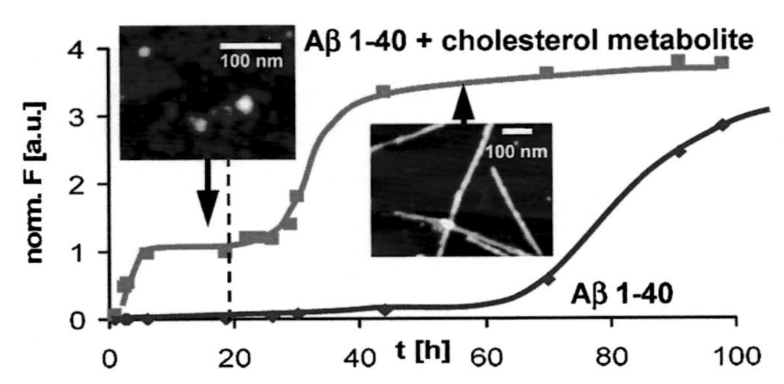

FIGURE 22.7 Two step process for the development of long term amyloid fibrils (right) from initial spherical misfolded A_β peptides (left). *Bieschke J, Zhang Q, Powers ET, Lerner RA, Kelly JW. Oxidative metabolites accelerate Alzheimer's amyloidogenesis by a two-step mechanism, eliminating the requirement for nucleation. Biochemistry 2005;44(13): 4977–83.*

4. DEMYELINATION DISEASES

4.1 Composition/Structure/Function of Myelin

Demyelination diseases are a family of related maladies of the nervous system in which the myelin sheath of neurons is damaged [28]. Since myelin is an essential component of the nervous system, its destruction impairs nerve impulse conduction (Chapter 18) leading to many adverse symptoms [29]. Myelin sheath is the protective covering surrounding nerves (Fig. 22.8). It is essentially a jelly roll of lipid-rich bilayers whose function is to protect and insulate the axon. The insulating properties of the myelin sheath allows an action potential to be transmitted up to 100 times faster in myelinated cells compared to unmyelinated cells. When the myelin sheath is damaged, nerve impulses slow or even stop, causing neurological disorders.

The myelin sheath was discovered in 1854 by German doctor and pathologist Rudolph Virchow (Fig. 22.9). Virchow was a true "Renaissance Man" as he was productive in many areas. Virchow is credited with bringing science to medicine, earning him the title "father of modern pathology." He is also considered to be one of the founders of "social medicine." In addition to his work in science and medicine, Virchow developed a substantial reputation in archeology. In 1879, he accompanied Heinrich Schliemann on his excavation of Homer's Troy!

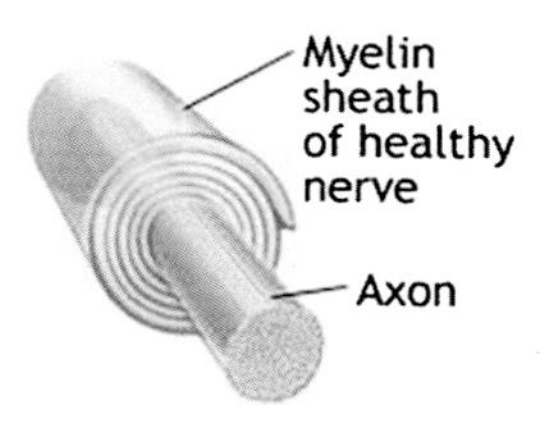

FIGURE 22.8 Diagram of a cross section through an axon showing the jelly roll-like structure of a myelin sheath. http://www.msreversed.com/efaoils.html.

FIGURE 22.9 Rudolph virchow, 1821–1902. http://www.findabout.net/rudolf-virchow.html.

The myelin sheath is considered to be the "simplest membrane" as it has by far the lowest percentage of protein by weight of any membrane (Chapter 1), and protein content directly reflects biochemical activity. The major function of the myelin sheath is simply to provide electrical insulation for the axon, a function that does not require biochemical activity, and hence many proteins. The myelin sheath is about 40% by weight water. The remaining dry mass by weight is 70–85% lipid and only 15–30% protein. Most biological membranes have protein contents of >50 to ~75%. Table 22.1 lists the lipid components of human myelin sheath [28]. Note that the myelin sheath is composed of the same basic lipids found in other membranes. There are no myelin-specific lipids. The primary myelin lipid is the glycolipid, galactocerebroside (Fig. 22.10).

TABLE 22.1 Composition of Human Myelin Sheath

Component	Percentage
OVERALL	
Protein	30
Lipid	70
MYELIN LIPID	
Cholesterol	27.7
Cerebroside	22.7
Sulfatide	3.8
Total galactolipids	27.5
Phosphatidylethanolamine	15.6
Phosphatidylcholine	11.2
Sphingomyelin	7.9
Phosphatidylserine	4.8
Phosphatidylinositol	0.6
Plasmalogens	12.3
Total phospholipids	43.1

Protein and lipid in percent dry weight; all others in percent total lipid.
Adopted from Love S. Demyelinating diseases. J Clin Pathol 2006;59(11):1151–59.

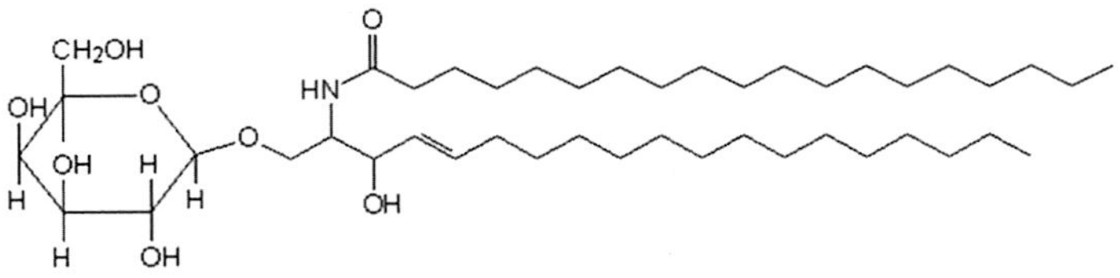

FIGURE 22.10 Structure of galactocerebroside. http://themedicalbiochemistrypage.org/krabbedisease.php.

There are only a handful of proteins in myelin, the major ones being myelin basic protein, myelin oligodendrocyte glycoprotein, and myelin proteolipid protein (lipophilin). The function of these three proteins is to establish and maintain myelin sheath structure. Therefore, they are structural proteins and not dynamic channels or enzymes. The major interest in these proteins concerns their potential roles in demyelination diseases.

4.2 Multiple Sclerosis

The decrease in nerve signal conduction caused by demyelination can produce deficiencies in sensation, movement, cognition, and many other functions depending on what nerves are affected [28]. Extreme cases can result in paralysis and even death. While it is known that the basis of these diseases is defects in the myelin sheath, the precise mechanism of demyelination is not clearly understood. Causative factors include genetics, environmental poisons, infectious diseases, and especially, autoimmune reactions [30]. Immune system T-cells and macrophages are found at the site of myelin sheath lesions, strongly implicating their involvement in disease.

The best known demyelination disease is multiple sclerosis (MS), a disease with no clear cause and no cure. MS occurs after the immune system attacks the myelin sheath causing inflammation and injury to the sheath and eventually the nerve itself [31,32]. Although the first description of MS is attributed to the French physician Jean-Martin Charcot (Fig. 22.11) in 1868, an earlier, and highly accurate water color painting of MS spinal cord lesions was published in 1838 by the Scottish physician Robert Carswell (Fig. 22.12) [33]. Carswell's painting is shown in Fig. 22.13. In addition to his description of MS, Charcot was the first to identify disintegration of bones in the foot and ankle of diabetics. This condition, now known as Charcot Foot, can lead to foot deformities, clubfoot, and eventually amputation. Since diabetics often have almost no feelings in their feet, they do not recognize the broken bones until it is too late. (The author of this book is diabetic and has had Charcot in both feet resulting in more than 10 broken bones.)

FIGURE 22.11 Jean-Martin Charcot, 1825–1893. Encyclopedia Britannica http://www.britannica.com/EBchecked/topic/106349/Jean-Martin-Charcot.

FIGURE 22.12 Robert carswell, 1793-1857. http://en.wikipedia.org/wiki/Robert_Carswell_(pathologist).

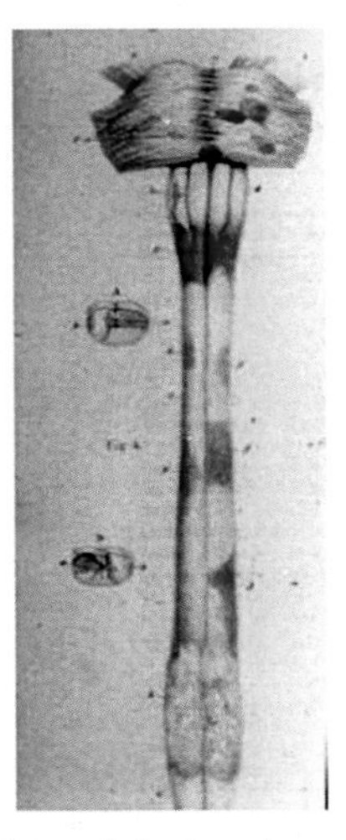

FIGURE 22.13 Lesions on the spinal cord noted during the autopsy of a multiple sclerosis (MS) patient as painted by Robert Carswell. The painting was published years later (in 1838) as Plate 4, Fig. 4 in Carswell's "Pathological Anatomy: Illustrations of the elementary forms of disease". http://www.ms-society.ie/pages/research/historical-overview/robert-carswell.

MS affects >2.5 million people worldwide and presently has no cure. MS symptoms can vary over a wide range from barely noticeable to completely debilitating. Search for the cause of MS remains intense. Consensus is that MS is an autoimmune disease where the body's immune system erroneously starts to attack its own neuronal myelin sheaths [31,32]. What causes this mistake is not known but it has been suggested that it could be triggered by a viral infection such as measles or herpes. MS is characterized by alternating cycles of remission and relapse. Unfortunately, remission is never complete and the sheath continues to get thinner and shorter as damage spreads.

Although no approaches have been really successful at curing MS, advances are continuously being made. At the lowest treatment level are combinations of diet and physical exercise used to manage the disease and reduce the symptoms. There are many examples of MS-modifying drugs that are designed to affect various aspects of the disease [31,32]. Potential MS drug targets include:

Remission and relapse: Beta interferon and glatiramer acetate (Copaxone) act by reducing the frequency of MS relapses through an unknown mechanism. It has been suggested that glatiramer acetate may resemble myelin basic protein and acts as a decoy diverting the autoimmune response away from the myelin sheath.

Immune cells: A common theme for MS drug development is to block the action of immune cells that attack the myelin sheath.

Inflammation: Antiinflammatory drugs are being developed to inhibit inflammation caused by the autoimmune response on myelin.

Stem Cells: There are two different ways that stem cells may be used to treat MS: use them to stop the immune system from attacking the myelin sheath; or use them to repair or regrow damaged myelin.

Although MS affects the most people and is the best known, many other important demyelination diseases exist. Leukodystrophies are a class of hereditary demyelination diseases [34]. Examples include Krabbe disease [35] and Alexander disease that have no treatments and affect infants. Most children afflicted with leukodystrophies die in early infancy. Another rare (affects ~1/100,000 people) but well-known autoimmune demyelination disease is Guillain-Barre syndrome [36]. This is a disease of the peripheral nervous system that is the most common cause of acute nontrauma-related paralysis. Guillain-Barre syndrome has claimed a number of famous victims including actor Andy Griffith.

5. AGING

5.1 Characteristics of Cellular Aging

Even the most devastating diseases only affect a portion of the human population. Aging, however, has a deleterious affect on everyone. In fact, a good case can be made that aging is the most intractable of all human maladies as it is a normal part of the human condition. Aging is a poorly understood, complex, multifaceted process that affects all 6.3×10^{13} cells in the human body. At the cellular level, aging results in a general decrease in metabolic activity and an accumulation of oxidized (nonfunctional) proteins, lipids, and mutated nucleic acids [37,38]. A measurable byproduct of aging is a granular yellow-brown pigment called lipofuscin (Fig. 22.14) [38,39]. As the name implies, lipofuscin has a high lipid content of oxidized unsaturated fatty acids originating from age-related damage to membranes and mitochondria.

5.2 Mitochondrial Theory of Aging

Through the years, many mostly overlapping theories of aging have been proposed [40,41]. The theory to be discussed here is referred to as "The Mitochondrial Theory of Aging"

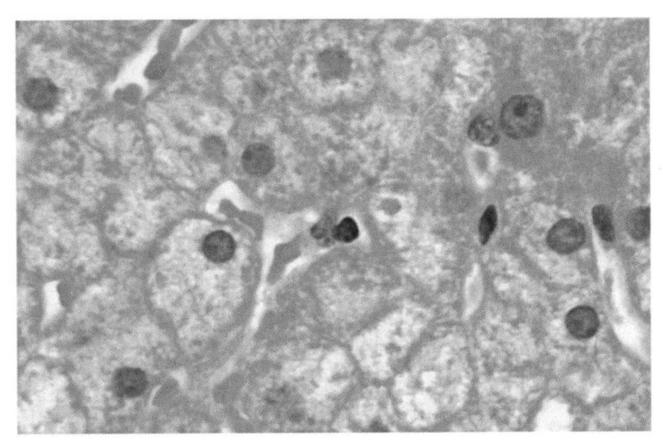

FIGURE 22.14 Lipofuscin particles in an aging cell. http://en.wikipedia.org/wiki/Lipofuscin.

FIGURE 22.15 Denham Harman, 1916–2014. http://www.the-scientist.com/?articles.view/articleNo/18085/title/Trying-To-Unlock-The-Mysteries-Of-Free-Radicals-And-Antioxidants/TheScientist.

proposed by Denham Harman (Fig. 22.15) in 1972 [42,43]. This theory is an extension of a more general theory of aging that was based on free radicals and was also proposed by Harman [44,45]. For his extensive work on aging, Denham Harman is often referred to as the "father of the free radical theory of aging."

From a human perspective, oxygen is a Jekyll and Hyde molecule. On the one hand, oxygen is necessary for the production of most ATP in aerobic organisms. Yet, if not carefully controlled, oxygen can become a deadly poison. Life is an ongoing conflict between oxygen's beneficial and deleterious properties, a conflict that everyone eventually loses! This battle is most noticeable in the mitochondria where the bioenergetic process of oxidative phosphorylation occurs (Chapter 18). With age, the deleterious properties of oxygen slowly gain an upper hand resulting in many age-related afflictions [37,38].

The age-related decrease in ATP-dependent metabolic functions and increase in undesirable oxidation byproducts form the foundation of the "Mitochondrial Theory of Aging" [37,38].

5.2.1 *Mitochondria*

It is well established that mitochondria house the components that drive aerobic bioenergetics (Chapter 18) and that mitochondrial structure and function change with age [43]. Histologically, aged mitochondria display an increase in size and disorganization with shortened cristae and an increase in matrix vacuolization. Associated with these physical alterations are diminished bioenergetics resulting from an increase in proton leakage across the inner mitochondrial membrane, a decrease in mitochondrial proton motive force (Chapter 18), a decrease in state 3/state 4 respiratory rates, a decline in the activities of Complexes I, II and IV (Chapter 18), and a decrease in membrane fluidity which reduces mobility and hence electron movement between the various Complexes. Aging also results in the accumulation of damaged electron transport proteins, mutated mitochondrial DNA (mtDNA), and peroxidized membrane lipids, the result of increased reactive oxygen species (ROS) [38,42,43].

5.2.2 *Oxidation*

Mitochondria are the major source of damaging ROS that become more prevalent with age, and the electron transport system consumes approximately 85% of the entire oxygen utilized by the cell. Therefore mitochondrial electron transport would be a logical place to look for a link between oxidation and aging. There are several ROS found in aging mitochondria, starting with oxygen (O_2) itself and extending to the one electron reduction product of O_2, superoxide ($\cdot O_2^-$) [46]. While oxygen is a di-radical, $\cdot O_2^-$ is an anion radical. Other examples of ROS include hydrogen peroxide (H_2O_2), hydroxyl radical ($HO^\cdot$), and hydroperoxyl radical ($HO_2^\cdot$). These ROS can all damage mitochondrial DNA, proteins, and membrane lipids, diminishing metabolic energy, a characteristic of aging. Damaged electron transport can lead to the production of even more ROS, accelerating the aging process. For example, even in isolated, unaged mitochondria, it has been reported that 0.1–2% of electrons passing through the electron transport chain, prematurely and incompletely reduce oxygen, generating $\cdot O_2^-$.

5.2.3 *Proteins and mitochondrial DNA*

Anything that disrupts the normal flow of electrons, the transmembrane electrochemical gradient (proton motive force), or ATP synthase, will result in diminished energy production. Oxidized or cross-linked electron transport cytochromes accumulate in aged mitochondria, further increasing ROS and decreasing cellular energy level. The affect of oxidation on mitochondrial lipids, proteins, and mtDNA has been nicely reviewed by Ames and coworkers [37,38].

The increased level of ROS also increases the mutation rate of mtDNA, altering stoichiometry of the bioenergetic protein components. mtDNA is particularly susceptible to mutation (~17 times more than nuclear DNA) due to its lack of histones and its location adjacent to the ROS-producing inner-mitochondrial membrane. Mutation-induced imbalances in the normal stoichiometry of electron transport components can result in leakage of electron flow to cytochrome oxidase, producing O_2^- and H_2O_2. It has been suggested that mtDNA mutations may even be sufficient to account for age-related deficits in mitochondrial bioenergetics.

5.2.4 Lipids

The potential for lipid peroxidation (the "peroxidizability index") for inner mitochondrial membrane lipids increases with age. This is of particular concern for the signature mitochondrial lipid, cardiolipin (CL) (Chapter 5). CL is a highly unusual phospholipid being essentially two phospholipids fused together with one large head group and four acyl chains (Fig. 5.18). CL is known to support the function of many mitochondrial proteins, including those involved in electron transport and oxidative phosphorylation, much better than other phospholipids. Decreased levels of CL, observed with aging, negatively affect the mitochondrial bioenergetic proteins [46]. In addition to providing the proper environment for optimal protein performance, CL is a major lipid component of the mitochondrial inner membrane where it helps establish proper bilayer fluidity that controls diffusion rates, and hence interaction between the electron transport complexes (Chapter 18). Age-dependent oxidative damage to mitochondrial membrane lipids decreases membrane fluidity. CL is enriched in α-linoleic acid ($18{:}2^{\Delta 9,12}$) (Chapter 4), maximizing its interaction with membrane proteins but concomitantly making it susceptible to peroxidation. CL has also been shown to help control transmembrane proton leakage that affects the proton motive force. With age, there is: a decrease in CL content, electron transport, and inner mitochondrial membrane surface area; production of smaller, sparser cristae; and an increase in mitochondrial fragility and CL peroxidation – all adversely affecting cell bioenergetics [47].

6. SUMMARY

While most diseases have at least some interaction with membranes, Chapter 22 discusses five of the best known and most debilitating diseases that are closely associated with membranes: Cystic Fibrosis (CF), Duchenne Muscular Dysrtophy (DMD), Alzheimer's Disease (AD), demyelination diseases, and aging.

Cystic Fibrosis (CF) is an autosomal recessive genetic disorder that is primarily known for its affect on lungs, resulting in fluid buildup, difficulty breathing, fatigue, frequent lung infections, and often death. The protein responsible for CF, CFTR, is an anion channel responsible for the flow of Cl^- in and out of cells. The open channel is the normal state, and channel blockage leads to CF.

Duchenne Muscular Dystrophy (DMD) is a fatal genetic muscle disorder whose mutated protein (dystrophin) is linked to the muscle cell plasma membrane of boys. DMD produces severe muscle mass loss and weakness resulting in the victim being wheelchair-bound by age 12 and dead by age 25. Dystrophin is a very large, rod-shaped protein that is a component of the Dystrophin-Glycoprotein Complex (DGC), linking the extracellular matrix to intracellular actin across the muscle plasma membrane. A mutation in dystrophin results in a decrease in stability of the DGC, decreased muscle fiber strength, increased muscle stiffness, and increased sarcolemmal deformability.

Alzheimer's Disease (AD) is the most common form of dementia. AD worsens as it progresses, is incurable, and invariably results in death. For over 100 years, AD has been characterized by the appearance, in the brain of abnormal clumps (known as amyloid plaques) and tangled bundles of fibers (known as neurofibrillary tangles). Despite intense investigation, the cause, progression, and cure of AD remain a mystery. A wide variety of often overlapping hypotheses have been suggested, but the final solution may well involve a complex mix of genetic, environmental, and lifestyle factors. Suggested hypotheses for AD include: genetic, viral, possible link to amyloid plaque development, reduction in acetylcholine synthesis, age-related myelin breakdown in the brain, hyperphosphorylation of the microtubule-associated protein *tau*, and formation of misfolded A_β and *tau* proteins. Finally, there have been a few hypothesis caused by either a generalized defect in plasma membrane structure–function, alteration in cell signaling initiated in lipid rafts, or amyloid protein (A_β) induction of neuron cytotoxicity through an A_β channel that allows external Ca^{2+} to flood into the cell at toxic levels.

Demyelination diseases are a family of related maladies of the nervous system in which the myelin sheath of neurons is damaged by the victim's own immune system. Demyelination diseases impair nerve impulse conduction, producing deficiencies in sensation, movement, cognition, terminating in paralysis, and even death. While the mechanism of demyelination is not clearly understood, instigating factors may include genetics, environmental poisons, infectious diseases, and especially autoimmune reactions. The best known demyelination disease is Multiple Sclerosis (MS), a disease with no clear cause and no cure. MS occurs after the immune system attacks its own myelin sheaths causing inflammation and injury to the sheath and eventually the nerve itself. MS is characterized by cycles of remission and relapse, however, remission is never complete and the myelin sheath continues to get thinner and shorter as damage spreads. Other demyelination diseases include: Krabbe disease, Alexander disease, and Guillain-Barre syndrome.

Aging is the most intractable of all human maladies as it is a normal part of the human condition. Aging is a poorly understood, complex, multifaceted process that affects all cells in the body. At the cellular level, aging results in a general decrease in metabolic activity and an accumulation of oxidized (nonfunctional) proteins, lipids, and mutated nucleic acids. Many theories of aging have been proposed. Of primary importance here is "The Mitochondrial Theory of Aging" that is based on several oxygen-generated free radicals that are grouped together as ROS. ROS-damage accumulates with age, particularly in the mitochondria, where the organelle's structure and function (eg, ATP production) is severely impacted. With age, there is accumulation of damaged electron transport proteins, mutated mtDNA, and peroxidized membrane lipids (particularly CL).

Chapter 23 will discuss two approaches to improving human health by utilizing aspects of membrane composition and structure. Discussed are how liposomes can be used for targeted drug delivery. Also, the "bad" dietary fatty acids (trans-fatty acids) will be contrasted with a "good" fatty acid (the omega-3 fatty acid docosahexaenoic acid), as they impact human health.

References

[1] Bush A, Alton EWFW, Davies JC, Griesenbach U, Jaffe A, editors. Cystic fibrosis in the 21st century. Progress in respiratory research, vol. 34. Basel, Switzerland: Karger: Medical and Scientific Publishers; 2006.

[2] Andersen DH. Cystic fibrosis of the pancreas and its relation to celiac disease: a clinical and pathological study. Am J Dis Child 1938;56:344–99.

[3] Rommens JM, Iannuzzi MC, Kerem B, et al. Identification of the cystic fibrosis gene: chromosome walking and jumping. Science 1989;245(4922):1059–65.

[4] Gadsby DC, Vergani P, Casanady L. The ABC protein turned chloride channel whose failure causes cystic fibrosis. Nature 2006;440(7083):477–83.

[5] Duchenne muscular dystrophy: MedlinePlus Medical Encyclopedia. Nlm.nih.gov.
[6] Hoffman E, Brown R, Kunkel L. Dystrophin: the protein product of the Duchenne muscular dystrophy locus. Cell 1987;51(6):919–28.
[7] Tennyson CN, Klamut HJ, Worton RG. The human dystrophin gene requires 16 hours to be transcribed and is cotranscriptionally spliced. Nat Genet 1995;9(2):184–90.
[8] Khurana TS, Davies KE. Pharmacological strategies for muscular dystrophy. Nat Rev Drug Discov 2003;2:379–90.
[9] NIH National Institute on Aging. Alzheimer's Disease Education and Referral Center. Alzheimer's Disease Fact Sheet; 2016.
[10] Katzman R, Bick KL. Alzheimer disease the changing view (Hardcover). Academic Press 2007; 2007. 387 pp.
[11] Tiraboschi P, Hansen LA, Thal LJ, Corey-Bloom J. The importance of neuritic plaques and tangles to the development and evolution of AD. Neurology 2004;62(11):1984–9.
[12] Bertram L, Tanzi RE. The genetics of Alzheimer's disease. Prog Mol Biol Transl Sci 2012;107:79–100.
[13] Itzhaki RF, Wozniak MA. Herpes simplex virus type 1 in Alzheimer's disease: the enemy within. J Alzheimer's Dis 2008;13(4):393–405.
[14] Francis PT, Palmer AM, Wilcock GK. The cholinergic hypothesis of Alzheimer's disease: a review of progress. J Neurol Neurosurg Psychiatr 1999;66(2):,137–147.
[15] Hardy J, Allsop D. Amyloid deposition as the central event in the aetiology of Alzheimer's disease. Trends Pharmacol Sci 1991;12(10):383–8.
[16] Holmes C. Long-term effects of Abeta42 immunisation in Alzheimer's disease: follow-up of a randomised, placebo-controlled Phase I trial. Lancet 2008;372(9634):216–23.
[17] Nikolaev A, McLaughlin T, O'Leary D, Tessier-Lavigne M. APP binds DR6 to cause axon pruning and neuron death via distinct caspases. Nature 2009;457(7232):981–9.
[18] Bertzokis G. Alzheimer's disease as homeostatic response to age-related myelin breakdown. Neurobiol Aging 2011;32(8):1341–71.
[19] Heneka MT, Nadrigny F, Regen T, Martinez-Hernandez A, Dumitrescu-Ozimek L, Terwel D, et al. Locus ceruleus controls Alzheimer's Disease pathology by modulating microglial functions through norephine. Proc Natl Acad Sci U S A 2010;107:6058–63.
[20] Chun W, Johnson GV. The role of *tau* phosphorylation and cleavage in neuronal cell death. Front Biosci 2007;12:733–56.
[21] Koren III J, Jinwal UK, Lee DC, Jones JR, Shults CL, Johnson AG, et al. Chaperone signalling complexes in Alzheimer's disease. J Cell Mol Med 2009;13(4):619–30.
[22] Arispe N, Diaz JC, Flora M. Efficiency of Histidine-associating compounds for blocking the Alzheimer's Aβ channel activity and cytotoxicity. Biophys J 2008;95:4879–89.
[23] Diaz JC, Simakova O, Jacobson KA, Arispe N, Pollard HB. Small molecule blockers of the Alzheimer Aβcalcium channel potently protect neurons from Aβ cytotoxicity. Proc Natl Acad U S A 2009;106:3348–53.
[24] Lukiw W. Alzheimer's disease (AD) as a disorder of the plasma membrane. Front Physiol General Comment 2013;16. Feb 2013.
[25] Zubenko GS, Cohen BM, Growdon J, Corkin S. Submembrane abnormality in Alzheimer's disease. Lancet July 1984;28:235.
[26] Sciacca MFM, Kotler SA, Brender JR, Chen J, Le D-K, Ramamoorthy A. Two-step mechanism of membrane disruption by Aβ through membrane fragmentation and pore formation. Biophys J 2012;103:702–10.
[27] Bieschke J, Zhang Q, Powers ET, Lerner RA, Kelly JW. Oxidative metabolites accelerate Alzheimer's amyloidogenesis by a two-step mechanism, eliminating the requirement for nucleation. Biochemistry 2005;44(13):4977–83.
[28] Love S. Demyelinating diseases. J Clin Pathol 2006;59(11):1151–9.
[29] Siegel GJ, Agranoff BW, Alberts RW, et al. Characteristic composition of myelin. In: Basic Neurochemistry: Molecular, cellular and medical aspects. 6th ed. Philadelphia: Lippincott-Raven; 1999.
[30] Zamvil SS, Steinman L. Autoimmune demyelinating disease. West J Med 1989;150(3):335–6.
[31] Compston A, Coles A. Multiple sclerosis. Lancet 2008;372(9648):1502–17.
[32] Compston A, Coles A. Multiple sclerosis. Lancet 2002;359(9313):1221–31.
[33] Carswell R. Pathological Anatomy: Illustrations of the elementary forms of disease. 1838. Plate 4, Fig. 4.
[34] NIH National Institute of Neurological Disorders and Stroke. NINDS Leukodystrophy Information Page. http://www.ninds.nih.gov/disorders/leukodystrophy/leukodystrophy.htm.

[35] Lee WC, Tsoi YK, Troendle FJ, et al. Single-dose intracerebroventricular administration of galactocerebrosidase improves survival in a mouse model of globoid cell leukodystrophy. FASEB J 2007;21(10):2520–7.
[36] MNT. What Is Guillain-Barre Syndrome? What Causes Guillain-Barre Syndrome? http://www.medicalnewstoday.com/articles/167892.php
[37] Ames BN, Shigenaga MK, Hagen TM. Oxidants, antioxidants, and the degenerative diseases of aging. Proc Natl Acad Sci U S A 1993;90:7915–22.
[38] Shigenaga MK, Hagen TM, Ames BN. Oxidative damage and mitochondrial decay in aging. Proc Nail Acad Sci U S A 1994;91:10771–8.
[39] Gaugler C. Lipofuscin. Stanislaus J Biochem Rev May 1997.
[40] Weiner BT, Timiras PS. Invited review: theories of aging. J Appl Physiol 2003;95(4):1706–16.
[41] Kunlin J. Modern biological theories of aging. Aging Dis 2010;1(2):72–4.
[42] Harman D. A biologic clock: the mitochondria? J Am Geriatr Soc 1972;20(4):145–7.
[43] Lee HC, Wei YH. Mitochondria and aging. Adv Exp Med Biol 2012;942:311–27.
[44] Harman D. Aging: a theory based on free radical and radiation chemistry. J Gerontol 1956;11(3):298–300.
[45] Harman D. Role of free radicals in mutation, cancer, aging and maintenance of life. Radic Res 1962;16:752–63.
[46] Devasagayam TPA, Tilak JC, Boloor KK, Sane KS, Ghaskadbi SS, Lele RD. Free radicals and antioxidants in human health: current status and future prospects. J Assoc Physicians India (JAPI) 2004;52:796.
[47] Paradies G, Petrosillo G, Paradies V, Ruggiero FM. Oxidative stress, mitochondrial bioenergetics, and cardiolipin in aging. Free Rad Biol Med 2010;48(10):1286–95.

CHAPTER

23

Membranes and Human Health

OUTLINE

1. LIPOSOMES AS DRUG DELIVERY AGENTS

The complexity of membrane structure and functions outlined in this book supports a wealth of potential targets to alleviate human afflictions. Indeed, drug companies have taken notice and made membranes a major area for future drug development. A thorough survey of these approaches is far too large to address here. Instead, a few representative examples are discussed.

1.1 Gout

For decades it has been appreciated that most life processes are in some way related to membranes. Many examples of clever experiments designed to test possible disease mechanisms of action adorn the literature. An early example by Gerald Weissmann (Fig. 23.1) in 1972 [1] typifies this approach. (Gerald Weissmann is also known for coining the term "liposome"). His

An Introduction to Biological Membranes
http://dx.doi.org/10.1016/B978-0-444-63772-7.00023-3

FIGURE 23.1 Gerald Weissmann, 1930–.

experiment involved understanding the molecular mechanism of gout. For centuries many famous historical figures suffered from gout. Included in this list were Henry VIII, Isaac Newton, Thomas Jefferson, Benjamin Franklin, Charles Darwin, and even "Sue", the famous Tyrannosaurus Rex skeleton. So what is responsible for gout? It is known that gout is related to a rich diet, perhaps one that is based on red meat dripping with cholesterol. Men "take the gout", but young women do not. However, postmenopausal women do get gout. Also, it is a historical fact that eunuchs (castrated males) do not "take the gout". Weissmann therefore proposed that gout was related to cholesterol and testosterone, but not estrogen. Gout is also associated with accumulation of uric acid crystals. Weissmann proposed that uric acid crystals attach to the lysosomal membrane containing cholesterol and testosterone, making the organelle very leaky to the sequestered hydrolytic enzymes. Once released from the lysosome, the enzymes destroy the cell, inducing gout. To test this hypothesis, Weissmann made "boy" (containing testosterone) and "girl" (containing estrogen) cholesterol-enriched liposomes. Upon the addition of uric acid the "boy" liposomes, but not the "girl" liposomes, became leaky to a sequestered solute. This clever experiment accounts for the basic characteristics of gout. Years later, Weissmann developed two liposome-encapsulated drugs (Abelcet and Myocet, see Table 23.1).

1.2 Liposomes

Soon after their discovery by Alec Bangham in 1961, it became evident that the basic properties of liposomes (Chapter 13) make them ideally suited for a plethora of medical applications [2,3]. Liposomes are tiny, sealed lipid vesicles that have: a sequestered aqueous compartment for water-soluble (hydrophilic) drugs; a surrounding, largely impermeable lipid bilayer membrane that can simultaneously house lipid-soluble (hydrophobic) drugs; and an external facing surface that can be modified with specific ligands (Fig. 23.2, [4]). Valuable liposome properties for drug delivery are listed in Table 23.1.

From this impressive list, it would appear that liposomes can be constructed for just about any drug delivery purpose. But are liposomes really a panacea? Unfortunately, nothing in life is as simple as one would hope. Upon testing liposomes as drug delivery agents, it was soon

TABLE 23.1 A List of Liposome Properties That Makes Them Valuable Assets for Drug Delivery Systems

1.	They can be made from thousands of natural and artificial lipids.
2.	They exist in lipid bilayers and so resemble biological membranes.
3.	They are generally not antigenic.
4.	They can be made in a wide range of sizes.
5.	They can be made with a range of different sequestered volumes.
6.	They can be unilamellar or multilamellar.
7.	They can be positive, negative, or neutral.
8.	They can be made with a wide range of permeability to sequestered solutes.
9.	They can simultaneously accommodate water-soluble and lipid-soluble drugs.
10.	They are nontoxic and biodegradable.
11.	Various ligands can be attached to their surfaces.
12.	Lectins, antigens, and receptors can be attached to their surfaces.
13.	There are many ways to make liposomes (Chapter 13).

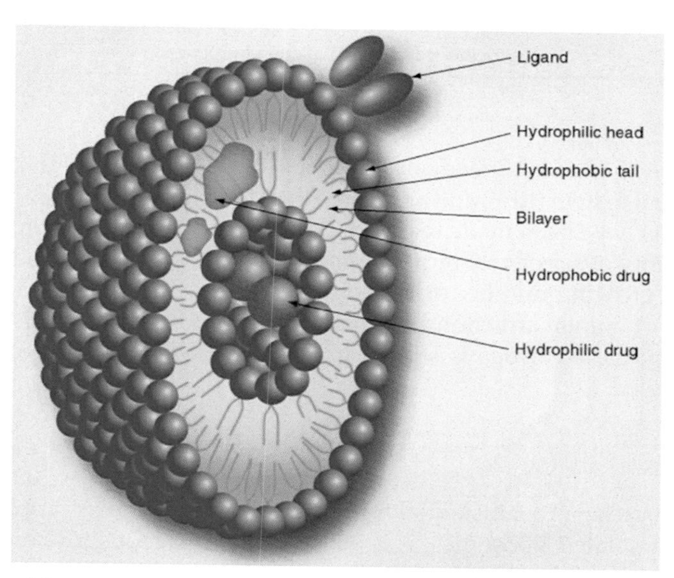

FIGURE 23.2 Drug delivery using liposomes. Liposomes are tiny, sealed lipid vesicles that have: a sequestered aqueous compartment for water-soluble (hydrophilic) drugs; a surrounding, largely impermeable lipid bilayer membrane that can simultaneously house lipid-soluble (hydrophobic) drugs; and an external facing surface that can be modified with specific ligands[4].

TABLE 23.2 Examples of Liposome-Sequestered Drugs Currently on the Market

Name	Drug carried	Disease targeted
Myocet	Doxorubicin	Breast cancer
Doxil, Caelyx	Doxorubicin	Several cancers
Lipodox	Doxorubicin	Several cancers
ThermoDox	Doxorubicin	Liver cancer
DaunoXome	Daunorubicin	Kaposi's sarcoma
Abelcet	Amphoteracin B	Fungal infections
AmBisome	Amphoteracin B	Fungal infections
Estrasorb	Estradiol	Menopausal therapy
Marqibo	Vincristine	Metastatic malignant melanoma
Visudyne	Verteporfin	Eye diseases
DepoCyt	Cytarabine	Meningitis
DepoDur	Morphine sulfate	Postoperative pain
Arikace	Amikacin	Lung infections
Lipoplatin	Cisplatin	Epithelial malignancies
LEP-ETU	Paclitaxel	Ovarian, breast and, lung cancer
Epaxal	Hepatitis A vaccine	Hepatitis A
Inflexal V	Influenza vaccine	Influenza

discovered that when injected intravenously, most liposomes are very rapidly taken up by cells of the reticuloendothelial system (RES) and are sent to the lysosome for destruction [5]. Through the years, many attempts have been made to avoid the RES and thus enhance liposome circulation time and the possibility of selectively targeting many parts of the human body.

There are now a growing number of liposome-sequestered drugs that are medically used to combat a variety of human afflictions and many more are in clinical trials. A representative list of liposome-sequestered drugs is shown in Table 23.2. A few selected applications are subsequently discussed.

1.3 Leishmaniasis

Whereas the RES presents a substantial hurdle for most intravenous applications of targeting with drug-encapsulated liposomes, it is actually advantageous for certain diseases. The best known of these is leishmaniasis, a serious, mostly tropical disease [6]. While rare in the United States, leishmaniasis has been reported in immigrants, returning tourists, and military personnel from the Persian Gulf. Leishmaniasis (also known as kala-azar, black fever, or Dumdum fever) is the second parasitic killer in the world (after malaria) and affects about

FIGURE 23.3 Carl Alving, 1939–.

12 million people worldwide, with 1.5 to 2 million new cases reported each year. Sadly, leishmaniasis is often fatal. Historically, the most effective anti-leishmaniasis drugs are antimonial compounds that unfortunately have about the same inherent toxicity as arsenates. The cure can be as bad as the disease!

In the late 1970s, Carl Alving (Fig. 23.3) proposed an unusual solution that was based on the life cycle of the leishmaniasis protozoan parasite [7]. The disease is spread by the bite of a female sandfly whereupon the protozoan is rapidly taken up by the victim's macrophages (part of the RES) where they multiply. Alving noticed that when liposomes were injected intravenously, they were rapidly taken up by the same RES macrophages. He reasoned that sequestering the antimonial drugs into liposomes would target the drugs directly to the parasite living in the macrophages, thus greatly decreasing the drug's systemic toxicity. Alving encapsulated the antimonial drugs, meglumine antimoniate and sodium stibogluconate into liposomes composed of phospholipids with di-saturated chains (dipalmitoyl phosphatidylcholine [DPPC]), cholesterol, and an anionic lipid. When encapsulated into liposomes, both drugs were more than 700 times more effective than either of the free (unencapsulated) drugs when tested in hamsters [7].

1.4 Development of Liposomes That Avoid the RES

The purpose of developing drug-sequestered liposomes is to increase the therapeutic index of the drug while minimizing its side effects. For most potential applications of drug-sequestered liposomes, being largely removed by the RES in a first pass through the circulatory system is a fatal flaw. Therefore countless attempts have been made to avoid the RES and thus enhance liposome circulation time [8,9]. The simplest method is to first inject empty liposomes (free of drug) to satiate the RES before the drug-encapsulated liposomes are injected. Most attempts to make long-circulating liposomes have involved changing liposomal properties by altering their lipid composition, size, and charge. As a general rule, small liposomes have longer circulation lifetimes than do larger liposomes [10]. Unfortunately small unilamellar vesicles (SUVs) have much smaller sequestered volumes than large unilamellar vesicles (LUVs) (Chapter 13), and so have diminished drug-carrying capacities. Even worse are the multilamellar vesicles (MLVs) that are both large and have a very limited sequestered aqueous space. Most of an MLV's sequestered space is occupied by lipid bilayers. In 1982, Senior and Gregoriadis [10] reported that saturated chain phosphatidylcholines (PCs) and sphingomyelin (SM) (both lipids have high main transition temperatures (T_ms), see Chapter 5) have longer circulating half-lives than unsaturated PCs. Cholesterol, a phospholipid-"condensing" sterol (Chapter 11), also increased liposome circulation time. On the other hand, positively charged liposomes are useless as drug carrying agents since they are toxic. A major advance in producing long-circulating liposomes came from studies on the surface of erythrocytes, cells that manage to avoid the RES for their entire life of ~127 days. These liposomes had their surfaces modified with gangliosides (particularly ganglioside GM1, discussed in Chapters 5 and 7) and sialic acid derivatives [11]. Therefore, the modified liposomes had surface properties similar to erythrocytes. GM-liposomes have additional advantages in being useful to deliver drugs by oral administration and being able to cross the blood–brain barrier. The explanation of GM1 in RES avoidance relates to its flexible sugar chain that occupies the space immediately adjacent to the liposome surface [9]. This space (periliposomal layer) makes it difficult for a macrophage to bind to a liposome.

1.5 Stealth Liposomes

The importance of the GM1 periliposomal layer led to the idea of replacing the ganglioside with a nontoxic synthetic polymer, such as polyethylene glycol (PEG). PEG-liposomes are usually referred to as "stealth liposomes" [12] due to their ability to avoid the RES and are heavily employed in many types of drug delivery systems. PEG is a linear polyether diol (Fig. 23.4(A)). An important feature of PEG is the ability to be readily synthesized from short to very long polymers, and polymer length affects circulation longevity. PEG is normally

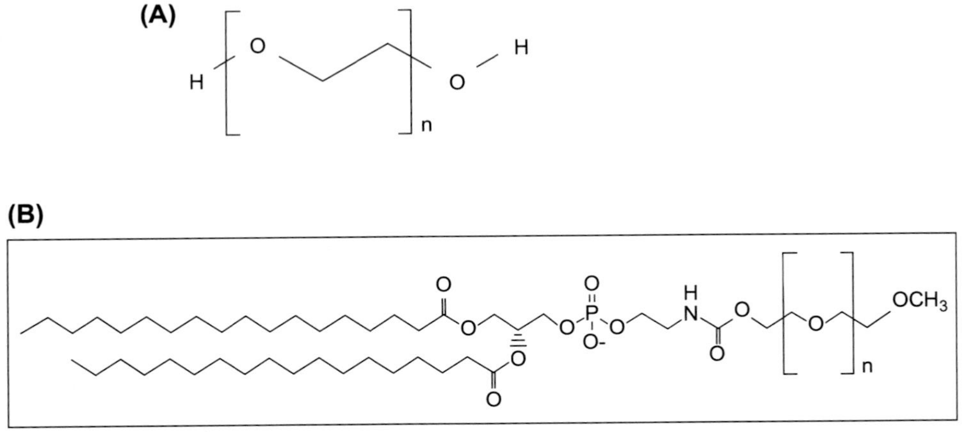

(A)

(B)

FIGURE 23.4 Structure of: (A) Polyethylene glycol and (B). Distearoylphosphatidylethanolamine–polyethylene glycol (DSPE–PEG) (18:0,18:0 N-succinyl-1,2-dioleoylphosphatidylethanolamine–polyethylene glycol (PE–PEG)).

anchored to the liposome by attachment to the primary amine head group of N-succinyl-1,2-dioleoylphosphatidylethanolamine (PE) (Fig. 23.4 (B)).

In one example, Allen et al. [13] made liposomes from SM/PC/cholesterol/DSPE–PEG where the PEG length varied from short to long. They reported increased circulation longevity for liposomes made from long length PEGs (PEG-1900 and PEG-5000) compared to short PEGs (PEG-750 and PEG-120). While PEG increases circulation longevity, by itself it cannot specifically target a particular cell. In fact, PEG actually hinders binding of the liposome to the delivery site. Most PEG-liposomes are further modified by attachment of "biomolecules" including monoclonal antibodies or fragments (immunoliposome), vitamins, lectins, specific antigens, peptides, growth factors, glycoproteins, carbohydrates, or other appropriate ligands [14] The ideal situation would be to create a drug-carrying liposome that would be invisible to the RES but would specifically bind to the diseased tissue or cell. The "biomolecule" is attached to the free end of the PEG in DSPE–PEG by a maleimide group. Transferrin is often the "biomolecule" of choice for specific delivery of anticancer drugs. Modified liposomes are also commonly used in diagnostic imaging and in vaccines (virosomes). Other synthetic polymers, including poly(vinyl pyrrolidone) and poly(acryl amide), have been tested as possible replacements for PEG, with some limited success.

1.6 Thermo-liposomes

It is now well accepted that by using systemic chemotherapy for solid tumors, it is almost impossible to achieve therapeutic drug levels without damaging healthy organs and tissues [15]. Hence, there is a very keen interest in liposomes as vehicles for drug delivery. It is believed that modified liposomes can meet the four basic requirements of a successful drug delivery system: "Retain (the sequestered drug), Evade (the RES), Target (the diseased cell), and Release (the drug at the appropriate location)" [16]. We have already discussed the role of liposomes in "Retain", "Evade", and "Target". Finally we will consider the issue of "Release", the process whereby the drug is rapidly deposited at the target site.

In the late 1970s, Yatvin et al. [17] pioneered the use of mild hyperthermia to release liposome-sequestered drugs at the site of solid tumors. They based their approach on the large increase in leakiness that liposomes exhibit when heated through their main lipid phase T_m [18]. Permeability of DPPC liposomes was discussed in Chapter 19. Liposomes in the gel state are poorly permeable while those in the liquid crystalline state exhibit considerable permeability. However, maximum permeability is achieved at the T_m, where equal amounts of gel and liquid crystalline state domains coexist. Domain interfaces are locations of extremely high permeability. By this methodology, temperature-sensitive (thermo-) liposomes are in the impermeable gel state at physiological temperature (37°C), but undergo a sharp phase transition a few degrees above this. At T_m they rapidly become leaky, dropping most of their sequestered drug load. The liposome T_m cannot be too high as mammalian cells start to show damage at ~42°C [19]. So how does one make liposomes with T_ms between ~40°C and 44°C? Many thermo-liposomes have as their major bilayer component, DPPC, since the T_m of this phospholipid is 41.3°C (Chapter 5).

In their initial report, Yatvin et al. [17] used, as a model system, neomycin-sequestered liposomes to affect *Escherichia coli* protein synthesis and cell survival in culture. Their initial experiments employed sonicated SUVs composed of DPPC–distearoylphosphatidylcholine (DSPC) (3:1). This formulation may seem strange since DSPC has a higher T_m (56°C) than does DPPC (41.3°C) and the liposomes would have a T_m of greater than 44°C. However, it is known that tightly curved SUVs exhibit permeability maxima ~4°C below their actual T_m. Therefore, the effective (permeability maxima) temperature was a very acceptable ~42°C. These investigators reported that *E. coli* protein synthesis was inhibited and cell killing enhanced by heating the neomycin-containing liposomes through their T_m. Maximum drug release and cell killing with the thermo-liposomes was between 42–46°C. This experiment established the potential of thermo-liposomes as useful drug carriers. A large number of drug-baring thermo-liposomes were soon developed and tested in animal models [20]. The first generation of thermo-liposomes concentrated on avoiding the RES by lipid composition and size manipulations. In a typical example, Lindner et al. [21] made 175 nm thermo-liposomes from DPPC, DSPC, and a novel lipid 1.2-dipalmitoyl-*sn*-glycero-3-phosphoglyceroglycerol (DPPGOG). These thermo-liposomes had a long circulation time and trapped drugs were released under mildly hyperthermic conditions (41–42°C).

Thermo-liposomes were taken a step further by making them "stealth" with attached PEG. For example, Needham et al. [15] developed a series of stealth thermo-liposome drug carriers to target solid tumors. The liposomes were made from various mixtures of DPPC, 1-palmitoyl-2-hydroxy-*sn*-glycero-3-phosphocholine (DPPGOG) (a lyso-PC), hydrogenated soy *sn*-glycero-3-phosphocholine (HSPC), cholesterol, and DSPE–PEG-2000. The sequestered drug was doxorubicin and the tumor a human squamous cell carcinoma xenograft line (FaDu). DPPC and cholesterol provided the gel state, monopalmitoyl phosphatidylcholine was responsible for the narrow phase transition (39°C to 40°C) and DSPE–PEG-2000 made the thermo-liposomes invisible to the RES. The liposomes proved to be very susceptible to mild hyperthermia where they rapidly (tens of seconds) released their drug load. Finally, the stealth thermo-liposomes are being further modified by adding components (biomolecules including monoclonal antibodies or fragments, vitamins, lectins, specific antigens, peptides, growth factors, glycoproteins, carbohydrates, or other appropriate ligands [14]) that will specifically target and bind the liposomes to the diseased cells.

A final aspect of delivering drugs to target cells or tissues via thermo-liposomes involves methods of generating localized, but limited, heat. This is referred to as "mild hyperthermia." The target tumor is locally heated by focusing electromagnetic or ultrasound energy on the tumor. In some cases radiofrequency electrodes are directly implanted into the tumor. More frequently, localized heating is achieved noninvasively by microwave antennas or ultrasound transducers (focused ultrasound) [22]. Upon heating, the circulating liposomes rapidly release their sequestered anticancer drug directly inside the tumor.

1.7 pH-Sensitive Liposomes

In the previous sections (Chapters 1–7) we have discussed some of the many challenges associated with the use of liposomes as drug delivery agents. Another problem involves the ultimate fate of the liposomes after they have been internalized into a cell through endocytosis (Chapter 17). Once internalized into an endosome, the liposomes and associated drugs are targeted to the lysosome where they are destroyed. In most cases it would be tremendously advantageous if the drug (or plasmid DNA or RNA) could be released from the liposome, and escape from the endosome into the cell's cytoplasm before encountering the lysosome. Early endosome studies showed that their interior pH is mildly acidic (pH is ~5), offering a possible target for liposome drug release [23]. This observation led to the development of a wide variety of "pH-sensitive liposomes" [24,25]. pH-sensitive liposomes are stable (nonleaky) at physiological pH (pH ~7.4), but become unstable (leaky) at low pH (pH ~5.0). The pH-sensitive liposomes must also avoid the RES, be stable in the presence of blood plasma, and have the ability to fuse with the cell's plasma membrane and endosome membrane in order to release their load into the cytoplasm.

Many pH-sensitive liposome compositions have been tested, but most have been mixtures of a lipid containing a pH titratable group and, as the bulk lipid, an unsaturated chain PE [26]. The titratable group is responsible for pH sensitivity and is often an acylated amino acid, a phospholipid derivative (usually a PE), a free fatty acid, a cholesterol derivative, or a double chain amphiphile. The earliest pH-sensitive liposomes [27] were composed of PE as the major lipid component and single-chain amphiphiles such as fatty acids or N-acyl amino acids as the pH-sensitive group. Unfortunately, these early liposomes were destroyed in the plasma before reaching the target cell. Although liposome stability was shown to be increased by incorporation of cholesterol, the sterol unfortunately decreased liposome fusion to the endosome membrane, creating yet another problem. Therefore an initial objective of a functional pH-sensitive liposome was to produce a cholesterol-free liposome that remained stable in the plasma yet still retained fusogenicity and pH sensitivity.

One popular paradigm for the design of pH-sensitive liposomes is based on lipid-anchored compounds that exist in two different conformations, one existing in mildly acidic conditions (the compound is protonated) and the other in neutral or slightly basic conditions (the compound is dissociated). Most of these compounds have been lipid-linked to homocysteine or succinate. The original 1980 report of a pH-sensitive liposome by Yatvin et al. [27] employed homocysteine linked to the membrane through a palmitic acid. Homocysteine exists in a cyclic structure that disrupts the liposomal membrane under acidic conditions but does not affect membrane permeability near neutral pH (Fig. 23.5(A)). These pH-sensitive liposomes were only useful between pH 7.4 and 6.0. It was predicted that they could be useful in targeting drugs to acidic tissues such as tumors and sites of inflammation or infection.

For the last 20 years or more, the most successful pH-sensitive liposomes have incorporated succinate. The first use of this type of pH-sensitive liposome was reported by Leventis et al. [28]. This group used a series of double chain amphiphiles, including 1,2-dioleoyl-3-succinylglycerol (DOSG) that, when combined with 1-palmitoyl-2-phosphatidylcholine (POPE) (16:0, 18:1 PE) at pH 7.4, formed nonleaky liposomes. Importantly, these liposomes became fusogenic and leaky under mildly acidic conditions, but were stable in serum. A few years later, Collins et al. [29] made a series of pH-sensitive liposomes from three similar diacylsuccinylglycerols (including DOSG) and PE. These workers advanced the earlier report [28] by attaching an anti-H2k^k antibody (isolated from the murine hybridoma cell line 11–4.1) to the N-hydroxysuccinimide ester of palmitic acid, thus converting the pH-sensitive liposome into a pH-sensitive "immunoliposome". Later additions attached a phosphate to the glycerol *sn*-3 position (Fig. 23.5 (B)). Finally, the phosphate was replaced by a PE and PEG was attached to the PE. Therefore the final product was a pH-sensitive, stealth, immunoliposome with tremendous cytotoxic potential. This illustrates the versatility of liposomes for drug targeting.

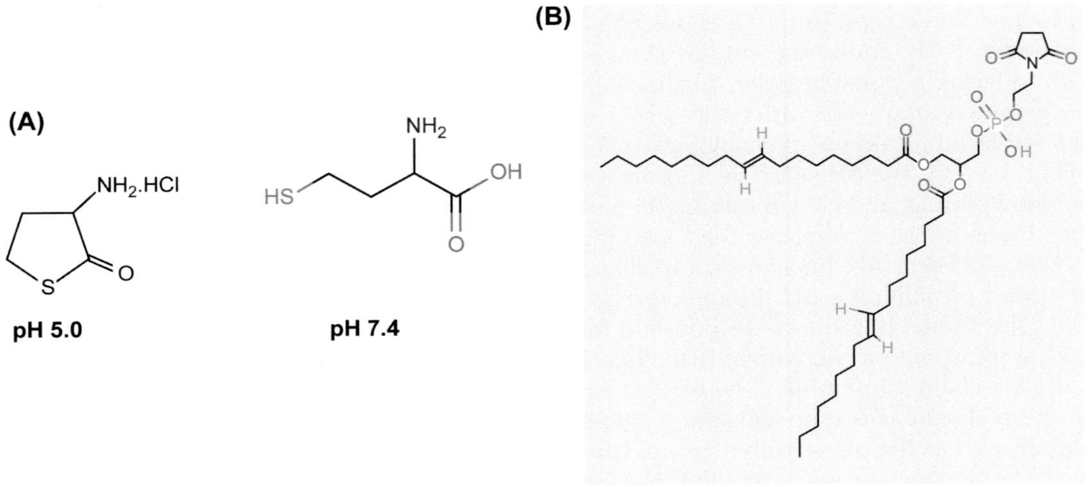

(A)

$NH_2.HCl$

pH 5.0

pH 7.4

(B)

FIGURE 23.5 (A) Structure of homocysteine at pH 5.0 and pH 7.4. The low pH cyclic conformation disrupts the liposome membrane, increasing leakiness. The neutral pH conformation does not affect liposome membrane permeability. (B) N-succinyl 1,2-dioleoyl-3-phosphoglycerol.

2. AFFECT OF DIETARY LIPIDS ON MEMBRANE STRUCTURE FUNCTION

The old adage "you are what you eat" certainly applies to the affect of dietary lipids on membrane structure and function. At the vanguard of dietary fatty acids are both, "bad guys" (*trans*-fatty acids, TFAs) and "good guys" (docosahexaenoic acid, DHA). Although many other lipids also affect membranes, this chapter will only contrast TFA and DHA, the alpha and omega of the fatty acid business, because of their topical interest in human health.

2.1 *Trans*-Fatty Acids

Dietary TFAs (*trans*-fatty acids) terrify the health-conscious public, and for good reason. Since the 1950s, TFAs have been closely linked to coronary heart disease and arteriosclerosis [30]. A 2004 report from the Harvard School of Public Health estimated that partially hydrogenated fat, the major dietary source of TFAs, may be responsible for between 30,000 and 100,000 premature coronary deaths per year in the United States alone [31]. To date, there have been numerous large epidemiology studies that have strongly supported this conclusion [32]. Much less certain are reports that TFAs may also be weakly carcinogenic [33] and may even affect normal brain function [34]. TFAs are now universally recognized to be "bad guys," but why? TFAs are very similar to *cis* fatty acids that are far more common and are essential for life. So why have TFAs turned to the dark side?

For millennia, TFAs have been a normal component of human membranes. For example, the sphingosine backbone of sphingolipids contains a *trans* double bond in its fixed hydrophobic chain. The simplest TFA, elaidic acid ($18{:}1^{\Delta 9t}$), is naturally present in ruminant fat, meat, and dairy products and so is commonly found at low levels in human membranes [35]. The fact that elaidic acid has been in the human diet for so long cannot account for the large increase in heart disease observed over the past century. It is doubtful that elaidic acid is responsible for the health problems. Instead, the culprit is likely to be partially hydrogenated plant oils. The first successful hydrogenation of plant oils was reported in 1897 [35]. An industrial process was developed to harden fluid plant oils and to decrease their susceptibility to oxidation, opening their application in food processing. Since then there has been a steady increase in the amount of TFAs appearing in the human diet and with it, a concomitant increase in heart disease. Partial hydrogenation of plant lipids produces a bewildering array of TFAs, the result of *cis* double bond reductions and migration up and down the chain and partial conversion of normal *cis* to deleterious *trans* double bonds. The process creates a wide range of geometric and positional fatty acid isomers. In recent years, the increase in heart disease has resulted in banning *trans*-fats from many parts of the world.

The question remains why *cis* fatty acids are essential for human health while TFAs are so harmful. One possible answer, related to the theme of this book, involves incorporation of unnatural TFAs into membrane phospholipids where they replace the natural *cis* lipids, thus altering the structure and function of membranes. All fatty acids, including TFAs, can be incorporated into phospholipids (Chapter 15) and thereby affect the hydrophobic interior of membranes [36,37].

A first approach of how *cis*-fatty acids and TFAs differ in their affect on membrane physical properties can be extracted from the main phase T_ms reported in Chapter 4, Tables 4.3 and 4.7. This data is summarized in Table 23.3, listing the T_ms and ΔT_ms for the 18-carbon series of fatty acids.

Both unsaturated fatty acids (oleic and elaidic) exhibit T_ms that are lower than that of saturated stearic acid. However, the change in T_ms (ΔT_m) is much larger for the *cis*-fatty acid than for the TFA. Therefore, at least by T_m analysis, although both elaidic and oleic acid have 18-carbons and a single $\Delta 9$ double bond, elaidic acid is more similar to the saturated stearic acid than it is to oleic acid. From the T_ms it can be concluded that *cis* double bonds have a larger affect on lipid packing than do *trans* double bonds that exhibit considerable saturated fatty acid properties.

Roach et al. [38] extended the T_m measurements on oleic versus elaidic acid to several PC model membranes (monolayers and bilayers). They compared the affect of oleic versus elaidic and linoleic versus linelaidic on homo-chain and hetero-chain PCs using molecular dynamics, lateral lipid packing, thermotropic phase behavior, fluidity, lateral mobility, and

TABLE 23.3 T_ms and ΔT_ms for the 18-carbon Series of Fatty Acids

Fatty acid	Designation	T_m (°C)	ΔT_m (°C)
Stearic	Saturated (18:0)	69.6	0
Oleic	*cis* ($18{:}1^{\Delta 9c}$)	16.2	−53.4
Elaidic	*trans* ($18{:}1^{\Delta 9t}$)	43.7	−25.9

permeability. In all cases, the *cis* unsaturated chains induced much larger membrane perturbations than did the *trans* unsaturated chains. Once again, the *trans* double bond acyl chains behaved more like a saturated chain than like a *cis* unsaturated chain. Corroborating this conclusion is the molecular dynamics simulations of Pasenkiewicz-Gierula and coworkers who could detect no significant difference between 16:0–18:1$^{\Delta 9c}$ PC and 16:0–18:1$^{\Delta 9t}$ PC at the aqueous interface [39], but did observe a difference between the two PCs within the bilayer hydrophobic interior [40]. They concluded that the *trans*-PC was more similar to the disaturated DMPC standard than to the *cis*-PC. A similar conclusion was obtained from a diet study of serum lipoprotein levels [41]. The effect of dietary *trans*-elaidic acid was more similar to saturated stearic acid than to *cis*-oleic acid. In addition there has been a plethora of membrane enzymes and receptors whose activity has been shown to decrease when TFAs replaced *cis* [42].

Although it is now clear that TFAs are incorporated into membrane phospholipids where they have the chance to affect many membrane properties, questions still abound. Are TFAs mistakenly identified as a saturated fatty acid and placed in the *sn*-1 chain position of a phospholipid, or are they recognized as an unsaturated fatty acid and placed in the *sn*-2 position? Either option seems possible. Emken et al. [43] reported that in human erythrocytes and platelets three times more elaidic acid than oleic acid accumulates in the *sn*-1 position of PCs. Since elaidic acid resembles a saturated fatty acid, its accumulation into the *sn*-1 position is not surprising and would likely have only a minimal affect on normal membrane structure and function [44]. Being a natural food product, elaidic acid is probably not responsible for heart disease. Instead, it is more likely that the enormous number of positional and geometric isomers found in partially hydrogenated oils are responsible. When incorporated into membrane phospholipids, TFAs must replace either existing saturated or *cis* unsaturated acyl chains. Supporting replacing *sn*-1 chain saturated fatty acids, Larque et al. [36] reported that as dietary TFA levels rise in liver microsomes and mitochondria, net saturated fatty acid levels drop. It has also been reported that dietary TFAs can replace natural DHA in the *sn*-2 position of brain membrane phospholipids, thus affecting the electrical activity of neurons and hence, brain function.

Although there are examples supporting TFAs replacing *sn*-1 chain saturated fatty acids and *sn*-2 chain unsaturated fatty acids, it is more likely that addition of polyunsaturated TFAs to the *sn*-2 position of phospholipids will alter membrane structure and function far more than elaidic acid in the *sn*-1 position [45]. However, other completely different possibilities exist. For example, it is possible that only one of the multitude of TFA isomers, existing at miniscule levels, is doing 99%+ of the harm. Identifying such a unique *trans* isomer will be a difficult endeavor.

2.2 Docosahexaenoic Acid

DHA (docosahexaenoic acid, 22:6(n-3)) (Chapters 4 and 10) is the longest (22 carbons) and most unsaturated (6 *cis* double bonds) fatty acid commonly found in mammalian membranes [46]. In recent years DHA (and other omega-3 fatty acids) has received a great deal of attention due to its reputed involvement in alleviating a wide variety of human afflictions [47] (listed in Table 23.4, [48]). This list, which spans the entire gambit of human disorders, can be roughly divided into six nonexclusive categories: heart disease, cancer, immune problems,

TABLE 23.4 A Partial List of Human Afflictions That Have Been Linked to Docosahexaenoic Acid

ADHD	Depression	Multiple sclerosis
Aggression	Dermatitis	Neurovisual development
Alcoholism	Diabetes	Nephropathy
Alzheimer's disease	Dyslexia	Periodontitis
Arthritis	Eczema	Phenylketonuria
Asthma	Fertility	Placental function
Atrial fibrillation	Gingivitis	Psoriasis
Autism	Heart disease	Respiratory diseases
Bipolar disorder	Hypersensitivity	Schizophrenia
Blindness	Inflammatory response	Sperm fertility
Blood clotting	Kidney disease	Suicide
Bone mineral density	Lupus	Ulcerative colitis
Brain development	Malaria	Visual acuity
Cancer	Methylmalonic acidaemia	Zellweger syndrome
Crohn's disease	Migraine headaches	Cystic Fibrosos Elasticity
Cystic fibrosis	Mood and behavior	

neuronal functions, aging, and "other" hard to categorize afflictions such as migraine headaches, malaria, and sperm fertility.

Historically, the primary source of DHA in the human diet has been oily, coldwater fish. In fact, the initial link of fish oils to a human health problem (ischaemic heart disease in Greenland Eskimos) originated in the pioneering work of Bang and Dyerberg from the 1970s [49]. DHA supplementation is currently in vogue. DHA, omega-3 fatty acid and fish oil capsules are commonly found in grocery and drug stores. DHA is also included in many infant formulas since it is known to accumulate in the brain and eyes of the fetus, where it affects early visual acuity and enhances neural development. DHA is also used as a component of parenteral (intravenous) and enteral (feeding tube) nutrition.

An obvious question is how such a simple molecule can affect so many seemingly unrelated processes. To accomplish this, DHA must be exerting its influence at some fundamental level that is common to many types of cells and tissues. Possible nonexclusive modes of action for DHA include: (1) eicosanoid biosynthesis; (2) protein activity through direct interaction; (3) protein activity through indirect interactions involving transcription factors; (4) lipid peroxidation products; (5) membrane structure and function; and (6) lipid raft structure and function [50,51]. Previous chapters in this book have discussed the enormous change in lipid physical properties that affect membrane structure and function occurring when a single double bond is added to a saturated acyl chain. Table 23.5 shows a list of these properties.

TABLE 23.5 Lipid Physical Properties Affecting Membrane Structure and Function That Occur When a Single Double Bond is Added to a Saturated Acyl Chain

Bilayer thickness	Lipid microdomain formation
Compressiblity	Lipid protein affinity
Elasticity	Membrane permeability
Fluidity	Membrane stability
Flip-flop rate	Packing free volume
Interaction with cholesterol	Phase preference
Lateral diffusion rate	Phase transition temperature
Lipid area/molecule	Susceptibility to peroxidation
Lipid packing	Susceptibility to phospholipases
Lipid–lipid interaction	"Squeeze out"

Since double bonds have such an enormous affect on membranes, it seems logical to predict that DHA with 6-double bonds might be the most influential membrane fatty acid of all. Indeed, many biophysical studies on lipid monolayer and bilayer properties have been reported [50,52,53]. The problem is that none of the listed properties are unique to DHA, but instead are characteristic of all polyunsaturated fatty acids. In most, but not all examples, DHA does exert a larger affect than other less unsaturated fatty acids, but this difference is probably not sufficient to account entirely for DHA's unusual health benefits. One possibility for DHA's molecular mode of action is in altering important cell signaling processes by affecting the composition, size, structure, and stability of lipid rafts. Initial investigations have demonstrated that DHA does indeed affect lipid raft structure and cell signaling [54–56].

It is likely that no single membrane property is sufficient to account for DHA's health benefits, but rather a combination of as yet unidentified DHA-affected properties is required. But what exactly are these properties? The answer to this conundrum will require further research into the molecular aspects of membrane structure and function. It can be predicted that in the near future the next major advance in membranes will involve understanding exactly what a lipid raft is at the molecular level. How many types of lipid rafts and lipid nonrafts are there and how do they control cellular events? Once the very fundamental questions about membrane structure is better understood, it should be possible to design new paradigms to benefit the human condition.

3. SUMMARY

The complexity of membrane structure and function outlined in this book suggests a wealth of potential targets to alleviate human afflictions. Chapter 23 investigates how liposomes can be modified for intravenous, targeted drug delivery. The liposomes must be nonleaky and able to avoid the RES (reticuloendothelial system). Several types of drug-carrying liposomes are discussed including stealth liposomes (coated by PEG, polyethylene-glycol), thermo-liposomes (made from lipid mixtures with a phase T_m of ~2–4°C above physiological), pH-sensitive liposomes (containing a titratable group, often derivatives of homocysteine or succinate), and several targeted liposomes (with specific antibodies, lectins, receptors, and vitamins attached to the liposome surface). Also discussed are the affects of dietary fatty acids on membrane structure and function as they influences human health. In this regard, harmful TFAs (*trans*-fatty acids) are contrasted with the beneficial omega-3 fatty acid, DHA (docosahexaenoic acid).

Chapter 24 will review events associated with apoptosis (programmed cell death). In sharp contrast to necrosis where cells die from acute injury, apoptosis occurs as a normal part of a cell's lifecycle and has advantages.

References

[1] Weissmann G, Rita GA. Molecular basis of gouty inflammation: interaction of monosodium urate crystals with lysosomes and liposomes. Nat New Biol 1972;240(101):167–72.
[2] Duzgunes N. Liposomes. Elsevier Academic Press; 2009. 369 pp.
[3] Chrai SS, Murari R, Ahmad I. Liposomes (a review): Part Two: drug delivery systems. BioPharm 2002:40–9.
[4] http://img.medscape.com/article/734/055/734055-fig3.jpg.
[5] Gregoriadis G, Ryman BE. Lysosomal localization of β-fructofuranosidase-containing liposomes injected into rats. Some implications in the treatment of genetic disorders. Biochem J 1972;129:123–33.
[6] Myler PJ, Fasel N, editors. Leishmania: After the genome. Norfolk, UK: Caister Academic Press; 2008. 306 pp.
[7] Alving CR, Steck EA, Chapman Jr WL, Waits VB, Hendricks LD, Swartz Jr GM, et al. Therapy of leishmaniasis: superior efficacies of liposome-encapsulated drugs. Proc Natl Acad Sci U S A 1978;75(6):2959–63.
[8] Immordino ML, Dosio F, Cattel L. Stealth liposomes: review of the basic science, rationale, and clinical applications, existing and potential. Intern J Nanomed 2006;1(3):297–315.
[9] Drummond DC, Meyer O, Hong K, Kirpotin DB, Papahadjopoulos D. Optimizing liposomes for delivery of chemotherapeutic agents to solid tumors. Pharm Rev 1999;51(4):691–744.
[10] Senior J, Gregoriadis G. Stability of small unilamellar liposomes in serum and clearance from the circulation: the effect of the phospholipid and cholesterol components. Life Sci 1982;30:2123–36.
[11] Gabizon A, Papahadjopoulos D. Liposome formulations with prolonged circulation time in blood and enhanced uptake by tumors. Proc Natl Acad Sci U S A 1988;85:6949–53.
[12] Lasic DD, Martin EJ, editors. Stealth liposomes. CRC Press, Taylor & Francis Group; 1995. 320 pp.
[13] Allen TM, Hansen C, Martin F, Redemann C, Yau-Young A. Liposomes containing synthetic lipid derivatives of poly(ethylene glycol) show prolonged circulation half-lives in vivo. Biochim Biophys Acta 1991;1066:29–36.
[14] Sapra P, Allen TM. Ligand-targeted liposomal anticancer drugs (Review). Prog Lipid Res 2003;42:439–62.
[15] Needham D, Anyarambhatla G, Kong G, Dewhirst MW. A new temperature-sensitive liposome for use with mild hyperthermia: characterization and testing in a human tumor xenograft model. Cancer Res 2000;60:1197–201.
[16] Needham D. Materials engineering of lipid bilayers for drug carrier function. MRS Bull 1999;24:32–40.
[17] Yatvin MB, Weinstein JN, Dennis WH, Blumenthal R. Design of liposomes for enhanced local release of drugs by hyperthermia. Science 1978;202:1290–3.
[18] Blok MC, van Deenen LLM, de Gier J. Effect of the gel to liquid crystalline phase transition on the osmotic behaviour of phosphatidylcholine liposomes. Biochim Biophys Acta 1976;433:1–12.
[19] Crile Jr G. Selective destruction of cancers after exposure to heat. Ann Surg 1962;156:404–7.
[20] Kong G, Dewhirst MW. Hyperthermia and liposomes: a review. Int J Hyperth 1999;15:345–70.
[21] Lindner LH, Eichhorn ME, Eibl H, Teichert N, Schmitt-Sody M, Issels RD, et al. Novel temperature-sensitive liposomes with prolonged circulation time. Clin Cancer Res 2004;10:2168–78.
[22] Koning GA, Eggermont AMM, Lindner LH, ten Hagen TLM. Hyperthermia and thermosensitive liposomes for improved delivery of chemotherapeutic drugs to solid tumors. Pharm Res 2010;27:1750–4.

[23] White J, Matlin K, Helenius A. Cell fusion by Semliki Forest, influenza, and vesicular stomatitis viruses. J Cell Biol 1981;89:674–9.
[24] Chu C-J, Szoka FC. pH-sensitive liposomes. J Liposome Res 1994;4:361–95.
[25] Drummond DC, Zignani M, Leroux I. Current status of pH-sensitive liposomes in drug delivery. Prog Lipid Res 2000;39(5):409–60.
[26] Connor J, Yatvin MB, Huang L. pH-sensitive liposomes: acid-induced liposome fusion. Proc Natl Acad Sci U S A 1984;81:1715–8.
[27] Yatvin MB, Kreuz W, Horowitz BA, Shinitzky M. pH-sensitive liposomes: possible clinical implications. Science 1980;210:1253–5.
[28] Leventis R, Diacovo T, Silvius JR. pH-dependent stability and fusion of liposomes combining protonatable double-chain amphiphiles with phosphatidylethanolamine. Biochemistry 1987;26:3267–76.
[29] Collins D, Litzinger DC, Huang L. Structural and functional comparisons of pH-sensitive liposomes composed of phosphatidylethanolamine and three different diacylsuccinylglycerols. Biochim Biophys Acta 1990;1025:234–42.
[30] Denke MA. Serum lipid concentrations in humans. In: Trans fatty acids and coronary heart disease risk. Am J Clin Nutr 1995;62:693S–700S.
[31] Ascherio A, Stampfer MJ, Willett WC. Trans fatty acids and coronary heart disease. Harv Sch Public Health 2004:1–8.
[32] Zaloga GP, Harvey KA, Stillwell W, Siddiqui R. Trans fatty acids and coronary heart disease. Nutr Clin Pract 2006;21:505–12.
[33] Slattery ML, Benson J, Ma KN, Schaffer D, Potter JD. Trans fatty acids and colon cancer. Nutr Cancer 2001;39:170–5.
[34] Phivilay A, Julien C, Tremblay C, Berthiaume L, Julien P, Giguere Y, et al. High dietary consumption of trans fatty acids decreases brain docosahexaenoic acid but does not alter amyloid-beta and tau pathologies in the 3xTg-AD model of Alzheimer's disease. Neuroscience 2009;159:296–307.
[35] Emken EA. Nutrition and biochemistry of trans and positional fatty acid isomers in hydrogenated oils. Annu Rev Nutr 1994;4:339–76.
[36] Larque E, Garcia-Ruiz PA, Perez-Llamas F, Zamora S, Gil A. Dietary trans fatty acids alter the composition of microsomes and mitochondria and the activities of microsome Δ6- fatty acid desaturase and glucose-6-phosphatase in livers of pregnant rats. J Nutr 2003;133:2526–31.
[37] Morgado N, Galleguillos A, Sanhueza J, Garrido A, Nieto S, Valenzuela A. Effect of the degree of hydrogenation of dietary fish oil on the trans fatty acid content and enzymatic activity of rat hepatic microsomes. Lipids 1998;33:669–73.
[38] Roach C, Feller SE, Ward JA, Shaikh SR, Zerouga M, Stillwell W. Comparison of cis and trans fatty acid containing phosphatidylcholines on membrane properties. Biochemistry 2004;43:6344–51.
[39] Murzyn K, Rog T, Jezierski G, Takaoka Y, Pasenkiewicz- Gierula M. Effects of phospholipid unsaturation on the membrane/water interface: a molecular simulation study. Biophys J 2001;81:170–83.
[40] Rog T, Murzyn K, Gurbiel R, Takaoka Y, Kusumi A, Pasenkiewicz-Gierula M. Effects of phospholipid unsaturation on the bilayer nonpolar region: a molecular simulation study. J Lipid Res 2004;45:326–36.
[41] Mensink RP, Katan MB. Effect of dietary trans fatty acids on high density and low density lipoprotein cholesterol levels in healthy subjects. N Engl J Med 1990;323:439–45.
[42] Alam SQ, Ren YF, Alam BS. Effect of dietary trans fatty acids on some membrane-associated enzymes and receptors in rat heart. Lipids 1989;24:39–44.
[43] Emken EA, Rohwedder WK, Dutton HJ, Dejarlais WJ, Adlof RO. Incorporation of deuterium-labeled cis- and trans-9-octadecenoic acids in humans: plasma, erythrocyte, and platelet phospholipids. Lipids 1979;14:547–54.
[44] Wolff RL, Entressangles B. Steady-state fluorescence polarization study of structurally defined phospholipids from liver mitochondria of rats fed elaidic acid. Biochim Biophys Acta 1994;1211:198–206.
[45] Lichtenstein AH. Dietary trans fatty acid. J Cardiopulm Rehab 2000;20:143–6.
[46] Salem NJ, Kim H-Y, Yergey JA. Docosahexaenoic acid: membrane function and metabolism. In: Simopolous AP, Kifer RR, Martin RE, editors. Health effects of polyunsaturated fatty acids in seafoods. New York: Academic Press; 1986. p. 319–51.
[47] Stillwell W, Shaikh SR, Lo Cascio D, Siddiqui RA, Seo J, Chapkin RS, et al. Docosahexaenoic acid: an important membrane-altering omega-3 fatty acid. In: Huamg JD, editor. Frontiers in nutrition research. New York, NY: NOVA Science Publishers; 2006. p. 249–71 [chapter 8].
[48] Stillwell W. Docosahexaenoic acid: a most unusual fatty acid. Chem Phys Lipids 2008;153:1–2.
[49] Dyerberg J, Bang HO, Hjorne N. Fatty acid composition of the plasma lipids in Greenland Eskimos. Am J Clin Nutr 1975;28:958–66.

[50] Stillwell W, Wassall SR. Docosahexaenoic acid: membrane properties of a unique fatty acid. Chem Phys Lipids 2003;126:1–27.
[51] Stillwell W. The role of polyunsaturated lipids in membrane raft function. Scand J Food Nutr 2006;50 (Suppl. 2):107–13.
[52] Salem Jr N, Litman B, Kim HY, Gawrisch K. Mechanisms of action of docosahexaenoic acid in the nervous system. Lipids 2001;36:945–59.
[53] Feller SE, Gawrisch K. Properties of docosahexaenoic acid-containing lipids and their influence on the function of rhodopsin. Curr Opin Struct Biol 2005;15:416–22.
[54] Stillwell W, Shaikh SR, Zerouga M, Siddiqui R, Wassall SR. Docosahexaenoic acid affects cell signaling by altering lipid rafts. Reprod Nutr Dev 2005;45:559–79.
[55] Kim W, Fan Y, Barhoumi R, Smith R, McMurray DN, Chapkin RS. n-3 polyunsaturated fatty acids suppress the localization and activation of signaling proteins at the immunological synapse in murine CD4+ T cells by affecting lipid raft formation. J Immunol 2008;181:6236–43.
[56] Shaikh SR. Diet-induced docosahexaenoic acid non-raft domains and lymphocyte function. 2010. Prostaglandins Leukotrienes Essent. Fat Acids 2010;28:159–64.

CHAPTER

24

Cell Death, Apoptosis

OUTLINE

1. INTRODUCTION

In previous chapters (Chapters 14–16), the many complex steps involved in the "birth" (biogenesis) of cell membranes was investigated. The fact that cells, and therefore cell membranes, do not live forever, implies there is likely a cell "death" processes common to all cells. There are two basic ways that a cell dies, by necrosis and by apoptosis [1]. Necrosis refers to traumatic cell death that results from acute cellular injury. Necrosis is uncontrolled, chaotic and detrimental. In sharp contrast is apoptosis or "programmed cell death" (PCD) that is actually beneficial. Apoptosis follows a complex set of prescribed processes and pathways and is therefore somewhat predictable [2].

2. NECROSIS

Necrosis is the premature death of living cells or tissues caused by external factors such as infection, toxins, or trauma [3]. Necrosis results in the destruction of many normal cell processes and structures. Membranes are disrupted resulting in hypoxia and loss of electron transport and oxidative phosphorylation, greatly reducing adenosine triphosphate (ATP)

An Introduction to Biological Membranes
http://dx.doi.org/10.1016/B978-0-444-63772-7.00024-5

levels, inducing metabolic collapse, cell swelling and rupture, leading to inflammation. Unlike apoptotic cells, necrotic cells cannot send a normal chemical signal to the immune system, thus preventing nearby phagocytes from locating and rapidly engulfing the dead cells. Therefore dead cell debris accumulates during necrosis resulting in damage to surrounding healthy cells and inflammation that if severe enough can become chronic.

3. APOPTOSIS

Apoptosis was discovered by the German scientist Carl Vogt (Fig. 24.1) in 1842 while studying tadpole development of the midwife toad. In 1972, the term apoptosis was introduced to the scientific literature by Kerr, Wyllie and Currie [4]. Apoptosis is a naturally occurring and orderly cause of cell death that does not produce inflammation. Apoptosis is a common process in multi-cellular organisms and is beneficial, likely essential, for their life [5,6]. It has been estimated that in a typical adult human, between 50 and 70 billion cells die each day from apoptosis [7]! This probably sounds like a lot of cells, and it is. In one year, an average child between the ages of 8 and 14 loses their body weight in cells due to apoptosis. Of course, the cell loss is balanced by an equal cell proliferation. This exemplifies the role of apoptosis; maintaining homeostasis between cell death rate and mitosis rate.

Apoptosis is also called programmed cell death (PCD) because it involves a well-defined series of biochemical events leading to controlled cell destruction and death. Balance is the key objective of apoptosis. Excessive apoptosis causes hypotropy (progressive degeneration of an organ or tissue caused by loss of cells), whereas insufficient apoptosis results in uncontrolled cell proliferation (eg, cancer). Apoptosis leads to a variety of characteristic morphological changes including membrane alterations (blebbing, loss of membrane asymmetry, and cell–cell attachment), cell shrinkage, nuclear fragmentation, chromatin condensation, DNA fragmentation, and apoptopic cell body formation. Unlike dead necrotic cells

FIGURE 24.1 Carl Vogt, 1817–1895. *File: Carl Vogt.jpg. From Wikimedia Commons, the free media repository http://commons.wikimedia.org/wiki/File:Carl_Vogt.jpg.*

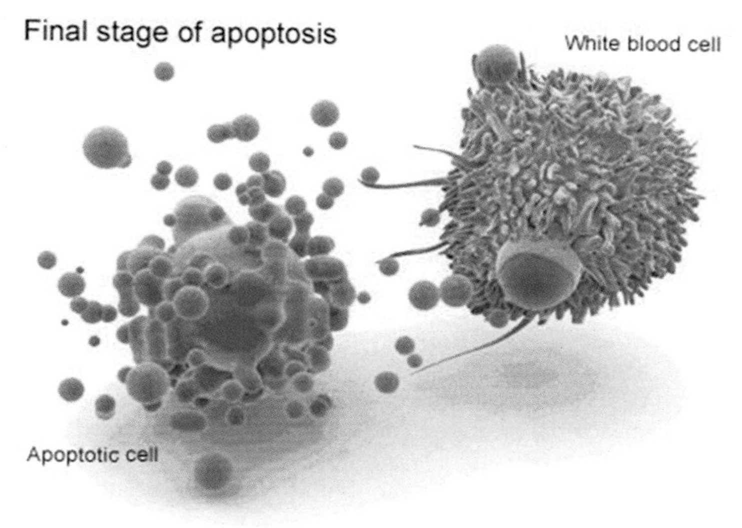

FIGURE 24.2 Depicted is the final stage of apoptosis where a white blood cell (phagocyte, macrophage) engulfs apoptotic cell bodies. Phagocytes rapidly engulf and remove apoptopic cell bodies before they can spread and damage adjacent, healthy cells. *SBI human apoptosis-related RNAi library. Phone 888 266-5066. http://www.systembio.com/rnai-libraries/pathway-focused/apoptosis/overview.*

that spill cellular debris, apoptopic cell bodies (Fig. 24.2) send out a chemical signal that attracts phagocytes. Phagocytes rapidly engulf and remove the apoptopic cell bodies before cell debris can leak out, damaging adjacent cells. Therefore necrosis damages many adjacent cells while apoptosis does not.

The process of apoptosis is extremely complex. It involves multiple steps and multiple pathways that are controlled by a diverse range of cell signals, and is inherent in every cell in the body. Only an abbreviated outline of apoptosis can be included in this short book chapter.

4. MECHANISMS OF APOPTOSIS

There are two basic types of apoptosis that are distinguished by where the triggering signal originates: extracellular triggers (extrinsic inducers) or intracellular triggers (intracellular inducers) [8].

Extracellular inducers: Extracellular inducers include: toxins, hormones, growth factors, nitric oxide, and cytokines. They must either cross the plasma membrane themselves or transduce a transplasma membrane response. In either case the plasma membrane is directly involved.

Intracellular inducers: Intracellular inducers include: glucocorticoids, heat, radiation, nutrient deprivation, viral infection, hypoxia, and increased intracellular calcium. They are produced in response to stress.

5. MITOCHONDRIAL ROLE IN APOPTOSIS

Apoptopic signals must first cause the initiation of regulatory proteins that instigate an apoptosis pathway, eventually leading to cell death. Apoptosis has two main methods of regulation, through mitochondria and by directly transducing the signal via adaptor proteins to advance the apoptosis mechanisms. Major targets of apoptopic signals are the two mitochondrial membranes and particularly the inner membrane (the cristae) [9,10]. The initial apoptotic signals may cause mitochondrial swelling through pore formation or by directly making the mitochondrial membrane leaky to sequestered apoptotic effectors. Among the compounds that leak from the mitochondria are the small proteins, small mitochondria-derived activator of caspases (SMAC) and cytochrome c [11]. It is the activated caspases that destroys the cell from the inside. Caspases (cysteine-aspartic proteases or cysteine-dependent aspartate-directed proteases) are a family of 12 proteases that play essential roles in apoptosis, necrosis, and inflammation. They are sometimes referred to as "executioner" proteins.

Under normal (nonapoptotic) conditions, an inhibitor of apoptosis proteins (IAPs) exists in the cytoplasm. IAPs turn off the apoptotic caspase pathways allowing the cell to live normally. Released SMAC and cytochrome c bind to IAPs, deactivating them and thereby allowing apoptosis to proceed. Therefore, apoptosis is indirectly regulated by mitochondrial permeability. The basic sequence of apoptotic events is:

$$\text{Apoptosis Signal} \rightarrow \text{SMAC/Cytochrome c} \rightarrow \text{IAC} \rightarrow \text{Caspases} \rightarrow \text{Apoptosis}$$

Cytochrome c is a most remarkable protein. It is quite small for a protein, being a single chain of 104 amino acids (MW 12,233 Da), covalently bound to a single heme group. Cytochrome c is loosely attached to the outer surface of the mitochondrial inner membrane (it is an extrinsic or peripheral protein, Chapter 6). When released from the mitochondria it is highly water-soluble ($\sim$100 g/L). Cytochrome c is also an example of a "moonlighting protein", having two major activities that are completely unrelated to one another. Cytochrome c's major function is as an essential oxidation–reduction component of the mitochondrial electron transport chain (Chapter 18). Cytochrome c transfers electrons between electron transport Complexes III (coenzyme Q-cytochrome c reductase) and IV (cytochrome c oxidase) (Chapter 18). Cytochrome c's second major function is its unrelated role in apoptosis [11].

6. EXTRACELLULAR INDUCERS OF APOPTOSIS

There are two basic types of mechanisms involved in extracellular (direct) induction of apoptosis: tumor necrosis factor (TNF)-induced and Fas–Fas (apoptosis stimulating fragment) ligand-mediated apoptosis. Both pathways transduce a signal across the plasma membrane of a cell targeted for apoptosis. Transduction for both processes involves members of the TNF receptor (TNFR) family. TNF, the major extracellular inducer of apoptosis, is a cytokine produced by activated macrophages. TNF binds to the cell surface transmembrane receptors, TNF-R1 and TNF-R2 that are commonly found on most cells in the human body. Binding of TNF to TNF-R1 stimulates the intermediate membrane protein TNF receptor-associated death domain

(TRADD) which in turn cleaves to form the Fas-associated death domain protein (FADD) on the plasma membrane inner leaflet (Fig. 24.3, Left). The extracellular (direct) inducer of apoptosis known as Fas-ligand (FasL) is generated by cleaving the membrane-bound inactive form of Fas-receptor by the external matrix metalloproteinase, MMP-7 (Fig. 24.3, Right). FasL then binds to its transmembrane receptor Fas-receptor (also known as Apo-1 or CD95). This receptor is also a member of the TNF family of receptors. The activated Fas-receptorisis (this should read) Fas-receptoris is cleaved cleaved generating FADD on the inner membrane leaflet. FADD is a central component in the apoptosis process. FADD then leads to the generation of pro-caspase 8 that activates other members of the caspase cascade [12] of proteases, triggering apoptosis through a multitude of cytoplasmic steps that will not be discussed here. In Fig. 24.3 that there is considerable similarity between the TNF and FasL pathways.

In addition to the caspases, many other proteins are involved in the steps between FADD and apoptosis. Some inhibit, while others promote apoptosis. A balance must be struck between proapoptotic (BAX, BID, BAK or BAD) and antiapoptic (Bcl-XI and Bcl-2) proteins that are all members of the Bcl-2 family of proteins. This balance is established on the outer mitochondrial membrane. Proapoptotic homodimers are required to make the mitochondrial membrane permeable for the release of caspase activators such as SMAC and cytochrome c.

The mechanism of apoptosis is very complex and involves scores of proteins. The various pathways involve considerable overlap and share several common components. Importantly, cells undergoing apoptosis do display several characteristic morphologies. These include:

1. The cells shrink and round-up due to breakdown of the cytoskeleton by the caspases.
2. The cytoplasm becomes dense with tightly packed organelles.
3. The cell's chromatin (chromatin is the combined cell's nuclear DNA and protein) condenses into compact patches against the nuclear envelope. This process, known as pyknosis, is a hallmark of apoptosis.
4. The nuclear envelope breaks up into several discrete chromatin bodies and the DNA is fragmented by endonucleases in a process known as karyorrhexis. The DNA fragments are short and regularly spaced in size giving a characteristic "laddered" appearance on agar gel after electrophoresis.
5. The plasma membrane develops many surface structures known as blebs (Fig. 24.4). A bleb is an irregular bulge in the plasma membrane caused by apoptosis-induced breakup of the cytoskeleton. As a result, the cytoskeleton is decoupled from the plasma membrane producing outward facing bulges called blebs. Furthermore, the blebs may pinch off from the cell taking a portion of the cytoplasm with them, generating apoptotic cell bodies (Fig. 24.2).
6. In the final stages of apoptosis, the plasma membrane loses its normal lipid asymmetry (Chapter 9). Particularly important is phosphatidylserine (PS). In nonapoptotic cells the plasma membrane outer leaflet is almost completely devoid of PS. This is the result of an ATP-dependent PS—flippase that immediately takes any PS that has flipped to the outer leaflet and flips it back to the inner leaflet [13,14]. During apoptosis this process is disrupted. The cause may be due to a loss of flippase activity, increased flop-flop associated with membrane disruption or the activity of a hypothetical scramblase that destroys all phospholipid asymmetry. In any case, the appearance of plasma membrane outer leaflet PS marks the apoptotic cell for phagocytosis by macrophages.

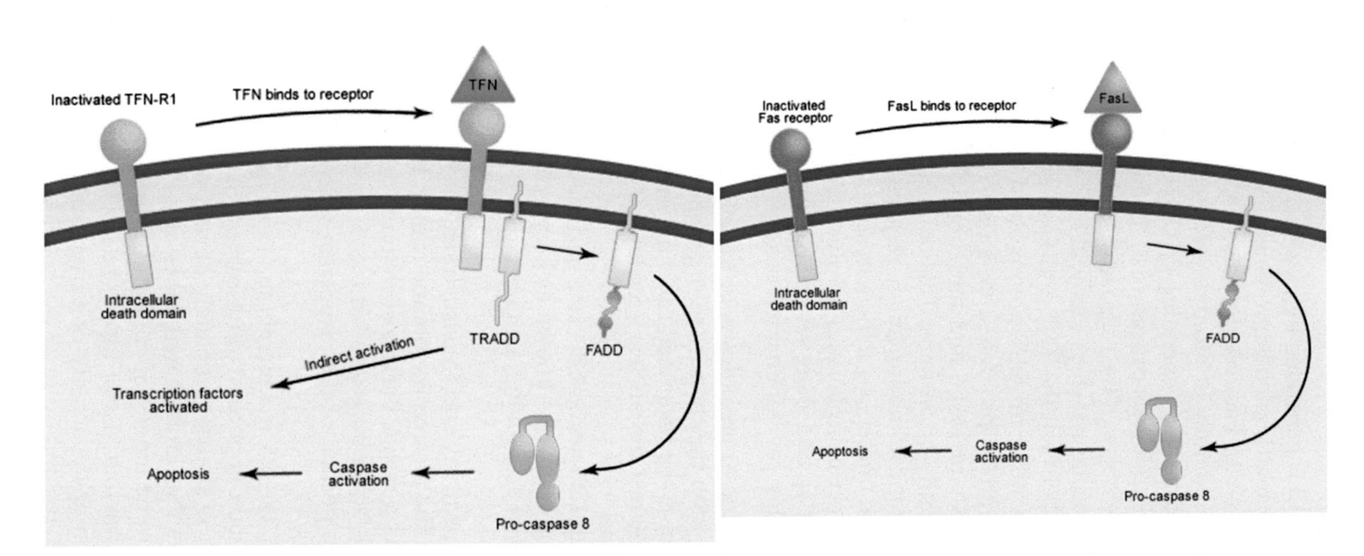

FIGURE 24.3 Overview of TNF (left) and Fas (right) signaling in apoptosis. *FADD*, Fas-associated death domain protein; *TNF-R1*, tumor necrosis factor-receptor 1; *TRADD*, tumor necrosis factor receptor-associated death domain. *Both sides of this figure come from the same source—Wikipedia Commons Adapted from Figs. 1-23 (TNF) and 1-22 (Fas),* Robbins Pathologic Basis of Disease, *6th edition. Author: Emma Farmer. Date: December 16, 2006. Permission: domain (own work). "I, the copyright holder of this work, release this work into the public domain. This applies worldwide. In some countries this may not be legally possible; if so:* I grant anyone the right to use this work ***for any purpose***, without any conditions, unless such conditions are required by law." *Left side TNF http://en.wikipedia.org/wiki/File:TFN-signalling.png; right side Fas http://en.wikipedia.org/wiki/File:Fas-signalling.png.*

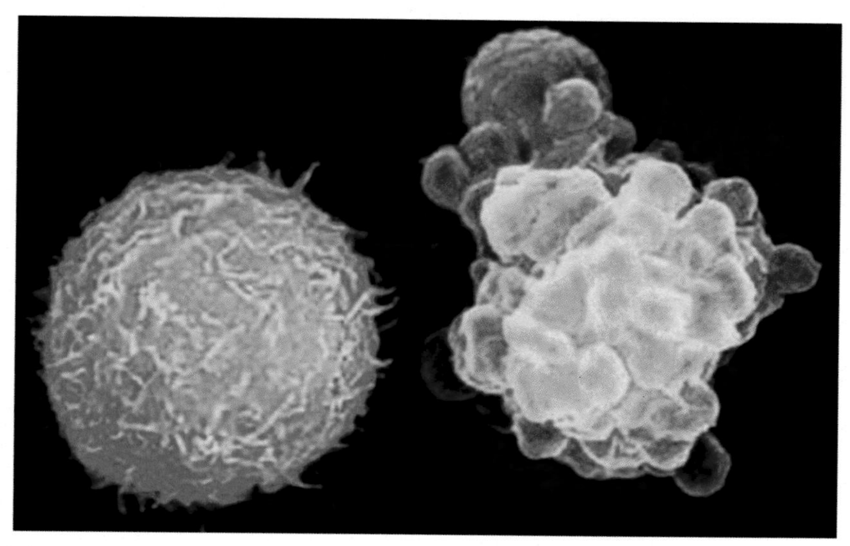

FIGURE 24.4 Apoptosis-induced membrane blebs. On the left is a healthy cell and on the right is a cell in late apoptosis showing many blebs. http://www.nephrology-uni-kiel.com/start-page/research/current-projects/programmed-cell-death/apoptosis/. *Clinic For Nephrology and Hypertension in Kiel, Germany.*

7. Finally, the cell breaks apart into apoptotic cell bodies that are engulfed by macrophages (Fig. 24.2). Apoptotic cell body removal is orderly, very fast, and leaves no trace of their prior existence.

7. SUMMARY

All living cells must die at some time. There are two very different ways that cells succumb, by necrosis and by apoptosis. Necrosis results from acute cellular injury and is traumatic, uncontrolled, chaotic, and detrimental. In sharp contrast is apoptosis or programmed cell death (PCD) that is actually beneficial. Apoptosis follows a complex set of prescribed processes and pathways and is therefore somewhat predictable. In necrosis, membranes are disrupted resulting in hypoxia, metabolic collapse, cell swelling and rupture, leading to accumulation of dead cell debris, inflammation and damage spreading to adjacent healthy cells. Apoptosis is a naturally occurring and orderly cause of cell death that does not produce inflammation. Apoptosis maintains the proper balance between cell death rate and mitosis rate. Insufficient apoptosis results in uncontrolled cell proliferation (eg, cancer). There are two basic types of apoptosis that are distinguished by where the triggering signal originates: extracellular triggers or intracellular triggers. Apoptosis has two main methods of regulation, through mitochondria and by directly transducing the signal via adaptor proteins. In the final stages of apoptosis the plasma membrane develops many surface structures in the form of irregular bulges known as blebs. The blebs may pinch off from the cell taking a portion of the cytoplasm with them, generating apoptotic cell bodies.

Chapter 25 presents a chronological compendium of many important membrane studies from ~450 BC to the present.

References

[1] Green D. Means to an end: apoptosis and other cell death mechanisms. Cold Spring Harbor (NY): Cold Spring Harbor Laboratory Press; 2011.
[2] Kanduc D, Mittelman A, Serpico R, Sinigaglia E, Sinha AA, Natale C, et al. Cell death: apoptosis versus necrosis (review). Int J Oncol 2002;21(1):165–70.
[3] Proskuryakov SY, Konoplyannikov AG, Gabai VL. Necrosis: a specific form of programmed cell death? Exp Cell Res 2003;283:1–16.
[4] Kerr JF, Wyllie AH, Currie AR. Apoptosis: a basic biological phenomenon with wide-ranging implications in tissue kinetics. Br J Cancer 1972;26:239–57.
[5] Potten P, Wilson J. Apoptosis: the life and death of cells. In: Developmental & cell biology series. Cambridge (UK): Cambridge University Press; 2014.
[6] Alberts B, Johnson A, Lewis J, Raff M, Roberts K, Walter P. Apoptosis: programmed cell death eliminates unwanted cells. In: Molecular biology of the cell. 5th ed. Garland Science; 2008. p. 1115.
[7] Karam JA. Apoptosis in carcinogenesis and chemotherapy. Netherlands: Springer; 2009.
[8] Reed JC. Mechanisms of apoptosis. Am J Pathol 2000;157(5):1415–30.
[9] Rolland S, Conradt B. The role of mitochondria in apoptosis induction in *Caenorhabditis elegans*: more than just innocent bystanders? Cell Death Differ 2006;13:1281–6.
[10] Wang C, Youle RJ. The role of mitochondria in apoptosis. Annu Rev Genet 2009;43:95–118.
[11] Jiang X, Wang X. Cytochrome C-mediated apoptosis. Annu Rev Biochem 2004;73:87–106.
[12] Loque SE, Martin SJ. Caspase activation cascades in apoptosis. Biochem Soc Trans 2008;36(Pt 1):1–9.
[13] Devaux PF. Phospholipid flippases. FEBS Lett 1988;234:8–12.
[14] Daleke DL. Phospholipid flippases. J Biol Chem 2007;282:821–5.
[15] SBI human apoptosis-related RNAi library, http://www.systembio.com/rnai-libraries/pathway-focused/apoptosis/overview.
[16] http://en.wikipedia.org/wiki/File:TFN-signalling.png.
[17] http://en.wikipedia.org/wiki/File:Fas-signalling.png.
[18] http://www.nephrology-uni-kiel.com/start-page/research/current-projects/programmed-cell-death/apoptosis/. Clinic For nephrology and hypertension in Kiel, Germany.

CHAPTER

25

Chronology of Membrane Studies

The study of membranes has a very long and rich history. Below is a compendium of many of the most important observations concerning membrane composition, structure, and function. Such a list must terminate several years before its compilation date since it takes some considerable time before it is clear which newer studies will have a lasting impact and which studies will reach a dead end.

Year	Discoverer	Discovery
~540 BC	Thales of Miletus	First to emphasize the fundamental importance of water.
450 BC	Hippo of Samos	Perceived life as water.
77	Pliny the Elder	Described the commonly used fishing trick of floating oil on water.
1650	Francis Glisson	First description of Rickets
1665	Robert Hooke	Early microscopist. Coined term "cell."
1769	F.P. de la Salle	Discovered cholesterol in gallstones.
1771	Luigi Galvani	Discovered that muscles of dead frog legs twitched when struck by an electrical spark.
1773	William Hewson	Discovered osmotic swelling and shrinking of erythrocytes. Proposed existence of a plasma membrane.
1774	Benjamin Franklin	First scientific lipid monolayer study.
1781	Henry Cavendish	Discovered the chemical composition of water.
1806	Vauquelin & Robiquet	Discovered the first amino acid, asparagine, from asparagus.
1823	Michel Chevreul	Discovered stearic acid and oleic acid in pork fat.
1828	Friedrich Wohler	Synthesized an organic molecule (urea) from an inorganic molecule (ammonium cyanate). Proved organic molecules are not "special."
1836	C.H. Schultz	Visualized erythrocyte plasma membrane with iodine as a stain. First to use the term "membrane."
1838	Robert Carswell	Drew multiple sclerosis spinal cord lesions.
1839	T. Schwann	Established the "Cell Theory."

(*Continued*)

An Introduction to Biological Membranes
http://dx.doi.org/10.1016/B978-0-444-63772-7.00025-7

Year	Discoverer	Discovery
1842	Carl Vogt	Discovered apoptosis.
	Crawford Long	First to use a general anesthetic (diethyl ether) in an operation.
1847	Theodore Gobley	Discovered lecithin [phosphatidylcholine].
	Von Bibra & Harless	Proposed the nonspecific mechanism of general anesthetic action.
1852	George Stokes	Discovered fluorescence.
1854	Rudolph Virchow	Discovered the myelin sheath.
1855	Karl von Nageli	Membrane is barrier to osmosis.
	Adolf Fick	Described what is now known as Fick's Laws of Diffusion.
1861	G.B.A. Duchenne	Described Duchenne Muscular Dystrophy.
1871	Hugo de Vries	Measured membrane permeability of ammonia and glycerol.
1877	Wilhelm Pfeffer	Proposed the first membrane theory. A membrane is thin and semipermeable.
1882	Elie Metchnikoff	Discovered phagocytosis.
1888	F. Reinitzer	Discovered the liquid crystalline phase.
	Herrmann Stillmark	Discovered lectins.
	Walter Nernst	Developed theory of electrical potentials (basis of electrophysiology) that explains ion flux across membranes.
1889	Agnes Pockels	Developed methodology for lipid monolayer studies while working in her kitchen.
1890	Lord Raleigh	Determined the size of triolein using lipid monolayer methodology.
1894	S.R. y Cajal	Proposed the "neuron doctrine."
1896	Jules Bordet	Isolated complement.
1898	Carmillo Golgi	Discovered the Golgi apparatus.
	Carl Benda	Introduced the term "mitochondria."
1899	Charles Ernest Overton	First true "membranologist." Membranes have a lipid-like barrier. Proposed passive and active transport.
	H.H. Myer & E. Overton	Origin of Myer–Overton Correlation for anesthetics.
1902	Julius Bernstein	Excitable cell membranes are selectively permeable to K^+.
1906	Alois Alzheimer	Described Alzheimer's disease.
1913	E. McCollum & M. Davis	Discovered vitamin A in butterfat and cod liver oil.
1917	Irving Langmuir	Credited with making lipid monolayers a "precise" science. His work led to the seminal 1925 Gorter–Grendel experiment.

Year	Discoverer	Discovery
1920	Latimer & Rodebush	Suggested H-bonding in water structure is the major driving force for membrane stabilization.
1921	Otto Loewi	Discovered the first neurotransmitter, acetylcholine.
1922	H.M. Evans & K.S. Bishop	Discovered vitamin E.
	E. McCollum	Identified vitamin D.
1924	Feulgen & Voit	Discovery of plasmalogens.
1925	Gorter & Grendel	The most important paper ever published on membranes. First experimental evidence for a lipid bilayer.
	Leathers & Raper	In their book, *The Fats*, they suggested phospholipids were essential components of membranes.
	H. Fricke	Used electrical impedance measurements to determine the thickness of an erythrocyte membrane.
	David Keilin	Discovered cytochromes.
1926	James Sumner	Isolated the first enzyme, urease.
	J.B. Perrin	Developed Fluorescence Polarization.
1929	Warren Lewis	Discovered pinocytosis.
	Henrik Dam	Discovered vitamin K.
1930	J.D. van der Waal Fritz London	London extended the van der Waal force to include induced dipole–dipole interactions.
1931	Ernst Ruska	Invented the electron microscope.
1934	Flaschenträger & Wolffersdorff	Discovered phorbol.
1935	Danielli & Davson	Proposed the "Pauci-Molecular model" for membrane structure. Based on the lipid bilayer.
1938	Izmailov & Shraiber	Discovered thin layer chromatography, a major technique for membrane lipid separation.
	Dorothy H. Anderson	First correlation of various aspects of Cystic Fibrosis.
1941	Fritz Lipmann	Proposed the bioenergetic function of adenosine triphosphate.
	M. Pangborn	Discovered cardiolipin.
1942	E. Klenk	Identified gangliosides from brain.
1945	Cole & Marmount	Developed voltage clamp technique.
	Porter, Claud & Pullam	First to observe the endoplasmic reticulum.

(*Continued*)

Year	Discoverer	Discovery
1946	Fritz Lipmann	Discovered coenzyme A.
1948	Linus Pauling	Proposed the α-helix.
1949	W.C. Griffin	Proposed hydrophile–lipophile balance parameter for detergents.
1950	A.J.P. Martin	Developed gas–liquid chromatography, the major technique used for fatty acid analysis.
1951	Oliver Lowery	Developed the first sensitive method to quantify proteins.
1952	Gunnar Blix	After 15 years of work on this sugar, Blix coined the term "sialic acid."
	Frederick Sanger	Sequenced the first protein, insulin.
1953	George Palade	Discovered Caveolae using electron microscopy.
1955	Lathe and Ruthven	Discovered size exclusion (gel filtration) chromatography.
	Christian de Duve	Discovered lysosomes.
	Eugene P. Kennedy	CDP-choline pathway for PC biosynthesis (Kennedy Pathway).
1956	Earl Sutherland	Discovered the first, second messenger, cAMP.
1957	J.D. Robertson	Proposed the "Unit membrane" model for membrane structure.
	Jens Skou	Discovered the plasma membrane Na^+/K^+ ATPase.
	Folch, Lees & Stanley	Developed method to extract lipids from membranes that is still in use today.
	Frederick Crane	Discovered Coenzyme Q (ubiquinone).
1960	Watson & O'Neill	Developed the technique of differential scanning calorimetry.
1961	Peter Mitchell	Proposed the Chemiosmotic Hypothesis for oxidative phosphorylation.
	A.D. Bangham	First production of liposomes.
1962	Mueller et al.	Made the first stable large lipid bilayer called a planar BLM.
	Kauzman & Tanford	Developed the Hydrophobic Effect theory that explains membrane stability.
1963	Palade & Farquahr	First description of Gap Junctions.
1964	R.T. Holman	Proposed omega nomenclature for fatty acids.
	H. Fernandez-Moran	First definitive evidence that a membrane protein (mitochondrial F_1 ATPase) is 100% asymmetrically distributed across a membrane.
	Saul Roseman	Discovered the PTS sugar transport system.
	Bernard Pressman	Valinomycin recognized as a potassium ionophore.
1966	Andre Jagendorf	"Acid Bath Experiment" supporting the Chemiosmotic hypothesis.
1967	Charles Pederson	Discovered crown ethers.

Year	Discoverer	Discovery
	Christian de Duve	Discovered peroxisomes.
1968	D. Zilversmit	Discovered phospholipid exchange proteins.
1969	Braun & Rodin	Discovered palmitoylation of a membrane protein.
	Huang	Reconstituted membrane protein into LUV. Important for transport studies.
1970	L. Frye & M. Edidin	First measurement of lateral diffusion in membranes.
	Hladky & Haydon	Discovered Gramicidin A transmembrane channel.
	Martin Rodbell	Introduced concept of G-proteins.
1971	Gunter Blobel	Proposed the "Signal Hypothesis."
1972	Singer & Nicolson	Proposed the "Fluid Mosaic Model" for membrane structure.
	Mark Bretscher	First report of partial lipid asymmetry in membranes.
	Denham Harman	Proposed the "Mitochondrial Theory of Aging."
	Kerr, Wyllie & Currie	Introduced the term apoptosis.
1974	M. Sinensky	Proposed Homeoviscous Adaptation.
	Morre & Mollenhauer	Introduced the "Endomembrane Concept."
1975	Henderson & Unwin	First electron microscopy-derived structure of a membrane protein, bacteriorhodopsin (has seven transmembrane alpha helices).
	Tomita & Marchesi	Sequenced glycophorin, first integral membrane protein sequenced.
	Barbara Pearse	Discovered clathrin.
	G. Goldstein	Discovered ubiquitin.
1976	J. Axelrod	Membrane lateral diffusion rates determined by FRAP.
	M. Low	Described GPI-anchored proteins.
	H. Ikezawa	
	Brown & Goldstein	Formulated Receptor-Mediated Endocytosis hypothesis for cholesterol metabolism.
	Neher & Sakmann	Developed patch clamp technique, able to measure single channel conductance.
1977	Demel et al.	Used DSC to determine the affinity of cholesterol for various phospholipids: SM > PS, PG > PC > PE.
	Yasutomi Nishizuka	Discovered protein kinase C.
	Werner Lowenstein	Measured Gap junction pore size.
1978	Kamiya et al.	Discovered first prenylated protein.
	Laskey et al.	Reported the first chaperone assists in assembling nucleosomes.

(*Continued*)

Year	Discoverer	Discovery
1982	Kyte & Doolittle	Propose hydropathy scale to predict orientation of an integral membrane protein.
	Karnovsky & Klausner	Proposed lipid microdomain concept, foreshadowing lipid rafts.
	Binnig & Rohrer	Invented atomic force microscopy.
	Aitken et al.	Discovered first myristoylated protein.
	Carr et al.	
1984	R.A. Schmidt	Reported that many proteins are anchored to membranes via long chain isoprenoids.
	Seigneuret & Devaux	Discovered the first ATP-dependent flippase.
1985	Diesenhofer et al.	First high resolution X-ray structure of a membrane protein, a bacterial reaction center.
1986	Akiharo Kusumi	Developed Single Particle Tracking to measure lateral movement of a single protein in a membrane.
1989	Tsui et al.	Sequenced the gene for Cystic Fibrosis.
1991	Sarah Spiegel	Sphingosine-1-phosphate is linked to cell signaling.
1992	Peter Agre	Discovered the water channel, aquaporin.
1997	Kai Simons	Proposed the "Lipid Raft" model for membrane structure.
1998	Rod MacKinnon	Determined X-ray crystallography structure of the K^+ channel.
2005	E. Fahy et al.	Published a comprehensive classification system for lipids based on lipidomics.

BLM, bimolecular lipid membrane; *cAMP*, cyclic adenosine monophosphate; *CDP*, cytidine diphosphate choline; *FRAP*, fluorescence recovery after photobleaching; *GPI*, glycosylphosphatidylinositol; *LUV*, large unilamellar vesicles; *PE*, phosphatidylethanolamine; *PG*, phosphatidylglycerol; *PS*, phosphatidylserine; *PTS*, phosphotransferase system; *SM*, sphingomyelin.

Index

'*Note:* Page numbers followed by "f" indicate figures, "t" indicate tables.'

B

C

D

E

F

G

H

I

M

N

O

P

Q

R

V

W

X

Printed in the United States
By Bookmasters